Lecture Notes in Computer S
Edited by G. Goos, J. Hartmanis and J. va

Springer
Berlin
Heidelberg
New York
Barcelona
Hong Kong
London
Milan
Paris
Singapore
Tokyo

Jean-Yves Girard (Ed.)

Typed Lambda Calculi and Applications

4th International Conference, TLCA'99
L'Aquila, Italy, April 7-9, 1999
Proceedings

Springer

Series Editors

Gerhard Goos, Karlsruhe University, Germany
Juris Hartmanis, Cornell University, NY, USA
Jan van Leeuwen, Utrecht University, The Netherlands

Volume Editor

Jean-Yves Girard
Institut de Mathématiques de Luminy, CNRS - UPR9016
163 Avenue de Luminy, case 907
F-13288 Marseille cedex 9, France
E-mail: girard@iml.univ-mrs.fr

Cataloging-in-Publication data applied for

Die Deutsche Bibliothek - CIP-Einheitsaufnahme

Typed lambda calculi and applications : 4th international conference / TLCA '99, L'Aquila, Italy, April 7 - 9, 1999. Jean-Yves Girard (ed.). - Berlin ; Heidelberg ; New York ; Barcelona ; Hong Kong ; London ; Milan ; Paris ; Singapore ; Tokyo : Springer, 1999
(Lecture notes in computer science ; Vol. 1581)
ISBN 3-540-65763-0

CR Subject Classification (1998): F.4.1, F.3, D.1.1

ISSN 0302-9743
ISBN 3-540-65763-0 Springer-Verlag Berlin Heidelberg New York

Typesetting: Camera-ready by author
SPIN 10703139 06/3142 – 5 4 3 2 1 0 Printed on acid-free paper

Preface

This volume represents the proceedings of the Fourth International Conference on Typed Lambda Calculi and Applications, TLCA'99, held in L'Aquila, on 7-9 April 1999.

It contains 25 contributions. Fifty were submitted, their overall quality was high, and selection was difficult. The Programme Committee is very grateful to everyone who submitted a paper. It also contains two papers introducing the "demos" of "tlca software", i.e. industrial products making use of typed lambda-calculi.

The tutorials on

- *Denotational semantics* by Thomas Ehrhard and John Longley, and
- *Intersection types* by Mario Coppo and Mariangiola Dezani are not included in this volume.

The editor wishes to thank the members of the Programme Committee and the Organizing Committee listed, for their hard work and support, with a special mention for Benedetto Intrigila. He also thanks Corrado Böhm for kindly accepting the task of delivering a banquet speech.

The editor also expresses his gratitude to all the referees listed on the next page, as well as to those who wish not to be listed for their essential assistance and time generously given.

Marseille, January 1999

Jean-Yves Girard

Programme Committee

S. Abramsky (Edinburgh)
J.-Y. Girard (Marseille) (Chair)
J.-L. Krivine (Paris)
S. Ronchi (Torino)
T. Streicher (Darmstadt)
P. Urzyczyn (Warszawa)
T. Coquand (Göteborg)
R. Hindley (Swansea)
J. Reynolds (Pittsburgh)
A. Scedrov (Philadelphia)
M. Takahashi (Tôkyô)

Organizing Committee

F. Corradini, A. Formisano, B. Intrigila, (Chair), **M.-C. Meo, M. Nesi, A. Pierantonio, I. Salvo, S. Sorgi**

(Dip. Matematica , L'Aquila and Dip.Informatica, La Sapienza, Roma)

Referees

Y. Akama	T. Altenkirch	F. Barbanera	H. Barendregt
O. Bastonero	S. Berardi	A. Berarducci	C. Berline
V. Bono	M. Bugliesi	F. Cardone	G. Castagna
I. Cervesato	G. Chen	R. Cockett	A. Compagnoni
M. Coppo	T. Crolard	P.-L. Curien	V. Danos
R. Davies	P. De Groote	U. De Liguoro	M. Dezani
R. Dickhoff	H. Geuvers	N. Ghani	P. Giannini
S. Guerrini	B. Harper	R. Hasegawa	H. Herbelin
M. Hofmann	K. Honda	R. Jagadeesan	T. Jim
Y. Kameyama	M. Kanovich	R. Kashima	Y. Kinoshita
T. Kurata	Y. Lafont	J. Laird	F. Lamarche
P. B. Levy	C. Mc Bride	M. Marz	R. Matthes
P.-A. Mellies	G. Mitschke	H. Nickau	S. Nishizaki
M. Parigot	C. Paulin	F. Pfenning	B. Pierce
A. Piperno	L. Regnier	J. Rehof	D. Remy
E. Ritter	K. Rose	L. Roversi	P. Roziere
P. Ruet	A. Schalk	C. Schüermann	P.J. Scott
R.A.G. Seely	P. Selinger	M. H. Sörensen	R. Statman
C. Stewart	I. Takeuti	M. Tameyama	C. Urban
J. Vauzeilles	M. Wehr	J. Wells	H. Yokouchi

Table of Contents

Invited demonstrations

Contributions

The Coordination Language Facility and Applications

Jean-Marc Andreoli

Xerox Research Centre Europe, 38240 Grenoble, France,
Jean-Marc.Andreoli@xrce.xerox.com,
http://www.xrce.xerox.com

Abstract. This short paper gives a quick overview of CLF, a distributed object coordination middleware, and two applications of that platform to workflow. The driving concepts behind CLF derive from a reflection on proof search in Linear Logic, and in particular, the systematic exploitation of its resource conscious nature.

1 CLF: A Coordination Middleware

CLF is born from a reflection on the application of Linear Logic to distributed object coordination. It exploits the resource-conscious nature of Linear Logic in the framework of the concurrent logic programming paradigm, where computations are identified with proof-search [And92]. Turning a theoretical model of resource manipulation into a concrete distributed object coordination middleware required two main steps:

- First, the notions of "resources" and "objects" had to be integrated. This was achieved through a modification of the traditional object model of computation, making plain the role of objects as resource managers.
- Second, the concurrent logic programming paradigm of proof search had to be adapted to this new object model. This was realized by a scripting language based on Linear Logic formulae to express coordination.

1.1 The CLF Object Model

The CLF object model enriches the traditional one by viewing objects as resource managers, thus separating, inside the object state, the resources themselves from their management state. Primitives are introduced to (i) inquire and negotiate objects capabilities in terms of resource availability, (ii) perform basic transaction operations over the resources of several objects (two-phase commit) and (iii) request resource insertion. This enriched interaction model (Figure 1) is characterized by a set of 8 interaction verbs (similar to KQML performatives) together with a protocol describing correct sequences of invocations of these verbs, and their intended meaning in terms of resource manipulations. Figure 2 gives an overview of the verbs and the protocol. The interface of a CLF object distinguishes between "CLF services", accessed through the CLF interaction protocol, and regular methods, accessed through the traditional request/answer protocol.

1.2 The CLF Coordination Scripting Facility

The CLF coordination scripting facility takes full advantage of the object model. It allows high-level declarative specifications of coordinated invocations of CLF object services. A coordination is viewed here as a complex block of inter-related manipulations (removal, insertion, etc.) of the resources held by a set of objects (called the participants of the coordination). CLF scripts describe, using Linear Logic formulae, the expected global behavior of such blocks in terms of resulting resource transformations, but abstracts away the detailed sequencing of invocations of the CLF interaction verbs required to achieve such a behavior. It is this abstraction feature which considerably simplifies the design and verification of coordination scripts and makes them highly platform independent and hence, portable. The abstract operational semantics of CLF scripts is given in terms of proof search. Currently, the fragment of Linear Logic used by the CLF scripting language is a small subset of LinLog [And92], which is itself a "complete" fragment of Linear Logic in terms of proof search (complete in the sense that proof search in full Linear Logic can be reduced without loss to proof search in LinLog). Extensions of CLF to larger fragments of LinLog are possible, and may lead to further refinements of the object model.

2 The Demonstration: Applications of the CLF

There are two ways to demonstrate a middleware tool such as CLF: (*i*) focus on the middleware platform itself, but this is rather aimed at a somewhat specialized audience (developers of distributed object-based applications); (*ii*) show applications (or rather prototype applications) which have been developed using the platform.

We propose to demonstrate here two prototype applications developed using CLF. The first one, called XFolder, is a lightweight workflow management system; the second one called XPect [AP98], is a generic electronic commerce broker.

2.1 XFolder: a Lightweight Workflow Management System

XFolder uses the metaphor of the well-known circulation folder envelope to organize lightweight workflow within an organization or across several independent organizations (with possible access restrictions between them). A circulation folder consists of a set of documents, enclosed in an envelope, and a route, usually specified on the envelope, and describing the expected path of the envelope through different services (or people) of the organization(s) and the tasks to be performed at each stop. Whenever a user gets hold of the envelope, s/he can perform the current task assigned to it (e.g. read, create, modify, annotate a document, sign a sheet, insert a memo etc.), and, possibly, modify the route (e.g. extend it or change some tasks), then forward it.

XFolder implements an electronic version of the traditional circulation folder, with additional features allowed by this "virtualization". The architecture of the system is described in Figure 3. The documents contained in the envelope are held in electronic form in heterogeneous document repositories implemented as CLF objects (the resources of which are the documents). The status of the individual folders (route, current active task in the route, assignment of tasks etc.) are held in a specific CLF object, the XFolder manager (the resources of which are the virtual envelopes). CLF scripts handle the notification of available tasks to each user, implement the task status transformation as tasks are performed, and take care of migrating documents across different repositories when needed (i.e. when a firewall or some access restriction prevents a document reference from being directly shared between users).

2.2 XPect: an Electronic Commerce Broker

XPect realises the functionality of a broker for electronic commerce. It handles the coordination of the different partners involved in an electronic commerce transaction: customers, bankers, providers, delivery providers etc. Basically, the customer submits a query to the broker, describing items of interest. The broker browses through the catalogs of the different providers to extract offers matching the query constraints (description of good, required options, price limits etc.). The user may then select a set of different offers for purchase. This is different from the "shopping basket" of traditional electronic commerce systems in the sense that the selected items are considered as a whole: either all of them are available at the condition of the offers, and the commercial transaction is continued, or the whole transaction is cancelled (but the customer can always resume the search phase). This atomic behavior is ensured even across independent providers (e.g. with queries of the form "24x36mm camera from provider A and a matching 50mm lens from provider B", or "10 hardcopies of a book X from bookshop A and a 24-hour delivery of the whole set from provider B").

XPect is implemented as a CLF application. The architecture of the system is described in Figure 4. The CLF objects involved are the providers, offering virtual or hard goods classified in catalogs, the financial services (credit cards, electronic cash etc.) and the customer management services. CLF scripts are used to implement the different phases of the electronic commerce transaction.

References

[And92] J-M. Andreoli: Logic programming with focusing proofs in linear logic. *Journal of Logic and Computation*, 2(3), 1992.

[AP98] J-M. Andreoli and F. Pacull: Distributed print on demand systems in the xpect framework. *Journal of Distributed and Parallel Databases*, 1998. To appear.

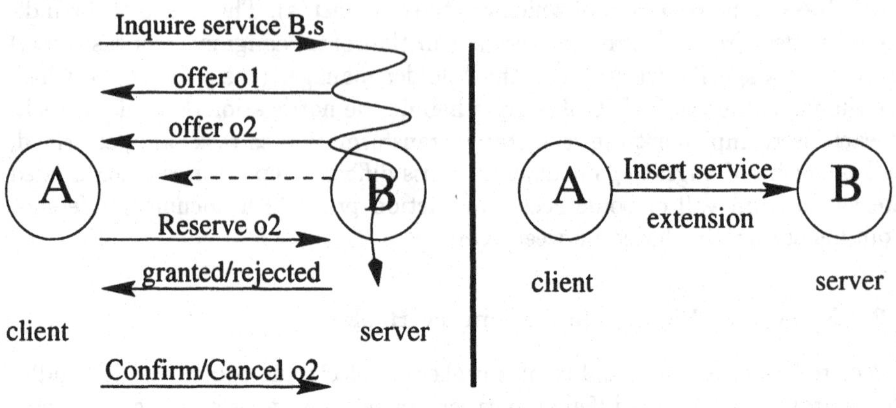

Fig. 1. The CLF interaction model

```
Inquire: input-tuple-pattern -> stream
Next:    stream -> [ action ; output-tuple ] or NO-MORE-VALUE
Kill:    stream -> void
Check:   action -> TRUE or FALSE

Reserve: action -> ACCEPT or SOFT-REJECT or HARD-REJECT
Confirm: action -> void
Cancel:  action -> void

Insert:  tuple -> void
```

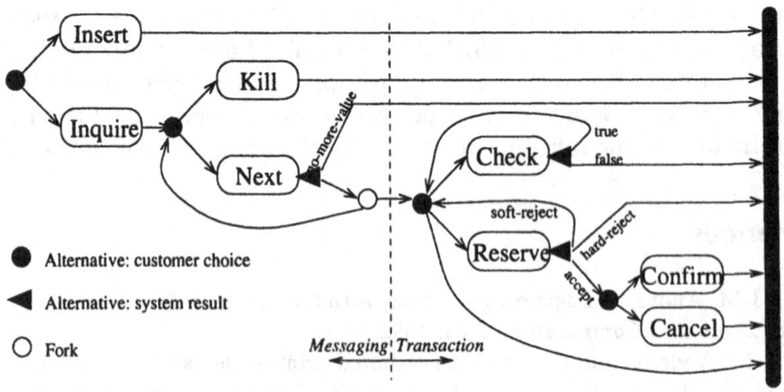

Fig. 2. The CLF protocol

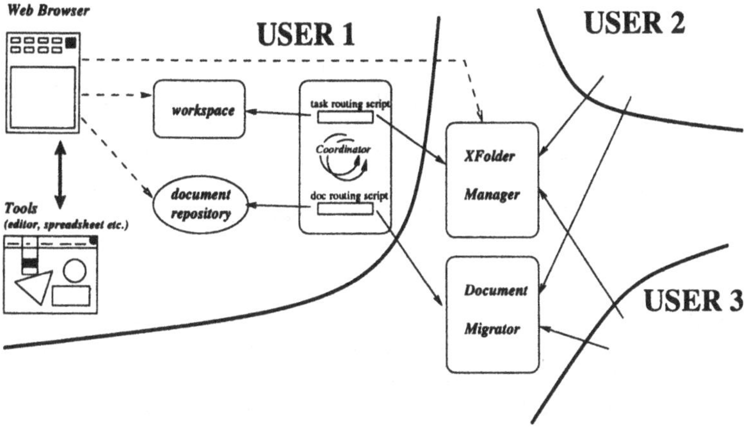

Fig. 3. Architecture of XFolder

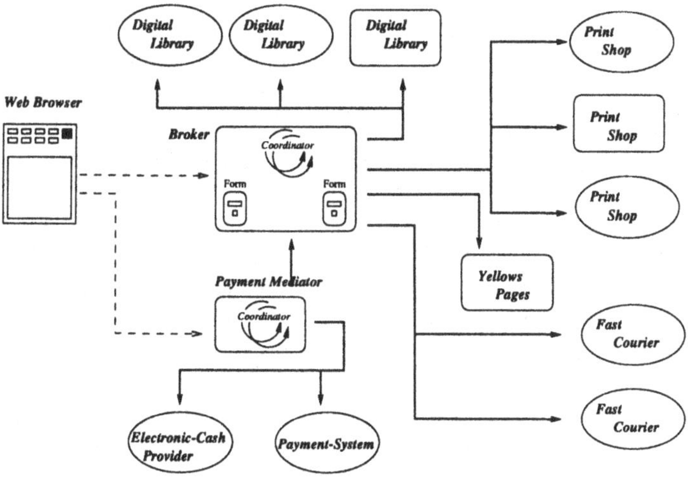

Fig. 4. Architecture of XPect

Legend:

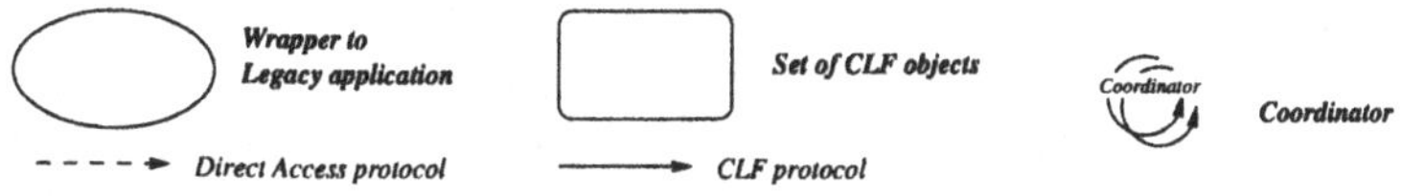

AnnoDomini in Practice: A Type-Theoretic Approach to the Year 2000 Problem

Peter Harry Eidorff Fritz Henglein Christian Mossin
Henning Niss Morten Heine B. Sörensen Mads Tofte

Dept. of Computer Science, Univ. of Copenhagen (DIKU) and Hafnium ApS
{phei,henglein,mossin,hniss,rambo,tofte}@diku.dk
http://www.diku.dk, http://www.hafnium.com

Abstract. AnnoDomini is a commercially available source-to-source conversion tool for finding and fixing Year 2000 problems in COBOL programs. AnnoDomini uses type-based specification, analysis, and transformation to achieve its main design goals: flexibility, completeness, correctness, and a high degree of safe automation.

1 Introduction

The *Year 2000 (Y2K) problem* refers to the inability of software and hardware systems to process dates in the 21st century correctly.[1] The problem arises from representing calendars years by their last two digits and thus restricting the range of representable years to 1900-1999. Starting some 40 years ago, this convention was established as one of numerous techniques for conserving precious memory space.

The most widespread *Year-2000-unsafe* date representation consists of six characters. It has two characters each for the day of the month, the month of the year, and the calendar year, often in the order year-month-day (YYMMDD). The string "981106", for example, represents November 6th, 1998. The problem, of course, is that no provision is made for representing years in the 21st century: "00" represents 1900, not 2000.

Since the year 2000, mistakenly represented as "00", comes before e.g. "99", comparison of two-digit years may produce unexpected results in the 21st century, and this may incur problems in operating with e.g. expiry dates. Similarly, arithmetical operations involving two-digit years may produce unexpected results, affecting e.g. interest calculations.

The Y2K problem affects countless systems at all levels: embedded systems, operating systems, applications and data bases that process or contain dates. Both its size and consequences are staggering. Cost estimates vary widely, but according to Capers Jones, "the costs of fixing the year 2000 problem appear to constitute the most expensive single problem in human history" [2, p. xxiii].

Updating application programs to become Year 2000 compliant usually involves a combination of *expansion* and *masking*. Expansion refers to expanding

[1] We adopt the convention of viewing the year 2000 as belonging to the 21st century.

unsafe two-digit years to four-digit years in applications, data bases, files, etc. Expansion can be expensive, however: it requires that not only application programs be changed, but also data bases, files and all other programs communicating dates. Masking denotes a variety of methods for extending two-byte year representations into the 21st century; e.g. *windowing*, *compression* and *encapsulation.* These techniques aim at extending the lifetime of existing data in data bases and files as well as screen and print maps into the 21st century. In windowing, for example, a pivot year determines whether a two-digit year belongs to the 20th or the 21st century. For example, with pivot 70, 79 represents 1979, and 41 represents 2041.

AnnoDomini[2] is a tool (and accompanying method) for finding and fixing Year 2000 problems in COBOL programs. It accommodates expansion as well as masking by source-to-source transformation of COBOL programs. The converted programs do not require special compiler support, but compile and execute in their existing operating environment.

AnnoDomini consists of three components: an analysis and conversion engine (60,000 lines of Standard ML), a graphical user interface (10,000 lines of Visual Basic), and IBM's Live Parsing Editor (a syntax-sensitive program editor). The three components are tightly integrated, as will be explained in what follows. AnnoDomini runs on Windows NT 4.0 and Windows 9X, and is commercially available from Computer Generated Solutions, Inc. (an IBM business partner)—see http://www.cgsinc.com and http://www.hafnium.com.

AnnoDomini uses type-based specification, analysis, and transformation to achieve its main design goals: flexibility, completeness, correctness, and a high degree of safe automation. The type-theoretic foundations of AnnoDomini have been described elsewhere [1]. In the present brief account we aim to demonstrate how AnnoDomini actually works in practice—emphasizing the role of types—although we shall ignore many practical issues, e.g., key fields with years, alignment of key fields, aliasing, editing characters, padding/truncation, justification, and usage.

2 The AnnoDomini Approach

In COBOL programs, dates are represented using the data types and operations of the source language: strings of characters and digits, and flat records. Their *intensional* interpretation as representations of dates is not explicit. The AnnoDomini approach is based on *reverse engineering* the programmer-intended date interpretations as *abstract types.* This is done in three conceptual phases: seeding, type checking, and conversion.

2.1 An Example COBOL Program Fragment

To illustrate the AnnoDomini approach, we consider the following fragment of a COBOL program.

[2] AnnoDomini is a registered trademark of Hafnium ApS.

```
77 CUR-DATE PIC 999999.
77 LRD      PIC 999999.
77 COLUMN   PIC 99.
   IF CUR-DATE > LRD PERFORM ISSUE-LAST-REMINDER.
   IF COLUMN < 80    PERFORM DISPLAY-STATUS.
```

The first three lines are declarations of three variables: CUR-DATE (containing six-digit data, signified by the six occurrences of 9), LRD (also containing six-digit data), and COLUMN (containing two-digit data).

The first statement invokes procedure ISSUE-LAST-REMINDER if CUR-DATE ("current date") is greater than LRD ("last reminder date").

The current date will most likely have form 000101 on January 1st, year 2000, so with a last reminder date of, say, December 31st, year 1999 (991231), no last reminder will ever be issued as a result of running the application in the year 2000 or later. This is not the desired behavior of the program.

The last statement invokes procedure DISPLAY-STATUS, provided the value contained in COLUMN is less than 80. This comparison has nothing to do with years and will continue to work in the 21st century.

2.2 Seeding

In the first phase of the AnnoDomini approach the user *seeds* the program with year (and possibly non-year) information. This is done by annotating variable declarations with *Type System 2000 (TS2K) types* that specify where years occur in them, if at all.

TS2K types are concatenations of the following different base types:

1. YYYY: four-digit year;
2. WW: two-digit, windowed year relative to a fixed pivot, by default 00;[3]
3. N: single non-year character;
4. -...- (n occurrences of -): n characters of unknown type.

For example, from the declarations alone in our example program we might guess that CUR-DATE is a six-digit date with a leading year, and that COLUMN is a column position at the terminal screen and hence unrelated to years. What LRD denotes, is not clear from the declaration alone. Thus, a seeded version of the example program might read:

```
*TS2K WWNNNN
 77 CUR-DATE PIC 999999.
*TS2K ------
 77 LRD      PIC 999999.
*TS2K NN
 77 COLUMN   PIC 99.
    IF CUR-DATE > LRD PERFORM ISSUE-LAST-REMINDER.
    IF COLUMN < 80    PERFORM DISPLAY-STATUS.
```

[3] With pivot 00, two-digit windowed years are the two-digit years of the 20th century.

Since COBOL comment lines start with * in column 7, the above TS2K type declarations are treated as comments by the COBOL compiler. However, to AnnoDomini they provide type information.

Seeding can be done *automatically* or *manually*. Automatic seeding works by scanning variable names in a program, including all the libraries it imports, and looking for matches according to both lexical and data description criteria. Informally, for each program variable the user asks: "Could this variable contain a calendar year, based on its name and its data description?" For example, a variable named `DEP-DAT` and occupying 6 bytes, might represent a six-digit date ("departure date"). Then again, it might not ("deposition data"). Automatic seeding is specified by a combination of lexical inclusion and exclusion criteria and a list of target date types. These specifications can be configured interactively, and they can be stored in separate files for future use. Automatic seeding is known to be quick, but also error-prone since it depends on nomenclature for variable names. AnnoDomini presents a list of all matches along with annotation suggestions, but does not automatically accept the results as bona-fide year annotations. Instead, it expects the user to explicitly accept or reject them, possibly after inspecting the variable declarations through a point-and-click interface.

Manual seeding works by systematically checking the *interfaces* of a program; e.g., data base, file, terminal and print map descriptions. In COBOL, these are typically localized in shared libraries that are copied into programs by `COPY` statements, COBOL's macro expansion and source library access mechanism. Manual seeding is less error-prone since it reduces guesswork. Since data base, file, and map descriptions need to be annotated only once, but are typically used by multiple programs, manual seeding need not be done for each program and is thus often a quite efficient and safe seeding method.

2.3 Type checking

In the second phase AnnoDomini *propagates* the seeding information to other data by *type inference*. In particular, types are propagated through comparisons and assignments. For instance, since our example program contains the statement

```
IF CUR-DATE > LRD PERFORM ISSUE-LAST-REMINDER.
```

the type of `CUR-DATE` is propagated to `LRD`, and AnnoDomini *suggests* that `LRD` be given the same type. As with seeding suggestions, the user accepts and rejects such suggestions through a point-and-click interface.

During propagation AnnoDomini also *checks* that the seeded and propagated types are *consistent* with each other. For example, based on its cryptic name, we might mistakenly have assumed that `LRD` is entirely composed of non-year data and have assigned the type `NNNNNN` to it, in which case the types of `CUR-DATE` and `LRD` would be inconsistent. In this case AnnoDomini signals a *type error*.

In general, type errors may stem from the following sources:

1. *Seeding error.* Seeding might be wrong; for instance, we might have assumed the incorrect type `NNNNNN` for `LRD`.

2. *Not a Year 2000 problem.* The type system does not allow both years and and non-years to occupy the same storage at different times, such as when printing both years and non-years through the same print buffer, as in the following program fragment.[4]

```
*TS2K WW
 77 CUR-YEAR  PIC 99.
*TS2K NN
 01 NON-YEAR  PIC 99.
*TS2K NN
 77 PRINT-BUF PIC 99.
    MOVE CUR-YEAR TO PRINT-BUF.
    MOVE NON-YEAR TO PRINT-BUF.
```

3. *Year 2000 problem.* The error might signal a Year 2000 problem or other questionable computations on dates, e.g. a hardwired conversion between four-digit years to two-digit years, as in the following program fragment in which the two last digits of a four-digit year are moved into a two-digit variable utilizing COBOL's alignment and truncation rules. A similar coercion in the other direction is adding 1900 to a two-digit year. Such hardwired coercions do not generally work in the 21st century.

```
*TS2K WW
 77 YEAR2 PIC 99.
*TS2K YYYY
 77 YEAR4 PIC 9999.
    MOVE YEAR4 TO YEAR2.
```

AnnoDomini does not attempt to guess what the cause of a type error is and how to eliminate it. It suggests a number of plausible *corrective actions*, however. These include changing the declarations of the variables involved in the type incorrect statement—the relevant option in case of a seeding error.

Two other forms of suggestions are to insert `ASSUME` and `COERCE` annotations. For instance, for the print buffer example, AnnoDomini suggests annotating the `MOVE` statements with an `ASSUME` annotation; e.g.,

```
*TS2K WW
 77 CUR-YEAR  PIC 99.
*TS2K NN
 77 NON-YEAR  PIC 99.
*TS2K NN
 77 PRINT-BUF PIC 99.
*TS2K ASSUME CUR-YEAR IS NN
    MOVE CUR-YEAR TO PRINT-BUF.
    MOVE NON-YEAR TO PRINT-BUF.
```

[4] The `MOVE` statement is COBOL's assignment statement.

The ASSUME annotation tells the type checker that CUR-YEAR should be treated as having type NN *in this statement only.* (This is dangerous, of course, and therefore requires an *explicit* annotation in the source code.)

The COERCE annotation is used to convert between different year formats. For instance, AnnoDomini suggest annotating the type incorrect statement MOVE YEAR4 TO YEAR2 with a COERCE statement, e.g.,

```
*TS2K COERCE YEAR4 TO WW BY D4TO2NO
      MOVE YEAR4 TO YEAR2.
```

The coercion D4TO2NO converts a four-digit year to a value with the same year in windowed representation. The COERCE annotation is similar to the ASSUME annotation: the former instructs in the above example AnnoDomini to regard YEAR4 as having type WW. The difference is that, in the conversion phase COERCE annotations are replaced by code performing the coercions, whereas ASSUME annotations have no run-time significance.

AnnoDomini also provides point-and-click access to the statements causing type errors and to the declarations of the variables occurring in the type incorrect statement for manual browsing and editing of the source code.

AnnoDomini issues *warnings* for all relational and arithmetic operations on two-digit years as well as for all relational and arithmetic operations for which there is insufficient type information to determine whether their operands contain years or not. This is a case where seeding is incomplete, with potentially dangerous consequences. The user is expected to check the warnings to determine whether they cover over any potential Year 2000 problems. They can also be eliminated by strengthening the seeding to resolve the operand types.

Seeding and type checking are repeated, possibly interchangeably, until *all* type errors are eliminated and the program is *type correct.*

2.4 Conversion

The third and final phase consists of *virtual conversion* and *actual conversion.* During virtual conversion the user specifies Year 2000-safe types for each variable. For example

```
*TS2K WWNNNN->YYYYNNNN
  77 CUR-DATE PIC 999999.
*TS2K WWNNNN->YYYYNNNN
 77 LRD       PIC 999999.
*TS2K NN
 77 COLUMN    PIC 99.
    IF CUR-DATE > LRD PERFORM ISSUE-LAST-REMINDER.
    IF COLUMN < 80    PERFORM DISPLAY-STATUS.
```

is a virtual conversion which specifies that CUR-DATE and LRD should be *expanded* from a six-digit to an eight-digit date representation. The actual conversion is then fully automatic, yielding the following program fragment:

```
*TS2K YYYYNNNN
 77 CUR-DATE PIC 99999999.
*TS2K YYYYNNNN
 77 DUE-DATE PIC 99999999.
*TS2K NN
 77 COLUMN PIC 99.
    IF CUR-DATE > DUE-DATE  PERFORM ISSUE-LAST-REMINDER.
    IF COLUMN < 80          PERFORM DISPLAY-STATUS.
```

Alternatively, a virtual conversion can be specified by changing the default pivot for windowing from 00 to, say, 70 (this does not require any change to the program). Actual conversion then yields, fully automatically:

```
*TS2K WWNNNN
 77 CUR-DATE PIC 999999.
*TS2K WWNNNN
 77 DUE-DATE PIC 999999.
*TS2K NN
 77 COLUMN PIC 99.
    MOVE CUR-DATE TO ARG-1 OF ARGUMENT OF LT70N4-PARAMS.
    MOVE DUE-DATE TO ARG-2 OF ARGUMENT OF LT70N4-PARAMS.
    CALL "LT70N4" USING LT70N4-PARAMS.
    IF RESULT OF LT70N4-PARAMS = '1'
                                 PERFORM ISSUE-LAST-REMINDER.
    IF COLUMN < 80  PERFORM DISPLAY-STATUS.
```

The first four program statements call the AnnoDomini library routine `LT70N4` which compares dates with leading two-digit windowed years relative to pivot 70 (COBOL's built-in operator `>` does not work since it does not take the pivot into account). The year-unrelated comparison `COLUMN < 80` is left as is.

Each variable can have its own year representation. AnnoDomini has built-in support for four-digit years and windowed two-digit years, Apart from these, it allows abstract, user-defined two-digit years. These are denoted `AA(`t`)`, where t is the name of a user-defined library, which must contain the required arithmetic and relational operations. These can be type-checked on a par with the built-in year types.

Actual conversion is fully automatic: at the push of a button, data declarations are expanded as desired, calls to the specified coercions are inserted, and arithmetic and relational operations involving two-digit years are replaced by calls to AnnoDomini's Year 2000-safe library routines.

3 Conclusion and Related Work

The decision to base AnnoDomini on types has had a number of advantages.

First, types are good for explicating the intention of data. For instance, types are good for distinguishing between years and non-years, e.g. between 80 as a two-digit year and 80 as a column position. Similarly, types are good for distinguishing between different types of years, e.g. 80 as the two-digit windowed year 1980 and 80 as the year 80 A.D.

Second, types are good for discovering Year 2000 problems; for instance, the comparison `CUR-YEAR < 80` is problematic if `CUR-YEAR` is a two-digit year, whereas `COLUMN < 80` is unproblematic if `COLUMN` denotes non-year information.

Third, types are good for guiding transformations. In particular, year-unrelated code can be left as is.

Fourth, many design choices are made simply and elegantly by casting our analysis as a type inference. For instance, how to report inconsistent usage of years in the program? This obviously becomes a type error.

Finally, we have been able to benefit greatly from the design and implementation of ML, and its underlying theory, in developing AnnoDomini. For instance, some of the main results concerning type inference in [1] were adopted from Hindley-Milner type inference, with some modifications.

There is a vast literature on type theory and type-based program analysis. There are also numerous Year 2000 tools. Very few of those are semantics-based, however, and of those only AnnoDomini appears to be type-based with integrated automatic analysis and conversion.

The value of working with type notions in software understanding and reengineering has been observed previously by O'Callahan and Jackson [3]. Van Deursen and Moonen [5] describe type inference rules for COBOL for classifying data into sets of data representations. Subtyping is interpreted as subsumption of value sets. Their system specifies type equivalences, and it allows subtyping steps at assignments. Intuitively, this specifies a flow-insensitive data flow analysis, refined by data flow sensitivity at assignments.

Independently of us, Ramalingam, Field and Tip have developed basically the same unification algorithm as used by AnnoDomini's type inference algorithm [4]. They also demonstrate how their algorithm is applicable to Year 2000 program analysis.

References

1. P. H. Eidorff, F. Henglein, C. Mossin, H. Niss, M.H.B Sörensen, and M. Tofte. Annodomini: From type theory to year 2000 conversion tool. In *Principles of Programming Languages*, January 1999.
2. C. Jones. *The Year 2000 Software Problem — Quantifying the Costs and Assessing the Consequences.* Addison-Wesley, ACM Press, 1998. ISBN 0-201-30964-5.
3. R. O'Callahan and D. Jackson. Lackwit: A program understanding tool based on type inference. In *Proc. 1997 International Conference on Software Engineering (ICSE '97), Boston, Massachusetts*, pages 338–348, May 1997.
4. G. Ramalingam, J. Field, and F. Tip. Aggregate structure identification and its application to program analysis. In *Principles of Programming Languages*, January 1999.
5. A. van Deursen and L. Moonen. Type inference for COBOL systems. To appear in Proc. 5th IEEE Working Conference on Reverse Engineering, Honolulu, Hawaii, October 1998.

Modules in Non-commutative Logic

V. Michele Abrusci

Università Roma Tre, Dipartimento di Filosofia,
via Magenta, 5 00185 – ROMA
abrusci@uniroma3.it

1 Introduction

The question we want to investigate was expressed by Girard in [3]:

> "Assume that I am given a program P [a *proof-net* Π], and that I cut it in two parts arbitrarily. I create two ... *modules*, linked together by their *border*. Can I express that my two modules are complementary [*orthogonal*], in other terms that I can branch them by identification over their common border? One would like to define the *type* of the modules as their branching instructions; these branching instructions should be such that they authorized the restoring of the original P [the *proof-net* Π]."

Girard in [3] gave the solution for the multiplicative fragment of linear logic (MLL); another deep investigation of this question for MLL has been given by Danos and Regnier in [2]. Here we present the first steps towards a solution of this question for the multiplicative fragment of non-commutative logic (MNL), which is a refinement of MLL and an extension of both MLL and the cyclic multiplicative linear logic. MNL was introduced in [1].

The lines of Girard's investigations [3] are the basis for our investigations, since they can be improved and adapted also for MNL.

In the following:

1. we define what a module is in MNL, i.e. what we obtain from a proof-net in MNL by splitting it arbitrarily, and we define what is a type of a module in MNL;
2. we define when two modules in MNL are *orthogonal*, and we prove the theorem: if two modules Π_1 and Π_2 in MNL are orthogonal, then $\Pi_1 \circ \Pi_2$ (i.e what we get by gluing their common border) is a proof-net in MNL.

In order to understand the paper, and the new questions posed to MNL with respect to MLL, we give a short summary about proof nets in MNL (called MNL proof-nets).

The formulas of MNL are built from atoms $(p, q, \ldots)$ and their orthogonal $(p^\perp, q^\perp, \ldots)$ by using the following binary connectives:

- multiplicative connectives: $\otimes$, $\wp$
- non-commutative connectives: $\odot$ (next), ∇ (sequential).

A MNL proof-structure is defined as usual by taking the following links:

- axiom-link (no premises; conclusions: $A^\perp$ and A),
- cut-link (no conclusions; premises: A and $A^\perp$),
- for each connective $\diamondsuit$, the $\diamondsuit$-link (conclusion: $A\diamondsuit B$; first premise: A; second premise: B).

A set of switches is defined for each link:

- for the axiom-link, one switch ($A^\perp\uparrow\rightarrow A\downarrow$, $A\uparrow\rightarrow A^\perp\downarrow$),
- for the cut-link, one switch ($A\downarrow\rightarrow A^\perp\uparrow$, $A^\perp\downarrow\rightarrow A\uparrow$),
- for the $\otimes$-link, the usual two switches $\otimes$-L and $\otimes$-R,
- for the $\wp$-link the usual two-switches $\wp$-L and $\wp$-R,
- for the $\odot$a unique switch: $\odot$-R ($(A\odot B)\uparrow\rightarrow A\uparrow$, $A\downarrow\rightarrow B\uparrow$, $B\downarrow\rightarrow(A\odot B)\downarrow$),
- for the ∇-link three switches: ∇-L ($(A\nabla B)\uparrow\rightarrow A\uparrow$, $B\downarrow\rightarrow B\uparrow$, $A\downarrow\rightarrow (A\nabla B)\downarrow$), ∇-R ($(A\nabla B)\uparrow\rightarrow B\uparrow$, $A\downarrow\rightarrow A\uparrow$, $B\downarrow\rightarrow(A\nabla B)\downarrow$), ∇-3 ($(A\nabla B)\uparrow\rightarrow A\uparrow$, $B\downarrow\rightarrow(A\nabla B)\downarrow$).

A switching for a MNL proof-structure Π is a function s such that, for every link l in Π, $s(l)$ is a switch for l.

Given a switching s for a MNL proof-structure Π, we can construct an oriented graph $s(\Pi)$. The set of the vertices of $s(\Pi)$ is:

$$\{A^x \mid A \text{ occurrence of formula in } \Pi \text{ and } x\in\{\uparrow,\downarrow\}\}$$

In $s(\Pi)$ there is the oriented edge $A^x\rightarrow B^y$ if, for some link l, $A^x\rightarrow B^y$ is given by $s(l)$, and, moreover, there is an edge between $C\downarrow$ and $C\uparrow$ for every conclusion C.

A trip in $s(\Pi)$ is a maximal path in $s(\Pi)$. The trips in $s(\Pi)$ may be:

- either cyclic, i.e. of the from $A^x\ldots A^x$, with $x\in\{\uparrow,\downarrow\}$; a *long trip* is a cyclic trip containing $B\uparrow$ and $B\downarrow$ for each occurrence of formula B in Π;
- or non cyclic; in this case the trip has the form $B\uparrow\ldots A\downarrow$, where B is the second premise of a ∇-link l such that $s(l)\nabla$-3 and A is the first premise of a ∇-link l' such that $s(l')\nabla$-3.

Π is a MNL proof-net iff: Π is a MNL proof-structure such that, for any switching s, there is exactly one cyclic trip T in $s(\Pi)$ and T:

- is *bilateral*, i.e. T does not contain the pattern $B^x\ldots C^y\ldots B^{\bar{x}}\ldots C^{\bar{y}}$, where $x,y\in\{\uparrow,\downarrow\}$ and $\bar{\uparrow}\downarrow$, $\bar{\downarrow}\uparrow$,
- contains all the conclusions of Π.

So, when Π is a MNL proof-net and s is a switching for Π, non-cyclic trips do not contain conclusions of Π.

Due to the possible presence of ∇-links, with switch ∇-3, the unique cycle in $s(\Pi)$, where Π is a MNL proof-net, is not necessarily a long trip. Due to the unique switch $\odot$-R for $\odot$-links, we can not conclude that every trip is bilateral, from the existence of a unique cycle T in $s(\Pi)$ for every switching s.

The existence of non-cyclic trips in $s(\Pi)$, and the requirement of the bilaterality of the unique cycle in $s(\Pi)$ for every switching s, are the main difficulties when we try to adapt the lines of Girard's investigations in [3] to MNL. The existence of non-cyclic trips leads to consider *partial permutations* (instead of total permutations), induced by the switchings; the need that a cyclic trip must contain all the conclusions leads to consider sets γ_s induced by each switching s; the requirement of the bilaterality of cyclic trips leads to introduce a more refined concept of orthogonality between permutations and to use relations δ_s and χ_s induced by each switching s.

2 Bordered MNL proof-structures, switchings and trips

Definition 1. $\langle \Pi; \Gamma \rangle$ *is a* bordered MNL proof-structure *iff:*

- *Π is a graph of occurrences of formulas of MNL, linked by occurrences of links of MNL; $\mid \Pi \mid$ is the set of all the occurrences of formulas of Π, and L_Π is the set of all occurrences of links in Π,*
- *for every $A \in \mid \Pi \mid$ there is at most a link $l \in L_\Pi$, such that A is a premise of l;* $\mathrm{Con}(\Pi)\{A \in \mid \Pi \mid \mid \forall l \in L_\Pi . Aisnotpremiseofl\}$;
- *for every $A \in \mid \Pi \mid$ there is at most a link $l \in L_\Pi$, such that A is a conclusion of l;* $\mathrm{Hyp}(\Pi)\{A \in \mid \Pi \mid \mid \forall l \in L_\Pi . Aisnotconclusionofl\}$;
- *Γ (the* border*) is a finite sequence of occurrences of formulas of Π such that:*
  - *for every $A \in \mathrm{Hyp}(\Pi)$, A is in Γ,*
  - *if A is in Γ, then $A \in \mathrm{Hyp}(\Pi) \cup \mathrm{Con}(\Pi)$;*
    $\mathrm{PCon}(\langle \Pi; \Gamma \rangle)\{A \in \mathrm{Con}(\Pi) \mid A \notin \Gamma\}$.

The elements of $\mathrm{Hyp}(\Pi)$ are called the hypothesis of Π. The elements of $\mathrm{Con}(\Pi)$ are called the conclusions of Π and those of $\mathrm{PCon}(\Pi)$ are called the proper conclusions of $\langle \Pi; \Gamma \rangle$. If Γ is empty, then $\langle \Pi; \emptyset \rangle$ is exactly a MNL proof-structure as defined in [1].

Definition 2. *Two bordered MNL proof-structures $\langle \Pi_1; \Gamma_1 \rangle$ and $\langle \Pi_2; \Gamma_2 \rangle$ are* compatible *iff:*

- *$\Gamma_1 \Gamma_2$,*
- *for every A in Γ_1, $A \in \mathrm{Con}(\Pi_1)$ iff $A \in \mathrm{Hyp}(\Pi_2)$, and $A \in \mathrm{Con}(\Pi_2)$ iff $A \in \mathrm{Hyp}(\Pi_1)$.*

Definition 3. *If $\langle \Pi_1; A_1 \dots A_n \rangle$ and $\langle \Pi_2; A_1 \dots A_n \rangle$ are compatible bordered MNL proof-structures, then $\Pi_1 \circ \Pi_2$ is the graph obtained from Π_1 and Π_2 by i dentifying of the occurrences A_i in Π_1 with the occurrences A_i in Π_2, for all $i \in \{1 \dots n\}$.*

Remark that, if Π is *commutative* (i.e. without $\odot$ and ∇), then the above definitions are those given by Girard in [3].

If $\langle \Pi_1; \Gamma \rangle$ and $\langle \Pi_2; \Gamma \rangle$ are compatible bordered MNL proof-structures, then $\Pi_1 \circ \Pi_2$ is an MNL proof-structure.

Definition 4. *s is a switching for a bordered MNL proof-structure $\langle \Pi; \Gamma \rangle$ iff s is a function such that, for every $l \in L_\Pi$, $s(l)$ is a switch for l.*

If s is a switching for a bordered MNL proof-structure $\langle \Pi; \Gamma \rangle$, then $s(\langle \Pi; \Gamma \rangle)$ is the oriented graph such that:

- $\mid s(\langle \Pi; \Gamma \rangle) \mid \{A^x \mid A \in \mid \Pi \mid \ textandx \in \{\uparrow, \downarrow\}\}$.
- *In $s(\langle \Pi; \Gamma \rangle)$ there is the oriented edge $A^x \longrightarrow B^y$ (where $x, y \in \{\uparrow, \downarrow\}$) if:*
  - *either, for some link $l \in L_\Pi$, $A^x \longrightarrow B^y$ is given by $s(l)$,*
  - *or AB is a proper conclusion of $\langle \Pi; \Gamma \rangle$, and both $x \downarrow$, and $y \uparrow$.*

Remark that in $s(\langle \Pi; \Gamma \rangle)$ there is no oriented edge

- from $A \downarrow$, when $A \in \mathrm{Con}(\Pi)$ and A is in Γ, or when A is the first premise of a ∇-link $l \in L_\Pi$ such that $s(l)\nabla$-3,
- from $A \uparrow$, when $A \in \mathrm{Hyp}(\Pi)$,
- to $A \downarrow$, when $A \in \mathrm{Hyp}(\Pi)$,
- to $A \uparrow$, when $A \in \mathrm{Con}(\Pi)$ and A is in Γ, or when A is the second premise of a ∇-link $l \in L_\Pi$ such that $s(l)\nabla$-3.

We shall use the following notations for $A \in \Gamma$: $A^{\mathrm{in}(\Pi)}$ is:

$$\begin{cases} A \uparrow \mathrm{if} A \in \mathrm{Con}(\Pi) \\ A \downarrow \mathrm{if} A \in \mathrm{Hyp}(\Pi) \end{cases}$$

and $A^{\mathrm{out}(\Pi)}$ is:

$$\begin{cases} A \downarrow \mathrm{if} A \in \mathrm{Con}(\Pi) \\ A \uparrow \mathrm{if} A \in \mathrm{Hyp}(\Pi) \end{cases}$$

$A^{\mathrm{in}(\Pi)}$ is the position of A *($A \downarrow$ or $A \uparrow$)* such that we can move in Π through A.

$A^{\mathrm{out}(\Pi)}$ is the position of A ($A \downarrow$ or $A \uparrow$) such that we can exit from Π through A.

Definition 5. *Let s be a switching for a bordered MNL proof-structure $\langle \Pi; \Gamma \rangle$. A* trip *in $s(\langle \Pi; \Gamma \rangle)$ is a maximal path in $s(\langle \Pi; \Gamma \rangle)$.*

Let s be a switching for a bordered MNL proof-structure $\langle \Pi; \Gamma \rangle$. Each trip in $s(\langle \Pi; \Gamma \rangle)$ belongs to exactly one of these classes:

- *cyclic trips* or *cycles*, i.e. trips of the form $A^x \dots A^x$, where $A \in \mid \Pi \mid$ and $x \in \{\uparrow, \downarrow\}$,

- Γ *trips*, i.e. the trips of the form $A^{\text{in}(\Pi)} \dots B^{\text{out}(\Pi)}$, where A, B are formulas in Γ,

- *critical trips*, i.e. trips of the form $C \uparrow \dots D^x$ or $A^y \dots B \downarrow$, where C is the second premise of a ∇-link $l \in L_\Pi$, such that $s(l)\nabla$-3 and B is the first premise of a ∇-link $l' \in L_\Pi$, such that $s(l')\nabla$-3.

3 The type of a bordered MNL proof-structure

Definition 6. $\sigma \in \text{perm}(\{1, \dots, n\})$ *iff* σ *is a (total or partial) permutation of* $\{1, \dots, n\}$, *i.e. is a (total or partial) injection from* $\{1 \dots, n\}$ *to* $\{1, \dots, n\}$.

Definition 7. *Let* $\langle \Pi; \Gamma \rangle$ *be a bordered MNL proof-structure, where* Γ *is* $A_1 \dots A_n$. *Let* s *be a switching for* $\langle \Pi; \Gamma \rangle$.

- p_s *is the (partial or total) permutation of* $\{1, \dots, n\}$, *defined by:*
  for every $i \in \{1, \dots, n\}$, $p_s(i)j$ *iff there is a* Γ*-trip in* $s(\langle \Pi; \Gamma \rangle)$ *of the form* $A_i^{\text{in}(\Pi)} \dots A_j^{\text{out}(\Pi)}$.
  Remark that, if $i \in \{1 \dots n\}$ *and there is a critical trip* $A_i^{\text{in}(\Pi)} \dots B \downarrow$ *in* $s(\langle \Pi; \Gamma \rangle)$, *then* $p_s(i)$ *is not defined.*

- $\beta_s \subseteq \{1 \dots n\}^4$ *is defined by:*
  for every $i, j, k, m \in \{1, \dots, n\}$, $\beta_s(i, j, k, m)$ *iff the* Γ*-trips* $A_i^{\text{in}(\Pi)} \dots A_{p_s(i)}^{\text{out}(\Pi)}$, $A_j^{\text{in}(\Pi)} \dots A_{p_s(j)}^{\text{out}(\Pi)}$, $A_k^{\text{in}(\Pi)} \dots A_{p_s(k)}^{\text{out}(\Pi)}$, $A_m^{\text{in}(\Pi)} \dots A_{p_s(m)}^{\text{out}(\Pi)}$ *exist in* $s(\langle \Pi; \Gamma \rangle)$ *and by linking these trips as follows*

$$A_i^{\text{in}(\Pi)} \dots A_{p_s(i)}^{\text{out}(\Pi)} \longrightarrow A_j^{\text{in}(\Pi)} \dots A_{p_s(j)}^{\text{out}(\Pi)} \longrightarrow$$
$$\longrightarrow A_k^{\text{in}(\Pi)} \dots A_{p_s(k)}^{\text{out}(\Pi)} \longrightarrow A_m^{\text{in}(\Pi)} \dots A_{p_s(m)}^{\text{out}(\Pi)}$$

  we get a bilateral trip (i.e. there is not a configuration $B^x \dots C^y \dots B^{\bar{x}} \dots C^{\bar{y}}$, *where* $x, y \in \{\uparrow, \downarrow\}$, $\bar{\uparrow} \downarrow$ *and* $\bar{\downarrow} \uparrow$*),*

- γ_s *is the subset of* $\{1, \dots, n\}$ *defined by:*
  $i \in \gamma_s$ *iff the trip* $A_j^{\text{in}(\Pi)} \dots A_{p_s(j)}^{\text{out}(\Pi)}$ *exists in* $s(\langle \Pi; \Gamma \rangle)$ *and in this trip there is at least one proper conclusion of* $\langle \Pi; \Gamma \rangle$,

- $\delta_s \subseteq \{1, \dots, n\}^2$ *is defined by:*
  for every $i, j \in \{1, \dots, n\}$, $\delta_s(i, j)$ *iff* $p_s(i)$ *and* $p_s(j)$ *are defined and there is* B *in* Π *such that* B^x *is in the* Γ*-trip from* $A_i^{\text{in}(\Pi)}$ *and* $B^{\bar{x}}$ *is in the* Γ*-trip from* $A_j^{\text{in}(\Pi)}$,

- $\chi_s \subseteq \{1,\ldots,n\}^3$ *is defined by:*
  for every $i,j,k \in \{1,\ldots,n\}$,$\chi_s(i,j,k)$ *iff* $p_s(i), p_s(j)$ *and* $p_s(k)$ *are defined and by linking these trips as follows*

$$A_i^{\mathrm{in}(\Pi)} \ldots A_{p_s(i)}^{\mathrm{out}(\Pi)} \longrightarrow A_j^{\mathrm{in}(\Pi)} \ldots A_{p_s(j)}^{\mathrm{out}(\Pi)} \longrightarrow A_k^{\mathrm{in}(\Pi)} \ldots A_{p_s(k)}^{\mathrm{out}(\Pi)}$$

  we get a bilateral trip (i.e. there is not a configuration $B^x \ldots C^y \ldots B^{\bar{x}} \ldots C^{\bar{y}}$, *where* $x,y \in \{\uparrow,\downarrow\}$, $\bar{\uparrow}\downarrow$, *and* $\bar{\downarrow}\uparrow$*).*

Remark that β_s, δ_s and $\chi_s$are related as showed in the following lemma.

Lemma 1. *Let s be a switching for* $\langle \Pi; A_1,\ldots,A_n\rangle$. *Let S be a cycle made of elements of* $\{1\ldots,n\}$. *Let us consider the following statements:*

1. *For every* $i,j,k,m \in \{1,\ldots,n\}$, *if* $i\ldots j\ldots k\ldots m$ *occurs in this order in S, then* $\beta_s(i,j,k,m)$,
2. *For every* $i,j,k,m \in \{1,\ldots,n\}$,
   - *if* $i\ldots j\ldots k$ *occurs in this order in S, then* $\chi_s(i,j,k)$, *and*
   - *if* $i\ldots j\ldots k\ldots m$ *occurs in this order in S, then either* $\delta_s(i,k)$ *does not hold, or* $\delta_s(j,m)$ *does not hold.*

Then, the statement in point 2 implies the one in point 1.

Proof. Let the statement 2 hold. Let **i** be the Γ-trip from A_i, **j** from A_j, **k** from A_k, and **m** from A_m. Let $i\ldots j\ldots k\ldots m$ occur in this order in S. Then, by the statement 2: $\chi_s(i,j,k)$, $\chi_s(i,k,m)$, $\chi_s(j,k,m)$ hold so that the possible pattern $B^x \ldots C^y \ldots B^{\bar{x}} \ldots C^{\bar{y}}$ in

$$\mathbf{i} \longrightarrow \mathbf{j} \longrightarrow \mathbf{k} \longrightarrow \mathbf{m}$$

may occur only when B^x is in **i**, C^y is in **j**, $B^{\bar{x}}$ is in **k**, $C^{\bar{y}}$ is in **m**, i.e. when $\delta_s(i,k)$ and $\delta_s(j,m)$ hold; but this is excluded by the statement 2.

Definition 8. *Let* $\sigma_1, \sigma_2 \in \mathrm{perm}(\{1\ldots n\})$. *Let* $\gamma_1, \gamma_2 \subset \{1\ldots n\}$. *Let* $\delta_1, \delta_2 \subset \{1\ldots n\}^2$. *Let* $\chi_1, \chi_2 \subset \{1\ldots n\}^3$.
$\langle \sigma_1, \gamma_1, \delta_1, \chi_1\rangle \perp \langle \sigma_2, \gamma_2, \delta_2, \chi_2\rangle$ *iff:*

- *the composition* $\sigma_1\sigma_2$ *contains exactly one cycle* T_1*; if* T_1 *is the cycle*

$$\sigma_1\sigma_2(i)\ldots(\sigma_1\sigma_2)^k(i) \ ,$$

  for some $i \in \{1\ldots n\}$ *and* $k \leq n$, *then call* T_2 *the following unique cycle of* $\sigma_2\sigma_1$*:*

$$\sigma_2(i), \sigma_2((\sigma_1\sigma_2)(i)), \ldots, \sigma_2((\sigma_1\sigma_n)^{k-1}(i)) \ ,$$

  and T the following cycle:

$$i, \sigma_2(i), (\sigma_1\sigma_2)(i) \ldots \sigma_2((\sigma_1\sigma_2)^{k-1}(i)), (\sigma_1\sigma_2)^k(i) \ .$$

- *T is bilateral, i.e. T does not contain the configuration $j \ldots k \ldots j \ldots k$,*

- *for every $j \in \{1 \ldots n\}$, if $j \in \gamma_1$ then j is in T_1, and if $j \in \gamma_2$ then j is in T_2,*

- *for every $j, k, m, h \in \{1 \ldots n\}$,*
  - *if $j \ldots k \ldots m \ldots h$ is a portion of T_1, then either $\delta_1(j, m)$ does not hold, or $\delta_1(k, h)$ does not hold,*
  - *if $j \ldots k \ldots m \ldots h$ is a portion of T_2, then either $\delta_2(j, m)$ does not hold, or $\delta_2(k, h)$ does not hold,*

- *for every $j, k, m \in \{1 \ldots n\}$,*
  - *if $j \ldots k \ldots m$ is a portion of T_1 then $\chi_1(j, k, m)$, and*
  - *if $j \ldots k \ldots m$ is a portion of T_2 then $\chi_2(j, k, m)$.*

Remark that $\langle \sigma_1, \gamma_1, \delta_1, \chi_1 \rangle \perp \langle \sigma_2, \gamma_2, \delta_2, \chi_2 \rangle$ iff $\langle \sigma_2, \gamma_2, \delta_2, \chi_2 \rangle \perp \langle \sigma_1, \gamma_1, \delta_1, \chi_1 \rangle$.

Definition 9. – *The set* $\mathrm{Dperm}(\{1 \ldots n\})$ *is*

$$\{\langle \sigma, \gamma, \delta, \chi \rangle \mid \sigma \in \mathrm{perm}(\{1 \ldots n\}), \gamma \subseteq \{1 \ldots n\}, \delta \subseteq \{1 \ldots n\}^2, \chi \subseteq \{1 \ldots n\}^3\}.$$

– *If $X, Y \subseteq \mathrm{Dperm}(\{1 \ldots n\})$ then*

$$X \perp Y \text{ iff } \forall x \in X, y \in Y . x \perp y \ .$$

Definition 10. *Let $\langle \Pi; \Gamma \rangle$ be a bordered MNL proof-structure.* $\mathrm{type}(\langle \Pi; \Gamma \rangle)$ $\{\langle p_s, \beta_s, \gamma_s \rangle \mid s$ *switching of* $\langle \Pi; \Gamma \rangle\}$.

Remark that if we restrict us to total permutations and we delete γ, δ and χ, the definition 9 is exactly the definition of orthogonality of permutations and between sets of permutations given by Girard in [3].

If Π is *commutative*, then, for every s, p_s is a total permutation.

The need for δ_s and χ_s comes from the condition that trips in MNL proof-nets must be bilateral. The need for γ_s comes from the condition that the unique cycle in MNL proof-nets, under a switching s, must contain all the conclusions.

So, when Π is commutative, we can disregard $\gamma_s, \delta_s, \chi_s$.

4 Orthogonal MNL modules

Definition 11. *$\langle \Pi; \Gamma \rangle$ is a MNL proof-module iff:*

- *$\langle \Pi; \Gamma \rangle$ is a bordered MNL proof-structure and Γ is not empty,*
- *for every switching s for $\langle \Pi; \Gamma \rangle$:*
  - *in $s(\langle \Pi; \Gamma \rangle)$ there is no cycle,*
  - *for every critical trip T in $s(\langle \Pi; \Gamma \rangle)$there is no proper conclusion in T.*

Remark that the only condition added to the definitions given by Girard in [3] is the last point in definition 11.

Definition 12. $\langle \Pi_1; \Gamma\rangle \bot \langle \Pi_2; \Gamma\rangle$ *iff:*

- $\langle \Pi_1; \Gamma\rangle$ *and* $\langle \Pi_2; \Gamma\rangle$ *are compatible MNL modules,*
- $\text{type}(\langle \Pi_1; \Gamma\rangle) \bot \text{type}(\langle \Pi_2; \Gamma\rangle)$.

Theorem 1. *If* $\langle \Pi_1; \Gamma\rangle \bot \langle \Pi_2; \Gamma\rangle$, *then* $\langle \Pi_1; \Gamma\rangle \circ \langle \Pi_2; \Gamma\rangle$ *is a proof net of MNL.*

Proof. Let s be an arbitrary switching for $\Pi_1 \circ \Pi_2$; we show that:

1. there is a cycle T^* in $s(\Pi_1 \circ \Pi_2)$,
2. T^* is bilateral,
3. T^* contains all the conclusions of $\Pi_1 \circ \Pi_2$,
4. T^* is the unique cycle in $s(\Pi_1 \circ \Pi_2)$.'

Observe that $ss_1 + s_2$, where s_1 is a switching for Π_1 and s_2 is a switching for Π_2, and the conclusions of $\Pi_1 \circ \Pi_2$ are the elements of $\text{PCon}(\langle \Pi_1; \Gamma\rangle) \cup \text{PCon}(\langle \Pi_2; \Gamma\rangle)$.

Let σp_{s_1} and τp_{s_2}, $\gamma_1 \gamma_{s_1}$, $\gamma_2 \gamma_{s_2}$, $\delta_1 \delta_{s_1}$, $\delta_2 \delta_{s_2}$, $\chi_1 \chi_{s_1}$, $\chi_2 \chi_{s_2}$.

Proof of 1. Since $\text{type}(\langle \Pi_1; \Gamma\rangle) \bot \text{type}(\langle \Pi_2; \Gamma\rangle)$ we get that $\langle \sigma, \beta_1, \gamma_1\rangle \bot \langle \tau, \beta_2, \gamma_2\rangle$. Take the unique cycle T_1 of $\sigma\tau$:

$$\sigma\tau(i), (\sigma\tau)^2(i), \ldots, (\sigma\tau)^k(i)i$$

where $i \in \{1 \ldots n\}$ and $k \leq n$, and consider the cycle T obtained by interpolating in T_1 the cycle T_2 in $\tau\sigma$:

$$i, \tau(i), \sigma\tau(i), \tau(\sigma\tau(i)), (\sigma\tau)^2(i), \ldots, \tau((\sigma\tau)^{k-1}(i)), (\sigma\tau)^k(i)i$$

T^* is:

$$A_i^{\text{in}(\Pi_2)} \ldots A_{\tau(i)}^{\text{out}(\Pi_2)} \ldots A_{\sigma\tau(i)}^{\text{in}(\Pi_2)} \ldots A_{\tau(\sigma\tau(i))}^{\text{out}(\Pi_2)} \ldots$$

$$A_{(\sigma\tau)^2(i)}^{\text{in}(\Pi_2)} \ldots A_{\tau((\sigma\tau)^{k-1}(i))}^{\text{out}(\Pi_2)} \ldots A_{(\sigma\tau)^k(i)}^{\text{in}(\Pi_2)}$$

Remark that $\text{out}(\Pi_1)\,\text{in}(\Pi_2)$ and $\text{in}(\Pi_1)\,\text{out}(\Pi_2)$.
T^* is a cyclic trip in $s(\Pi_1 \circ \Pi_2)$ since, by definition of σ and τ,

$$A_{\tau(i)}^{\text{out}(\Pi_2)} \ldots A_{\sigma\tau(i)}^{\text{in}(\Pi_2)}, A_{\tau(\sigma\tau(i))}^{\text{out}(\Pi_2)} \ldots A_{(\sigma\tau)^2(i)}^{\text{in}(\Pi_2)}, \ldots, A_{\tau((\sigma\tau)^{k-1}(i))}^{\text{out}(\Pi_2)} \ldots A_{(\sigma\tau)^k(i)}^{\text{in}(\Pi_2)}$$

are trips in $s_1(\langle \Pi_1; \Gamma\rangle)$ and

$$A_i^{\text{in}(\Pi_2)} \ldots A_{\tau(i)}^{\text{out}(\Pi_2)}, A_{\sigma\tau(i)}^{\text{in}(\Pi_2)} \ldots A_{\tau(\sigma\tau(i))}^{\text{out}(\Pi_2)}, A_{(\sigma\tau)^{k-1}(i)}^{\text{out}(\Pi_2)} \ldots A_{\tau((\sigma\tau)^{k-1}(i))}^{\text{in}(\Pi_2)}$$

are trips in $s_2(\langle \Pi_2; \Gamma\rangle)$.

Proof of 2. First remark that T is bilateral so that also the sequence of the occurrence of A_i^x (with $i \in \{1 \ldots n\}$ and $x \in \{\uparrow, \downarrow\}$) in T^* is bilateral; i.e. the border is arranged inside T^* in a bilateral way.

Suppose that in T^* there is a non bilateral pattern

$$B^x \dots C^y \dots B^{\bar{x}} \dots C^{\bar{y}}$$

$(x, y \in \{\uparrow, \downarrow\})$.

First case: $B \in\mid \Pi_1 \mid$ and $C \in\mid \Pi_1 \mid$.
In this case, $B^x, C^y, B^{\bar{x}}, C^{\bar{y}}$ are in some Γ-trips of $\sigma_1(\langle \Pi_1; \Gamma \rangle)$. If these occurrences are distributed in at most three Γ-trips, then we find three Γ-trips:

$$A_j^{\mathrm{in}(\Pi_1)} \dots A_{\sigma(j)}^{\mathrm{out}(\Pi_1)}, A_k^{\mathrm{in}(\Pi_1)} \dots A_{\sigma(k)}^{\mathrm{out}(\Pi_1)}, A_m^{\mathrm{in}(\Pi_1)} \dots A_{\sigma(m)}^{\mathrm{out}(\Pi_1)}$$

such that the order of the occurrences of j, k, m in T_1 is $j \dots k \dots m$, and

$$A_j^{\mathrm{in}(\Pi_1)} \dots A_{\sigma(j)}^{\mathrm{out}(\Pi_1)} \longrightarrow A_k^{\mathrm{in}(\Pi_1)} \dots A_{\sigma(k)}^{\mathrm{out}(\Pi_1)} \longrightarrow A_m^{\mathrm{in}(\Pi_1)} \dots A_{\sigma(m)}^{\mathrm{out}(\Pi_1)}$$

contains the pattern$B^x, C^y, B^{\bar{x}}, C^{\bar{y}}$; but then not $\chi_1(j, k, m)$, in contradiction with type$(\langle \Pi_1; \Gamma \rangle) \perp$ type$(\langle \Pi_2; \Gamma \rangle)$.

Let B^x be in the Γ-trips from $A_j^{\mathrm{in}(\Pi_1)}$, C^y be in the Γ-trips from $A_k^{\mathrm{in}(\Pi_1)}$, $B^{\bar{x}}$ be in the Γ-trips from $A_m^{\mathrm{in}(\Pi_1)}$, $C^{\bar{y}}$ be in the Γ-trips from $A_h^{\mathrm{in}(\Pi_1)}$. In T_1 the order of the occurrences j, k, m, h is just:

$$j \dots k \dots m \dots h$$

but $\delta_1(j, m)$ and $\delta_1(k, h)$ in contradiction with type$(\langle \Pi_1; \Gamma \rangle) \perp$ type$(\langle \Pi_2; \Gamma \rangle)$.

Second case: $B \in\mid \Pi_2 \mid$ and $C \in\mid \Pi_2 \mid$. Analogous to the first one.

Third case: $B \in\mid \Pi_1 \mid$ and $C \in\mid \Pi_2 \mid$.

In this case B^x and $B^{\bar{x}}$ are in some Γ-trips of $s_1(\langle \Pi_1; \Gamma \rangle)$, say B^x is in the Γ-trip from

$$A_j^{\mathrm{in}(\Pi_1)}$$

and $B^{\bar{x}}$ is in the Γ-trip from

$$A_m^{\mathrm{in}(\Pi_1)} \ ;$$

C^y and $C^{\bar{y}}$ are in some Γ-trips of $s_2(\langle \Pi_2; \Gamma \rangle)$, say C^y is in the Γ-trip from

$$A_h^{\mathrm{in}(\Pi_2)}$$

and $C^{\bar{y}}$ is in the Γ-trip from

$$A_k^{\mathrm{in}(\Pi_2)} \ .$$

In T the order of the occurrences of j, m, k, h is:

$$j \dots k \dots m \dots h \ .$$

Consider the portion of T^*:

$$A_j^{\mathrm{in}(\Pi_1)} \dots B^x \dots A_{\sigma(j)}^{\mathrm{out}(\Pi_1)} \dots A_k^{\mathrm{in}(\Pi_2)\,\mathrm{out}(\Pi_1)}$$

and let r be such that $A_r^{\text{out}(\Pi_1)}$ is in this portion after B^x, whereas $A_r^{\text{in}(\Pi_1)}$ is not in this portion (such an r exists by obvious considerations.) Now, consider the portion:

$$A_j^{\text{in}(\Pi_1)} \ldots B^x \ldots A_r^{\text{out}(\Pi_1)} \ldots A_m^{\text{in}(\Pi_1)\,\text{out}(\Pi_2)} \ldots B^{\bar{x}} \ldots A_{\sigma(m)}^{\text{out}(\Pi_1)} \; .$$

rm, because, otherwise there would be the pattern:

$$A_r^{\text{in}(\Pi_1)} \ldots B^x \ldots A_r^{\text{out}(\Pi_1)} \ldots B^{\bar{x}} \; ,$$

or the pattern:

$$B^x \ldots A_r^{\text{out}(\Pi_1)} \ldots B^{\bar{x}} \ldots A_r^{\text{in}(\Pi_1)}$$

with only formulas belonging to Π_1, and this is excluded by the first case. Therefore the portion of T^* from $A_j^{\text{in}(\Pi_1)}$ to $A_{\tau(h)}^{\text{out}(\Pi_2)}$ looks as follows:

$$A_j^{\text{in}(\Pi_1)} \ldots B^x \ldots A_r^{\text{out}(\Pi_1)} \ldots A_k^{\text{in}(\Pi_2)} \ldots C^y \ldots A_{\tau(k)}^{\text{out}(\Pi_2)} \ldots$$
$$\ldots\ldots A_r^{\text{in}(\Pi_1)} \ldots B^{\bar{x}} \ldots A_h^{\text{in}(\Pi_2)} \ldots C^{\bar{y}} \ldots A_{\tau(h)}^{\text{out}(\Pi_2)} \; .$$

Restrict our attention to Γ-trips of $s_2(\langle \Pi_2; \Gamma \rangle)$ which occur in the here above portion of T^*:

$$A_r^{\text{in}(\Pi_2)} \ldots A_{\tau(r)}^{\text{out}(\Pi_2)} \tag{1}$$
$$A_k^{\text{in}(\Pi_2)} \ldots C^y \ldots A_{\tau(k)}^{\text{out}(\Pi_2)} \tag{2}$$
$$A_{\tau^{-1}(r)}^{\text{in}(\Pi_2)} \ldots A_r^{\text{out}(\Pi_2)} \tag{3}$$
$$A_h^{\text{in}(\Pi_2)} \ldots C^{\bar{y}} \ldots A_{\tau(h)}^{\text{out}(\Pi_2)} \tag{4}$$

but (1)⟶(2)⟶(3)⟶(4) is not bilateral since it contains the pattern

$$A_r^{\text{in}(\Pi_2)} \ldots C^y \ldots A_r^{\text{out}(\Pi_2)} \ldots C^{\bar{y}} \; ;$$

with only formulas belonging to Π_2, and this contradicts the second case.

Fourth case: $B \in\mid \Pi_2 \mid$ and $C \in\mid \Pi_1 \mid$. Analogous to the third one.

Proof of 3. Let C be a proper conclusion of a $\langle \Pi_1; \Gamma \rangle$. Since $\langle \Pi_1; \Gamma \rangle$ is an MNL module C is not in a critical trip of $s_1(\langle \Pi_1; \Gamma \rangle)$, so that C is in a Γ-trip, say in the Γ-trip

$$A_j^{\text{in}(\Pi_1)} \ldots A_{\sigma(j)}^{\text{out}(\Pi_1)} \; .$$

So $j \in \gamma_1$.

Since type$(\langle \Pi_1; \Gamma \rangle) \perp$ type$(\langle \Pi_2; \Gamma \rangle)$,$j$ is in T so that

$$A_j^{\text{in}(\Pi_1)} \ldots A_{\sigma(j)}^{\text{out}(\Pi_1)}$$

is in T^* which contains C.

If C is a proper conclusion of $\langle \Pi_2; \Gamma \rangle$ the proof is analogous.

Proof of 4. Let S be a cycle in $s(\langle \Pi_1; \Gamma \rangle \circ \langle \Pi_2; \Gamma \rangle)$. S can not be a cycle in $s_1(\langle \Pi_1; \Gamma \rangle)$ or a cycle in $s_2(\langle \Pi_2; \Gamma \rangle)$ since $\langle \Pi_1; \Gamma \rangle$ are $\langle \Pi_2; \Gamma \rangle$ MNL modules. So S must contain some formulas of the border Γ. If S contains A_j^x, where A_j is in Γ, and A_j^x is in T^*, then cycle S is equal to T^*. Otherwise, as S is a cycle, we can extract from S another cycle in $\sigma\tau$. But this would contradict type$(\langle \Pi_1; \Gamma \rangle) \perp$ type$(\langle \Pi_2; \Gamma \rangle)$.

References

1. V.M. Abrusci and P. Ruet. Non-commutative logic I: the multiplicative fragment. Università di Roma Tre and McGill University Preprints, 1998.
2. V. Danos and L. Regnier. The structure of multiplicatives. *Archive for Mathematical Logic*, 28:181 - 203, 1989.
3. J.-Y. Girard. Multiplicatives. In *Logic and Computer Science: new trends and applications (Lolli Editor)*, pages 11 - 34. Rendiconti del Seminario Matematico dell'Università e Politecnico di Torino,1988.

Elementary Complexity and Geometry of Interaction
(Extended Abstract)

Patrick Baillot[1] and Marco Pedicini[2]

[1] Institut de Mathématiques de Luminy (Marseille), C.N.R.S.
`baillot@iml.univ-mrs.fr`
[2] Istituto per le Applicazioni del Calcolo "Mauro Picone" (Roma), C.N.R.
`marco@iac.rm.cnr.it`

Abstract. We introduce a geometry of interaction model given by an algebra of clauses equipped with resolution (following [Gir95a]) into which proofs of Elementary Linear Logic can be interpreted. In order to extend geometry of interaction computation (Execution) to more programs of the algebra than just those coming from proofs, we define a variant of Execution (called Weak Execution). Its application to any program of clauses is shown to terminate with a bound on the number of steps which is elementary in the size of the program. We establish that Weak Execution coincides with standard Execution on programs coming from proofs.

Geometry of interaction (GOI) was introduced by Girard ([Gir88a]) as a semantics of computation which: on the one hand, contrary to denotational semantics interprets explicitly the dynamics of computation and handles finite objects; on the other hand expresses this dynamic by more *mathematical* means than syntactical rewriting.

The EXECUTION operation is the mathematical tool inside the model used to interpret the cut-elimination process. This operation is not always defined and sufficient conditions have been given which ensure termination of the computation : in the case of second-order Linear Logic ([Gir88a, Gir95a]) and of untyped lambda-calculus [MR91]), operators coming from the syntax do satisfy such conditions (a nilpotency condition for instance in the case of LL). Various frameworks have been used to describe GOI models: bounded operators on Hilbert spaces ([Gir88a, DR95]), partial applications ([Dan90, Reg92]) and clauses ([Gir95a]). This latter point of view is the one we adopt here.

Elementary Linear Logic (ELL), as Light Linear Logic (LLL), is a variant of Linear Logic in which the rules introducing exponentials have been modified (cf. [Gir95b]) in order to limit the size explosion of proofs during normalization. It is obtained by removing the two principles : $!A \vdash A$ and $!A \vdash !!A$; contraction and weakening are kept unchanged. We consider here a version of ELL without additive connectives and where introduction of the modality ! is handled through a (multi-)functorial promotion rule (called t-promotion, see [Ped96]), which offers the advantage of having simple proof-nets. A proof-net has two main parameters: its *size* (say the number of edges) and its *depth* (maximal nesting of the boxes it contains). The number of steps of its normalization is bounded by a function of the size which is elementary: the expression of this function is an exponential tower whose height only depends on the depth (see [Ped96]).

A drawback of ELL (as well as of LLL) is the lack of a specific semantics of proofs, though a semantics of provability has been given by Kanovitch et al. ([KOS97]). We address this problem from the angle of a GOI semantics.

Achievements and limits of the present work. We present here an algebra of clauses along the lines of [Gir95a] with a kind of depth-preservation property analogous to that of ELL (section 1). Execution is defined through resolution and the operators are certain sets of clauses; a comparison of these operators with Prolog programs can be found in [Gir95a], section 2.3. In addition to usual EXECUTION we define a *weak execution* (section 2) which amounts to giving up the computation of certain products of the EXECUTION (products yielding a deadlock when one restrains the depth).

A size and a depth are defined for general operators respectively as the number of clauses and the arity of the predicates of the terms (all predicates have the same arity). Our main result is then that WEAK EXECUTION always terminates (there is no need for nilpotency sufficient condition for instance) and that the depth being fixed, the number of steps of the computation is bounded by a function of the size of the program which is elementary (proved in section 2). Clearly speaking, in this setting we can bound in advance the run-time of a program provided we know its size and depth. Therefore the intrinsic elementary bound obtained in ELL by logical means has been extended to a semantical ground.

Yet this WEAK EXECUTION presents a serious drawback as it is not in general an associative operation ... Still at least one inclusion is obtained instead of the expected equality (we call this property sub-associativity): the result of *global* EXECUTION is included in the result of any *modular* EXECUTION (see section 3 for a precise statement).

In this abstract we only sketch the main proofs; complete proofs can be found in [BP99].

Acknowledgments. Authors wish to thank Jean-Yves Girard for important suggestions and for pointing out the crucial lemma 2.

1 Resolution Algebra

We need first to introduce clauses and resolution. We then recall the definition of the algebra of clauses given in [Gir95a] before describing the particular algebra we consider in this work : the *layered algebra of clauses*.

A *term language* T is built over variables and a set of symbols of functions; its elements will be denoted t, u. Let $\{P_i\}_{i \in I}$ be a set of predicate symbols given together with their arity; the *language of atoms* $\mathcal{L}$ built over T and this set of predicates is the set of $P_i(t_1, \ldots, t_n)$, where n is the arity of P_i and the t_j's belong to T.

We say that two terms (or atoms) e and e' are *comparable* when there is a substitution θ defined over the variables in e and e' such that $e\theta = e'\theta$.

If e, e' are comparable then there exists a *most general unifier* (*m.g.u.*) i.e. a substitution θ_0 such that for every unifier θ there is a θ' such that $\theta = \theta_0\theta'$. If e and e' are not comparable, we say that they are *orthogonal*: $e \perp e'$.

A *clause* ϕ in the language $\mathcal{L}$ is a sequent

$$\forall x_0, \ldots, x_d. \, (P_i(t_0, \ldots, t_m) \vdash P_j(u_0, \ldots, u_n)) ,$$

where $P_i(t_0, \ldots, t_m)$ and $P_j(u_0, \ldots, u_n)$ are atoms of L with the same variables $x_0, \ldots, x_d$. We will omit to write the quantification.

The *head* of the clause ϕ is the atom $\mathtt{head}(\phi) = P_i(t_0, \ldots, t_m)$, its *tail* is the atom $\mathtt{tail}(\phi) = P_j(u_0, \ldots, u_n)$.

We introduce also a formal clause 0. Let C denote the set of clauses over $\mathcal{L}$.

A substitution θ *acts* on a clause ϕ by : $\phi\theta = \mathtt{head}(\phi)\theta \vdash \mathtt{tail}(\phi)\theta$.

Definition 1 (Resolution). *Given two clauses ϕ and ϕ' we can assume they have disjoint variables (by choosing appropriate instantiations). If $\mathtt{tail}(\phi)$ is comparable with $\mathtt{head}(\phi')$ and θ is their m.g.u. we define the* resolution *of the two clauses as the clause*

$$\phi \cdot \phi' = \mathtt{head}(\phi)\theta \vdash \mathtt{tail}(\phi')\theta$$

Otherwise, if $\mathtt{tail}(\phi)$ and $\mathtt{head}(\phi')$ are not comparable: $\phi \cdot \phi' = 0$.

We fix by convention that the resolution of the clause zero with any other clause is zero; this implies that resolution is associative.

A clause ϕ is said to be a *projection* (resp. a *null-square*) if $\phi^2 = \phi$ (resp. $\phi^2 = 0$), which is equivalent to $\mathtt{head}(\phi) = \mathtt{tail}(\phi)$ (resp. $\mathtt{head}(\phi) \perp \mathtt{tail}(\phi)$).

Definition 2 (Resolution Algebra). *Let $\lambda^*(\mathcal{L})$ be the set of all finite formal linear combinations $\sum \alpha_i\phi_i$ where the scalars α_i belong to $\mathbb{C}$ and the clauses ϕ_i to C. The set $\lambda^*(\mathcal{L})$ is equipped with*

- *a structure of complex vector space,*
- *a structure of complex algebra, the multiplication being extended by bilinearity from resolution:* $\sum \alpha_i\phi_i \sum \beta_j\phi'_j = \sum \alpha_i\beta_j(\phi_i \cdot \phi'_j)$,
- *a unit w.r.t. multiplication :* $\sum_{i\in I} P_i(x_0, \ldots, x_n) \vdash P_i(x_0, \ldots, x_n)$,
- *an anti-involution defined by* $(\sum \alpha_i\phi_i)^* = \sum \overline{\alpha_i}\phi_i{}^*$ *where* $\phi^* := \mathtt{tail}(\phi) \vdash \mathtt{head}(\phi)$.

A norm can be introduced in order to get a $\mathbb{C}^*$-algebra, see [Gir95a].

Another way to write a combination of clauses is as $\sum \alpha(\phi)\phi$, where the sum is taken over C and α is an application from the set of clauses C to $\mathbb{C}$ such that $\alpha^{-1}(\mathbb{C}\backslash\{0\})$ is finite. We will use this notation when it is more convenient.

If $U = \sum \alpha_i\phi_i$ and $V = \sum \beta_i\phi_i$ are two elements with coefficients in $\mathbb{N}$, we write $U \subseteq V$ if for all i, $\alpha_i \leq \beta_i$.

Definition 3 (Execution Formula). *A* wiring *is a finite sum of clauses $\sum \phi_i$ such that for $i \neq j$: $\mathtt{head}(\phi_i) \perp \mathtt{head}(\phi_j)$ and $\mathtt{tail}(\phi_i) \perp \mathtt{tail}(\phi_j)$.*

A loop *is a pair of wirings (U, σ) such that σ is hermitian (i.e. $\sigma^* = \sigma$),*

A loop converges *when σU is nilpotent, i.e. when $(\sigma U)^n = 0$ for some n. The* execution *of the loop (U, σ) is then the element*

$$\mathtt{Ex}_\sigma(U) := U(1 - \sigma U)^{-1} = U \sum_{k=0}^{n} (\sigma U)^k$$

and the result of the execution is given by

$$\mathtt{Result}_\sigma(U) := (1 - \sigma^2)\mathtt{Ex}_\sigma(U)(1 - \sigma^2).$$

Remark 1. Another way to write the execution is directly as a sum of clauses:

$$\mathtt{Ex}_\sigma(U) = \sum_{\substack{\phi_0 \in U \\ \phi_i \in \sigma U, 1 \leq i \leq k \\ k \leq n}} \phi_0 \cdot \phi_1 \cdots \phi_k$$

Now we specify the particular language we are going to consider. The terms of T are built over a set of unary symbols of function $\{p, q, r, s\}$; therefore such terms have exactly one free variable, and $t[x]$ will denote a term with free variable x. The *length* $|t|$ of t is the number of symbols of function appearing in it.

Remark 2. Notice that as T is defined over unary symbols of function, if two terms t and u are unifiable then their m.g.u. θ leaves at least one of the two terms unchanged (up to renaming of its variable). For any pair of terms (t, u), only one of the following cases can occur: $t \perp u$ or $t \leq u$ or $t \geq u$, where $t \leq u$ means that u is the unchanged term.

We will consider a family of symbols of predicate $\{P_i\}_{i \in \{1,\dots,m\}}$ of same arity[1] $(d+1)$. Let $T^d \cdot m$ denote the set of atoms defined this way.

The set C^d is the set of clauses:

$$\phi = \forall x_0, \dots, x_d. \left(P_i(t_0[x_0], \dots, t_d[x_d]) \vdash P_j(u_0[x_0], \dots, u_d[x_d])\right),$$

where $\mathtt{head}(\phi)$ and $\mathtt{tail}(\phi)$ belong to $T^d \cdot m$. Notice that t_k and u_k $(0 \leq k \leq d)$ are required to have the same free variable.

We call *layered algebra* the algebra of clauses defined over C^d and we denote it by $\lambda^*(T^d \cdot m)$. From now on this is the algebra we consider.

2 Weak Execution

A *word* of clauses w is a finite sequence of clauses $w = (\phi_1, \dots, \phi_n)$ with $\phi_i \in C^d$, and the *product clause* is $\phi_1 \cdot \phi_2 \cdots \phi_n$. A *sub-product* of the word w is the product clause of a word $(\phi_i, \dots, \phi_j)$ for some $i \leq j \leq n$.

Given a clause $\phi = P(t_0, \dots, t_d) \vdash P'(u_0, \dots, u_d)$ its *width* is defined as $||\phi|| := \sup\{|t_k|, |u_k| \,/\, 0 \leq k \leq d\}$. The width of a word w is simply given by $||w|| := \sup_{1 \leq i \leq n}(||\phi_i||)$. The *cardinality* of w is: $N(w) := \#\{\phi_i \mid 1 \leq i \leq n\}$.

[1] The choice of $d+1$ is done to keep the same notations when we interpret proof-structures, see section 4.

Example : Consider in $\mathcal{C}^0$ the clause $\phi = P(x) \vdash P(rx)$ and let w_n be the word $(\phi, \ldots, \phi)$ of length n. In that case we have $||w_n|| = ||\phi|| = 1$, $N(w_n) = 1$ and the product of w_n is the clause $P(x) \vdash P(r^n x)$.

Definition 4 (Acyclicity). *A clause $\phi = P(t_0, \ldots, t_d) \vdash P'(u_0, \ldots, u_d)$ is an* acyclic clause *if $P \neq P'$, or ($P = P'$ and there exists $k \leq d$ such that for every $i < k$ we have $u_i = t_i$ and $u_k \perp t_k$).*

An acyclic word *(resp.* strictly acyclic word*) is a word $(\phi_1, \ldots, \phi_n)$ such that every sub-product ψ is either an acyclic clause or a projection (resp. an acyclic clause).*

Example : Consider in $\mathcal{C}^1$ the clauses $\phi_1 = P(sx_0, x_1) \vdash P(rx_0, x_1)$, $\phi_2 = P(rx_0, rrx_1) \vdash P(rx_0, sx_1)$ and $\phi_3 = P(x_0, sx_1) \vdash P(rx_0, rsx_1)$. Each of them is an acyclic clause. The word $w = (\phi_1, \phi_2, \phi_3)$ has a non-null product but is not acyclic since its subproduct $\phi_2 \cdot \phi_3 = P(rx_0, rrx_1) \vdash P(rrx_0, rsx_1)$ is not an acyclic clause (though it is a null-square).

We now introduce a restricted form of execution over strictly acyclic words of clauses. Contrarily to usual execution we define it not only for converging loops but for any pair of combination (U, σ); theorem 1 will establish the fact that this definition always makes sense (the sum is finite).

Definition 5 (Weak Execution). *Given a pair of combinations (U, σ) denote $U = \sum \alpha(\phi)\phi$ and $\sigma U = \sum \gamma(\phi)\phi$. Its* weak execution *is defined as:*

$$\mathtt{Ex}^{\dagger}_{\sigma}(U) = \sum_{(\phi_0, \phi_1 \ldots, \phi_n) \in A} \alpha(\phi_0)(\prod_{i=1}^{n} \gamma(\phi_i))\, \phi_0 \cdot \phi_1 \cdot \cdots \cdot \phi_n$$

$$\textit{where } A := \left\{ (\phi_0, \phi_1 \ldots, \phi_n) \,\middle|\, \begin{array}{l} \alpha(\phi_0) \neq 0, \gamma(\phi_i) \neq 0 \textit{ when } i \neq 0, \textit{ and} \\ \textit{the word } (\phi_0, \ldots, \phi_n) \textit{ is strictly acyclic} \end{array} \right\}.$$

As a particular case, given a loop (U, σ) its weak execution *is*

$$\mathtt{Ex}^{\dagger}_{\sigma}(U) = \sum_{(\phi_0, \phi_1 \ldots, \phi_n) \in A'} \phi_0 \cdot \phi_1 \cdot \cdots \cdot \phi_n$$

$$\textit{where } A' := \left\{ (\phi_0, \phi_1 \ldots, \phi_n) \,\middle|\, \begin{array}{l} \phi_0 \in U, \phi_i \in \sigma U \textit{ when } i \neq 0, \textit{ and} \\ \textit{the word } (\phi_0, \ldots, \phi_n) \textit{ is strictly acyclic} \end{array} \right\}.$$

The result of the weak execution is in that case defined as

$$\mathtt{Result}^{\dagger}_{\sigma}(U) := (1 - \sigma^2)\mathtt{Ex}^{\dagger}_{\sigma}(U)(1 - \sigma^2).$$

Remark 3. Note that in cases where $\mathtt{Ex}_{\sigma}(U)$ makes sense ((U, σ) is a loop and σU is nilpotent), we have that $\mathtt{Ex}^{\dagger}_{\sigma}(U) \subseteq \mathtt{Ex}_{\sigma}(U)$.

Our first goal is to show that we can bound the width of the product clause of an acyclic word (proposition 1). In the case of a strictly acyclic word this implies that the length of the word cannot exceed a certain bound (depending on the number and the width of the clauses) without yielding zero as result. This bound will be expressed as an exponential tower of height d (proposition 2).

Proposition 1. *Given an acyclic word $w = (\phi_1, \dots \phi_n)$ with non-null product, we have the following inequality[2]: $||\phi_1 \cdot \dots \cdot \phi_n|| \leq L(||w||, N(w), d)$, where L is defined by*

$$L(a,b,d) := 2^{2^{\cdot^{\cdot^{2^{4ab(d+1)^2}}}}}$$

and the height of the exponential tower is d (for d=0 we get the exponent $8ab$).

This proposition will be proved further. The result relies of course on the fact that w is strictly acyclic. Otherwise given a fixed width (of word) and cardinality one might exhibit non acyclic words whose products are of arbitrary big width: see for instance the first example given where for any n, $||w_n|| = 1$, $N(w_n) = 1$ and the product of w_n is $P(x) \vdash P(r^n x)$ whose width is n.

Definition 6. *Given three integers $l, N \geq 1$ and s we define*

$$B(l,N,s) := 2^{2^{\cdot^{\cdot^{2^{9 \cdot lN(s+1)^2}}}}}$$

where the height of the exponential tower is $s+1$.

Proposition 2. *Given a strictly acyclic word w with non-null product and such that $||w|| \geq 1$, its length is bounded by $B(||w||, N(w), d)$.*

We give now the main result: weak execution always terminates and can be computed in an *elementary* number of resolution steps. We state it first for a loop and then give the result for an arbitrary pair of combinations:

Theorem 1. *Let (U, σ) be a loop and let us fix the variables $N = \#\sigma U$ and $k = 1 + \max\{||\phi|| \,/\, \phi \in \sigma U\}$. We have*

$$\mathbf{Ex}^{\dagger}_{\sigma}(U) = \sum_{\substack{(\phi_0,\dots,\phi_n) \in A' \\ n \leq B(k,N,d)}} \phi_0 \cdot \phi_1 \cdot \dots \cdot \phi_n$$

where A' is the set given in definition 5.

More generally, (U, σ) being simply a pair of combinations let us denote $U = \sum \alpha(\phi)\phi$, $\sigma U = \sum \gamma(\phi)\phi$. We define then as before $N = \#\{\phi, \gamma(\phi) \neq 0\}$ and $k = 1 + \max\{||\phi|| \,/\, \gamma(\phi) \neq 0\}$. We have:

[2] Note that $||\phi_1 \cdot \dots \cdot \phi_n||$ should not be confused with $||(\phi_1, \dots, \phi_n)||$, the former being the width of a clause and the latter the width of the word.

$$\mathrm{Ex}^{\dagger}_{\sigma}(U) = \sum_{\substack{(\phi_0,\ldots,\phi_n)\in A \\ n\leq B(k,N,d)}} \alpha(\phi_0)(\prod_{i=1}^{n}\gamma(\phi_i))\ \phi_0\cdot\phi_1\cdot\dots\cdot\phi_n$$

where A is the set given in definition 5.

Proof (Proof of theorem 1). Let $w = (\phi_0,\ldots,\phi_n)$ be a word in the set A with $n \geq 1$. Let us denote by w' the word $(\phi_1,\ldots,\phi_n)$ on σU. We have $N(w') \leq N$ and $||w'|| \leq k$. Now, if $n > B(k,N,d)$ then $n > B(||w'||, N(w'), d)$ and we know by proposition 2 that w' (and consequently w) has a null product.

Therefore the sum in $\mathrm{Ex}^{\dagger}_{\sigma}(U)$ can be restricted to the words of A such that $n \leq B(k,N,d)$. □

Let us introduce a few more notations on clauses and words of clauses. To each predicate symbol P_i of our set we associate d predicate symbols, one for each arity $k+1$ in $\{1,\ldots,d\}$; we will denote them all by P_i as anyway in atoms the arity of the predicate will be made explicit by the number of terms. For $0 \leq k \leq d-1$ we denote by $T^k \cdot m$ the language built from T and the family of predicates of arity $k+1$ and $\mathcal{C}^k$ is defined as before from $T^k \cdot m$.

Given a clause $\phi = P(t_0,\ldots,t_d) \vdash P'(u_0,\ldots,u_d)$ of $\mathcal{C}^d$ and $0 \leq k \leq d-1$, its k-th *layer* is the clause of $\mathcal{C}^0$ $[\phi]_k := P(t_k) \vdash P'(u_k)$ and its k-th *truncation* is the clause $[\phi]_{(0,k)} := P(t_0,\ldots,t_k) \vdash P'(u_0,\ldots,u_k)$ of $\mathcal{C}^k$.

The k-th layer of a word $w = (\phi_1,\ldots,\phi_n)$ is $[w]_k = ([\phi_1]_k,\ldots,[\phi_n]_k)$; similarly its k-th truncation is $[w]_{(0,k)} = ([\phi_1]_{(0,k)},\ldots,[\phi_n]_{(0,k)})$.

We define width of atoms by: $|P(t_0,\ldots,t_k)| = \sup\{|t_i|/0 \leq i \leq k\}$.

Proof (Proposition 1). We prove the proposition by means of an intermediate inequality, namely we will prove by induction on d the following one:

$$||\phi_1\cdot\dots\cdot\phi_n|| \leq L'(||w||, N(w), d) \tag{1}$$

where $L'(a,b,s)$ is defined inductively by:

$$\begin{cases} L'(a,b,0) = 2ab \\ L'(a,b,s+1) = 2ab2^{4(s+1)L'(a,b,s)} \end{cases} \tag{2}$$

Then the announced result will be obtained as a consequence. Next lemmas give the result for $d = 0$. Until it is differently specified we consider clauses in $\mathcal{C}^0$.

Lemma 1. *Given two clauses ϕ and ψ,*

1. *if* $\mathtt{tail}(\phi) \geq \mathbf{head}(\psi)$ *then* $|\mathbf{head}(\phi\psi)| = |\mathbf{head}(\phi)|$,
2. *if* $\mathtt{tail}(\phi) \leq \mathbf{head}(\psi)$ *then* $|\mathbf{head}(\phi\psi)| \leq |\mathbf{head}(\psi)| + |\mathbf{head}(\phi)|$.

Remark 4. Given a word $w = (\phi_1,\ldots,\phi_n)$ with non-null product, let us denote $\{j_1,\ldots,j_m\} = \{j \geq 2 \mid \mathtt{tail}(\phi_1\cdot\dots\cdot\phi_{j-1}) < \mathbf{head}(\phi_j)\}$. By induction over the integer m we deduce from the previous lemma the following inequality:

$|\mathbf{head}(\phi_1\cdot\dots\cdot\phi_n)| \leq |\mathbf{head}(\phi_1)| + \sum_{i=1}^{m}|\mathbf{head}(\phi_{j_i})|$; *analogously for the* $\mathtt{tail}(\phi_1\cdot\dots\cdot\phi_n)$.

Lemma 2. *An acyclic word $w = (\phi_1, \ldots, \phi_n)$ with non null product denoted by $\psi := \phi_1 \cdot \cdots \cdot \phi_n$, satisfies*

$$||\psi|| \leq ||w||(N(w) + 1).$$

Proof. In order to get contradiction assume $||\psi|| > ||w||(N(w)+1)$. In that case either $|\mathtt{head}(\psi)| > ||w||(N(w) + 1)$ or $|\mathtt{tail}(\psi)| > ||w||(N(w) + 1)$. Suppose for instance that we are in the first situation (the second case is handled in a completely symmetric way). By remark 4, using the same notations we have that $|\mathtt{head}(\phi_1 \cdot \cdots \cdot \phi_n)| \leq (m+1)||w||$; then we have $m \geq N(w) + 1$, so there exist $i_1 < i_2$ such that $\phi := \phi_{j_{i_1}} = \phi_{j_{i_2}}$.

We claim that the sub-product $\phi_{j_{i_1}} \cdot \cdots \cdot \phi_{j_{i_2}-1}$ gives a cyclic clause, hence the contradiction with the acyclicity of w. Indeed: let us denote $\Pi' := \phi_1 \cdot \cdots \cdot \phi_{j_{i_1}-1}$ and $\Pi'' := \phi_{j_{i_1}+1} \cdot \cdots \cdot \phi_{j_{i_2}-1}$; then we have $\mathtt{tail}(\Pi' \cdot \phi \cdot \Pi'') < \mathtt{head}(\phi)$. So as $\mathtt{tail}(\phi \cdot \Pi'') \leq \mathtt{tail}(\Pi' \cdot \phi \cdot \Pi'')$, we get $\mathtt{tail}(\phi \cdot \Pi'') < \mathtt{head}(\phi)$. Moreover, from $\mathtt{head}(\phi) \leq \mathtt{head}(\phi \cdot \Pi'')$, we deduce $\mathtt{tail}(\phi \cdot \Pi'') < \mathtt{head}(\phi \cdot \Pi'')$ and we are done. □

This lemma ends the base case of induction ($d = 0$) since

$$||w||(N(w)+1) \leq 2||w||N(w) = L'(||w||, N(w), 0), \quad \text{as } N(w) \geq 1.$$

In order to get the step of induction, we need a few intermediary results about products of clauses.

Lemma 3. *Let us consider a word $w = (\phi_1, \ldots, \phi_n)$ with non null product; the product of w induces a unique substitution family $(\sigma_1^0, \ldots, \sigma_n^0)$ such that σ_i^0 is defined on the variable of ϕ_i and*

$$\begin{aligned} \phi_1 \cdot \cdots \cdot \phi_n &= \mathtt{head}(\phi_1)\sigma_1^0 \vdash \mathtt{tail}(\phi_n)\sigma_n^0, \\ \textit{and} \quad \mathtt{tail}(\phi_i)\sigma_i^0 &= \mathtt{head}(\phi_{i+1})\sigma_{i+1}^0 \ \textit{when}\ 1 \leq i \leq n-1. \end{aligned}$$

Moreover, every substitution family $(\sigma_1, \ldots, \sigma_n)$ such that σ_i is defined on the variable of ϕ_i and satisfying:

$$\mathtt{tail}(\phi_i)\sigma_i = \mathtt{head}(\phi_{i+1})\sigma_{i+1} \ \textit{when}\ 1 \leq i \leq n-1 \tag{3}$$

can be obtained from $(\sigma_1^0, \ldots, \sigma_n^0)$ by means of a substitution θ such that $(\sigma_1, \ldots, \sigma_n) = (\sigma_1^0\theta, \ldots, \sigma_n^0\theta)$.

These properties are proved by induction over the length n of the word.

Remark 5. Note that for $2 \leq i \leq n-1$ we have $\mathtt{tail}(\phi_i\sigma_i^0) = \mathtt{head}(\phi_{i+1}\sigma_{i+1}^0)$ and that this term is equal either to $\mathtt{tail}(\phi_1 \cdot \cdots \cdot \phi_i)$ if $\mathtt{tail}(\phi_1 \cdot \cdots \cdot \phi_i) \geq \mathtt{head}(\phi_{i+1} \cdot \cdots \cdot \phi_n)$, or to $\mathtt{head}(\phi_{i+1} \cdot \cdots \cdot \phi_n)$ otherwise.

Lemma 4. *Let us consider a word $w = (\phi_1, \ldots, \phi_n)$ with non null product; the product of w induces a word $w' = (\phi'_1, \ldots, \phi'_n)$ such that $\phi'_i = \phi_i\sigma_i^0$ $(1 \leq i \leq n)$ with σ_i^0 as in lemma 3. Then for every i and h, we have that the subproduct $\phi_i \cdot \cdots \cdot \phi_{i+h}$ is a projection if and only if the corresponding sub-product of w', $\phi'_i \cdot \cdots \cdot \phi'_{i+h}$ is a projection.*

If $\phi_i \cdot \cdots \cdot \phi_{i+h}$ is a null-square then $\phi'_i \cdot \cdots \cdot \phi'_{i+h}$ is a null-square.

We establish now the induction step of inequality (1). Assume the inequality is true for any acyclic word in $\mathcal{C}^k$ with $k \leq d$, and take a word $w = (\phi_1, \dots, \phi_n)$ over $\mathcal{C}^{(d+1)}$. Consider for every layer $[w]_k$ the induced family of substitutions: $(\sigma_1^k, \dots, \sigma_n^k)$. Let w' be the word obtained by applying in w the substitution family to every layer $k \leq d$ and by freezing variables by means of newly introduced symbols of constants a_k: i.e. $[\phi_i']_k = [\phi_i]_k \sigma_i^k \langle a_k/x_k \rangle$ for $0 \leq k \leq d$ and $[\phi_i']_{d+1} = [\phi_i]_{d+1}$.

Notice that in w' variables remain only in the last layer $d+1$, so we can consider w' built over clauses of $\mathcal{C}^{(0)}$ with the first d layers constituting the predicate (we enlarge our set of predicates).

Let us show that w' is an acyclic word: we take a sub-word $(\phi_i', \dots, \phi_{i+h}')$ and its product $\phi' := \phi_i' \cdot \dots \cdot \phi_{i+h}'$; we denote the corresponding sub-product in w by $\phi := \phi_i \cdot \dots \cdot \phi_{i+h}$. By lemma 4, if the layer $[\phi]_k$ is a projection then $[\phi']_k$ is a projection too and if $[\phi]_k$ is a null-square then $[\phi']_k$ is a null-square. Combined with the fact that ϕ is an acyclic clause (definition 4), this implies that ϕ' is an acyclic clause.

So w' is an acyclic word in $\mathcal{C}^0$ and by establishing $N(w')$ and $||w'||$ we obtain the following inequality:

$$||\phi_1' \cdot \dots \cdot \phi_n'|| \leq L'(||w'||, N(w'), 0). \tag{4}$$

As the width of a word doesn't depend upon predicates appearing in its clauses and terms in w' are equal to terms in the last layer of w, we have $||w'|| \leq ||w||$. By definition $N(w')$ is the number of distinct clauses in w'; in order to find it we can calculate the number of all possible instances of terms in w'.

Remark 5 tells us that $\mathtt{tail}([\phi_i']_k)$ is equal to $\mathtt{tail}([\phi_1 \cdot \dots \cdot \phi_i]_k \langle a_k/x_k \rangle$ or $\mathtt{head}([\phi_{i+1} \cdot \dots \cdot \phi_n]_k \langle a_k/x_k \rangle$ (similarly for $\mathtt{head}([\phi_i']_k)$). Moreover we have

$$||[\phi_1 \cdot \dots \cdot \phi_i]_k|| \leq ||[\phi_1 \cdot \dots \cdot \phi_i]_{(0,k)}||.$$

Since by induction hypothesis: $||[\phi_1 \cdot \dots \cdot \phi_i]_{(0,k)}|| \leq L'(||[w]_{(0,k)}||, N([w]_{(0,k)}), k)$, we can apply the inequalities $N([w]_{(0,k)}) \leq N(w)$ and $||[w]_{(0,k)}|| \leq ||w||$, and we get

$$||[\phi_1 \cdot \dots \cdot \phi_i]_{(0,k)}|| \leq L'(||w||, N(w), k).$$

$$\text{Similarly} \quad ||[\phi_{i+1} \cdot \dots \cdot \phi_n]_{(0,k)}|| \leq L'(||w||, N(w), k).$$

Finally, we obtain: $||[\phi_i']_k|| \leq L'(||w||, N(w), k)$, and the number of all possible unary terms built in our language[3] of length at most $L'(||w||, N(w), k)$ is bounded by $4^{L'(||w||, N(w), k)+1}$.

We are now able to bound the number of clauses $N(w')$ in w': the number of possibilities for the choice of the head and tail predicates is bounded by $N(w)$;

[3] as the number of symbols of function in our language is 4, the number of terms of length k is 4^k, and the number of terms of length at most l is $\sum_{k=0}^{l} 4^k \leq 4^{l+1}$.

at the level k the number of possibilities for the head and tail terms is bounded by $4^{2(L'(||w||,N(w),k)+1)}$. Therefore we have:

$$N(w') \leq N(w) \prod_{k=0}^{d} 4^{2(L'(||w||,N(w),k)+1)} = N(w) \prod_{k=0}^{d} 2^{4(L'(||w||,N(w),k)+1)}.$$

By substitution of quantities $N(w')$ and $||w'||$ in (4) we have

$$\begin{aligned} ||\phi'_1 \cdot \dots \cdot \phi'_n|| &\leq L'(||w'||, N(w'), 0) = 2||w'||N(w') \\ &\leq 2||w||N(w) \prod_{k=0}^{d} 2^{4(L'(||w||,N(w),k)+1)} \\ &\leq 2||w||N(w)2^{4(d+1)L'(||w||,N(w),d)} = L'(||w||, N(w), d+1) \quad (5) \end{aligned}$$

We used the following inequality $L'(a,b,k) + 2 \leq L'(a,b,d)$ for $k \leq d-1$.

We therefore get $||[\phi_1 \cdot \dots \cdot \phi_n]_{d+1}|| \leq L'(||w||, N(w), d+1)$, and by induction hypothesis we have: $||[\phi_1 \cdot \dots \cdot \phi_n]_{(0,d)}|| \leq L'(||w||, N(w), d)$. Since $L'(||w||, N(w), d) \leq L'(||w||, N(w), d+1)$, we get:

$$||\phi_1 \cdot \dots \cdot \phi_n|| \leq L'(||w||, N(w), d+1).$$

This ends our proof for the induction step and the inequality (1) is established.

Using inequalities $\alpha \leq 2^{\alpha}$ and $x + y \leq xy$ whenever $x \geq 2$ and $y \geq 2$ one easily checks that: $L'(a,b,d) \leq L(a,b,d)$. This way we infer from inequality (1) proposition 1. □

The proof of proposition 2 uses proposition 1 and is in the same spirit (see [BP99]).

3 Sub-Associativity of Weak Execution

Let $\vdash \Gamma, \Delta, \Delta'$ be a sequent such that in Δ and Δ' formulas can be assembled dually in pairs $(B, B^{\perp})$. We consider the algebra $\lambda^*(\Delta, \Delta', \Gamma)$ built, as in section 1, using the language T and the family of predicates of arity $(d+1)$, $\{P_A\}_{A \in \Delta, \Delta', \Gamma}$. Let:

$$\sigma_{\Delta;\Delta',\Gamma} = \sum_{B \in \Delta} P_B(x_0, \dots, x_d) \vdash P_{B^{\perp}}(x_0, \dots, x_d)$$

We denote $\sigma_{\Delta;\Delta',\Gamma}$ by σ and $\sigma_{\Delta';\Delta,\Gamma}$ by τ, so that $\sigma + \tau = \sigma_{\Delta,\Delta';\Gamma}$.

Proposition 3 (Sub-associativity of weak execution). *Let U be a wiring of $\lambda^*(\Delta, \Delta', \Gamma)$ and σ and τ defined as above; we have:*

$$\mathtt{Result}^{\dagger}_{\sigma+\tau}(U) \subseteq \mathtt{Result}^{\dagger}_{\tau}(\mathtt{Result}^{\dagger}_{\sigma}(U)).$$

Remark 6. The equality is false in general, which contrasts with usual execution and the expected modularity of a valuable computation process. Still, as far as we are dealing with loops coming from proofs, associativity is valid since we will prove in the sequel that weak execution and ordinary execution coincide on such loops.

4 Interpretation of ELL Proof-Structures

We consider elementary linear logic with *t*-promotion and without additives and quantifiers. The sequent calculus is given in [Ped96]; the rules are as in multiplicative exponential linear logic but for dereliction which is not included and for promotion which is replaced by *t*-promotion: from $\vdash A, \Delta$ infer $\vdash !A, ?\Delta$. We now give the corresponding definition of *ELL proof-structures.* As usual there is a translation of proofs into proof-structures, yielding *ELL proof-nets.*

4.1 ELL Proof-Structures

Definition 7. *The* ELL proof-structures *are graphs with boxes whose edges are labelled by (multiplicative exponential) LL formulas; they are defined inductively together with their* depth *by:*

- *A proof-structure of depth 0 is a labeled graph R built over the nodes:*
  - *Axiom and cut: an* **ax** *(axiom) node has no premise and two conclusions labeled by dual formulas A and $A^\perp$; a* **cut** *node has two premises labeled by dual formulas (*cut formulas*) and no conclusion; we consider axioms labeled by atomic formulas to simplify the definition of the interpretation.*
  - *Multiplicative nodes: a $\otimes$ node (resp. a $⅋$ node) has two premises labeled by A and B and one conclusion $A \otimes B$ (resp. $A ⅋ B$).*
- *if $R_1, \ldots, R_n$ are proof-structures of maximal depth d then a graph R built from $R_1, \ldots R_n$ using the preceding nodes and the following exponential nodes is a proof-structure of depth d: a* **?c** *(contraction) node has two premises labeled by $?A$ and one conclusion labeled by $?A$; a* **?w** *(weakening) node has no premise and one conclusion labeled by a formula $?A$;*
- *if R is a proof-structure of depth d, the box containing R and with conclusions as on the figure is a proof-structure of depth $d + 1$.*

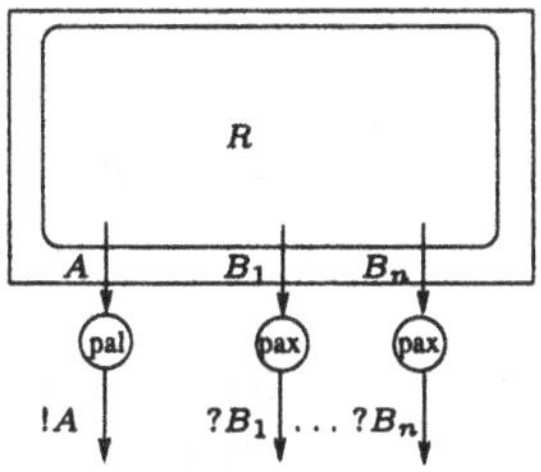

The depth *of an edge is the number of boxes it is contained in.*

We consider in proof-structures oriented paths crossing multiplicative and exponential nodes either from a premise to the conclusion or from the conclusion to a premise, and axiom nodes (resp. cut nodes) from a conclusion (resp. premise) to the other conclusion (resp. premise). A path is *up* (resp. *down*) if it only crosses nodes from conclusion to premise (resp. premise to conclusion).

The *length* of a path is the number of edges it goes through. If γ_1 is a path ending upwards (resp. downwards) with an edge conclusion (resp. premise) of a node N and γ_2 starts upwards (resp. downwards) with an edge premise (resp. conclusion) of N, we denote by $\gamma_1;\gamma_2$ their concatenation.

An *elementary path* of R is a path going upwards from a conclusion or a cut node to an axiom and then downwards to a conclusion or a cut node; we denote their set by $\mathcal{P}_e(R)$. A *constant-depth path* of R is a path of R which doesn't cross any box node, axiom node or cut node and starting upwards with a premise of box node or downwards with a conclusion of box node. The depth of such a path is the number of boxes of R it is contained in.

A proof-structure R gives a multiset Γ of conclusion formulas and a multiset Δ of cut formulas (associated dually in couples $(B, B^{\perp})$ by cut nodes). The language we consider is $T^d \cdot m$ where d is the depth of the proof-structure R and m is the cardinality of Γ, Δ. Predicates are indexed by formulas in Γ, Δ. The wiring part U_R of the loop interpreting R will be obtained by interpreting each elementary path of R by a clause.

4.2 Interpretation of a Proof-Structure

Representation of a constant-depth path by a term. As they don't cross axiom or cut nodes, constant-depth paths are up or down. We only consider constant-depth paths which don't visit any weakening node; this is enough to give the interpretation of proof-structures.

We associate to such a path γ of depth i a term $t_\gamma[x_i]$; we define this interpretation below in the case of a path oriented up by induction on the length of the path. In the case of a down path the interpretation t_γ is that of the reverted up path (orientation will be taken into account when we introduce the clauses...).

- if γ is reduced to an edge premise of a box node, then $t_\gamma = x_i$,
- otherwise we can write $\gamma = \gamma_1;\gamma_2$ where γ_2 is reduced to an edge premise of a multiplicative or a contraction node:
  - if γ_2 is the left (resp. right) premise of a multiplicative node then $t_\gamma = t_{\gamma_1}[px_i/x_i]$ (resp. $t_\gamma = t_{\gamma_1}[qx_i/x_i]$).
  - if γ_2 is the left (resp. right) premise of a contraction node then $t_\gamma = t_{\gamma_1}[rx_i/x_i]$ (resp. $t_\gamma = t_{\gamma_1}[sx_i/x_i]$).

Representation of an elementary path by a clause. If γ is an elementary path of the proof-structure R of depth d, it can be decomposed as:

$$\gamma = \gamma_i;\gamma_{i+1};\ldots;\gamma_j;\gamma'_j;\gamma'_{j-1};\ldots;\gamma'_k$$

where $0 \leq i, j \leq d$ and the path γ_l (resp. γ'_l) for $i \leq l \leq j$ (resp. $k \leq l \leq j$) is a constant-depth up path (resp. down path) of depth l.

Let A (resp. A') be the beginning (resp. ending) conclusion or cut formula. Their respective depths (i.e. the depths of their edges) are i and k. The clause $W(\gamma)$ interpreting the path γ (the *weight* of the path) is given by:

$$P_A(x_0, .., x_{i-1}, t_{\gamma_i}, .., t_{\gamma_j}, x_{j+1}, .., x_d) \vdash P_{A'}(x_0, .., x_{k-1}, t_{\gamma'_k}, .., t_{\gamma'_j}, x_{j+1}, .., x_d)$$

The ELL proof-structure R is interpreted by the loop (U_R, σ_R) with:

$$U_R = \sum_{\gamma \in \mathcal{P}_e(R)} W(\gamma)$$

$$\sigma_R = \sum_{B \in \Delta} P_B(x_0, \ldots, x_d) \vdash P_{B^\perp}(x_0, \ldots, x_d).$$

5 Weak Execution of Proof-Nets

In this section we prove that for every proof net R the associated loop (U, σ) satisfies: $\mathtt{Result}_\sigma(U) \subseteq \mathtt{Result}^\dagger_\sigma(U)$. The equality $\mathtt{Result}_\sigma(U) = \mathtt{Result}^\dagger_\sigma(U)$ follows then by remark 3.

First we will prove a proposition (4) and then we will derive this result as a corollary (1). Let us give before a few definitions.

A *balanced path* of R is a path starting upwards in a conclusion of R or downwards in a cut premise, ending downwards in a conclusion of R or in a cut premise. An *elementary balanced path* γ of R is a balanced path crossing at most one cut node, so that :

- if γ crosses no cut node it is an elementary path and its weight is given in the previous section;
- if it crosses a cut node from the premise B to the premise $B^\perp$ then it can be decomposed in the path just crossing the cut with weight $\sigma_0 = P_B(x_0, \ldots, x_d) \vdash P_{B^\perp}(x_0, \ldots, x_d)$ and in an elementary path γ_0 with weight $W(\gamma_0)$, so its weight is $W(\gamma) = \sigma_0 \cdot W(\gamma_0)$.

Any balanced path γ can be written as a concatenation of elementary balanced paths: $\gamma = \gamma_0; \ldots; \gamma_n$ and its weight is given by the product

$$W(\gamma) = W(\gamma_0) \cdot \cdots \cdot W(\gamma_n).$$

Definition 8. *We say a clause $\phi = P(t_0, \ldots, t_d) \vdash P'(u_0, \ldots, u_d)$ is* cyclic at depth $k \leq d$ *if:* **(1)** $P = P'$, **(2)** *for all $i < k$, $t_i = u_i$,* **(3)** *$t_k \neq u_k$ and t_k and u_k are comparable.*

We say the clause is cyclic at depth $+\infty$ if it is a projection.

We need three intermediary lemmas:

Lemma 5. *Let R be a proof-net and γ be a balanced path of R such that $W(\gamma)$ is non-null and cyclic at depth k. Then γ crosses at least one cut in R at depth lower than k.*

A *special cut w.r.t. a path* γ is an exponential cut σ such that γ crosses σ but doesn't cross any cut below the auxiliary ports of the box associated to the !-premise of σ (special cuts have been introduced by Regnier and Danos in [Reg92], [Dan90]). We use a variant of the "special cut lemma" stated in [Reg92]:

Lemma 6. *Let γ be a path of a proof-net R. If γ crosses only exponential cuts at depth lower than k and at least one, then R has a special exponential cut w.r.t. γ at depth lower than k.*

Lemma 7. *Let R be a proof-net and γ be a balanced path of R such that $W(\gamma)$ is cyclic at depth k. Assume σ is a cut of R at depth lower than k and crossed by γ which is either a multiplicative or axiom cut or a special exponential cut w.r.t. γ. Let R' be the proof-net obtained from R by reducing σ. Then R' has a balanced path γ' such that $W(\gamma')$ is non-null and cyclic at depth k.*

Proposition 4. *Given a proof net R and γ a balanced path of non-null weight, the clause $W(\gamma)$ associated to γ is acyclic.*

Proof. In order to get contradiction assume the proof-net R has a balanced path φ of non-null weight cyclic at depth k. By lemma 5 this implies that φ crosses at least one cut in R at depth lower than k. The idea is then to reduce progressively all the cuts at depth lower than k crossed by φ in such a way that at each step we keep in the corresponding proof-net a path satisfying the hypothesis. Now in order to do so we need to consider a particular strategy of reduction:

- if there is a multiplicative cut at depth lower than k crossed by the path, then we reduce it,
- otherwise, if all cuts crossed by the path at depth lower than k are exponential then we choose a special cut w.r.t. the path and reduce it.

We build a sequence (R_i, δ_i) of pairs of a proof-net and a path in it satisfying the property : $W(\delta_i)$ is non-null and cyclic at depth lower than k. Put $R_0 = R$ and $\delta_0 = \varphi$. Now assume the sequence has been defined up to rank $i \geq 0$. By lemma 5, δ_i crosses at least one cut in R_i at depth lower than k. If it crosses a multiplicative or axiom cut σ at depth lower than k take for R_{i+1} the proof-net obtained from R by reducing σ ; then by lemma 7 we know that R_{i+1} has a path satisfying the hypothesis which we take as δ_{i+1}. Otherwise lemma 6 ensures that δ_i has a special exponential cut σ at depth lower than k and this is the cut we choose.

This way we build an infinite sequence (R_i, δ_i) of pairs of a proof-net and a path in it with these properties. This sequence contradicts the strong normalization property of ELL. □

From this proposition, we derive the two following corollaries:

Corollary 1. *Let (U, σ) be the loop associated to a proof-net R; we have:*

$$\texttt{Result}^{\dagger}_{\sigma}(U) = \texttt{Result}_{\sigma}(U).$$

Corollary 2. *Let $(U, \sigma + \tau)$ be the loop associated to a proof-net R, then:*

$$\texttt{Result}^{\dagger}_{\sigma}(\texttt{Result}^{\dagger}_{\tau}(U)) = \texttt{Result}^{\dagger}_{\sigma+\tau}(U) = \texttt{Result}_{\sigma+\tau}(U).$$

Conclusion and Perspectives: Broadly speaking, our aim is to define a setting as large – and as simple – as possible for elementarily bounded computations. Weak Execution satisfies the complexity requirement with respect to programs of clauses of our algebra but (partially) fails to fulfill the modularity requirement. We are looking for a sufficient condition on programs which would ensure this modularity/associativity property for a larger class of programs than those coming from proof-nets. One direction under exploration (suggested in [Gir95b]) is that of an untyped calculus whose computations would be performed in the algebra through Weak Execution.

References

[BP99] P. Baillot and M. Pedicini. Elementary complexity and geometry of interaction. Prepublication, Quaderni dell'Istituto per le Applicazioni del Calcolo "M.Picone" di Roma, IAC-CNR n. 21-998, January 1999.

[Dan90] V. Danos. *La Logique Linéaire appliquée à l'étude de divers processus de normalisation (principalement du λ-calcul).* PhD thesis, Université Paris VII, Juin 1990.

[DR95] V. Danos and L. Regnier. Proof nets and the Hilbert space. In J.-Y. Girard, Y. Lafont, and L. Regnier, editors, *Advances in Linear Logic*, pages 307–328. Cambridge University Press, 1995. London Mathematical Society Lecture Note Series 222, Proceedings of the 1993 Workshop on Linear Logic, Cornell University, Ithaca.

[Gir88a] J.-Y. Girard. Geometry of interaction I: an interpretation of system *F*. In Ferro and al, editors, *Proceedings of A.S.L. Meetings*, Padova, 1988. North-Holland.

[Gir88b] J.-Y. Girard. Geometry of interaction II: Deadlock free algorithms. In Martin-Löf and Mints, editors, *Proceedings of COLOG'88*, volume 417 of *Lecture Notes in Computer Science*, pages 76–93. Springer Verlag, 1988.

[Gir95a] J.-Y. Girard. Geometry of interaction 3: the general case. In J.-Y. Girard, Y. Lafont, and L. Regnier, editors, *Advances in Linear Logic*, pages 329–389. Cambridge University Press, 1995. London Mathematical Society Lecture Note Series 222, Proceedings of the 1993 Workshop on Linear Logic, Cornell University, Ithaca.

[Gir95b] J.-Y. Girard. Light linear logic. In D. Leivant, editor, *Proceedings of the International Workshop on Logic and Computational Complexity (LCC'94)*, volume 960 of *LNCS*, pages 145–176, Berlin, GER, October 1995. Springer.

[KOS97] M. I. Kanovich, M. Okada, and A. Scedrov. Phase semantics for light linear logic (extended abstract). In *Mathematical foundations of programming semantics (Pittsburgh, PA, 1997)*, volume 6 of *Electron. Notes Theor. Comput. Sci.* Elsevier, Amsterdam, 1997.

[MR91] P. Malacaria and L. Regnier. Some results on the interpretation of λ-calculus in operator algebras. In *Proceedings of the Sixth Annual Symposium on Logic in Computer Science*, pages 63–72. IEEE Computer Society Press, 1991.

[Ped96] M. Pedicini. Remarks on elementary linear logic. In *A Special Issue on the "Linear Logic 96, Tokyo Meeting"*, volume 3 of *Electronic Notes in Theoretical Computer Science*, Amsterdam, The Netherlands, 1996. Elsevier.

[Reg92] L. Regnier. *Lambda-Calcul et réseaux.* PhD thesis, Université Paris VII, 1992.

Quantitative Semantics Revisited *
(Extended Abstract)

Nuno Barreiro** and Thomas Ehrhard

Institut de Mathématiques de Luminy
C.N.R.S. U.P.R. 9016
{nbar,ehrhard}@iml.univ-mrs.fr

Abstract. In the coherence space semantics of linear logic, the webs of the spaces interpreting the exponentials may be defined using multi-cliques (multisets whose supports are cliques) instead of cliques. Inspired by the quantitative semantics of Jean-Yves Girard, we give a characterization of the morphisms of the co-Kleisly category of the corresponding comonad (this category is cartesian closed and, therefore, is a model of intuitionistic logic). It turns out that these morphisms are the convex and multiplicative functions mapping multicliques to multicliques. This characterization is achieved via a normal form theorem, which associates a trace to each such map.

Introduction

The notion of *stable* function has been introduced by Berry for the purpose of modeling functional programming languages like PCF [1]. In the framework of *dilators* (functors acting on ordinals), Girard discovered independently stability as a condition allowing for a finitary representation of these functors. He applied the same idea to the denotational semantics of system F (see [3]) and this led him to the crucial observation that this semantics (which is an extension of Berry's semantics of PCF) can be described in the framework of *qualitative domains*, and even in the one of *coherence spaces*, which are particular qualitative domains. Berry actually developed his semantics in the framework of dI-domains (Scott domains satisfying some further properties). Coherence spaces are very particular dI-domains which define a sub-cartesian-closed category of the category of dI-domains and stable functions.

A coherence space is a symmetric and reflexive unlabelled graph (its *web* is the set of vertices; two vertices which are related are said to be *coherent*). The cliques of this graph are the elements of the corresponding dI-domain (singletons correspond to prime elements, finite cliques to compact elements).

* This work was partially supported by HCM Project CHRX-CT93-0046 "Typed Lambda-Calculus".

** On leave from the Departamento de Informática da Faculdade de Ciências da Universidade de Lisboa with partial financial support from JNICT (Programa de Mobilidade de Recursos Humanos BD813).

The space of stable functions from a coherence space X to a coherence space Y can in turn be described as a coherence space Z through traces: if f is a stable function from X to Y, the trace of f is the set of all couples (x_0, b) where b is a vertex of Y and x_0 is a finite clique of X minimal such that $b \in f(x_0)$. This leads to the idea that the function space operation (which corresponds to implication through the Curry-Howard isomorphism) is not atomic. It can be decomposed in two operations: (set) exponential and linear functions space.

The exponential $\underset{s}{!}X$ of a coherence space X has as web the set of all finite cliques of X, and the linear function space $X' \multimap Y$ of two coherence spaces X' and Y has as web the cartesian product of the webs of X' and Y. There is a natural isomorphism between the space of stable functions from the cliques of X to those of Y and the cliques of $\underset{s}{!}X \multimap Y$.

These two operations have logical counterparts which are made explicit as logical connectives in linear logic ([4, 8, 6] describe the coherence space semantics of linear logic).

Van de Wiele observed that alternative definitions of the exponential operation on coherence spaces are available. More specifically, from a categorical viewpoint, the exponential is an endofunctor on the category of coherence spaces and linear maps, and this functor has an additional structure of *comonad* satisfying some further requirements (the image of a coherence space by this functor has a canonical structure of commutative comonoid, see [2]). These properties do not characterize the exponential in a unique way. Van de Wiele proposed in particular a version of this operation where the web of $\underset{m}{!}X$, the multiset exponential of X, is the set of all finite *multisets* of the web of X whose support is a clique. From a categorical viewpoint, this exponential is extremely natural: the image of a coherence space by this functor is the *free* commutative comonoid on this coherence space.

This multiset exponential gives rise to a semantics of linear logic where the cliques of $\underset{m}{!}X \multimap Y$ may also be viewed as functions acting on the cliques of X. But these multiset morphisms are not characterized by their applicative behavior on cliques, in sharp contrast with the set semantics. Hence, a very natural question arises: can we, in an uniform way, associate to each coherence space X some space $\bar{X}$ in such a way that each clique of $\underset{m}{!}X \multimap Y$ may be seen as a function preserving some structure from $\bar{X}$ to $\bar{Y}$, *and conversely*?

This paper provides a positive and natural answer to this question, inspired by a work of Girard who, before introducing qualitative domains, and already guided by his dilators intuitions, considered in [5] a *quantitative* semantics of λ-calculus where the interpretation of a term takes into account the number of times a value is used in a computation. Actually, in that semantics, these "numbers" are sets, and morphisms are functors acting on families of sets, preserving directed limits, pullbacks and kernels. The nice feature of this semantics is that, like dilators and stable functions, these morphisms admit a "normal form theorem" (relying on a "formal series" representation).

Simplifying this approach, we replace each of the arbitrary sets of quantitative semantics by a natural number (the simplification is thus twofold: first we restrict to finite sets, and second, we restrict to their cardinality). This leads to associating to a coherence space X the set $\bar{X}$ (that we shall denote by $\mathcal{M}(X)$) of all the multicliques of X. We establish an isomorphism between the cliques of $\underset{m}{!}X \multimap Y$ and the functions from $\mathcal{M}(X)$ to $\mathcal{M}(Y)$ which preserve finite products of compatible multicliques, and satisfy a convexity criterion (these properties, together, imply Scott continuity). If convexity has hardly a domain-theoretic counterpart, the product preservation property clearly corresponds to the standard meets preservation property of stable functions. In particular, for multiset supports, the two preservation properties trivially coincide (the meet of the supports is the support of the product).

1 Preliminaries on multisets

We denote by $\mathbb{N}^+$ the set of non zero natural numbers.

Let S be a set. A multiset μ of S is a function mapping each element a of S to a natural number, the *multiplicity* of a in μ. We denote by $|\mu|$ the set $\{a \,/\, \mu(a) \neq 0\}$, which we call *support* of μ. We denote a multiset by an enumeration (delimited by square brackets) of the elements of its support, each as many times as its multiplicity in the multiset. We denote by 0 the multiset whose support is the empty set. A multiset whose support is a finite set is called a finite multiset. Let $a \in S$, we also denote by a the multiset whose support is $\{a\}$ and in which the multiplicity of a is 1.

Observe that, since multisets are functions to natural numbers, the sum, product and exponentiation of multisets are well-defined (in a pointwise manner), and similarly, the standard order on natural numbers induces a (partial) order on multisets that we shall denote by $\leqslant$.

Let μ be a multiset of S. Observe that, if $\mu^2 = \mu$ then, for every a in S, $\mu(a)$ is 0 or 1. We shall represent sets using these multisets. They enjoy the following immediate property.

Lemma 1. *Let S be a set. Let μ be a multiset of S such that $\mu^2 = \mu$. For any multisets ρ_1 and ρ_2 of S such that $|\rho_1|, |\rho_2| \leqslant \mu$,*

$$\text{if } \rho_1\rho_2 = 0 \text{ then } (\mu + \rho_1)(\mu + \rho_2) = \mu + \rho_1 + \rho_2 \,.$$

2 Coherence semantics

We shall give a brief review of coherence semantics. We begin with the definition of a coherence space.

Definition 1. *A* coherence space *X is a pair $(|X|, \frown_X)$ where $|X|$ is a countable set (the* web *of X, whose elements are called* points *of $|X|$) and $\frown_X$ is a symmetric and reflexive binary relation on $|X|$. Two elements of $|X|$ that are in this relation are said to be* coherent. *Otherwise they are said to be* incoherent.

A clique *of X is a subset x of $|X|$ such that, for any $a_1, a_2 \in x$, $a_1 \frown_X a_2$. A* multiclique *of X is a multiset μ of $|X|$ such that $|\mu|$ is a clique of X.*

We denote by $\frown_X$ and call *strict coherence relation* of X the relation obtained from $\bigcirc_X$ by removing the diagonal. We denote by 1 the coherence space whose web is the singleton $\{*\}$. The set of cliques of X, that we denote by $\mathcal{C}(X)$, is a qualitative domain ordered by the inclusion order. The set of multicliques of X, which is ordered by $\leqslant$, is denoted by $\mathcal{M}(X)$.

Definition 2. *Let X and Y be coherence spaces. A* linear map *from X to Y is a function f from $\mathcal{C}(X)$ to $\mathcal{C}(Y)$ such that:*

- *for any $x \in \mathcal{C}(X)$, $f(x) = \bigcup_{a \in x} f(\{a\})$;*
- *for any $x_1, x_2 \in \mathcal{C}(X)$, if $x_1 \cup x_2 \in \mathcal{C}(X)$ then $f(x_1 \cap x_2) = f(x_1) \cap f(x_2)$.*

We denote by Coh the category whose objects are the coherence spaces and whose morphisms are the linear maps.

Let X and Y be coherence spaces. The *linear implication* of X and Y is defined by $|X \multimap Y| = |X| \times |Y|$ and

$$\begin{aligned}(a_1, b_1) \bigcirc_{X \multimap Y} (a_2, b_2) \quad \text{if} \quad & a_1 \bigcirc_X a_2 \Rightarrow b_1 \bigcirc_Y b_2 \\ \text{and} \quad & a_1 \frown_X a_2 \Rightarrow b_1 \frown_Y b_2 \,.\end{aligned}$$

There is a bijective correspondence between the linear maps from X to Y and the cliques of $X \multimap Y$. To any linear map f from X to Y, we associate its *trace* $\mathcal{T}(f) \in \mathcal{C}(X \multimap Y)$, defined by

$$\mathcal{T}(f) = \{(a, b) \,/\, b \in f(\{a\})\} \,.$$

Reciprocally, to any clique t of $X \multimap Y$, we associate the linear map $\mathcal{F}(t)$ from X to Y, defined, for any clique x of X, by

$$\mathcal{F}(t)(x) = \{b \,/\, \exists a \in x \quad (a, b) \in t\} \,.$$

We shall indifferently use the same symbol to denote a linear map and its trace.

We recall now the definitions of the set exponentials and the multiset exponentials. Let X be a coherence space.

- Set *of course*. The points of $|\underset{s}{!}X|$ are the finite cliques of X. Two finite cliques x_1 and x_2 are coherent in $\underset{s}{!}X$ if, for any $a_1 \in x_1$ and $a_2 \in x_2$, $a_1 \bigcirc_X a_2$.
- Multiset *of course*. The points of $|\underset{m}{!}X|$ are the finite multicliques of X. Two finite multicliques μ_1 and μ_2 are coherent in $\underset{m}{!}X$ if $|\mu_1|$ and $|\mu_2|$ are coherent in $\underset{s}{!}X$.

We shall write ! to refer to both $\underset{s}{!}$ and $\underset{m}{!}$. The operation ! is a functor from Coh to itself which, furthermore, is endowed with a comonad structure. We denote by coK(!) the co-Kleisli category of the comonad !, which is a cartesian closed category (CCC) and, hence, is a model of intuitionistic logic (and also of PCF). The objects of this category are the coherence spaces and, given X and Y any coherence spaces, a morphism from X to Y is a clique in $!X \multimap Y$.

Whenever a category has a terminal object, a *point* of an object in that category is a morphism from the terminal object to the object in question. Then, given any coherence space X, a point of X in coK(!) is a morphism from the terminal object of coK(!) (the coherence space whose web is the empty set) to X. This means that a point of X in coK(!) is a clique of X. Let X and Y be coherence spaces. The evaluation map of the CCC yields a canonical notion of application of a morphism from X to Y to a point of X in coK(!), thus getting a point of Y in coK(!). In other terms, the cliques of $!X \multimap Y$ send cliques of X to cliques of Y in a canonical way. Let us focus on the set exponentials.

Definition 3. *Let X and Y be coherence spaces. A* stable map *from X to Y is a function f from $\mathcal{C}(X)$ to $\mathcal{C}(Y)$, such that f is monotone, continuous (commutation to directed unions) and, for any cliques x_1 and x_2 of X,*

$$x_1 \cup x_2 \in \mathcal{C}(X) \quad \Rightarrow \quad f(x_1 \cap x_2) = f(x_1) \cap f(x_2)\,.$$

Stable maps enjoy the following normal form theorem: let X and Y be coherence spaces and f be a stable map from X to Y; let $x \in \mathcal{C}(X)$ and $b \in f(x)$; then

- there is a $x_0 \subseteq x$ such that x_0 is finite and $b \in f(x_0)$;
- if x_0 is chosen minimal w.r.t. inclusion, then it is unique.

There is a bijective correspondence between the stable maps from X to Y and the cliques of $\underset{\cdot}{!}X \multimap Y$. To every stable map f from X to Y we associate, by use of the normal form theorem, its trace $\mathcal{T}(f) \in \mathcal{C}(\underset{\cdot}{!}X \multimap Y)$, defined by

$$\mathcal{T}(f) = \{(x_0, b) \,/\, x_0 \in \mathcal{C}(X) \;\wedge\; b \in f(x_0) \;\wedge\; \forall y \subseteq x_0\, (b \in f(y) \Rightarrow y = x_0)\}\,.$$

Reciprocally, to any clique t of $\underset{\cdot}{!}X \multimap Y$ we associate, by use of the evaluation map of coK($\underset{\cdot}{!}$), the stable map $\mathcal{F}(t)$ from X to Y, defined, for any clique x of X, by

$$\mathcal{F}(t)(x) = \{b \,/\, \exists x_0 \in |\underset{\cdot}{!}X| \;\; (x_0, b) \in t \;\wedge\; x_0 \subseteq x\}\,.$$

In fact, $\mathcal{F}$ is a functor from coK($\underset{\cdot}{!}$) to the category whose objects are the coherence spaces and whose morphisms are the stable maps; $\mathcal{T}$ is a functor going in the opposite direction. And it holds that, for any stable map f from X to Y and for any clique t of $\underset{\cdot}{!}X \multimap Y$,

$$\mathcal{F}(\mathcal{T}(f)) = f \quad \text{and} \quad \mathcal{T}(\mathcal{F}(t)) = t\,.$$

We shall indifferently use the same symbol to denote a stable map and its trace.

Observe that all linear maps are stable. Furthermore, given any coherence spaces X and Y, there is a bijective correspondence between the linear maps from X to Y and the stable maps from X to Y such that

- $f(\emptyset) = \emptyset$;
- if $x_1 \cup x_2 \in \mathcal{C}(X)$ then $f(x_1 \cup x_2) = f(x_1) \cup f(x_2)$.

Stable maps are naturally ordered by the set inclusion of traces. This order, denoted by $\leqslant_B$, is called *stable order* or *Berry order* and has the following functional counterpart: let X and Y be coherence spaces; let f and g be stable maps from X to Y; then

$$f \leqslant_B g \quad \text{iff} \quad \forall x, y \in \mathcal{C}(X)\,(x \cup y \in \mathcal{C}(X) \Rightarrow f(x \cap y) = f(x) \cap g(y))\,.$$

For more information on the coherence spaces denotational semantics of linear logic, we refer to [4, 8, 6].

Let us now go back to the multiset exponentials. Let X and Y be coherence spaces. By use of the evaluation map of $\mathrm{coK}(\underset{m}{!})$, we associate to each clique t of $\underset{m}{!}X \multimap Y$ a function from $\mathcal{C}(X)$ to $\mathcal{C}(Y)$, sending a clique x of X to the following clique of Y:

$$\{b \,/\, \exists \mu_0 \in |\underset{m}{!}X| \;\; (\mu_0, b) \in t \;\wedge\; |\mu_0| \subseteq x\}\,.$$

But this kind of functions does not enjoy a normal form theorem allowing to recover the underlying structure of clique. In fact, several cliques may correspond to the same function. Indeed, if we take the two cliques $\{([*], *)\}$ and $\{([*, *], *)\}$ of $\underset{m}{!}1 \multimap 1$, one may easily check that, by use of the evaluation map of $\mathrm{coK}(\underset{m}{!})$, they have the same functional behavior. This means that the category $\mathrm{coK}(\underset{m}{!})$ does not have enough points to sort its morphisms (it is not well pointed).

3 Convex and multiplicative maps

We start by the definitions.

Definition 4. *Let X, Y be coherence spaces. A* convex map *from X to Y is a function f from $\mathcal{M}(X)$ to $\mathcal{M}(Y)$ such that, for any $\mu, \nu, \rho \in \mathcal{M}(X)$ obeying $\mu + \rho, \nu + \rho \in \mathcal{M}(X)$,*

$$\mu \leqslant \nu \quad \Rightarrow \quad f(\mu + \rho) + f(\nu) \leqslant f(\mu) + f(\nu + \rho)\,.$$

Definition 5. *Let X, Y be coherence spaces. A* multiplicative map *from X to Y is a function f from $\mathcal{M}(X)$ to $\mathcal{M}(Y)$ such that, for any $\mu, \nu \in \mathcal{M}(X)$,*

$$\mu + \nu \in \mathcal{M}(X) \quad \Rightarrow \quad f(\mu\nu) = f(\mu)f(\nu)\,.$$

In fact, our main purpose in this section will be to prove that convexity and multiplicativity, together, imply continuity. We first observe the following.

Lemma 2. *Let X, Y be coherence spaces and f a function from $\mathcal{M}(X)$ to $\mathcal{M}(Y)$. If f is convex and multiplicative, then f is monotone.*

We shall now place us in the conditions of Lemma 2. Since the referred lemma yields that f is monotone, then, for any $\mu, \nu, \rho \in \mathcal{M}(X)$ such that $\mu + \rho \in \mathcal{M}(X)$ and $\nu + \rho \in \mathcal{M}(Y)$, we have that

$$f(\mu) \leqslant f(\mu + \rho) \quad \text{and} \quad f(\nu) \leqslant f(\nu + \rho)\,.$$

Therefore, f is convex and multiplicative if, and only if, f is monotone, multiplicative and obeys, for any $\mu, \nu, \rho \in \mathcal{M}(X)$ such that $\mu + \rho, \nu + \rho \in \mathcal{M}(X)$,

$$\mu \leqslant \nu \quad \Rightarrow \quad f(\mu + \rho) - f(\mu) \leqslant f(\nu + \rho) - f(\nu) .$$

This means that the increment of f grows with its argument, whence the term *convex.*

Let X be a coherence space. Observe that $\mathcal{M}(1) = \mathbb{N}$ and, then, it makes sense to speak of the multiplicativity and the convexity of the functions from $\mathcal{M}(X)$ to $\mathbb{N}$. We shall now state some of their properties.

Lemma 3. *Let X be a coherence space and f a function from $\mathcal{M}(X)$ to $\mathbb{N}$, convex and multiplicative. Let μ be an infinite multiclique of X such that $\mu^2 = \mu$ and $f(\mu) = 1$. Given any $\rho \leqslant \mu$, if $f(\mu + \rho) > 1$, then there exists $a \in \rho$ such that $f(\mu + a) > 1$.*

Proof. Observe that, since $\rho \leqslant \mu$, the multiset ρ is actually a set (that is $\rho^2 = \rho$).

We shall first consider the case in which ρ is finite. Since $\rho^2 = \rho$ we can write $\rho = a_1 + \ldots + a_n$ with $a_i \neq a_j$, for any $i, j \in \{1, \ldots, n\}$ such that $i \neq j$. Iterating Lemma 1 and using the fact that f is multiplicative, we get

$$f(\mu + \rho) = \prod_{i=1}^{n} f(\mu + a_i) .$$

But, by hypothesis, $f(\mu + \rho) > 1$ and thus, for some $i \in \{1, \ldots, n\}$, we have $f(\mu + a_i) > 1$, which proves the lemma for the finite case.

Let us now assume that ρ is an infinite set. Since $|X|$ is countable, so is ρ. Let $(a_i)_{i \in \mathbb{N}^+}$ be an enumeration without repetitions of ρ. Let ν be the multiset such that $|\nu| = \rho$ and $\nu(a_i) = i$ for every $i \in \mathbb{N}^+$.

For every $N \in \mathbb{N}^+$ we define the multiset ν_N by

$$\nu_N(a) = \begin{cases} N & \text{if } 1 \leqslant \nu(a) \leqslant N \\ \nu(a) & \text{otherwise} . \end{cases}$$

so that clearly $|\nu_N| = \rho$. One may easily check that, for any $N \in \mathbb{N}^+$, $\nu_N \geqslant N\rho$, and then, since, by Lemma 2, f is monotone, it holds that $f(\mu + \nu_N) \geqslant f(\mu + N\rho)$.

We shall now prove, by an induction on $N \in \mathbb{N}^+$, that $f(\mu + N\rho) \geqslant N$. For $N = 1$, this property holds as we have assumed $f(\mu + \rho) > 1$. Let us now assume, as inductive hypothesis, that $f(\mu + N\rho) \geqslant N$. Since $\mu \leqslant \mu + \rho$, the convexity of f yields that

$$f(\mu + N\rho) + f(\mu + \rho) \leqslant f(\mu) + f((\mu + \rho) + N\rho) .$$

By hypothesis $f(\mu) = 1$ and $f(\mu + \rho) > 1$, which entails that

$$f(\mu + N\rho) + 2 \leqslant 1 + f(\mu + (N + 1)\rho) .$$

We finally apply the inductive hypothesis to get that $f(\mu + (N + 1)\rho) \geqslant N + 1$.

We have shown that, for any $N \in \mathbb{N}^+$, $f(\mu + \nu_N) \geqslant f(\mu + N\rho)$ and $f(\mu + N\rho) \geqslant N$. Hence, it holds that

$$\text{for any } N \in \mathbb{N},\ \ f(\mu + \nu_N) \geqslant N\,.$$

Let us define, for every $N \in \mathbb{N}^+$, the following multisets

$$\hat{\nu}_N = \sum_{1 \leqslant i \leqslant N} a_i \quad \text{and} \quad \check{\nu}_N = \sum_{i > N} a_i$$

which are such that $\hat{\nu}_N \check{\nu}_N = 0$ and $\rho = \hat{\nu}_N + \check{\nu}_N$. Since $\rho^2 = \rho$ and $|\nu| = |\nu_N| = \rho$, one may easily check that $\nu\rho = \nu$ and $\nu_N \rho = \nu_N$, which entails

$$\nu = \nu\hat{\nu}_N + \nu\check{\nu}_N \quad \text{and} \quad \nu_N = \nu_N\hat{\nu}_N + \nu_N\check{\nu}_N\,.$$

Since we have that $(\nu\hat{\nu}_N)(\nu\check{\nu}_N) = 0$ and $(\nu_N\hat{\nu}_N)(\nu_N\check{\nu}_N) = 0$, we are in the conditions of Lemma 1 and we get:

$$\mu + \nu = \mu + \nu\hat{\nu}_N + \nu\check{\nu}_N = (\mu + \nu\hat{\nu}_N)(\mu + \nu\check{\nu}_N)\,;$$

$$\mu + \nu_N = \mu + \nu_N\hat{\nu}_N + \nu_N\check{\nu}_N = (\mu + \nu_N\hat{\nu}_N)(\mu + \nu_N\check{\nu}_N)\,.$$

By inspecting the definition of ν_N one immediately deduces that, for every $N \in \mathbb{N}^+$, $\nu\check{\nu}_N = \nu_N\check{\nu}_N$, and then, trivially, $f(\mu + \nu\check{\nu}_N) = f(\mu + \nu_N\check{\nu}_N)$. Let us suppose that, for every $N \in \mathbb{N}^+$, we have $f(\mu + \nu\hat{\nu}_N) = f(\mu + \nu_N\hat{\nu}_N)$. Then, since f is multiplicative, for every $N \in \mathbb{N}^+$ it holds that $f(\mu + \nu) = f(\mu + \nu_N)$. But we have shown that, for any $N \in \mathbb{N}^+$, $f(\mu + \nu_N) \geqslant N$, and this entails that, for every $N \in \mathbb{N}^+$, $f(\mu + \nu) \geqslant N$, which is in contradiction with the fact that $f(\mu + \nu)$ is finite.

Then there exists $N_0 \in \mathbb{N}^+$ such that $f(\mu + \nu\hat{\nu}_{N_0}) \neq f(\mu + \nu_{N_0}\hat{\nu}_{N_0})$. But $\nu_{N_0}\hat{\nu}_{N_0} = N_0\hat{\nu}_{N_0}$, hence $\nu\hat{\nu}_{N_0} \leqslant N_0\hat{\nu}_{N_0}$, so, given that f is monotone with $f(\mu) = 1$, we have $f(\mu + N_0\hat{\nu}_{N_0}) > 1$. Let us define $\rho_0 = N_0\hat{\nu}_{N_0}$, which is finite, since $\hat{\nu}_{N_0}$ is finite. Furthermore, we know that $|\hat{\nu}_{N_0}| \leqslant |\nu_{N_0}| = \rho$, which entails $|\rho_0| \leqslant \rho$. But, by hypothesis, $\rho \leqslant \mu$ and, therefore, $|\rho_0| \leqslant \mu$.

Since $\mu \leqslant \mu + \rho_0$, the convexity of f yields that

$$f(\mu + (\rho_0 + |\rho_0|)) + f(\mu + \rho_0) \leqslant f(\mu) + f((\mu + \rho_0) + (\rho_0 + |\rho_0|))\,.$$

But f is monotone and, therefore, $f(\mu + (\rho_0 + |\rho_0|)) \geqslant f(\mu + \rho_0)$ which entails, since $f(\mu + \rho_0) > 1$, that $f(\mu + (\rho_0 + |\rho_0|)) > 1$. Then, the convexity inequality above implies

$$1 + f(\mu + \rho_0) < f(\mu) + f(\mu + 2\rho_0 + |\rho_0|)\,,$$

which amounts to $f(\mu + \rho_0) < f(\mu + 2\rho_0 + |\rho_0|)$. Observe that

$$(\mu + \rho_0)(\mu + |\rho_0|) = \mu^2 + \mu|\rho_0| + \rho_0\mu + \rho_0|\rho_0| = \mu + 2\rho_0 + |\rho_0|\,,$$

hence

$$f(\mu + \rho_0) < f(\mu + \rho_0) f(\mu + |\rho_0|)\,,$$

which amounts to $f(\mu + |\rho_0|) > 1$. Since ρ_0 is finite and $|\rho_0| \leqslant \mu$, we are back to the finite case we have treated in the first place, and this ends the proof. □

The previous lemma is the key to continuity. Indeed, it shows that when the output of f increases, it is possible to find at least one point in $|X|$ which is responsible for a non zero part of that increase. But, due to multiplicativity, there can be only finitely many such points. And we have the following lemma.

Lemma 4. *Let X be a coherence space and f a function from $\mathcal{M}(X)$ to $\mathbb{N}$, convex and multiplicative. Let μ be an infinite multiclique of X such that $\mu^2 = \mu$ and $f(\mu) = 1$. Then, there exists $\mu_0 \leqslant \mu$ such that μ_0 is finite and $f(\mu_0) = 1$.*

4 From cliques to functions

Let X be a coherence space and μ, μ_0 be multicliques of X, such that μ_0 is finite. Then, given that, for any $n \in \mathbb{N}$, $n^0 = 1$, the product $\prod_{a \in |X|} \mu^{\mu_0}(a)$ is well-defined and we have

$$\prod_{a \in |X|} \mu^{\mu_0}(a) = \prod_{a \in |\mu_0|} \mu^{\mu_0}(a) .$$

Furthermore, this product is equal to 0 if, and only if, there exists $a \in |\mu_0|$ such that $\mu^{\mu_0}(a) = 0$, or, in an equivalent way, $\mu(a) = 0$ and $\mu_0(a) \neq 0$ (since $0^0 = 1$), which means that $|\mu_0| \nleqslant \mu$. And this shows that

$$\prod_{a \in |X|} \mu^{\mu_0}(a) \neq 0 \quad \text{iff} \quad |\mu_0| \leqslant \mu .$$

Let Y be a coherence space and $t \in \mathcal{C}(\underset{m}{!} X \multimap Y)$. Let $(\mu_0, b), (\nu_0, b) \in t$ such that

$$\prod_{a \in |X|} \mu^{\mu_0}(a) \neq 0 \quad \text{and} \quad \prod_{a \in |X|} \mu^{\nu_0}(a) \neq 0 .$$

Then, we have that $|\mu_0| \leqslant \mu$ and $|\nu_0| \leqslant \mu$, and thus $\mu_0 \frown_{\underset{m}{!} X} \nu_0$. But $(\mu_0, b) \frown_{\underset{m}{!} X \multimap Y} (\nu_0, b)$, which entails that $\mu_0 = \nu_0$. Therefore, for any $\mu \in \mathcal{M}(X)$ and $b \in |Y|$, there is *at most* one multiclique μ_0 of X such that

$$(\mu_0, b) \in t \quad \text{and} \quad \prod_{a \in |X|} \mu^{\mu_0}(a) \neq 0 .$$

This entails that, for any $\mu \in \mathcal{M}(X)$, the sum

$$\sum_{(\mu_0, b) \in t} \Big(\prod_{a \in |X|} \mu^{\mu_0}(a) \Big) b$$

is, trivially, well-defined.

We shall now prove that this sum is a multiclique of Y. Let $b_1, b_2 \in |Y|$ such that there are $\mu_1, \mu_2 \in \mathcal{M}(X)$ obeying $(\mu_1, b_1), (\mu_2, b_2) \in t$ and

$$\prod_{a \in |X|} \mu^{\mu_1}(a) \neq 0 \quad \text{and} \quad \prod_{a \in |X|} \mu^{\mu_2}(a) \neq 0 .$$

Then, we have that $|\mu_1|, |\mu_2| \leqslant \mu$, and thus $\mu_1 \bigcirc_{\underset{m}{!}X} \mu_2$. Therefore, since $(\mu_1, b_1) \bigcirc_{\underset{m}{!}X \multimap Y} (\mu_2, b_2)$, it holds that $b_1 \bigcirc_Y b_2$.

We are now in conditions of giving the following definition.

Definition 6. *Let X and Y be coherence spaces and $t \in \mathcal{C}(\underset{m}{!}X \multimap Y)$. We define $\mathcal{F}(t)$, a function from $\mathcal{M}(X)$ to $\mathcal{M}(Y)$, as follows:*

$$\mathcal{F}(t)(\mu) = \sum_{(\mu_0, b) \in t} \Big(\prod_{a \in |X|} \mu^{\mu_0}(a) \Big) b .$$

And we have the following proposition.

Proposition 1. *$\mathcal{F}$ is a functor from $coK(\underset{m}{!})$ to the category whose objects are the coherence spaces and whose morphisms are the convex and multiplicative maps.*

5 The normal form theorem

We shall start this section by a simple exercise.

Lemma 5. *Let f be a function from $\mathbb{N}^+$ to $\mathbb{N}^+$, monotone and multiplicative. Then there is a natural number, k, such that, for any $n \in \mathbb{N}^+$, $f(n) = n^k$.*

Proof. Let $n \in \mathbb{N}^+$ such that $n > 1$. For any $p \in \mathbb{N}$ we define $q_p \in \mathbb{N}$ in the following way

$$q_p = \left\lfloor p \frac{\log n}{\log 2} \right\rfloor , \qquad \text{which obeys } \; 2^{q_p} \leqslant n^p < 2^{q_p + 1} .$$

Using the fact that f is monotone and multiplicative we get that

$$f(2)^{q_p} \leqslant f(n)^p \leqslant f(2)^{q_p+1} .$$

We apply the logarithm function to the inequalities above and we get

$$q_p \log 2 \leqslant p \log n \leqslant (q_p + 1) \log 2 ,$$

$$q_p \log f(2) \leqslant p \log f(n) \leqslant (q_p + 1) \log f(2) .$$

Since $n \neq 1$, we then have that

$$\frac{q_p \log f(2)}{(q_p + 1) \log 2} \leqslant \frac{\log f(n)}{\log n} \leqslant \frac{(q_p + 1) \log f(2)}{q_p \log 2}$$

which is valid for any $p \in \mathbb{N}^+$. Therefore, we can take the limit of the inequality when p goes to the infinity and we obtain, since q_p goes to the infinity with p,

$$\frac{\log f(n)}{\log n} = k \qquad \text{where } \; k = \frac{\log f(2)}{\log 2}$$

and then we clearly have that, for any $n \in \mathbb{N}^+$ such that $n > 1$, $f(n) = n^k$.

But f is multiplicative which means, in particular, that $f(1)^2 = f(1)$ and, since $f(1) \in \mathbb{N}^+$, $f(1) = 1$. Then, for any $n \in \mathbb{N}^+$,

$$f(n) = n^k .$$

For any $n \in \mathbb{N}^+$, we must have $n^k \in \mathbb{N}^+$, which implies that $k \in \mathbb{N}$ (this is a known result of number theory). □

And, with this lemma, we are able to prove the normal form theorem.

Theorem 1. *Let X and Y be coherence spaces. Let f be a convex and multiplicative map from X to Y. Let $\nu \in \mathcal{M}(X)$ and $b \in |f(\nu)|$. Then, there is an unique $\mu_0 \in \mathcal{M}(X)$ such that $|\mu_0| \leqslant \nu$, μ_0 is finite and, for any $\mu \in \mathcal{M}(X)$,*

$$\mu + \mu_0 \in \mathcal{M}(X) \quad \Rightarrow \quad f(\mu)(b) = \prod_{a \in |X|} \mu^{\mu_0}(a) .$$

Proof. Let f_b be the function from $\mathcal{M}(X)$ to $\mathbb{N}$ defined, for any $\mu \in \mathcal{M}(X)$, by $f_b(\mu) = f(\mu)(b)$. Observe that, since f is convex and multiplicative, so is f_b.

It is easy to prove that that $f_b(|\nu|) = |f_b(\nu)|$ and, given that, by hypothesis, $b \in |f(\nu)|$, we have that $f_b(|\nu|) = 1$. Then, by Lemma 4, there is a $\nu_0 \leqslant |\nu|$ such that ν_0 is finite and $f_b(\nu_0) = 1$. Observe that $\nu_0^2 = \nu_0$.

If ν_0 is minimal, it is unique (by multiplicativity of f); we suppose it is the case.

For every $a \in \nu_0$, let f_b^a be the function from $\mathbb{N}^+$ to $\mathbb{N}^+$ defined, for any $n \in \mathbb{N}^+$, by

$$f_b^a(n) = f_b(\nu_0 + (n-1)a) .$$

Since f_b is convex and multiplicative, then, for every $a \in \nu_0$, f_b^a is convex and multiplicative, and therefore, by Lemma 2, f_b^a is also monotone. Lemma 5, then, yields that, for every $a \in \nu_0$, there is a $k_a \in \mathbb{N}$ such that, for any $n \in \mathbb{N}^+$,

$$f_b^a(n) = n^{k_a} .$$

Moreover, the minimality of ν_0 and the convexity of f entail that, for all $a \in \nu_0$, the natural number k_a is different from 0.

Let $\mu \in \mathcal{M}(X)$ such that $\nu_0 \leqslant \mu$. By multiplicativity of f,

$$f_b(\nu_0\mu) = f_b(\nu_0)f_b(\mu) = f_b(\mu) .$$

As $\nu_0 \leqslant \mu$,

$$\nu_0\mu = \nu_0 + \sum_{a \in \nu_0} (\mu(a) - 1)\, a .$$

Then, iterating Lemma 1, we get that

$$\nu_0 + \sum_{a \in \nu_0} (\mu(a) - 1)\, a = \prod_{a \in \nu_0} (\, \nu_0 + (\mu(a) - 1)a \,)$$

and by multiplicativity of f_b again, we have

$$f_b(\nu_0\mu) = \prod_{a\in\nu_0} f_b(\,\nu_0 + (\mu(a) - 1)a\,)$$

and, equivalently,

$$f_b(\mu) = \prod_{a\in\nu_0} f_b^a(\mu(a))\,,$$

which finally yields that

$$f_b(\mu) = \prod_{a\in\nu_0} \mu(a)^{k_a}\,.$$

Let $\mu_0 \in \mathcal{M}(X)$ be defined by $\mu_0(a) = k_a$ if $a \in \nu_0$ and $\mu_0(a) = 0$ otherwise, so that $|\mu_0| = \nu_0$. By construction we have that $|\mu_0| \leqslant \nu$, μ_0 is finite and, for every $\mu \in \mathcal{M}(X)$,

$$|\mu_0| \leqslant \mu \Rightarrow f(\mu)(b) = \prod_{a\in|X|} \mu^{\mu_0}(a)\,.$$

Now we need to check that, for every $\mu \in \mathcal{M}(X)$ such that $\mu + \mu_0 \in \mathcal{M}(X)$ and $|\mu_0| \not\leqslant \mu$, it holds that $f(\mu)(b) = 0$.

Let μ be such a multiclique and let us assume that $f(\mu)(b) \neq 0$. Then, we have shown that there is a finite multiclique $\mu_0' \in \mathcal{M}(X)$ such that $|\mu_0'| \leqslant \mu$, and, for every $\rho \in \mathcal{M}(X)$,

$$|\mu_0'| \leqslant \rho \Rightarrow f(\rho)(b) = \prod_{a\in|X|} \rho^{\mu_0'}(a)\,.$$

Clearly, $\mu_0 + \mu_0' \in \mathcal{M}(X)$. Let $(a_i)_{i=1,\ldots,n}$ be an enumeration without repetitions of the clique $|\mu_0 + \mu_0'|$ and let $\xi \in \mathcal{M}(X)$ be given by $|\xi| = |\mu_0 + \mu_0'|$, and $\xi(a_i) = p_i$, the i-th prime number.

Since $|\mu_0| \leqslant \xi$ and $|\mu_0'| \leqslant \xi$, we have

$$\prod_{a\in|X|} \xi^{\mu_0}(a) = \prod_{a\in|X|} \xi^{\mu_0'}(a)\,,$$

that is

$$\prod_{i=1}^{n} p_i^{\mu_0(a_i)} = \prod_{i=1}^{n} p_i^{\mu_0'(a_i)}\,.$$

Hence $\mu_0 = \mu_0'$, and we have a contradiction.

The same argument shows that the multiclique μ_0 whose existence is stipulated by the theorem is unique.

□

Once we have the normal form theorem, we may define the trace of a convex and multiplicative map.

Definition 7. *Let X and Y be coherence spaces and f a convex and multiplicative map from X to Y. The* trace *of f, $\mathcal{T}(f)$, is the following clique of $\underset{m}{!}X \multimap Y$:*

$$\mathcal{T}(f) = \{(\mu_0, b) /\ \mu_0 \in \mathcal{M}(X) \ \wedge\ \mu_0 \text{ is finite} \ \wedge\ b \in |f(\mu_0)| \ \wedge \\ \forall \mu \in \mathcal{M}(X)\ (\mu + \mu_0 \in \mathcal{M}(X) \ \Rightarrow\ f(\mu)(b) = \prod_{a \in |X|} \mu^{\mu_0}(a))\}.$$

Finally, one may easily prove the following proposition, which expresses in a categorical way the bijective correspondence between the morphisms of $\mathrm{coK}(\underset{m}{!})$ and the convex and multiplicative maps.

Proposition 2. *$\mathcal{T}$ is a functor going in the opposite direction of $\mathcal{F}$. And it holds that, for any multiplicative and convex map f from X to Y and for any clique t of $\underset{m}{!}X \multimap Y$,*

$$\mathcal{F}(\mathcal{T}(f)) = f \quad \text{and} \quad \mathcal{T}(\mathcal{F}(t)) = t.$$

6 Some remarks

We have seen that linear maps are particular stable maps. There is a similar result for convex and multiplicative maps.

Proposition 3. *Let X and Y be coherence spaces. Let f be a convex and multiplicative map from X to Y. Then, f is a linear map iff, for every $\mu, \nu \in \mathcal{M}(X)$,*

$$\mu + \nu \in \mathcal{M}(X) \Rightarrow f(\mu + \nu) = f(\mu) + f(\nu).$$

The Berry order for convex and multiplicative functions is, obviously, defined as follows.

Definition 8. *Let X and Y be coherence spaces. The Berry order on the convex and multiplicative functions from $\mathcal{M}(X)$ to $\mathcal{M}(Y)$, which we denote by $\leqslant_B$, is defined, for any such functions f and g, by:*

$$f \leqslant_B g \quad \text{iff} \quad \mathcal{T}(f) \subseteq \mathcal{T}(g).$$

And, as for the set exponentials, the Berry order has a functional counterpart.

Proposition 4. *Let X and Y be coherence spaces. Let f and g be convex and multiplicative functions from $\mathcal{M}(X)$ to $\mathcal{M}(Y)$. Then*

$$f \leqslant_B g \quad \text{iff} \quad \forall \mu, \nu \in \mathcal{M}(X)(\mu + \nu \in \mathcal{M}(X) \Rightarrow f(\mu\nu) = f(\mu)g(\nu)).$$

Our approach presents a discrepancy which may have puzzled the reader. The traces of our morphisms are *cliques*, but as functions, these morphisms act on *multicliques*. A natural generalization would be to allow arbitrary multicliques as traces of morphisms. Furthermore, such an approach would also be closer to Girard's quantitative semantics, since the monomials of his "formal series" (which correspond to the elements of our traces) have coefficients.

Given a multiclique τ of $\underset{m}{!}X \multimap Y$ and a multiclique μ of X, how could we generalize Definition 6 to define $\mathcal{F}(\tau)(\mu)$ as a multiclique of Y? The answer is, obviously,

$$\mathcal{F}(\tau)(\mu) = \sum_{(\mu_0, b) \in |\underset{m}{!}X \multimap Y|} \tau(\mu_0, b)\Big(\prod_{a \in |X|} \mu^{\mu_0}(a)\Big)\, b\,.$$

Indeed, in the particular case where τ is a clique, this formula coincides with the formula of Definition 6. These functions are convex but fail to be multiplicative, and we have to introduce a notion of weak multiplicativity in order to characterize them.

In the complete version of the paper we consider the category $\mathrm{Coh}(\mathbb{N}^+)$, whose objects are the coherence spaces. In that category, a morphism from a coherence space X to a coherence space Y is a multiclique τ of $\underset{m}{!}X \multimap Y$ such that $|\tau|$ is a linear map. This corresponds to the parameterization of Coh by the monoid $(\mathbb{N}^+, 1, \times)$, in the same way it is done in [7].

We finally prove that the co-Kleisly category of the comonad $\underset{m}{!}$ in $\mathrm{Coh}(\mathbb{N}^+)$ is isomorphic to the category whose objects are the coherent spaces and the morphisms are the convex and weakly multiplicative functions. Observe that, in that co-Kleisly category, a point is a multiclique and, therefore, since the morphisms act on multicliques, the category is well pointed.

References

1. Gérard Berry. Stable models of typed lambda-calculi. In *Proceedings of the 5th International Colloquium on Automata, Languages and Programming*, number 62 in Lecture Notes in Computer Science. Springer-Verlag, 1978.
2. Gavin Bierman. What is a categorical model of intuitionistic linear logic? In *Proceedings of Typed Lambda Calculus and Applications*, number 902 in Lecture Notes in Computer Science. Springer-Verlag, 1995.
3. Jean-Yves Girard. The system F of variable types, fifteen years later. *Theoretical Computer Science*, 45:159–192, 1986.
4. Jean-Yves Girard. Linear logic. *Theoretical Computer Science*, 50:1–102, 1987.
5. Jean-Yves Girard. Normal functors, power series and the λ-calculus. *Annals of Pure and Applied Logic*, 37:129–177, 1988.
6. Jean-Yves Girard. Linear logic: its syntax and semantics. In Jean-Yves Girard, Yves Lafont, and Laurent Regnier, editors, *Advances in Linear Logic*, volume 222 of *London Mathematical Society Lecture Note Series*. Cambridge University Press, 1995.
7. Jean-Yves Girard. On denotational completeness. To appear in Theoretical Computer Science, 1998.
8. Jean-Yves Girard, Yves Lafont, and Paul Taylor. *Proofs and types*. Cambridge University Press, 1989.

Total Functionals and Well-Founded Strategies
(Extended Abstract)

Stefano Berardi and Ugo de'Liguoro

Dipartimento di Informatica, Università di Torino,
Corso Svizzera 185, 10149 Torino, Italy
{stefano,deligu}@di.unito.it

Abstract. In existing game models, total functionals have no simple characterization neither in term of game strategies, nor in term of the total set-theoretical functionals they define. We show that the situation changes if we extend the usual notion of game by allowing infinite plays. Total functionals are, now, exactly those having a tree-strategy in which all branches end in a last move, winning for the strategy. Total functionals now define (via an extensional collapse) all set-theoretical functionals. Our model is concrete: we used infinite computations only to have a nice characterization of totality. A computation may be infinite only when the input is a discontinous functional; in practice, never.

1 Introduction

Games and strategies have emerged as useful tools to model interaction, with applications both to logic and to the theory of higher type functionals.

We address the problem of characterizing total functionals in game theoretic models. A natural conjecture is that a functional is total if and only if it is the extensional counterpart of some winning well-founded strategy. This would mean that a total functional can always be described via strategies whose plays eventually end, after finitely many steps, in some move by the Player, which Opponent cannot reply to.

We prove, however, that this is the case only (and exactly) for Tait-definable functionals, and that some interesting computable total functionals have infinite branches in any strategy defining them. This calls for a generalization of the notion of play to ordinal sequences of moves (possibly of transfinite length), and for a proper notion of winning strategy. Later, we will remark that infinite plays arise only in the application of a functional to some discontinous functional. Hence transfinite plays are relevant to have a nice characterization of total maps, but they cannot arise in practice.

In the literature game theoretic concepts have been proposed to construct models of lambda calculi, by extensionally collapsing certain sets of strategies. There have been two proposals: the first one is based on the idea of history-free strategies [3]; according to the second one players move depending on "views" of the play: these are called dialog games and innocent strategies, as defined in [10, 11].

In [7] an apparently different notion of game, originally introduced by Novikoff, is used to give an intuitionistic explanation of the classical notion of truth. As it will be explained in sections 2 and 3 of the present paper, dialog games and Novikoff-Coquand games are closely related: the former can be obtained from the latter by distinguishing between question and answer moves, and by imposing Gandy's "no dangling question" condition (no computation may end before all its sub-computations ended).

In all cases quoted above strategies produce either finite plays, or non-terminated plays of length ω. This is not necessary, at least in the case of strategies depending on views (called "innocent strategies" in [10]), since a generalization of dialog games to plays of transfinite length has been achieved in [5]. As we pointed out in the abstract, in this way all total set-theoretical functionals become naturally definable via strategies in which all branches end (maybe after infinitely many steps) in a last move, winning for the strategy.

We do not loose concreteness of the game interpretation: transfinite plays may arise only as the effect of the application to discontinuous arguments. Yet, transfinite branches are necessary even to represent some computable functionals.

To substantiate this claim, we provide two type 3 examples of functionals, taken from Kreisel Realization model of the Analysis. They require strategies with transfinite branches; but, if their arguments are hereditarily continuous functionals, the resulting play is always finite, and it is recursive if the arguments are.

The plan of the paper is as follows. In section 2 we introduce the basic definitions of transfinite dialog games. Then, in section 3, we specialize games to functional games. In section 4 we characterize total functionals, as promised. In the same section, we characterize total functionals definable via well-founded strategies as the Tait-continuous functionals. Finally, in section 5, we prove that this class does not contain even all "computable" total functionals: in particular certain type 3 realizers for Classical Second Order Arithmetic cannot be described via well-founded strategies.

Because of lack of space, almost all proofs have been omitted.

2 Games with transfinite plays

In this section we introduce Coquand's notion of game, as generalized in [5].

We want games able to model computation consisting of questions/answers (or dialogues) between two process. The first question is the input value, its answer is the output value, and it ends the dialogue. During the dialogue, processes alternate: each process answers to some previous question of the other process. The answer may be another question (concerning the value of a subcomputation); or it may be the final value of a (sub)computation.

We fix a trivial example we will use through the paper. Let $F : (N \to N) \to N$, and $f : N \to N$. Assume $f(0) = a, f(1) = b, f(2) = c$. We will describe the computation of $F(f) = f(0) + f(1) + f(2)$ as a dialogue between a process F and a process f. First, f asks "$F(f) =?$" (asks F for the value of $F(f)$). F answers by asking "$f(x) =?$" (by asking f for the value of $f(x)$ in $x = 0$; f, in turn, answers "$x =?$" (asks F for the input value x). F answers by "$x = 0!$" (by sending an input value 0 to f); now f answers F's original question, by "$f(x) = a!$" (by returning the output value a of $f(x)$ in $x = 0$).

The same questions and answers are used to compute $f(1)$ and $f(2)$. Eventually, F may answers f's first question: "$F(f) =?$", by returning $(a + b + c)$. This ends the dialogue.

We will model processes by players, whose goal is always to provide an answer to other player's questions. The first player unable to answer looses. Game rules fix a possible set of answers to each question. Computations are represented by plays which follow the rules of the game. A winning strategy will model a total functional, while a strategy which may loose will model a partial functional. We will define strategies at the end of this next section. Before we will formally define Coquand's games and plays.

Definition 1. *A* game *is a 5-ple* $G = \langle A, B, M, R, m_0\rangle$ *such that:*

1. *A, B are the names of the first and the second player;*
2. *M is a set, whose elements are the moves of G;*
3. *$R \subseteq M \times M$ is the set of rules of G: $\langle m, m'\rangle \in R$, also written mRm', reads as "m' is a legal reply to move m";*
4. *$m_0 \in M$ is the starting move.*

We assume the relation R having finite depth: there exists $k < \omega$ such that, if $m_0Rm_1 \cdots m_{n-1}Rm_n$, then $n \leq k$.

In our example, A and B are the processes f and F. M is the set of possible questions and answers between any two $F : (N \to N) \to N$, and $f : N \to N$, that is: $F(f) =?, F(f) = i!, f(x) =?, f(x) = j!, f(x) = k!$. We list now a coding for the elements of M.

1. $m_0 = ?\varepsilon$ is $F(f) =?$, the first question of the game, of f about the value of $F(f)$. *2.* The possible answers of F to $?\varepsilon$ are: the answer $!i$, or $F(f) = i!$" consisting of the output value $i \in N$ for F, and another question, $?1$, or $f(x) =?$, of F to f, about the value of $f(x)$. *3.* The possible answers of f to $?1$ are: the answer $!1.j$, or $f(x) = j!$, consisting of the output value $j \in N$ of $f(x)$, and the question $?1.1$, or $x =?$, of f to F, about the value of its input x. *4.* The only possible answers of F to $?1.1$ are $?1.1.k$, or $x = k!$, consisting of a value $k \in N$ for x. (In the next section, we will describe more in general a coding for the elements of M).

The relation $R(m, m')$ on M, or "game rule", describes the set of all m' which are a correct answer to m: in our case, according to what said, we have $R(?\varepsilon, !i), R(?\varepsilon, ?1), R(?1, !1.j), R(?1, ?1.1), R(?1.1, !1.1.k)$. The height of R is finite (equal to 3).

The next step will be to introduce first "generic" plays, and then specialize them to the particular notion of play we will use: "Novikoff plays".

Definition 2. *A* generic play *of the game* U *above is a triple* $p = \langle I, r, m_{(.)} \rangle$ *such that:*

1. *I, called the carrier set, is a non-empty well-order (total and well-founded), with minimum 0_I. Its elements are the indexes of the moves of the play p.*
2. *$r : I - \{0_I\} \to I$ is a map, such that $r(i) < i$ for all $i \in I$. r is called the* replay map*; $r(i)$ denotes (the index of) the move to which the move with index i answers to. Thus, $r(0)$ is undefined.*
3. *$m_{(.)} : I \to M$ is a map, associating to each index $i \in I$ a move $m_i \in M$ of the play, having such index. We ask moreover that $R(m_i, m_{r(i)})$ (that whenever a move answers to another one, then it is a correct answer to it)*

In our example, the whole play has 14 moves, and index set $I = \{0 \ldots 13\}$. The moves are: $m_0 = ?\varepsilon$ (or $F(f) = ?$), $m_1 = ?1$ (or $f(x) = ?$), $m_2 = ?1.1$ (or $x = ?$), $m_3 = !1.1.0$ (or $x = 0!$), $m_4 = !1.a$ (or $f(x) = a!$), ... The last move is $m_13 = !(a+b+c)$, or $F(f) = (a+b+c)!$. The reply map r keeps track to which move answers each move: we may check that its values are: $r(1) = 0, r(2) = 1, r(3) = 2, r(4) = 1, \ldots$ (the move 4 provides the value of $f(x)$ in $x = 0$, hence it answers to the move 1). Remark that $r(13) = 0$ (the last move provides the value of the whole computation, hence it answers to the move 0).

We will now define a map turn : $I \to \{A, B\}$, telling which player is on turn at a given step. Since $r(i) < i$ for $i > 0$, we have $r^n(i) = 0$ for a unique $n \in N$. The player on turn on 0 is A by the rules of the game, and the player on turn on $r(i)$ is the opponent of the player on turn on i. Thus, we may define turn as follows: $\mathsf{turn}(i) = A$ if the first n such that $r^n(i) = 0$ is even, and $\mathsf{turn}(i) = B$ if such an n is odd.

The last step is to restrict the set of plays we allow by introducing the notion of visibility. Visibility models the *memory* of the computation (which past moves may be used by a player to decide the next move, or which moves may be answered). We follows Novikoff and Coquand, and we decide to assume that each move between a question in $j = r(i)$ and its answer in i are invisible for the player who got the answer. The reason is that we think of the moves in $]j, i[$ as a subcomputation, with input the question in j, and output the answer in i. And we want to model any computation by a "black box", with only visible points the input and the output, as real computations are. Thus, the player who sent the input in j and received the output in i should see nothing else in between.

Let $U = \mathsf{turn}(k)$. We may express Novikoff-Coquand by requiring: *1.* each segment $[0, i[$ of the play is split into a partition made of segments $[r(k), k]$ ($r(k)$ = question of U, k = answer of his opponent); *2.* the only visible moves, by U from i, are the endpoints $\{r(k), k\}$ of such segments; *3.* $r(i) = k$ for the last point k of one of such segments. This latter requirement means that U, in i, replies to some visible answer of his opponent. We will now formalize the idea above into definition of Novikoff play.

Definition 3. – *We associate to any $i \in I$ a segment by $S(i) = [0, 0]$ if $i = 0$, $S(i) = [r(i), i]$ if $i > 0$. We call $S(i)$ an R-segment: it is the segment of moves between the move i answers to (if any), and i itself.*

- *We say that* $\{S(k) | k \in V\}$ *is a "black box structure" over* I *if it is a* partition *of* I. *We call the set* V *above, consisting of the last points of the segments* $S(k)$, *a* visibility set *over* I.
- *We say that* $p = \langle I, r, m_{(.)} \rangle$ *is a* Novikoff play *if there is a map* $V(.) : I \to \wp(I)$ *such that, for all* $i \in I$, $V(i)$ *is a visibility set over* $[0, i[$ *and* $r(i) \in V(i)$.

Starting from the sets $V(i)$, we may formalize the visibility predicate $\mathsf{Vis}(U, \xi, \zeta)$ (to be read "ζ is visible by player U at ξ"), by $\mathsf{Vis}(\mathsf{turn}(\xi), \xi, \zeta) \Leftrightarrow \zeta \in V(\xi) \vee \zeta r(V(\xi) - \{0\})$ and, if $U = \mathsf{turn}(r(\xi)) \neq \mathsf{turn}(\xi)$, $\mathsf{Vis}(\mathsf{turn}(\xi), \xi, \zeta) \Leftrightarrow \zeta = r(\xi) \vee \mathsf{Vis}(U, r(\xi), \zeta)$. The first definition expresses that $V(\xi) \cup r(V(\xi) - \{0\})$ is the set of endpoints of the "black box structure" associated to ξ and to the player on turn on ξ. The second definition expresses the fact that no move in $]r(\xi), \xi[$ is visible by the the player U on turn on $r(\xi)$. This is because the segment $[r(\xi), \xi]$ starts by a question by U, and ends by the answer of the other player. Thus, according to our assumptions, its interior is invisible by U.

The *view* of U on p at ξ is the set

$$\mathsf{view}(U, p, \xi) = \{\zeta \mid \mathsf{Vis}(U, \xi, \zeta)\}.$$

The main result about Novikoff plays is the following (proved in [5]):

Theorem 1.
Let p be any Novikoff play. Then all one-step extensions of p have, in their last move, the same visibility set and the same player on turn.

Because of 1, if a play p of length α can be extended, it makes sense to speak of the player on turn at α-th step: abusing notation we simply write $\mathsf{turn}_p(\alpha)$.

The theorem 1 is easy to prove when I has a successor length, but difficult when I has a limit length. Herbelin [8] remarked that the case length$(I) = \omega$ is elementary equivalent to Tait's normalization result for ω-logic. As an easy corollary, the visibility assignment $V(.) : I \to \wp(I)$ such that $r(i) \in V(i)$ for all $i > 0$, if it exists, it is unique; and $V(i), \mathsf{turn}(i)$ are uniquely determined by r restricted to $[0, i[$. Thus, in principle, we could just say that a play is Novikoff, without quoting the map $V(.) : I \to \wp(I)$, since this map is unique.

Our example of play is a Novikoff play. We will now write down, for each move, a row with all visibility informations for the player on turn. Moves visible by the player on turn will be marked "v", or "**v** " for the moves of his opponent, forming the visibility set. Invisible moves will be marked "i". We call the process F "P" (for "Player"), and process f "O" (for "Opponent").

	turn	Move	Coding of the move	r	0	1	2	3	4	...	...	13
0	O	$F(f) =?$	$?\varepsilon$	-	i	i	i	i	i	..	..	i
1	P	$f(x) =?$	?1	0	**v**	i	i	i	i	..	..	i
2	O	$x =?$	?1.1	1	v	**v**	i	i	i	..	..	i
3	P	$x = 0!$	!1.1.0	2	**v**	v	**v**	i	i	..	..	i
4	O	$f(x) = a!$	$!1.a$	1	v	**v**	v	**v**	i	..	..	i
...	...	...	...	..	..	..	..	..	..	..	..	i
...	...	...	...	..	..	..	..	..	..	..	..	i
13	P	$F(f) = a + b + c!$	$!(a + b + c)$	0	**v**	v	i	i	**v**	..	..	i

Remark that move 13 cannot see, for instance, the moves 2, 3. The reason is that such moves are in the interior of the R-segment $[1,4]$, that is, of the subcomputation with question $f(x) =?$ and answer $f(x) = a!$. Thus, moves 2, 3 are, for the player on turn on move 13, inside a "black box", hence invisible.

In the case of finite pre-plays, we may prove that the set view is the visibility set of view-strategies (called "innocent" in [10]), having a simple inductive definition:

$$\mathsf{view}(U,p,i) = \begin{cases} \{i-1\} \cup \mathsf{view}(U,p,i-1) & \text{if } \mathsf{turn}(i) = U \\ \{r(i)\} \cup \mathsf{view}(U,p,r(i)) & \text{if } \mathsf{turn}(i) \neq U. \end{cases}$$

This is the standard notion of visibility in dialog games: it is defined in this way both in [10, 11] and in [7]. the case of plays of possibly transfinite length has been considered for the first time in [5], from which we borrow the axiomatic definition of Vis. Definition above does not tell, explicitly, who is the player on turn at a limit point $\lambda \in I$, nor his views. The main theorem 1, however, states that r restricted to $[0, \lambda[$ uniquely determine the turn and the view at point λ.

This ends the introduction of Novikoff plays. In the remaining of this section, we will introduce strategies. In the next section, we will use them to model functionals.

To define strategies, concepts and terminology about certain parts of plays are in order. First, if $\xi \in I$ then $p\lceil\xi$ (a prefix play of p) is the (pre) play whose carrier set is $[0,\xi[$, whose $r, m_{(.)}$ are the restrictions to $[0,\xi[$ of those of p. More in general if $J \subseteq I$ then $p\lceil J$ is the structure $\langle J, r', m'_{(.)}\rangle$ where $r', m'_{(.)}$ come from $r, m_{(.)}$, by restricting them to J.

Given a play p we can choose J such that $p\lceil J$ is closed under the reply function and has the structure of a play, but it is not such for trivial reasons: e.g. because its first move is not m_0, or it is played by P. To define the notion of subplay without being too restrictive we introduce the notion of play morphism (see also [10]).

Definition 4. *If p and q are (pre) plays, with carrier sets I, J, then $\varphi : p \to q$ is a* play morphism *if it consists of a pair of maps $\langle\varphi_0, \varphi_1\rangle$ such that $\varphi_0 : I \to J$ is strictly increasing and $\varphi_1 : \{O, P\} \to \{O, P\}$ is identity or exchange, and for all $\xi < \alpha$:*

$$\mathsf{turn}_q(\varphi_0(\xi)) = \varphi_1(\mathsf{turn}_p(\xi)), \qquad r_q(\varphi_0(\xi)) = \varphi_0(r_p(\xi)).$$

The image $\varphi[p]$ in q is a subplay *of q.*

The subplay $\varphi[p]$ of q has the same structure of p, and its reply and turn functions are $r_q\lceil\varphi_0[\alpha]$ and $\mathsf{turn}_q\lceil\varphi_0[I]$ (where $\varphi_0[I]$ is the image of I in J via φ_0).

Proposition 1. *If $\varphi[p]$ is a subplay of q, then $I = \varphi_0[\text{length}(p)]$ is such that:*

1. *if $\xi, \zeta \in I$ are such that $\xi < \zeta$ and there exists no $\eta \in I$ such that $\xi < \eta < \zeta$, then $\mathsf{turn}(\xi) \neq \mathsf{turn}(\zeta)$;*
2. *$I \neq \emptyset$ and $r[I \setminus \{\min(I)\}] \subseteq I$;*
3. *for any $\xi \in I$, if $I' = \{\zeta \in I \mid \zeta < \xi\}$ then $I' \cap \mathsf{view}(\mathsf{turn}(\xi), q, \xi)$ is cofinal in I'.*

Vice versa, if $I \subseteq \text{length}(q)$ satisfies the above conditions, then $q\lceil I$ is a subplay of q.

A pre-play is *U-cut free*, for $U \in \{O, P\}$ if

$$\xi > 0 \wedge \mathsf{turn}(\xi) \neq U \Rightarrow \xi = r(\xi) + 1,$$

namely if the opponent of U is forced to reply to the last move of U.

U-cut free (pre) plays is the terminology of [7]. If a pre-play has finite length then the previous definition is a generalization of [11], definition 3.1.3. Observe that in a U-cut free pre-play, U is the unique player allowed to play at limit points.

Any view determines a subplay (but not vice versa), i.e. any non empty $I = \mathsf{view}(U, q, \xi)$ satisfies the conditions of 1. Such a $q\lceil I$ is a U-cut free play which, with overloaded terminology, we call the *U-view of q at ξ*. Also $I \cup \{\xi\}$ determines a subplay $q\lceil(I \cup \{\xi\})$, which we call "large U-view".

We say that player U is *deterministic* on a play p if for all $\xi, \zeta < \text{length}(p)$, if $\mathsf{turn}(\xi) = \mathsf{turn}(\zeta) = U$ and $p\lceil\mathsf{view}(U, p, \xi)$ isomorphic to $p\lceil\mathsf{view}(U, p, \zeta)$ (i.e., that they are the same up to renaming of the elements of the carrier sets) then the lare U-views of ξ, ζ are isomorphic, too. A play p is a *deterministic play* if both players are deterministic on p.

Definition 5. *A* strategy s *for player U over a game U (shortly an U-strategy) is a tree (i.e. a prefix closed set) of U-cut free plays of U such that, for all $p \in s$ with $\alpha = \text{length}(p)$:*

1. *if $\mathsf{turn}(\alpha) = U$ then there is at most one $q \in s$ of length $\alpha + 1$ such that p is a prefix of q;*
2. *if $\mathsf{turn}(\alpha) \neq U$ (hence α is a successor) then for any $m \in M$ which is a legal reply to $p_{\alpha-1}$, i.e. such that $p_{\alpha-1}Rm$, there exists $q \in s$ of length $\alpha + 1$ such that p is a prefix of q, $q_\alpha = m$ and $r_q(\alpha) = \alpha - 1$.*

Player U follows the strategy s in the play q if for all $\xi < \text{length}(q)$ the large U-view p of q at ξ belongs to s, up to renaming of the carrier set. Clearly U follows some strategy in q if and only if U is deterministic on q.

The main consequence of Theorem 1 w.r.t. strategies is the cut-elimination theorem:

Theorem 2 (Cut-elimination [5]). *Let s be a P-strategy and t an O-strategy such that the heights of s and t are bounded above by some infinite regular ordinal κ. Then there exists a unique play p of maximal length such that P and O follow the strategies s and s' respectively, and* $\text{length}(p) = \alpha + 1 < \kappa$.

This play has successor length, hence it has a last move; the player who did the last move won. Therefore any two strategies s and t, for Player and Opponent respectively, determine a winning player.

3 Sequential functionals of finite type

The present section specializes dialog games to games and strategies representing functionals. In this case the role of Player is to show that a functional F_s, associated to the strategy s, is defined against the arguments $F_{t_1}, \ldots, F_{t_k}$: if s wins against $t_1, \ldots, t_k$ then either some t_i misses a move or the resulting play has a last move $!v$ such that $F_s(F_{t_1}, \ldots, F_{t_k}) = v$. Therefore winning strategies (i.e. strategies such that the player who follows them is always able to play a move, when on turn) naturally induce total functionals.

We base our treatment on [11]. Admittedly formalizations based on the categorical semantics of linear logic, as it is the case of [6, 2, 3, 1, 9], have the advantage of being compositional with respect to the type structure, which is not the case of the present one. However the actual description of strategies seems more direct in a formulation which does not make use of the decomposition of the function space bifunctor into linear implication and the comonad "!". Perhaps the best thing would be a compromise between the two, which is still on demand.

Let $\Gamma = \{\gamma_0, \gamma_1, \ldots\}$ be a set of ground types, and $\mathsf{T}(\Gamma)$ be the set of simple types over Γ. We fix an interpretation of types in Γ as a set of values $V = \bigcup\{V_\gamma \mid \gamma \in \Gamma\}$.

Any type has the form $\tau = \tau_1 \to (\cdots \to (\tau_k \to \gamma)) \in \mathsf{T}(\Gamma)$, and is abbreviated by $(\tau_1, \ldots, \tau_k \to \gamma)$. The set of *occurrences* of τ, $Occ(\tau)$ is defined inductively: $\varepsilon \in Occ(\tau)$ and $\tau_\varepsilon = \gamma$; if $1 \le i \le k$ and $a \in Occ(\tau_i)$ then $i.a \in Occ(\tau)$ and $\tau_{i.a} = (\tau_i)_a$.

To each type τ it is associated a game G_τ as follows.

Definition 6. *For $\tau \in \mathsf{T}(\Gamma)$, G_τ is the game $\langle M_\tau, R_\tau, ?\varepsilon\rangle$ where:*

1. $M_\tau = \{?a, !a.v \mid a \in Occ(\tau, v \in V_\gamma,$ *for γ last atom in* $\tau_a\}$;
2. *R_τ is the least binary relation over M_τ such that:*

   $(R1)$ $\quad a.i \in Occ(\tau) \Rightarrow ?aR_\tau?a.i,$
   $(R2)$ $\quad a \in Occ(\tau) \wedge \tau_a = \gamma \wedge v \in V_\gamma \Rightarrow ?aR_\tau!a.v.$

In M_τ moves of the form $?a$ are queries for the output value of a functional of type τ_a, applied to all its arguments; moves of the form $!a.v$ are the corresponding answers.

Definition 7. *A* functional play *(henceforth simply a play) over the game G_τ is a deterministic play p over it such that*

$$(F) \quad p_\xi = !v \wedge r(\xi) < \zeta < \xi \wedge p_\zeta = ?a \Rightarrow \exists \zeta', v'.\ \zeta' < \xi \wedge p_{\zeta'} = !v' \wedge r(\zeta') = \zeta.$$

(F) imposes that an answer replies to the last unanswered question (the "no dangling condition" of [10]). By $(R1)$-$(R2)$ only queries can be replied to.

Let $\langle p, r_p\rangle$ be a play and $\langle p', r_{p'}\rangle$ a subplay (of any other play). By $p * p'$ we indicate the partially defined operation of concatenating p with p': $p * p'$ is defined and equal to $\langle q, r_q\rangle$ if length(q) = length(p) + length(p'), $q_\xi = p_\xi$ if $\xi <$ length(p), p'_ζ if ξ = length$(p) + \zeta$, and finally

$$r_q(\xi) = \begin{cases} r_p(\xi) & \text{if } \xi < \text{length}(p) \\ \text{length}(p) + r_{p'}(\zeta) & \text{if } \xi = \text{length}(p) + \zeta > \text{length}(p) \\ \text{length}(p) - 1 & \text{if } \xi = \text{length}(p) \text{ is a successor, and} \\ & p_{\text{length}(p)-1} R q_\xi \end{cases}$$

If some of the above conditions cannot be satisfied, $p * p'$ is undefined. If it is defined we set turn_q as the function determined by r_q.

Let p be a play of type τ_i, for $1 \le i \le k$, and $\tau = (\tau_1, \dots, \tau_k \to \gamma)$. Then we may construct a play $p^{(i)} = \langle ?\varepsilon\rangle * p'$ of type τ by adding a first move $?\varepsilon$ and by transforming each move over τ_i into the corresponding move over τ: so p' is the (sub) play obtained from p by changing any question of the form $?a$ into a question of the form $?i.a$. Because of the definition of concatenation, the first move of p' replies to $?\varepsilon$, which implies that players on p' are interchanged with respect to p (indeed, for all $\xi <$ length(p), p_ξ corresponds to $p^{(i)}_{1+\xi}$, so that $r^n_p(\xi) = 0$ if and only if $r^{n+1}_{p^{(i)}}(1+\xi) = 0$: in particular, if ξ is limit, then $1 + \xi = \xi$ so that players are exchanged also at limit points); therefore, if p is a P-view of a play of type τ_i, then $p^{(i)}$ is an O-view of a play of type τ. Finally, if s is a strategy of type τ_i then we set $s^{(i)} = \{p^{(i)} \mid p \in s\}$.

Proposition 2. *Let $\tau = (\tau_1, \dots, \tau_k \to \gamma)$ and $s_1, \dots, s_k$ be P-strategies of type $\tau_1, \dots, \tau_k$. Then*

$$(s_1, \dots, s_k)^O = \bigcup_{i=1}^{k} s_i^{(i)}$$

is an O-strategy of type τ, and any such a strategy arises in this way.

Because of this proposition there is no theoretical loss in concentrating on P-strategies, henceforth called simply strategies. An immediate consequence of this and of 2 is that given some P-strategy s of type $(\tau_1, \dots, \tau_k \to \gamma)$ and the P-strategies $s_1, \dots, s_k$ of type $\tau_1, \dots, \tau_k$ it is uniquely determined the play $p = s \bullet (s_1, \dots, s_k)^O$ of maximal length in which P and O follow s and $(s_1, \dots, s_k)^O$ respectively.

A functional play is *terminated* if it has a move answering to the first move $?\varepsilon$. This move is necessarily the last one, by $(F1)$. If such $s \bullet (s_1, \dots, s_k)^O$ is terminated by the move $!v$ then write

$$s[s_1, \dots, s_k] = v.$$

$s[s_1, \dots, s_k]$ is undefined otherwise. By $s[s_1, \dots, s_k] \simeq t[t_1, \dots, t_h]$ we mean they are either both defined and equal, or both undefined.

The functional interpretation of strategies depends on the following fact. For each type τ define the binary relation $\sim_\tau$ among strategies of type τ inductively as follows:

- $s \sim_\gamma s' \Leftrightarrow s = s'$;
- $s \sim_{(\tau_1,\ldots,\tau_k\to\gamma)} s' \Leftrightarrow$
$$\forall s_1, s'_1, \ldots, s_k, s'_k. \bigwedge_{i=1}^{k} s_i \sim_{\tau_i} s'_i \Rightarrow s[s_1, \ldots, s_k] \simeq s'[s'_1, \ldots, s'_k].$$

Then, if $s_i \sim_{\tau_i} s'_i$ for $1 \le i \le k$ and s is a strategy of type $(\tau_1, \ldots, \tau_k \to \gamma)$, $s[s_1, \ldots, s_k] \simeq s[s'_1, \ldots, s'_k]$.

The type structure of the *Hereditarily Sequential Functionals*[1], HSF, is defined as follows. To each type τ it is associated a set HSF^τ of functionals, and to each strategy s of type τ a functional $F_s \in \mathsf{HSF}^\tau$. Set $F_{\langle ?\varepsilon\rangle} = \bot$ and $F_{\tilde{v}} = v$, where $\tilde{v} = \langle ?\varepsilon, !v\rangle$. If s is a strategy of type $\tau = (\tau_1, \ldots, \tau_k \to \gamma)$ then $F_s : \mathsf{HSF}^{\tau_1} v \cdots \mathsf{HSF}^{\tau_k} \to \mathsf{HSF}^\gamma$ is the functional

$$F_s(F_{s_1}, \ldots, F_{s_k}) = s[s_1, \ldots, s_k] \text{ if defined.}$$

Finally $\mathsf{HSF}^\tau = \{F_s \mid s \text{ is a strategy of type } \tau\}$, in particular $\mathsf{HSF}^\gamma = (V_\gamma)_\bot$.

The structure HSF is a type frame. To see this we need a definition of application between strategies of higher type, namely an operation $App(s,t) = s[t]$ where, if s is some strategy of type $\sigma \to \tau$ and t of type σ, $s[t]$ is a strategy of type τ.

Let p be a play of type $(\tau_1, \ldots, \tau_k \to \gamma)$: q is the subplay of p on the *i-th component* if it is the maximal subplay of p such that any question of q but the first one has the shape $?i.a$.

If p is a play of type $\tau = (\tau_1, \ldots, \tau_k \to \gamma)$, then we may construct a play $p|\sigma$ of type $\sigma = (\tau_2, \ldots, \tau_k \to \gamma)$, by restricting p to the moves not in the first component. Take $I = \{\zeta < \text{length}(p) \mid \forall a, b.\ p_\zeta \neq ?1.a \wedge p_{r(\zeta)} \neq ?1.b\}$: then $p\lceil I$ is a subplay of p and there exists a play q of type σ and a play morphism φ such that $\varphi[q] = p\lceil I$, $q_\zeta = ?(j-1).a$ whenever $p_{\varphi_0(\zeta)} = ?j.a$, $q_\zeta = p_{\varphi_0(\zeta)}$ else, and φ_1 is the identity. r_q is fully determined by φ and r_p.

Proposition 3. *Let s be a strategy of type $\tau = (\tau_1, \ldots, \tau_k \to \gamma)$. Consider $\sigma = (\tau_2, \ldots, \tau_k \to \gamma)$ and some strategy t of type τ_1. Define $s[t]$ as the set of all P-cut free plays p' such that for some play p of type τ:*

1. *p' is a P-view of $p|\sigma$;*
2. *P follows s on p;*
3. *if q is the subplay of p on the first component then O follows $t^{(1)}$ on q.*

Then $s[t]$ is a strategy of type $\sigma = (\tau_2, \ldots, \tau_k \to \gamma)$, such that, for all strategies $t_2, \ldots, t_k$ of type $\tau_2, \ldots, \tau_k$

$$s[t][t_2, \ldots, t_k] \simeq s[t, t_2, \ldots, t_k].$$

By this the functional application is simply defined by: $F_s(F_t) = F_{s[t]}$.

[1] We give to this structure the same name as in [11], but they are different since our HSF properly includes the structure considered by Nickau.

4 Well-founded total functionals

In this section and in the next one we restrict our attention to type structures over $\mathsf{T}(N) = \mathsf{T}(\{N\})$, namely to simple types with ground type N. We also fix $V_N = \omega$.

A P-strategy s is *winning* if P always wins against any O-strategy, by following s. It is *strongly winning* if any $p \in s$ has some extension $q \in s$ won by P. A strongly winning strategy is winning, but not vice versa: indeed a winning strategy may include plays lost by P which simply cannot be a P- view of any play against some O-strategy. Strongly winning strategies are *complete*: by Theorem 1 any play of limit length can be extended; on the other hand in a P-cut free play just P may play at limit points; therefore if s is a winning strategy and p is a P-cut free play of limit length λ, then $p \in s$ if and only if $p\lceil\xi \in s$ for all $\xi < \lambda$.

Winning strategies are related to total functionals: $F_s \in \mathsf{HSF}^{(\tau_1,\ldots,\tau_k\to\gamma)}$ is *total* if for all total $F_{s_1},\ldots,F_{s_k}$ there exists $n \in V_N$ such that

$$F_s(F_{s_1},\ldots,F_{s_k}) \simeq n.$$

Theorem 3. *F_s is total if and only if s is strongly winning.*

The proof of the last theorem depends on the fact that any strategy s is included in some strongly winning strategy (possibly of transfinite height). This implies that any partial object in HSF has a total extension within HSF: this should be contrasted with the Scott continuous functionals, where e.g. Plotkin continuous existential quantifier is maximal (w.r.t. the pointwise ordering) but not total (see [12]). The same remark applies to the PCF definable functionals: indeed (our) HSF is a larger model than the extensional collapse of innocent strategies.

Because of the existence of transfinite plays and of strategies of transfinite height, any functional in the type frame HTF of the *Hereditarily Total Functionals* (the full type hierarchy over $V_N = \omega$) is an object of HSF[2].

Theorem 4. *For all type τ and $F \in$ HTF there exists a winning strategy s of the same type such that $F = F_s$.*

If κ is an infinite regular ordinal and s is a strategy of height $\leq \kappa$ (recall that the height of a tree T is the first ordinal α such that for all sequence $x \in T$, length$(x) < \alpha$), we say that it is a *κ-strategy*: an ω-strategy is then a well-founded tree. A functional $F_s \in \mathsf{HSF}^\tau$ is *well-founded* if there exists an ω-strategy s such that $F = F_s$. The following Corollary is an immediate consequence of the Cut-Elimination Theorem 2 and of the definition of totality.

[2] Strictly speaking any object of HTF turns out to be the restriction to total functionals of some object of HSF, as the latter may have partial functionals in its domain. In the sequel we shall not enter into such details, and we will consider HTF as a subframe of HSF

Corollary 1. *Total well-founded functionals from* HSF *are closed under application.*

Let TWF be the type frame of *Total Well-founded Functionals.*

Theorem 5. TWF *is a model of simply typed* λ*-calculus.*

Well-founded functionals embody the idea of functionals determined by finite amounts of information about their arguments: the same idea at the basis of Kleene-Kreisel countable functionals and of Scott continuous functionals. In the final part of this section we characterize the well-founded total functionals using a generalization to all types, due to Tait, of Brouwer's notion of continuity for type 2 functionals.

Definition 8. *The* Tait Continuous Functionals, TCF, *is the least type frame over* T(N) *such that:*

1. TCF^N *is the set of natural numbers;*
2. TCF *contains the combinators* $\mathbf{S}, \mathbf{K}, \mathbf{I}$ *at all (suitable) types;*
3. *if* $\{F_n \mid n \in \omega\} \subseteq \mathsf{TCF}^\tau$ *then the functional* $F(n) = F_n$ *(also denoted by* $\lambda n.\ F_n$*) is in* $\mathsf{TCF}^{(N \to \tau)}$ *(the* ω*-rule).*

Recursive Tait-continuous functionals, which are obtained from Definition 8 by asking in the third clause that the set $\{F_n \mid n \in \omega\}$ is recursive, are total functionals (this is a consequence of Tait cut-elimination theorem for the ω-logic). That TCF is a subframe of HTF will be a consequence of the proof that TCF and TWF actually coincide.

It is not difficult to show that TCF $\subseteq$ TWF, since by Theorem 5 it suffices to prove the closure of TWF under the ω-rule. Suppose that $F_n = F_{s_n}$ for all n and take s as the prefix closure of the set of all P-cut free plays p of type $(N \to \tau)$ such that $p = \langle ?\varepsilon, ?1, !n\rangle * q$, and q is obtained from some $q' \in s_n$ by substituting each move of the form $?i.a$ by $?(i+1).a$. Then s is a strategy of type $(N \to \tau)$, and $F_s = \lambda n.\ F_n$.

To prove that TCF $\supseteq$ TWF the following lemma is needed (compare with [11] Theorem 3.3.6). If T is a tree then $T_{\langle x\rangle} = \{y \mid \langle x\rangle * y \in T\}$ is an immediate subtree of T; a proper subtree of T is either an immediate subtree or a proper subtree of some immediate subtree of T. Recall that well-founded trees admit an inductive definition: T is well-founded if all immediate subtrees of T are such.

Lemma 1. *Let* s *be an* ω*-strategy of type* $(\tau_1, \ldots, \tau_k \to N)$ *such that* $s \neq \tilde{n}$ *for any* n*. Then there exist* $1 \leq i \leq k$ *and the* ω*-strategies* $s_1, \ldots, s_{n_i}$ *(where* $\tau_i = (\sigma_1, \ldots, \sigma_{n_i} \to N)$*) and a family of* ω*-strategies* $\{s'_m\}_{m \in \omega}$ *such that, for all strategies* $t_1, \ldots, t_k$ *of type* $\tau_1, \ldots, \tau_k$*, if* $t_i[s_1[t_1, \ldots, t_k], \ldots, s_{n_i}[t_1, \ldots, t_k]] \simeq m$ *then* $s[t_1, \ldots, t_k] \simeq s'_m[t_1, \ldots, t_k]$*. Moreover* $s_1, \ldots, s_{n_i}$ *and each* s'_m *are isomorphic to proper subtrees of* s*.*

Theorem 6. *The well-founded functionals are exactly the Tait-continuous functionals, namely* TWF $=$ TCF*.*

Proof. Let $F = F_s$ be a well-founded functional of type $(\tau_1, \ldots, \tau_k \to N)$. If $s = \tilde{n}$ then $F_s = \lambda x_1 \cdots x_k.n$ and it is trivially Tait-continuous. Otherwise, by induction over the well founded tree s and by Lemma 1, there exist $G_1 = F_{s_1}, \ldots, G_{n_i} = F_{s_{n_i}}$ and $G'_m = F_{s'_m}$ for each $m \in \omega$ which are Tait-continuous and such that, if $F_i(G_1(F_1, \ldots, F_k), \ldots, G_{n_i}(F_1, \ldots, F_k)) = m$ then $F(F_1, \ldots, F_k) = G'_m(F_1, \ldots, F_k)$. Therefore

$$F(F_1, \ldots, F_k) = (\lambda m.G'_m(F_1, \ldots, F_k))(F_i(G_1(F_1, \ldots, F_k), \ldots, G_{n_i}(F_1, \ldots, F_k)))$$

is Tait-continuous as it is obtained applying the ω-rule to a combination of $F_1, \ldots, F_k$ and of constants for Tait-continuous functionals.

□

5 Computable non well-founded functionals

Given any $F \in \mathsf{HTF}^{((N \to N) \to N)}$, there exists $f, g \in \mathsf{HTF}^{(N \to N)}$ such that

$$f(F(f)) \neq g(F(g)) \quad (1)$$
$$F(f) = F(g) \quad (2)$$

Indeed for any ordinal ξ let h_ξ be the characteristic function of $X_\xi = \{F(h_\zeta) \mid \zeta < \xi\}$. By a cardinality reasoning there exists a minimal $\alpha < \omega_1$ such that $X_{\alpha+1} = X_\alpha$; therefore $h_\alpha(F(h_\alpha)) = h_{\alpha+1}(F(h_\alpha)) = 1$. Since $F(h_\alpha) \in X_{\alpha+1} = X_\alpha$ there exists a (unique) $\beta < \alpha$ such that $F(h_\alpha) = F(h_\beta)$. If $h_\beta(F(h_\beta)) = 1$ then $X_\beta = X_{\beta+1} = X_\alpha$ contradicting the minimality of α, so that $h_\beta(F(h_\beta)) \neq h_\alpha(F(h_\alpha))$: now set $f = h_\alpha$ and $g = h_\beta$.

The construction of f, g is uniform in F, so that there exist two total functionals Φ, Ψ of type $(((N \to N) \to N), N \to N)$ such that $f = \Phi(F)$ and $g = \Psi(F)$ satisfy (1), (2). If F is continuous (w.r.t. the product topology over $\mathsf{HTF}^{(N \to N)} = \omega^\omega$) then $\alpha < \omega$. In this case it is easily proved that $\Phi(F)(n) = m$ and $\Psi(F)(n) = m$ are predicates recursive in F. In this sense Φ and Ψ are "computable" type 3 functionals.

By Theorem 4 Φ, Ψ are objects of HSF. More explicitly a strategy for Φ is the least prefix closed set of P-cut free plays of type $(((N \to N) \to N), N \to N)$ including plays of the following two forms (using the symbolic notation):

$$\langle \Phi(F, x) =?, F(f) =?, F(f) = n_0, \ldots, F(f) =?, F(f) = n_\eta, \text{ (for all } \eta < \xi)$$
$$F(f) =?, f(y) =?, y =?, y = m, f(y) = h_\xi(m)\rangle$$

which accounts for the computation of $F(h_\xi)$, and

$$\langle \Phi(F, x) =?, F(f) =?, F(f) = n_0, \ldots, F(f) =?, F(f) = n_\alpha,$$
$$x =?, x = n, \Phi(F, x) = h_\alpha(n)\rangle.$$

which yields the value of $\Phi(F, x)$. In the second line, as in the informal definition of Φ, α is the minimum ordinal such that $n_\alpha = n_\beta$ for a (unique) $\beta < \alpha$. The definition of a strategy for Ψ is similar, but the last move in the second case is $\Psi(F, x) = h_\beta(n)$.

These strategies are both ω_1-strategies, where ω_1 is the first uncountable ordinal. Next we prove that Φ, Ψ have no ω-strategy.

Theorem 7. *Φ and Ψ are not well-founded functionals.*

The proof uses two Lemmas. By $F \subseteq G$ it is meant graph inclusion.

Lemma 2. *Let $F \in \mathsf{HSF}^{((N\to N)\to N)}$ be partial injective, X be the range of F, $x \notin X$ and $f \in \mathsf{HTF}^{(N\to N)} \subseteq \mathsf{HSF}^{(N\to N)}$: then there exists $G \in \mathsf{HSF}^{((N\to N)\to N)}$ partial injective such that* $\mathrm{Rng}(G) \subseteq X \cup \{x\}$, $F \subseteq G$ *and* $f \in \mathrm{Dom}(G)$.

Lemma 3. *Let $\{s_n \mid n \in \omega\}$ be a family of winning ω-strategies of type $(((N \to N) \to N) \to N)$, and $X \subseteq \omega$ an infinite set. Then there exists $F \in \mathsf{HSF}^{((N\to N)\to N)}$ partial injective with range X s.t. $F_{s_n}(F)$ is defined for all n.*

Proof of Theorem 7. Toward a contradiction suppose that $\Phi = F_s$ and $\Psi = F_t$, for some (winning) ω-strategies s, t. Then there exist winning ω-strategies s_n and t_m associated to $\Phi_n = \lambda F.\ \Phi(F, n)$ and $\Psi_m = \lambda F.\ \Phi(F, m)$ respectively. Let us abbreviate by $\theta_{\langle n,m\rangle}$ a strategy for the functional $\Theta_{\langle n,m\rangle} \in \mathsf{HSF}^{(((N\to N)\to N)\to N}$ such that

$$\Theta_{\langle n,m\rangle}(G) = \langle \Phi(G)(n), \Psi(G)(m)\rangle,$$

where $\langle \_, \_\rangle$ is a surjective pairing function over the natural numbers. Of course $\theta_{\langle n,m\rangle}$ can be constructed from s_n and t_m in such a way that it is an ω-strategy. Being $\Theta_{\langle n,m\rangle}$ a total functional, $\theta_{\langle n,m\rangle}$ is winning by 3.

By Lemma 3, given any infinite $X \subseteq \omega$ and $\langle i, j\rangle \notin X$ we can find $F \in \mathsf{HSF}^{((N\to N)\to N)}$ partial injective with range $\subseteq X$ such that $\Theta_{\langle n,m\rangle}(F)$ is defined for all n, m, which implies that $f = \Phi(F)$ and $g = \Psi(F)$ are total functions, since $\langle f(n), g(m)\rangle \simeq \Theta_{\langle n,m\rangle}(F)$ for all n, m.

Applying Lemma 2 twice we find U partial injective such that $F \subseteq G$, $X \cup \{i, j\}$ is the range of U and $f, g \in \mathrm{Dom}(G)$. Let H be any total extension of U: then $\Phi(F) \subseteq \Phi(G) \subseteq \Phi(H)$, and, as $f = \Phi(F)$ is total, $\Phi(H) = f$. Similarly $\Psi(H) = g$.

By the absurd hypothesis $f(H(f)) \neq g(H(g))$ and $H(f) = H(g)$. ¿From $H(f) = G(f)$ and $H(g) = G(g)$ it follows $G(f) = G(g)$, hence $f = g$ since U is injective: a contradiction.

□

6 Concluding remarks

Although well-founded functionals are a natural structure, they do not capture the idea of (relative) computable functionals at type 3 and higher. This may be of minor interest as soon as one is concerned with λ-calculus models, but becomes relevant when dealing with the constructive analysis of classical proofs, and with program extraction. Indeed the functionals Φ, Ψ can be shown to be natural realizers of the no-counterexample of the comprehension axiom scheme for classical second order arithmetic, and have been found following methods introduced in [4].

The fact that they are not well-founded may appear not surprising as they are set theoretic functionals, defined also on discontinuous type 2 arguments (i.e. non continuous w.r.t. the product topology on type 1 objects), as it is needed if they have to build "no-counterexamples" against any possible candidate as a counterexample. However they have the robust property, as argued in the previous section, to yield finite plays on continuous (namely well-founded) arguments, which are effectively computable if the arguments are recursive. Actually Φ, Ψ are examples of a large class of functionals enjoying this property, which, we think, deserves further investigation.

References

1. S. Abramsky, "Semantics of Interaction", in *Semantics and Logics of Computation*, A. Pitts and p. Dybjer eds., Cambridge University Press 1997, 1-31.
2. S. Abramsky, R. Jagadeesan, "Games and full completeness for multiplicative linear logic", *Journal of Symbolic Logic* 59 (2), 1994, 543-574.
3. S. Abramsky, R. Jagadeesan, P. Malacaria, "Full abstraction for PCF", Proceedings of TACS'94, *Springer Lecture Notes in Computer Science* 789, 1994, 1-15.
4. S. Berardi, M. Bezem, T. Coquand, "On the Constructive Content of the Axiom of Choice", *Journal of Symbolic Logic*, to appear.
5. S. Berardi, T. Coquand, "Transfinite Games", September 1996.
6. A. Blass, "A game semantics for linear logic", *Annals of Pure and Applied Logic* 56, 183-220.
7. T. Coquand, "A Semantics of Evidence for Classical Arithmetic", *Journal of Symbolic Logic* 60, 1995, 325-337.
8. H. Herbelin. Séquents qu'on calcule. Ph.D. thesis, Univeristy of Paris VII, 1995.
9. J.M.E. Hyland, "Game Semantics", in *Semantics and Logics of Computation*, A. Pitts and p. Dybjer eds., Cambridge University Press 1997, 131-184.
10. J.M.E. Hyland, C.-H.L. Ong, "On full abstraction for PCF", available by ftp at ftp://ftp.comlab.ox.ac.uk/pub/Documents/techpapers/Luke.Ong/ as pcf.ps.gz, 1994.
11. H. Nickau, *Hereditarily Sequential Functionals: A Game- Theoretic Approach to Sequentiality*, Shaker Verlag, Achen 1996.
12. G. Plotkin, "Full Abstraction, Totality and PCF", available by ftp at ftp://ftp.lfcs.ed.ac.uk/pub/gdp/ as Totality.ps.gz, 1997.

Counting a Type's Principal Inhabitants
(Extended Abstract)

Sabine Broda and Luís Damas

DCC & LIACC, Universidade do Porto

Abstract. We present a Counting Algorithm that computes the number of λ-terms in β-normal form that have a given type τ as a principal type and produces a list of these terms. The design of the algorithm follows the lines of Ben-Yelles' algorithm for counting normal (not necessarily principal) inhabitants of a type τ.

1 Introduction

In [2], Ben-Yelles presented a Counting Algorithm, also described in [3], which given a type τ computes the number of λ-terms in β-normal form that can receive type τ in $\mathbf{TA}_\lambda$. For each type τ the algorithm decides in a finite number of steps whether the number of closed β-normal forms with type τ is finite or infinite, computes this number in the finite case, and lists all relevant terms in both cases. Related to this is the problem of counting the number of β-normal forms that have a given type τ as a principal type. As pointed out in ([3], p. 127), this problem is still open and in this paper we present a Counting Algorithm which solves this case. Analogous to Ben-Yelles' algorithm, our algorithm for counting (and listing) principal normal inhabitants of a type τ is based on the following facts. First, it is sufficient to look for a special kind of principal normal inhabitants of τ, called long terms. Second, there are integers $0 < \mathbf{d_p}(\tau) < \mathbf{D_p}(\tau)$ such that the cardinality of the set of principal normal inhabitants of τ depends directly on the number of long principal normal inhabitants of τ with depth in $[0; \mathbf{d_p}(\tau)[$ and on the number of those with depth in $[\mathbf{d_p}(\tau); \mathbf{D_p}(\tau)[$, where depth is a measure on the structure of a λ-term in β-normal form. Finally, Ben-Yelles defined in [2] a Search Algorithm (others are in [6], [7] and [5]), that given a type τ and any integer $d \geq 0$ can be used to compute all (a finite number) long normal inhabitants of a type τ with depth $\leq d$. Thus, using any principal-type checking algorithm (for example in [3]) it is possible to compute the long principal normal inhabitants of τ with depth $\leq d$. Thus the problem of counting principal normal inhabitants of a type τ is essentially solved by computing the long principal normal inhabitants of τ with depth $< \mathbf{D_p}(\tau)$.

In section 2 we describe the Counting Algorithm for normal principal inhabitants based on the existence of $\mathbf{d_p}(\tau)$ and $\mathbf{D_p}(\tau)$ and on Ben-Yelles' Search Algorithm. In section 3 we obtain a characterization of long normal principal inhabitants, that will give us a better insight on principal deductions for long terms in β-normal form and will thereby enable us in section 4 to establish and prove the correctness of the limits $\mathbf{d_p}(\tau)$ and $\mathbf{D_p}(\tau)$.

2 The Counting Algorithm

We use standard notation from [1] and [3]. Type-variables (atoms) are denoted by "a,b,c,..."and arbitrary types are denoted by lower-case Greek letters. It has been pointed out in [3] that it is equivalent to count typed or untyped inhabitants of a type τ. In sake of simplicity, we restrict this paper to the untyped case. A term M has a bound-variable clash iff M contains an abstractor $\underline{\lambda x}$ and a (free, bound or binding) occurrence of x that is not in its scope. Note that for any λ-term M exists a λ-term N without bound-variable clashes and such that $M =_\alpha N$. In this paper we will only consider λ-terms without bound-variable clashes.

Definition 1. *A type-assignment is an expression of the form $M : \tau$, where M is a λ-term and τ is a type. The type τ is the predicate and M is the subject of the type-assignment. A type-context or basis Γ is any finite, perhaps empty, set of type-assignments with distinct variables as subjects. If $\Gamma = \{x_1, \rho_1, \ldots, x_m : \rho_m\}$ define $Subjects(\Gamma) = \{x_1, \ldots, x_m\}$. A* **TA***-formula is any expression of the form $\Gamma \vdash M : \tau$, where M is a term, Γ a type-context and τ a type.*

In the following we describe a system to assign types to λ-terms in β-nf.

Definition 2. *Given a λ-term M in β-nf, a type τ and a context Γ, we say that $M : \tau$ is derivable from Γ, and write $\Gamma \vdash M : \tau$ if the formula $\Gamma \vdash M : \tau$ can be produced by the following rules.*

$$(axiom)\ \frac{}{\Gamma \vdash x : \alpha} \qquad (if\ x : \alpha \in \Gamma)$$

$$(app)\ \frac{\Gamma \vdash M_1 : \alpha_1 \quad \ldots \quad \Gamma \vdash M_n : \alpha_n}{\Gamma \vdash x M_1 \ldots M_n : \beta} \qquad (if\ x : \alpha_1 \to \ldots \to \alpha_n \to \beta \in \Gamma,\ n \geq 1)$$

$$(abs)\ \frac{\Gamma, x : \alpha \vdash M : \beta}{\Gamma \vdash \lambda x.M : \alpha \to \beta}$$

A **TA***-deduction Δ of $\Gamma \vdash M : \tau$ (where M denotes a β-nf) is a tree of* **TA***-formulae, those at the tops of branches being axioms and those below being deduced from those immediately above them by a rule ((app) or (abs)) and with bottom formula $\Gamma \vdash M : \tau$.*

Proposition 3. *Given a β-nf M, a basis Γ and type τ such that $\Gamma \vdash M : \tau$, there is exactly one deduction Δ of $\Gamma \vdash M : \tau$.*

Proof Straightforward. •

Definition 4. *Let M be a β-nf and (Γ, τ) a pair such that $\Gamma \vdash M : \tau$. We say that M is long with respect to (Γ, τ) iff for every formula of the form $\Gamma' \vdash x M_1 \ldots M_n : \alpha$, $n \geq 0$, in the unique deduction of $\Gamma \vdash M : \tau$, α is an atom. We* *call M a normal inhabitant of a type τ iff M is in β-nf and $\vdash M : \tau$ and denote the set of normal inhabitants of a type τ by* $\mathsf{Nhabs}(\tau)$*. The set of principal normal inhabitants of a type τ is called* $\mathsf{Nprinc}(\tau)$*. The set of normal inhabitants of τ which are long with respect to $(\emptyset, \tau)$ is called* $\mathsf{Long}(\tau)$*. The set of long principal normal inhabitants is called* $\mathsf{Lprinc}(\tau)$*.*

Thus,

$$\mathsf{Lprinc}(\tau) = \mathsf{Long}(\tau) \cap \mathsf{Nprinc}(\tau) \subseteq \mathsf{Long}(\tau) \subseteq \mathsf{Nhabs}(\tau).$$

The (finite) set of all terms obtained by η-reducing a λ-term M is called the η-family of M and denoted by $\{M\}_\eta$. It has been shown (cf. [3]) that the η-families of the long normal inhabitants of τ partition $\mathsf{Nhabs}(\tau)$ into non-overlapping finite subsets, each η-family containing just one long member. From this and from the two following results (in [3]) we conclude, that counting $\mathsf{Nprinc}(\tau)$ corresponds essentially to counting $\mathsf{Lprinc}(\tau)$.

Lemma 5 (Completeness of Long(τ); Ben-Yelles 1979). *Every normal inhabitant of τ can be η-expanded to a long normal inhabitant of τ. And this long inhabitant is unique (modulo $=_\alpha$); i.e.*

$$\{M, N \in Long(\tau) \text{ and } M =_\eta N\} \quad \Longrightarrow \quad M =_\alpha N.$$

Lemma 6 (in Hindley'97). *Let M^+ be the unique member of $Long(\tau)$ to which M η-expands. Then,*

$$M \in \mathsf{Nprinc}(\tau) \quad \Longrightarrow \quad M^+ \in \mathsf{Nprinc}(\tau).$$

Hence, if $M \in \mathsf{Nprinc}(\tau)$, then $M^+ \in \mathsf{Lprinc}(\tau)$. We conclude that $\mathsf{Nprinc}(\tau) = \emptyset$ iff $\mathsf{Lprinc}(\tau) = \emptyset$ and that $\mathsf{Nprinc}(\tau)$ is infinite iff $\mathsf{Lprinc}(\tau)$ is. Furthermore,

$$\mathsf{Nprinc}(\tau) \subseteq \bigcup_{M \in \mathsf{Lprinc}(\tau)} \{M\}_\eta.$$

Hence, our algorithm will focus on long normal principal inhabitants and, following Ben-Yelles' algorithm, the searching will be done in order of increasing depth of terms.

Definition 7. *The depth of a λ-term M in β-nf is defined as follows and denoted by $Depth(M)$.*

i. $Depth(y) = Depth(\lambda x_1 \ldots x_m.y) = 0;$
ii. $Depth(\lambda x_1 \ldots x_m.yM_1 \ldots M_n) = 1 + max_{1\le j\le n} Depth(M_j),$ *if* $n > 0.$

In [2] Ben-Yelles defined an algorithm, called Search Algorithm, that given a composite type τ, i.e. non-atomic (note that atomic types have no inhabitants at all), produces a sequence $\mathcal{A}(\tau,0), \mathcal{A}(\tau,1), \mathcal{A}(\tau,2), \ldots$ of finite sets of expressions, called nf-schemes, such that each member of $\mathcal{A}(\tau, d+1)$, which is a λ-term, is a closed β-nf with depth d. More precisely, one has the following, where $\mathsf{Long}(\tau, d)$ denotes the set of long normal inhabitants of τ with depth $\le d$.

Theorem 8 (Search Theorem for $\mathsf{Long}(\tau)$, Ben-Yelles'79). *The Search Algorithm accepts as input any composite type τ and outputs a finite or infinite sequence of sets $\mathcal{A}(\tau,d)$ $(d = 0,1,2,\ldots)$ such that for all $d \geq 0$,*

i. each member of $\mathcal{A}(\tau,d)$ is a closed nf-scheme with type τ and long with respect to $(\emptyset,\tau)$, and is either

  (a) a proper nf-scheme with depth d, or

  (b) a λ-term with depth $d-1$;

ii. $\mathcal{A}(\tau,d)$ is finite;

iii. $\mathsf{Long}(\tau,d) \subseteq \mathcal{A}(\tau,0) \cup \ldots \cup \mathcal{A}(\tau,d+1)$;

iv. if we call the set of all λ-terms in $\mathcal{A}(\tau,d)$ "$\mathcal{A}_{terms}(\tau,d)$", then

$$\mathsf{Long}(\tau) = \bigcup_{d \geq 0} \mathcal{A}_{terms}(\tau,d).$$

Now, and analogous to the Counting Algorithm for $\mathsf{Long}(\tau)$, the Counting Algorithm for $\mathsf{Nprinc}(\tau)$ is based on the fact that $\mathsf{Lprinc}(\tau)$ is infinite iff it has some member whose depth lies between two integers $\mathbf{d_p}(\tau)$ and $\mathbf{D_p}(\tau)$, that can be computed from τ. Furthermore, if $\mathsf{Lprinc}(\tau)$ has no member with depth in $[\mathbf{d_p}(\tau);\mathbf{D_p}(\tau)[$, then $\mathsf{Lprinc}(\tau)$ is finite or empty according as the number of long principal inhabitants of τ with depth $< \mathbf{d_p}(\tau)$ is finite or zero.

Definition 9. *The total number of occurrences of type-variables in a type τ will be denoted by $|\tau|$ and is defined as follows*

$$|a| = 1, \qquad |\rho \to \sigma| = |\rho| + |\sigma|.$$

The number of distinct type-variables occurring in τ will be denoted by $||\tau||$. Furthermore, if τ is a type let

$$\mathbf{d_p}(\tau) = |\tau| \qquad \textit{and} \qquad \mathbf{D_p}(\tau) = |\tau|^4.$$

In section 4 we will prove the following.

Theorem 10. *For any type τ, there is*

i. $\mathsf{Lprinc}(\tau) = \emptyset$ iff $\mathsf{Lprinc}(\tau)$ has no member with depth $< \mathbf{D_p}(\tau)$;

ii. $\mathsf{Lprinc}(\tau)$ is infinite iff it has a member M with $\mathbf{d_p}(\tau) \leq depth(M) < \mathbf{D_p}(\tau)$;

iii. $\mathsf{Lprinc}(\tau)$ is finite iff all its members with depth $< \mathbf{D_p}(\tau)$ have depth $< \mathbf{d_p}(\tau)$.

Thus, we have the following algorithm to count $\mathsf{Nprinc}(\tau)$.

Counting Algorithm for $\mathsf{Nprinc}(\tau)$ **11** *If* τ *is an atom,* $\mathsf{Nprinc}(\tau)$ *is empty. If* τ *is composite, apply the Search Algorithm to* τ *and compute* $\mathcal{A}_{terms}(\tau, d)$ *for* $d = 0, \ldots, \mathbf{D_p}(\tau)$. *Determine the set* $\mathcal{A}_p$ *of all* λ*-terms in* $\mathcal{A}_{terms}(\tau, 0) \cup \ldots \cup \mathcal{A}_{terms}(\tau, \mathbf{D_p}(\tau))$ *that are principal inhabitants of* τ *(using any algorithm for checking principal types).*

Case I *If* $\mathcal{A}_p = \emptyset$, *then* $\mathsf{Nprinc}(\tau) = \emptyset$.

Case II *If* $\mathcal{A}_p$ *has a member with depth* $\geq \mathbf{d_p}(\tau)$, *then* $\mathsf{Nprinc}(\tau)$ *is infinite. Apply the Search Algorithm to enumerate* $\mathcal{A}_{terms}(\tau, d)$ *for* $d = 0, 1, 2, \ldots$, *outputting for each of these sets its members which are principal inhabitants of* τ *as well as the members of their* η*-families that are principal inhabitants.*

Case III *If all members in* $\mathcal{A}_p$ *have depth* $< \mathbf{d_p}(\tau)$, *then* $\mathsf{Nprinc}(\tau)$ *is finite. Output all members of* $\mathcal{A}_p$ *as well as the members of their* η*-families that are principal inhabitants.*

3 Principal type inference for long normal λ-terms

In this section we introduce the typing system $\mathbf{TA_{pln}}$ which will give us a better insight on deductions of principal types for long normal inhabitants, and will thus enable us to prove the Shrinking and Stretching Lemmas for $\mathsf{Lprinc}(\tau)$, which have Theorem 10 as a consequence.

Definition 12. *The system* $\mathbf{TA_{pln}}$ *has an infinite set of axioms and three deduction rules as follows.*

$$(axiom)\ \frac{}{\Gamma \vdash^{pln} x : a \parallel \emptyset}\quad (\text{if } x : a \in \Gamma)$$

$$(app)\ \frac{\Gamma \vdash^{pln} M_1 : \alpha_1 \parallel \phi_1 \quad \ldots \quad \Gamma \vdash^{pln} M_n : \alpha_n \parallel \phi_n}{\Gamma \vdash^{pln} xM_1 \ldots M_n : b \parallel \emptyset}$$

$$(\text{if } x : \alpha_1 \to \ldots \to \alpha_n \to b \in \Gamma \text{ and } n \geq 1)$$

$$(abs)\ \frac{\Gamma, x : \alpha \vdash^{pln} M : \beta \parallel \phi}{\Gamma \vdash^{pln} \lambda x.M : \alpha \to \beta \parallel \emptyset}$$

$$(U)\ \frac{\Gamma \vdash^{pln} M : a \parallel \emptyset}{\Gamma \vdash^{pln} M : b \parallel (a, b)}$$

A $\mathbf{TA_{pln}}$*-deduction* Δ *is a tree of* $\mathbf{TA_{pln}}$*-formulae, those at the tops of branches being axioms and those below being deduced from those immediately above them by a rule. The bottom formula in* Δ *is called its conclusion; if it is* $\Gamma \vdash^{pln} M : \tau \parallel \phi$, *we call* Δ *a deduction of* $\Gamma \vdash^{pln} M : \tau$.

In the following we are going to define a transformation-algorithm that given the unique **TA**-deduction Δ of a formula $\Gamma \vdash M : \tau$, where M is a β-nf which is long with respect to (Γ, τ), and a pair (Γ', τ'), obtained from (Γ, τ) by renaming occurrences of variables (note that different occurrences of variables may have been substituted by different variables), constructs a $\mathbf{TA_{pln}}$-deduction Δ' of $\Gamma' \vdash^{pln} M : \tau'$, such that several arrows occurring in Δ' are possibly marked with a $\star$. In the first step an unmarked version of Δ' will be constructed bottom-up from Δ as follows.

- if $\Gamma \vdash M : \tau$ is an axiom, i.e. $x : a \in \Gamma$, $M = x$ and $\tau = a$, then $x : a_k \in \Gamma'$ and $\tau' = a_l$. Take $\Gamma' \vdash^{pln} x : a_l \parallel (a_k, a_l)$ as the bottom formula of Δ' (U-rule) and precede it by $\Gamma' \vdash^{pln} x : a_k \parallel \emptyset$ (axiom).
- if $\Gamma \vdash M : \tau$ was obtained by the (app)-rule from $\Gamma \vdash M_1 : \alpha_1, \ldots, \Gamma \vdash M_n : \alpha_n$, thus $M = xM_1 \ldots M_n$ and $\tau = b$, then there is a type-assignment $x : \alpha_1' \rightarrow \ldots \rightarrow \alpha_n' \rightarrow b_k$ in Γ' and $\tau' = b_l$. Take $\Gamma' \vdash^{pln} xM_1 \ldots M_n : b_l \parallel (b_k, b_l)$ as the bottom formula of Δ' (U-rule), precede it by $\Gamma' \vdash^{pln} xM_1 \ldots M_n : b_k \parallel \emptyset$ (app) and precede this formula by the deductions $\Delta_1', \ldots, \Delta_n'$ constructed from the **TA**-deductions $\Delta_1, \ldots, \Delta_n$ of $\Gamma \vdash M_1 : \alpha_1, \ldots, \Gamma \vdash M_n : \alpha_n$ and pairs $(\Gamma', \alpha_1'), \ldots, (\Gamma', \alpha_n')$.
- if $\Gamma \vdash M : \tau$ was obtained by the (abs)-rule, i.e. $M = \lambda x.N$ and $\tau = \alpha \rightarrow \beta$, from $\Gamma, x : \alpha \vdash N : \beta$, then take $\Gamma' \vdash^{pln} \lambda x.N : \alpha' \rightarrow \beta' \parallel \emptyset$ as the bottom-rule in Δ' (abs) and precede it by the deduction Δ_1' constructed from the deduction Δ_1 of $\Gamma, x : \alpha \vdash N : \beta$ and pair $(\Gamma' \cup \{x : \alpha'\}, \beta')$.

In the second step we mark arrows, starting top-down from axioms as follows.

- No arrows are marked for axioms.
- If $\Gamma \vdash^{pln} xM_1 \ldots M_n : b \parallel \emptyset$ results from $\Gamma \vdash M_1 : \alpha_1 \parallel \phi_1, \ldots, \Gamma \vdash M_n : \alpha_n \parallel \phi_n$ by the (app)-rule, then we mark $x : \alpha_1 \rightarrow \ldots \rightarrow \alpha_n \rightarrow b$ in Γ as follows: $x : \alpha_1 \rightarrow^\star \ldots \rightarrow^\star \alpha_n \rightarrow^\star b$ (no arrows are marked in $\alpha_1, \ldots, \alpha_n$).
- If $\Gamma \vdash^{pln} \lambda x.M : \alpha \rightarrow \beta \parallel \emptyset$ was obtained from $\Gamma, x : \alpha \vdash^{pln} M : \beta \parallel \phi$ by the (abs)-rule, then we mark the following arrow: $\Gamma \vdash^{pln} \lambda x.M : \alpha \rightarrow^\star \beta \parallel \emptyset$.
- Finally, no arrows are marked in formulae obtained by the U-rule.

Definition 13. *The indexed counterpart $(\Gamma_{\mathtt{i}}, \tau_{\mathtt{i}})$ of a pair (Γ, τ) is obtained by successively indexing all occurrences of type variables and arrows in (Γ, τ).*

Note 1. If $(\Gamma_{\mathtt{i}}, \tau_{\mathtt{i}})$ is the indexed counterpart of (Γ, τ), then $||\tau|| \leq |\tau| = |\tau_{\mathtt{i}}| = ||\tau_{\mathtt{i}}||$.

Example 14. The pair

$$(\{z : ((a_1 \rightarrow_1 b_1) \rightarrow_2 c_1 \rightarrow_3 c_2) \rightarrow_4 d_1\}, d_2)$$

is the indexed counterpart of $(\{z : ((a \rightarrow b) \rightarrow c \rightarrow c) \rightarrow d\}, d)$. The **TA**-deduction of

$$z : ((a \rightarrow b) \rightarrow c \rightarrow c) \rightarrow d \vdash z(\lambda xy.y) : d$$

is

$$\begin{array}{ll}
z : ((a \to b) \to c \to c) \to d, x : a \to b, y : c \vdash y : c & \\
\hline
z : ((a \to b) \to c \to c) \to d, x : a \to b & \vdash \lambda y.y : c \to c \\
\hline
z : ((a \to b) \to c \to c) \to d & \vdash \lambda xy.y : (a \to b) \to c \to c \\
\hline
z : ((a \to b) \to c \to c) \to d & \vdash z(\lambda xy.y) : d
\end{array}$$

and the corresponding $\mathbf{TA_{pln}}$-deduction of $z : ((a_1 \to_1 b_1) \to_2 c_1 \to_3 c_2) \to_4 d_1 \vdash z(\lambda xy.y) : d_2$

is

$$\begin{array}{l}
z : ((a_1 \to_1 b_1) \to_2 c_1 \to_3 c_2) \to_4 d_1, x : a_1 \to_1 b_1, y : c_1 \vdash y : c_1 \parallel \emptyset \\
\hline
z : ((a_1 \to_1 b_1) \to_2 c_1 \to_3 c_2) \to_4 d_1, x : a_1 \to_1 b_1, y : c_1 \vdash y : c_2 \parallel (c_1, c_2) \\
\hline
z : ((a_1 \to_1 b_1) \to_2 c_1 \to_3 c_2) \to_4 d_1, x : a_1 \to_1 b_1 \vdash \lambda y.y : c_1 \to_3^\star c_2 \parallel \emptyset \\
\hline
z : ((a_1 \to_1 b_1) \to_2 c_1 \to_3 c_2) \to_4 d_1 \vdash \lambda xy.y : (a_1 \to_1 b_1) \to_2^\star c_1 \to_3 c_2 \parallel \emptyset \\
\hline
z : ((a_1 \to_1 b_1) \to_2 c_1 \to_3 c_2) \to_4^\star d_1 \vdash z(\lambda xy.y) : d_1 \parallel \emptyset \\
\hline
z : ((a_1 \to_1 b_1) \to_2 c_1 \to_3 c_2) \to_4 d_1 \vdash z(\lambda xy.y) : d_2 \parallel (d_1, d_2)
\end{array}$$

Definition 15. *Given a type τ and a binary relation Φ defined over the set of type-variables in τ, let C_Φ be the set of equivalence classes of the reflexive, symmetric and transitive closure of Φ.*

Lemma 16. *Consider any finite non-empty set A and n binary relations over A*

$$\Phi_1 \subseteq \ldots \subseteq \Phi_n$$

such that $C_{\Phi_i} \neq C_{\Phi_j}$ for $1 \leq i \neq j \leq n$. Then $n \leq \#A$.

Proof Straightforward •

The following result is a consequence of observing that the algorithm for computing the principal pair of a λ-term only introduces arrows required by the typing rules and only unifies two variables if this is absolutely required by the term structure. Note that the binary relation Φ corresponds directly to the connection relation for TA-figures in [4].

Proposition 17. *Let M be a λ-term in normal form, Γ a type-context and τ a type such that, $Subjects(\Gamma) = FV(M)$, $\Gamma \vdash M : \tau$ and M is long with respect to (Γ, τ). Consider the indexed counterpart $(\Gamma_{\mathtt{i}}, \tau_{\mathtt{i}})$ of (Γ, τ). Let $\Delta_{(\Gamma_{\mathtt{i}}, \tau_{\mathtt{i}})}$ be the* $\mathbf{TA_{pln}}$*-deduction of $\Gamma_{\mathtt{i}} \vdash^{pln} M : \tau_{\mathtt{i}}$ constructed from the unique* **TA***-deduction of $\Gamma \vdash M : \tau$ and let Φ be the set of all binary pairs in $\Delta_{(\Gamma_{\mathtt{i}}, \tau_{\mathtt{i}})}$. Then,*

(Γ, τ) is a principal pair for M
iff
for each type-variable a in (Γ, τ) there is one equivalence class in C_Φ containing exactly all indexed occurrences of a in $(\Gamma_{\mathtt{i}}, \tau_{\mathtt{i}})$ and all indexed arrows have a marked occurrence in $\Delta_{(\Gamma_{\mathtt{i}}, \tau_{\mathtt{i}})}$.

Proof It is straightforward to show that whenever $(a,b) \in \Phi$, then a and b are indexed versions of the same type-variable. Now, suppose that (Γ,τ) is no principal pair for M, i.e. there exists a (more general) pair (Γ_0,τ_0) such that $\Gamma_0 \vdash M : \tau_0$, $\Gamma_0^\star = \Gamma$ and $\tau_0^\star = \tau$ for some substitution $\star$, but (Γ,τ) is no variant of (Γ_0,τ_0). If (Γ_0,τ_0) has the same structure as (Γ,τ), but at least two type-variables in (Γ_0,τ_0) are given the same name, say a, in (Γ,τ), then, considering the observation made at the beginning of the proof, it is easy to see that there are at least two equivalence classes in C_Φ containing indexed occurrences of a in $(\Gamma_\mathtt{i},\tau_\mathtt{i})$. If, on the other hand, one variable in (Γ_0,τ_0) has a composite type $\alpha \to \beta$ in (Γ,τ), then the corresponding occurrences of the arrow will never be marked in $\Delta_{(\Gamma_\mathtt{i},\tau_\mathtt{i})}$. This follows from the fact that the **TA**-deduction of $\Gamma \vdash M : \tau$ is a copy of the **TA**-deduction of $\Gamma_0 \vdash M : \tau_0$ where all type-variables a are substituted by $\star(a)$. Analysing the transformation algorithm, one sees that the same holds for the corresponding $\mathbf{TA_{pln}}$-deductions.

Conversely, suppose that M is a β-nf, Γ a type-context and τ a type such that $\Gamma \vdash M : \tau$, M is long with respect to (Γ,τ) and $FV(M) \subseteq \mathit{Subjects}(\Gamma)$. Furthermore, consider $(\Gamma_\mathtt{i},\tau_\mathtt{i})$, $\Delta_{(\Gamma_\mathtt{i},\tau_\mathtt{i})}$ and Φ as before. Let $\Phi' \supseteq \Phi$ be any binary relation over the variables in $(\Gamma_\mathtt{i},\tau_\mathtt{i})$ such that there is some type-variable, say d, such that there are at least two different equivalence classes $\{d_{i_1},\ldots,d_{i_m}\}$ and $\{d_{j_1},\ldots,d_{j_n}\}$ in $C_{\Phi'}$ containing indexed occurrences of d. We prove by induction on the structure of M that, if we substitute in $\Delta_{(\Gamma_\mathtt{i},\tau_\mathtt{i})}$ the occurrences $d_{j_1},\ldots,d_{j_n}$ by $c_{j_1},\ldots,c_{j_n}$, for some new type-variable c, then we obtain a new $\mathbf{TA_{pln}}$-deduction Δ'_c that corresponds to the **TA**-deduction Δ_c of $\Gamma_c \vdash M : \tau_c$, where (Γ_c,τ_c) is obtained from (Γ,τ) by substitution of the occurrences of d corresponding to $d_{j_1},\ldots,d_{j_n}$ by c. Then (Γ_c,τ_c) is a more general pair such that $\Gamma_c \vdash M : \tau_c$ and consequently (Γ,τ) is no principal pair for M.

If $M = x$, then $(\Gamma,\tau) = (\{x : a\}, a)$ and the result holds vacuously, since $C_\Phi = \{\{a_1,a_2\}\}$.

If $M = xM_1 \ldots M_n$, then $\tau = b$, $x : \alpha_1 \to \ldots \alpha_n \to b \in \Gamma$ and $\Gamma \vdash xM_1 \ldots M_n : b$ results from $\Gamma \vdash M_1 : \alpha_1, \ldots, \Gamma \vdash M_n : \alpha_n$ by the $(\to E)$-rule. From the induction hypothesis we conclude that $\Gamma_c \vdash M_1 : \alpha_1^c, \ldots, \Gamma_c \vdash M_n : \alpha_n^c$. If $\Delta'_{c,1},\ldots,\Delta'_{c,n}$ are the corresponding $\mathbf{TA_{pln}}$-deductions, then Δ'_c has the form

$$\frac{\dfrac{\Delta'_{c,1}\cdots\Delta'_{c,n}}{\Gamma_c \vdash^{pln} xM_1\ldots M_n : b_k^c \parallel \emptyset}}{\Gamma_c \vdash^{pln} xM_1\ldots M_n : b_l^c \parallel (b_k^c, b_l^c).}$$

Thus $(b_k, b_l) \in \Phi \subseteq \Phi'$ belong to the same class in C_Φ and the corresponding occurrences in (Γ^c,τ^c) are occurrences of the same type-variable. Hence, $\Gamma^c \vdash xM_1 \ldots M_n : b^c$ can be inferred from $\Gamma_c \vdash M_1 : \alpha_1^c, \ldots, \Gamma_c \vdash M_n : \alpha_n^c$ by the (app)-rule and Δ'_c corresponds to the **TA**-deduction Δ_c of $\Gamma_c \vdash xM_1 \ldots M_n : b^c$.

The result is straightforward for $M = \lambda x.N$.

Thus we showed, that whenever (Γ,τ) is a principal pair for M then for each type-variable a in (Γ,τ) there is one equivalence class in C_Φ containing exactly all indexed occurrences of a in $(\Gamma_\mathtt{i},\tau_\mathtt{i})$. It remains to show that all (indexed) arrows have a marked occurrence in $\Delta_{(\Gamma_1,\tau_1)}$. We proceed by induction on the structure of M.

If $M = x$, then the result holds vacuously.

For $M = xM_1 \dots M_n$ let (Γ, b) be a principal pair of $xM_1 \dots M_n$ obtained from $\Gamma \vdash M_1 : \alpha_1, \dots, \Gamma \vdash M_n : \alpha_n$ by the (app)-rule and $x : \alpha_1 \to \dots \to \alpha_n \to b \in \Gamma$. Then there are $(\Gamma_1, \alpha_1^0), \dots, (\Gamma_n, \alpha_n^0)$ respectively principal pairs of $M_1, \dots, M_n$, such that $\Gamma = \Gamma_1^\star \cup \dots \cup \Gamma_n^\star \cup \{x : v_1 \to \dots \to v_n \to v\}^\star$, where $\star$ is the substitution resulting from the unification of the types assigned to variables in $\Gamma_1, \dots, \Gamma_n$ as well as in $\{x : v_1 \to \dots \to v_n \to v\}$ ($v_1, \dots, v_n, v$ are new type-variables) and such that $\alpha_1 = \star(v_1) = \star(\alpha_1^0), \dots, \alpha_n = \star(v_n) = \star(\alpha_n^0)$ and $b = \star(v)$. Now, by the induction hypothesis, all occurrences of arrows in $\Gamma_1, \dots, \Gamma_n, \alpha_1^0, \dots, \alpha_n^0$ are marked in the $\mathbf{TA_{pln}}$-deductions corresponding to the **TA**-deductions of $\Gamma_1 \vdash M_1 : \alpha_1, \dots, \Gamma_n \vdash M_n : \alpha_n$. Thus, the arrows resulting from the unification and corresponding to arrows in $\Gamma_1, \dots, \Gamma_n, \alpha_1^0, \dots, \alpha_n^0$ will also be marked in the $\mathbf{TA_{pln}}$-deduction that corresponds to the **TA**-deduction of $\Gamma \vdash xM_1 \dots M_n : b$. Finally, note that the n main arrows in $\alpha_1 \to \dots \to \alpha_n \to b$ will be marked too.

The case $M = \lambda x.N$ is trivial, since if $\Gamma \vdash \lambda x.N : \alpha \to \beta$ results from $\Gamma, x : \alpha \vdash N : \beta$ by the (abs)-rule and $(\Gamma, \alpha \to \beta)$ is a principal pair for $\lambda x.N$, then $(\Gamma \cup \{x : \alpha\}, \beta)$ is a principal pair for N and the result follows almost directly from the induction hypothesis. •

Example 18. Although $C_\Phi = \{\{a_1\}, \{b_1\}, \{c_1, c_2, \}, \{d_1, d_2\}\}$ we conclude that $(\{z : ((a \to b) \to c \to c) \to d\}, d)$ is no principal pair for $z(\lambda xy.y)$, since $\to_1$ has no marked occurrence in the $\mathbf{TA_{pln}}$–deduction of $z : ((a_1 \to_1 b_1) \to_2 c_1 \to_3 c_2) \to_4 d_1 \vdash z(\lambda xy.y) : d_2$.

4 Correctness

We begin this section with several definitions and results (mostly from [3]) on the structure of types and terms, that we will need later on in order to prove Theorem 10. Note that every type τ can be written uniquely in the form

$$\tau_1 \to \dots \to \tau_m \to e,$$

where e is an atom and $m \geq 0$. Iff $m \geq 1$ we call τ a composite type.

Definition 19. *The significant subtypes or s-subtypes of a type $\tau = \tau_1 \to \dots \to \tau_m \to e$, where e is an atom and $m \geq 0$, are defined recursively as follows.*

- *τ is an s-subtype of τ;*
- *every s-subtype of one of $\tau_1, \dots, \tau_m, e$ is an s-subtype of τ.*

A proper s-subtype of τ is an s-subtype $\neq \tau$. Particular occurrences of s-subtypes of τ are also called s-components of τ and are distinguished by underlining their names. An s-component of a type τ is defined to be positive or negative as follows.

- $\underline{\tau}$ *is a positive s-component of* τ;
- *if* $\tau = \tau_1 \to \ldots \to \tau_m \to e$, *then* $\underline{\tau}_1, \ldots, \underline{\tau}_m$ *are negative s-components of* τ *and* $\underline{e}$ *is a positive s-component of* τ;
- *if* $\tau = \tau_1 \to \ldots \to \tau_m \to e$ *and if* $\underline{\rho}$ *is an s-component of one of* $\tau_1, \ldots, \tau_m$, *then* $\underline{\rho}$ *is a positive or negative s-component of* τ *according as it is a negative or positive s-component of* $\tau_1, \tau_2, \ldots$ *or* τ_m.

Definition 20. *If* $\underline{\rho}$ *is a composite s-component of a type* τ *and* $\rho = \rho_1 \to \ldots \to \rho_n \to a$ $(n \geq 1)$, *the s-components* $\underline{\rho}_1, \ldots, \underline{\rho}_n$ *are called the premises of* $\underline{\rho}$ *and* $\underline{a}$ *is called the conclusion or tail-component of* $\underline{\rho}$.

An s-component of τ *is called a subpremise or subtail of* τ *according as it is a premise or tail of another s-component of* τ.

Definition 21. *If* τ *is composite,* $\mathsf{NSS}(\tau)$ *is the set of all finite sequences* $< \sigma_1, \ldots, \sigma_n >$ $(n \geq 1)$ *such that* τ *contains a positive composite s-component with form* $\sigma_1 \to \ldots \to \sigma_n \to a$ *for some atom a. Each member of* $\mathsf{NSS}(\tau)$ *is called a negative subpremise-sequence.*

Every non-atomic λ-term X can be expressed uniquely in the form

$$X = \lambda x_1 \ldots x_m.vY_1 \ldots Y_n, \qquad (m + n \geq 1).$$

The head and arguments of X are respectively $\underline{v}$ and $\underline{Y}_1, \ldots, \underline{Y}_n$.

Definition 22. *A subargument of a* λ*-term* X *is a component that is an argument of* X *or an argument of a proper component of* X. *If* $\underline{Z}$ *is a subargument of a* λ*-term* X, *the argument-branch from* $\underline{X}$ *to* $\underline{Z}$ *is the sequence* $< \underline{Z}_0, \underline{Z}_1, \ldots, \underline{Z}_k >$, $(k \geq 1)$, *such that* $\underline{Z}_0 = \underline{X}$ *and* $\underline{Z}_i$ *is an argument of* $\underline{Z}_{i-1}$ *for* $i = 1, \ldots, k$, *and* $\underline{Z}_k = \underline{Z}$. *It is called unextendable iff* $\underline{Z}$ *is an atom or abstracted atom. Its length is* k *(not* $k+1$*).*

Definition 23. *Let* Δ *be a* $\mathbf{TA_{pln}}$*-deduction of* $\Gamma \vdash^{pln} M : \tau$, *let* $\underline{Z}$ *be a subargument of* M; *say*

$$Z = \lambda x_1 \ldots x_m.yZ_1 \ldots Z_n \qquad (m, n \geq 0)$$

and let $\Gamma_Z \vdash^{pln} Z : \alpha_1 \to \ldots \alpha_m \to a \parallel \emptyset$ *be the node in* Δ *which corresponds to* $\underline{Z}$. *The Initial Abstractors' Types sequence* $\mathsf{IAT}(\underline{Z})$ *is defined to be*

$$\mathsf{IAT}(\underline{Z}) =< \alpha_1, \ldots, \alpha_m >$$

and has length m. The Initial Abstractors sequence $\mathsf{IA}(Z)$ *is the (possibly empty) sequence*

$$\mathsf{IA}(Z) =< x_1, \ldots, x_m > .$$

Lemma 24. *Let Δ be a deduction of $\vdash M : \tau$, where M is a long β-nf with respect to $(\emptyset, \tau)$ and let Δ' be the corresponding* $\mathbf{TA_{pln}}$*-deduction of* $\vdash^{pln} M : \tau_{\mathtt{i}}$. *Let $\underline{Z}$ be a subargument of M, and let $\Gamma_Z \vdash^{pln} Z : \sigma \parallel \emptyset$, with $\sigma = \sigma_1 \to \ldots \to \sigma_k \to s$, be the node in Δ' which corresponds to $\underline{Z}$. Then*

i. if σ is an atom, $\mathsf{IAT}(\underline{Z}) = \emptyset$;
ii. if σ is composite, $\mathsf{IAT}(\underline{Z}) \in \mathsf{NSS}(\tau_{\mathtt{i}})$.

Proof Part *i.* is trivial. For part *ii.*, we show by induction on the depth of $\underline{Z}$ in M, i.e. the length of the argument-branch from M to $\underline{Z}$, that whenever $x : \alpha \in \Gamma_Z$, then α occurs as a negative s-component in $\tau_{\mathtt{i}}$ and σ (composite) occurs as a positive s-component of τ. In fact, suppose that $\underline{Z}$ has depth 1 in M, i.e. $M = \lambda x_1 \ldots x_m.vY_1 \ldots Y_n$ with $m, n \geq 1$ and $\underline{Z} = Y_i$ for some $1 \leq i \leq n$. Then $\tau_{\mathtt{i}} = \alpha_1 \to \ldots \to \alpha_m \to a$ and the node in Δ' that corresponds to $\underline{Z}$ is $x_1 : \alpha_1, \ldots, x_m : \alpha_m \vdash^{pln} Z : \sigma \parallel \emptyset$ (note that σ is composite, so this node results from the (abs)-rule). By definition, $\alpha_1, \ldots, \alpha_m$ are negative s-components of $\tau_{\mathtt{i}}$. On the other hand, there is $v = x_j$ for some $1 \leq j \leq m$ and α_j is of the form $\beta_1 \to \ldots \to \beta_n \to b$ with $\beta_i = \sigma$. Thus, σ occurs as a negative s-component of a negative s-component of $\tau_{\mathtt{i}}$ and occurs consequently as a positive s-component of $\tau_{\mathtt{i}}$.

The induction step is mostly a repetition of the previous argument.

Thus, if σ is composite, then σ occurs as a positive s-component of $\tau_{\mathtt{i}}$ and by the definition of $\mathsf{NSS}(\tau_{\mathtt{i}})$ follows $\mathsf{IAT}(\underline{Z}) = <\sigma_1, \ldots, \sigma_k> \in \mathsf{NSS}(\tau_{\mathtt{i}})$. •

It has been shown in [3] that whenever τ is a composite type, then $\#(\mathsf{NSS}(\tau)) \leq |\tau| - 1$. Thus, if $(\emptyset, \tau_{\mathtt{i}})$ is the indexed counterpart of $(\emptyset, \tau)$, one has $\#(\mathsf{NSS}(\tau_{\mathtt{i}})) \leq |\tau_{\mathtt{i}}| - 1 = |\tau| - 1$.

Lemma 25. *If τ is composite and $(\emptyset, \tau_{\mathtt{i}})$ is the indexed counterpart of $(\emptyset, \tau)$, then*

$$\#(\mathsf{NSS}(\tau_{\mathtt{i}})) \leq |\tau| - 1. \bullet$$

The proofs of the following two lemmas follow closely the schemes of Ben-Yelles' proofs of corresponding results for $\mathsf{Long}(\tau)$, from which they differ essentially in the justification of the encountered limits $\mathbf{d_p}(\tau)$ and $\mathbf{D_p}(\tau)$. As in the original case ($\mathsf{Long}(\tau)$), the construction of a term with smaller depth (or greater in the case of the Stretching Lemma) is done by substitution of a subterm by another subterm with smaller (greater) depth. But in the case of principal inhabitants one has to be more careful choosing these subterms, leading thus to greater limits, in order to guarantee the preservation of principality.

Lemma 26 (Shrinking Lemma). *If* $\mathsf{Lprinc}(\tau)$ *has a member M with depth* $\geq \mathbf{D_p}(\tau)$, *then*

i. there exists $M^* \in \mathsf{Lprinc}(\tau)$ *with* $Depth(M) - |\tau|^3 \leq Depth(M^*) < Depth(M)$;
ii. there exists $N \in \mathsf{Lprinc}(\tau)$ *with* $\mathbf{D_p}(\tau) - |\tau|^3 \leq Depth(N) < \mathbf{D_p}(\tau)$.

Proof For part *i.* consider a λ-term $M \in \mathsf{Lprinc}(\tau)$ with depth $d \geq \mathbf{D_p}(\tau)$ and without bound-variable clashes. Let $\Delta_{(\emptyset,\tau_1)}$ be the $\mathbf{TA_{pln}}$-deduction of $\vdash^{pln} M : \tau_1$ constructed from the unique **TA**-deduction of $\vdash M : \tau$ and let Φ be the set of all binary pairs in $\Delta_{(\emptyset,\tau_1)}$. It follows from Proposition 17 that for each type-variable a in τ there is one equivalence class in C_Φ containing exactly all indexed occurrences of a in τ_1 and that all (indexed) arrows in τ_1 are marked in $\Delta_{(\emptyset,\tau_1)}$. In the following we are going to construct a term of depth $< Depth(M)$ which has $(\emptyset, \tau)$ as a principal pair. It has been shown in [3] that $d = Depth(M)$ is the maximum of the lengths of all the argument-branches in X. Thus M has at least one argument-branch with length d and in order to reduce the depth of M it is necessary to shrink all these branches. Let $< N_0, \ldots, N_d >$ be any such branch, with

$$N_i = \lambda x_{i,1} \ldots x_{i,m_i}.y_i P_{i,1} \ldots P_{i,n_i} \qquad (m_i, n_i \geq 0).$$

Let

$$\Gamma_i \vdash^{pln} N_i : \rho_{i,1} \to \ldots \to \rho_{i,m_i} \to a_i \parallel \emptyset$$

be the node in $\Delta_{(\emptyset,\tau_1)}$ that corresponds to $\underline{N}_i$, for $i = 0, \ldots, d$. Thus, $\mathsf{IAT}(\underline{N}_i) =< \rho_{i,1}, \ldots, \rho_{i,m_i} >$. For $i = 0, \ldots, d$ let $\underline{B}_i$ be the body of $\underline{N}_i$, i.e. $B_i = y_i P_{i,1} \ldots P_{i,n_i}$ and let $\Gamma_{B_i} \vdash^{pln} B_i : a_{b_i} \parallel \emptyset$ be the node in $\Delta_{(\emptyset,\tau_1)}$ that corresponds to $\underline{B}_i$. Furthermore, let Φ_i and $\mathcal{I}_i$ be respectively the set of binary pairs and of indexes of marked arrows in the subtree of $\Delta_{(\emptyset,\tau_1)}$ with bottom formula $\Gamma_{B_i} \vdash^{pln} B_i : a_{b_i} \parallel \emptyset$. Then, $\mathcal{I}_i \subseteq \mathcal{I}_{i-1}$ and $\Phi_i \subseteq \Phi_{i-1}$ for $i = 1, \ldots, d$.

As in [3] we define a sequence of integers $d_0, d_1, \ldots, d_n$ as follows: $d_0 = 0$ and d_{j+1} is the least $i > d_j$ such that $\mathsf{IAT}(\underline{N}_i)$ differs from all $\mathsf{IAT}(\underline{N}_{d_0}), \mathsf{IAT}(\underline{N}_{d_1}), \ldots, \mathsf{IAT}(\underline{N}_{d_j})$. Obviously, one has $n \leq d$ as well as $0 = d_0 < d_1 < \ldots < d_n \leq d$. Furthermore, for $0 \leq i \leq d$, $\mathsf{IAT}(\underline{N}_i)$ is identical to one of the $n+1$

$$\mathsf{IAT}(\underline{N}_{d_0}), \ldots, \mathsf{IAT}(\underline{N}_{d_n}),$$

which are all distinct and by 24 are either empty or members of $\mathsf{NSS}(\tau_1)$. Hence, by lemma 25

$$n + 1 \leq 1 + \#\mathsf{NSS}(\tau_1) \leq 1 + |\tau| - 1 = |\tau|.$$

For $i = 0, \ldots, n$ define the following non-empty sets, called IAT-intervals, as follows:

$$\begin{aligned} \mathbb{I}_j &= \{d_j, d_j + 1, \ldots, d_{j+1} - 1\} \quad 0 \leq j \leq n-1 \\ \mathbb{I}_n &= \{d_n, d_n + 1, \ldots, d\}. \end{aligned}$$

If $\mathbb{I}_j$ contains two numbers p, $p+r$ such that $r \geq 1$ and B_p and B_{p+r} have the same type (i.e. $a_{b_p} = a_{b_{p+r}}$), $C_{\Phi_p} = C_{\Phi_{p+r}}$ and $\mathcal{I}_p = \mathcal{I}_{p+r}$, we shall call $< p, p+r >$ a tail-repetition. It will be called minimal iff there is no other tail-repetition $< p', q' >$ with $p \leq p' < q' \leq p + r$. It follows that each IAT-interval $\mathbb{I}_j$ without a tail-repetition must have $\leq |\tau|^3$ members as well as $r \leq |\tau|^3$. In fact, there are $|\tau|$ distinct atoms in τ_1, there are at most $|\tau|$ distinct equivalence classes corresponding to $\Phi_0 \supseteq \Phi_1 \supseteq \ldots \supseteq \Phi_d$ (cf. Lemma 16), as well as at most $|\tau|$ distinct sets of indexes among $\mathcal{I}_0 \supseteq \ldots \supseteq \mathcal{I}_d$ (note that there are exactly $|\tau|$

arrows in τ_1). Thus, if none of the $n+1$ IAT-intervals contained a tail-repetition, then the branch would have $\leq |\tau|^4$ members. But the branch has $d+1$ members and

$$d+1 = Depth(M)+1 \geq \mathbf{D_p}(\tau)+1 > |\tau|^4.$$

Hence at least one IAT-interval contains a tail-repetition.

Now let I_j be the last interval containing a minimal tail-repetition, say $< p, p+r >$. Suppose that v is a variable that occurs free in B_{p+r} with type-assignment $v : \alpha \in \Gamma_{B_{p+r}}$. Since M is closed we conclude that $v \in \mathsf{IA}(\underline{N}_0) \cup \ldots \cup \mathsf{IA}(\underline{N}_{p+r})$. Furthermore, by the definition of IAT-intervals, $\alpha \in \mathsf{IAT}(\underline{N}_{d_q})$ for some $q \leq j \leq p < p+r$. Hence, there is some type-assignment $v' : \alpha \in \Gamma_{B_p}$.

Now let B'_{p+r} be a term obtained from B_{p+r} by substituting all free variables v in B_{p+r} by some variable v' as above (possibly itself if $v : \alpha \in \Gamma_{B_p}$). Finally let M' be obtained from M by replacing B_p by B'_{p+r}. Note that $\Gamma_{B_p} \subseteq \Gamma_{B_{p+r}}$, i.e. $\Gamma_{B_{p+r}} = \Gamma_{B_p} \cup \Gamma_r$ for some Γ_r with $\Gamma_{B_p} \cap \Gamma_r = \emptyset$ and such that no variable in $Subjects(\Gamma_r)$ occurs in B'_{p+r}. Consider the subtree Δ_{p+r} of $\Delta_{(\emptyset,\tau_1)}$ with bottom formula $\Gamma_{B_{p+r}} \vdash^{pln} B_{p+r} : a_{b_{p+r}} \parallel \emptyset$. Let Δ'_{p+r} be obtained from Δ_{p+r} by substituting every term variable v by the corresponding $v' \in Subjects(\Gamma_{B_p})$ and erasing in the bases of the formulae all type-assignments for variables in $Subjects(\Gamma_r)$. Then Δ'_{p+r} has bottom formula $\Gamma_{B_p} \vdash^{pln} B'_{p+r} : a_{b_{p+r}} \parallel \emptyset$. Now consider the tree obtained from $\Delta_{(\emptyset,\tau_1)}$ by substituting its subtree with bottom formula $\Gamma_{B_p} \vdash^{pln} B_p : a_{b_p} \parallel \emptyset$ by Δ'_{p+r} and by substituting B_p in all nodes below by B'_{p+r}. Then, it is straightforward to prove, that the resulting tree Δ' is a $\mathbf{TA_{pln}}$-deduction of $\vdash^{pln} M' : \tau_1$ which corresponds to a $\mathbf{TA}$-deduction of $\vdash M' : \tau$ and such that for the set Φ' of binary pairs in Δ' one has $C_{\Phi'} = C_{\Phi}$ and all arrows marked in $\Delta_{(\emptyset,\tau_1)}$ are also marked in Δ'. Thus, we conclude that M' is a principal long inhabitant of τ. On the other hand, in a branch in M $r \leq |\tau|^3$ arguments have been removed. Thus $d - |\tau|^3 \leq Depth(M') \leq d$. If $Depth(M') < d$ let $M^\star = M'$. Otherwise, repeat shortening branches of length d until there are none left and define $M^\star$ to be the first term produced by this procedure whose depth is less than d. Then $d - |\tau|^3 \leq Depth(M^\star) \leq d$. For part *ii.* it is sufficient to repeat *i.* and take the first output with depth $< \mathbf{D_p}(\tau)$. •

Lemma 27 (Stretching Lemma). *If* $\mathsf{Lprinc}(\tau)$ *has a member* M *with depth* $\mathbf{d_p}(\tau)$*, then*

i. there exists $M^\star \in \mathsf{Lprinc}(\tau)$ *with* $Depth(M^\star) \geq Depth(M)+1$*;*
ii. $\mathsf{Lprinc}(\tau)$ *is infinite.*

Proof Part *ii.* follows from *i.* by repetition. The construction of $M^\star$ in *i.* is identical to the one in [3] for the Stretching Lemma for $\mathsf{Long}(\tau)$ (not $\mathsf{Lprinc}(\tau)$). Choose any argument branch $< N_0, \ldots, N_d >$ of length $d = Depth(M) \geq |\tau|$ and, as in the proof of the Shrinking Lemma, let $\underline{B}_i$ be the body of $\underline{N}_i$ for $i = 0, \ldots, d$. Let $\Gamma_{B_i} \vdash B_i : a_{b_i} \parallel \emptyset$ be the corresponding node in $\Delta_{(\emptyset,\tau_1)}$. Since $d+1 > |\tau| = ||\tau_1||$, there are $0 \leq p < p+r \leq d$ such that $\underline{B}_p$ and $\underline{B}_{p+r}$ have the same type $a = a_{b_p} = a_{b_{p+r}}$. Define $M^\star$ to be the result of replacing $\underline{B}_{p+r}$ in M by a copy of $\underline{B}_p$, after (to avoid clashes) changing the names of the variables $x_1, \ldots, x_m$ in this copy to $x'_1, \ldots, x'_m$, where $\{x_1 : \alpha_1, \ldots, x_m : \alpha_m\} = Subjects(\Gamma_{B_p} \setminus \Gamma_{B_{p+r}})$. Then M^* has an argument branch with length $d + r > d$. In order to see that τ is a principal type of M^*, consider the tree Δ'_{p+r} obtained from the subtree of $\Delta_{(\emptyset,\tau_1)}$ with bottom formula $\Gamma_{B_p} \vdash B_p : a \parallel \emptyset$ by first replacing in each node all occurrences of $x_1, \ldots, x_m$ respectively by $x'_1, \ldots, x'_m$ and then adding $\{x_1 : \alpha_1, \ldots, x_m : \alpha_m\}$ to the bases in all nodes. Note that Δ'_{p+r} has bottom formula $\Gamma_{B_{p+r}} \vdash B'_{p+r} : a \parallel \emptyset$. Finally, let Δ' be the tree obtained from $\Delta_{(\emptyset,\tau_1)}$ by replacing its subtree with bottom formula $\Gamma_{B_{p+r}} \vdash B_{p+r} : a \parallel \emptyset$ by Δ'_{p+r}. It is straightforward to see that Δ' is the $\mathbf{TA_{pln}}$-deduction corresponding to the **TA**-deduction of $\vdash M^* : \tau$ and that all binary pairs and marked arrows in $\Delta_{(\emptyset,\tau_1)}$ occur in Δ'. Thus we conclude that $M^* \in \mathsf{Lprinc}(\tau)$. •

Whenever τ is a composite type, there is $|\tau| \geq 2$, thus

$$\mathbf{d_p}(\tau) = |\tau| < |\tau|^4 - |\tau|^3 = \mathbf{D_p}(\tau) - |\tau|^3 < \mathbf{D_p}(\tau).$$

Hence, Theorem 10 follows as a Corollary.

Acknowledgments

The work presented in this paper has been partially supported by funds granted to *LIACC* through *Programa do Financiamento Plurianual, Fundação para a Ciência e Tecnologia* and *Programa PRAXIS*.

References

1. H. Barendregt. Lambda calculi with types. In Abramsky, Gabbay, and Maibaum, editors, *Background: Computational Structures*, volume 2 of *Handbook of Logic in Computer Science*, pages 117–309. Oxford Science Publications, 1992.
2. C.-B. Ben-Yelles. *Type-assignment in the lambda-calculus; syntax and semantics*. PhD thesis, Mathematics Dept., University of Wales Swansea, UK, 1979.
3. J. R. Hindley. *Basic Simple Type Theory*. Cambridge Tracts in Theoretical Computer Science. Cambridge University Press, 1997.
4. S. Hirokawa. Principal types of BCK-lambda-terms. *Theoretical Computer Science*, 107:253–276, 1993.
5. M. Takahashi, Y. Akama, and S. Hirokawa. Normal proofs and their grammar. In M. Hagiya and J. Mitchell, editors, *Theoretical Aspects of Computer Software (TACS'94)*, volume 789 of *LNCS*, pages 465–493. Springer Verlag, 1994.
6. M. Zaionc. The set of unifiers in typed λ-calculus as regular expression. In J.-P. Jouannaud, editor, *Rewriting Techniques and Applications*, volume 202 of *LNCS*, pages 430–440. Springer Verlag, 1985.
7. M. Zaionc. Mechanical procedure for proof construction via closed terms in typed λ-calculus. *J. Automated Reasoning*, 4:173–190, 1988.

Useless-Code Detection and Elimination for PCF with Algebraic Data Types
(Extended Abstract)

Ferruccio Damiani

Dipartimento di Informatica, Università di Torino,
Corso Svizzera 185, 10149 Torino (Italy)
damiani@di.unito.it

Abstract. We present a non-standard type assignment system and simplifications mappings for detecting and removing *useless-code* in simply typed functional programs with algebraic datatypes and recursive functions. We characterize two classes of useless-code: the *dead-code*, that is code that is never executed under the lazy-call-by-name evaluation, and the *minimum-information-code*, that is code that contributes to the computation only with a minimum amount of constant information.

1 Introduction

Useless-code analysis for functional programming languages has been mainly studied in the context of *logical frameworks*, like Coq [9], to remove useless-code from functional programs extracted from formal proofs (see [12] for an introduction to the subject). In fact, programs extracted from proofs usually contain large parts that are useless for the computation of the final result and some sort of simplification is mandatory. To this aim various simplification techniques have been proposed in the last ten years (e.g. [16, 1, 5, 2, 6, 8, 3]). More in general useless-code elimination is worthwhile during compilation (see, for instance, [4]). Let us look at a couple of examples of useless-code detection and elimination.

Example 1. Let $M = (\lambda x^{\mathsf{int}}.3)P$ where P is a term of type int. Since x is never used in the body of the λ-abstraction $F = \lambda x^{\mathsf{int}}.3$, we have that the value of M can be computed without using P, which is therefore useless-code. In fact, in a *lazy-call-by-name* language (like Miranda, Haskell and Clean), M behaves like the term $M' = (\lambda x^{\mathsf{int}}.3)\mathsf{d}$, where d is a place-holder for the useless-code removed. Note that it is indeed possible to simplify the useless-code in a more substantial way, i.e. by removing the useless pair $\langle$formal parameter x, actual parameter d$\rangle$ and replacing M with the body of the λ-abstraction (the constant 3).

In the following we will call *dead-code* the useless-code that, like P above, is never executed under the lazy-call-by-name evaluation strategy. The next (more complex) example introduces another class of useless-code, called *minimum-information-code*, that has been characterized by Berardi and Boerio in [3].

Example 2. Let intList = data X.Nil ‖ Cons(int, X) be the datatype of the list of integers, and bool = data X.True ‖ False be the datatype of booleans. Consider the term $N = \lambda x^{\mathsf{int}}.GPA$, where G : int → intList → int = $\lambda y^{\mathsf{int}}.\lambda z^{\mathsf{intList}}$.case z of { Nil to Q_1 ‖ Cons(h, t) to Q_2 } and A : intList = case B of { True to Cons(E_1, L_1) ‖ False to Cons(E_2, L_2) }, for some terms P, Q_1, Q_2, B, E_1, L_1, E_2 and L_2. It is easy to see that Q_1 is useless-code[1], in particular dead-code, so it could be replaced by a place-holder d. Let z, h and t not occur in Q_2, then we have that E_1, L_1, E_2, L_2 are dead-code, and can therefore replaced by place-holders $\mathsf{d}'_1, \mathsf{d}''_1, \mathsf{d}'_2, \mathsf{d}''_2$. Suppose also that the variable y does not occur in Q_2, but only in Q_1 (that has been removed), then also P is dead-code and it can be replaced by a place-holder d_3.

We have removed a lot of *dead-code* but, also after the above simplifications, in the term N there is other useless-code, since the subexpression A is an example of *minimum-information-code*: it contributes to the computation only by providing the Cons data-constructor of intList. Quoting from Berardi and Boerio [3] we say that the subexpression A is minimum-information-code since *"we use only the first symbol of its output, and this first symbol is always the same"*. Minimum-information-code can be simplified by exploiting the (minimum) information that it provides. For the program N above this amounts to replacing the minimum-information-code A by the term $A' = \mathsf{Cons}(\mathsf{d}_1, \mathsf{d}_2)$, which provides the same (minimum) information, and replacing the term G (which uses the minimum-information-code A) by the term $G' = \lambda y^{\mathsf{int}}.\lambda z^{\mathsf{intList}}.Q_2$, obtaining the simplified term $N' = \lambda x^{\mathsf{int}}.G'\mathsf{d}_3 A'$.

Note that replacing A by A' may have changed the termination behaviour of N (even w.r.t. the lazy-call-by-name evaluation strategy[2]): minimum-information-code is useless-code, but is *not* dead-code, and simplifying it may change the termination property of the program (i.e. it may happen that the simplified program converges while the original one diverges).

Moreover after turning G to G' we have that A' is dead-code[3] and it can be replaced by a place-holder d_4. Note that A is not dead-code in N, but: *by simplifying the minimum-information-code A in N (i.e. by turning G to G' and A to A') we have obtained a term N' containing* new *dead-code.* By this second simplification step we obtain the term: $N'' = \lambda x^{\mathsf{int}}.G'\mathsf{d}_3\mathsf{d}_4$, that does not contain useless-code (supposing that Q_2 does not).

Also in this case (as in Example 1) it is possible to simplify the term in a clever way by eliminating the place-holders for the dead-code removed. I.e. the application $G'\mathsf{d}_3\mathsf{d}_4$ can be replaced by the body, Q_2 of the function G'. After this last "cosmetic" simplification we obtain the term: $N''' = \lambda x^{\mathsf{int}}.Q_2$, which is significantly simpler than the original one. We remark that the transformation

[1] Since A (the actual value of the formal parameter z) either diverges or converges to a term of the form Cons(E, L) (for some E, L), so Q_1, which is the first branch of the case examining A, is never executed.

[2] Suppose that, for some values of the parameter x, the term B is divergent.

[3] Remember that the variable z does not occur in Q_2.

ρ ::= int	(integers)
$\mid \rho \to \rho$	(functions)
$\mid \mu$	(datatypes)
μ ::= $\mathsf{data}\, X.\mathbf{dc}_1 \| \cdots \| \mathbf{dc}_m$	(datatype, $m \geq 0$, $Con(\mathbf{dc}_i) = Con(\mathbf{dc}_j)$ iff $i = j$, X bound in each $\mathbf{dc}_i$)
$\mathbf{dc}$ ::= $\mathsf{C}(\Diamond_1, \ldots, \Diamond_a)$	(data-clause, $a \geq 0$, each $\Diamond_i$ is either a type or the variable X)

Fig. 1. PCFD types

of N in N''' cannot be performed by (partially) evaluating the program N, since the value of the subexpression B may depend on the formal parameter x.

In Section 2 of this paper we introduce the programming language we are dealing with and its operational semantics. Section 3 briefly shows how program properties can be represented by partial equivalence relations on a term model of the programming language. In Section 4 we describe the language of non-standard types (that we call *evaluation types*) and its semantics. Section 5 presents an evaluation type assignment system and a program simplification based on the information provided by the evaluation types assigned to a program and to its subexpressions. Related work is considered in Section 6.

2 The language PCFD

In this section we introduce a simple functional programming language and its operational semantics. The acronym PCFD stands for "Programming Computable Functions with lazy algebraic Datatypes", since this language is a dialect of the language PCF [14] obtained by adding algebraic datatypes. For more details see, for instance, Pitts [13] and Gordon [10]. The set of PCFD types is defined assuming as *ground* type the set of integers, int. Types are ranged over by ρ, σ, τ (with superscripts and subscript when necessary), and *algebraic datatypes* (*datatypes* in the following) are ranged over by μ.

Definition 1 (PCFD types). *The language of* types $(\mathbf{T})$ *is defined by the grammar in Fig. 1 where, in a datatype* $\mu = \mathsf{data}\, X.\mathbf{dc}_1 \| \cdots \| \mathbf{dc}_m$*, the type variable* X *is bound by the* data*-binder and, for each clause* $\mathbf{dc}_i = \mathsf{C}_i(\Diamond_{i,1}, \ldots, \Diamond_{i,a_i})$*, each* $\Diamond_{i,j}$ *is either a type* $\rho \in \mathbf{T}$ *or the type variable* X *representing the datatype being defined. The function* $Con(\mathbf{dc})$ *is defined as:* $Con(\mathsf{C}(\Diamond_1, \ldots, \Diamond_a)) = \mathsf{C}$.

Sometimes we will use, as meta-notation, parametrized definitions like $\mathsf{list}(X_1) = \mathsf{data}\, X.\mathsf{Nil} \| \mathsf{Cons}(X_1, X)$, and $\mathsf{pair}(X_1, X_2) = \mathsf{data}\, X.\mathsf{Pair}(X_1, X_2)$. For every type $\rho, \rho_1, \rho_2 \in \mathbf{T}$, $\mathsf{list}(\rho) \in \mathbf{T}$ and $\mathsf{pair}(\rho_1, \rho_2) \in \mathbf{T}$ are the types "list of ρ elements" and "pair having a ρ_1 as first element and a ρ_2 as second element", respectively. We remark that in our language it is not possible to define polymorphic type constructors like $\mathsf{list}(X_1)$ above: type parameters are only a convenient

$$(\text{Var}) \vdash x^\rho : \rho \qquad (\text{Con}) \vdash \mathsf{k}^\rho : \rho \qquad (\text{Fix}) \frac{\vdash M : \rho}{\vdash \mathsf{fix}\, x^\rho.M : \rho}$$

$$(\to \text{I}) \frac{\vdash M : \sigma}{\vdash \lambda x^\rho.M : \rho \to \sigma} \qquad (\to \text{E}) \frac{\vdash M : \rho \to \sigma \quad \vdash N : \rho}{\vdash MN : \sigma}$$

$$(\text{Data}_{1 \leq i \leq m}) \frac{\vdash P_1 : \sigma_1 \quad \cdots \quad \vdash P_{a_i} : \sigma_{a_i}}{\vdash \mathsf{C}_i^\mu(P_1, \ldots, P_{a_i}) : \mu}$$

where $\mu = \mathsf{data}\, X.\mathsf{C}_1(\diamondsuit_{1,1}, \ldots, \diamondsuit_{1,a_1}) \,\|\, \cdots \,\|\, \mathsf{C}_m(\diamondsuit_{m,1}, \ldots, \diamondsuit_{m,a_m})$

$$\forall j \in \{1, \ldots, a_i\}.\sigma_j = \begin{cases} \mu, & \text{if } \diamondsuit_{i,j} = X \\ \diamondsuit_{i,j}, & \text{if } \diamondsuit_{i,j} \in \mathbf{T} \end{cases}$$

$$(\text{Case}) \frac{\vdash P : \mu \quad \vdash Q_1 : \tau \quad \cdots \quad \vdash Q_m : \tau}{\vdash \mathsf{case}\, P\, \mathsf{of}\, \{\mathsf{C}_1(x_{1,1}, \ldots, x_{1,a_1})\, \mathsf{to}\, Q_1 \,\|\, \cdots \,\|\, \mathsf{C}_m(x_{m,1}, \ldots, x_{m,a_m})\, \mathsf{to}\, Q_m\} : \tau}$$

where $m \geq 1, \forall i \in \{1, \ldots, m\}$ the variables $x_{i,1}, \ldots, x_{i,a_i}$ are distinct,
$\mu = \mathsf{data}\, X.\mathsf{C}_1(\diamondsuit_{1,1}, \ldots, \diamondsuit_{1,a_1}) \,\|\, \cdots \,\|\, \mathsf{C}_m(\diamondsuit_{m,1}, \ldots, \diamondsuit_{m,a_m})$, and
$\forall i \in \{1, \ldots, m\}$, the program variables $x_{i,1}, \cdots, x_{i,a_i}$

- have the types specified by the data-clause $\mathsf{C}_i(\diamondsuit_{i,1}, \ldots, \diamondsuit_{i,i_m})$
- may occur free in Q_i
- are bound by the left-hand-side of the case-clause $\mathsf{C}_i(x_{i,1}, \ldots, x_{i,a_i})$ to Q_i

Fig. 2. Rules for PCFD term formation (system $\vdash_\mathbf{T}$)

notation, and we need a separate definition for each particular instance of the parameter.

PCFD terms are defined from a set of typed *term constants* ($\mathcal{K} = \{0^{\mathsf{int}}, 1^{\mathsf{int}}, \ldots, -^{\mathsf{pair(int,int)\to int}}, +^{\mathsf{pair(int,int)\to int}}, \ldots, =^{\mathsf{pair(int,int)\to bool}}, <^{\mathsf{pair(int,int)\to bool}}, \ldots, \mathsf{not}^{\mathsf{bool\to bool}}, \mathsf{and}^{\mathsf{pair(bool,bool)\to bool}}, \mathsf{or}^{\mathsf{pair(bool,bool)\to bool}}\}$, ranged over by k), including the usual operations involving the datatype of booleans, $\mathsf{bool} = \mathsf{data}\, X.\mathsf{True} \,\|\, \mathsf{False}$, and a set $\mathcal{V}$ of typed *term variables* (ranged over by $x^\rho, y^\sigma, \ldots$). PCFD terms, ranged over by M, N, ..., are defined as follows.

Definition 2 (PCFD terms). *We write $\vdash_\mathbf{T} M : \rho$, and say that M is a term of type ρ, if $\vdash M : \rho$ is derivable by the rules in Fig. 2.*

Let $\Lambda_\mathbf{T}$ be the set of PCFD terms, i.e. $\Lambda_\mathbf{T} = \{M \mid \vdash_\mathbf{T} M : \rho$ for some type $\rho\}$, and $\Lambda_\mathbf{T}^\mathsf{c}$ be the set of the *closed* terms, i.e. $\Lambda_\mathbf{T}^\mathsf{c} = \{M \mid M \in \Lambda_\mathbf{T}$ and $\mathrm{FV}(M) = \emptyset\}$. The process of evaluating a program is specified in a standard way by giving a *structural operational semantics* (see [15,11]) in the form of an inductively defined *evaluation relation*, $M \Downarrow K$, where M is a closed term and K is a closed term in *weak head normal form* (*w.h.n.f.*), i.e. an element of the set of *values* $\mathbf{V}_\mathbf{T} = \mathcal{K} \cup \{\lambda x^\rho.N \mid \lambda x^\rho.N \in \Lambda_\mathbf{T}^\mathsf{c}\} \cup \{\mathsf{C}^\mu(P_1, \ldots, P_a)^\rho \mid \mathsf{C}^\mu(P_1, \ldots, P_a)^\rho \in \Lambda_\mathbf{T}^\mathsf{c}\}$. We assume that any functional constant has a type of the shape either $\rho_1 \to \rho_2$ or $\mathsf{pair}(\rho_1, \rho_2) \to \rho_3$, for some $\rho_1, \rho_2, \rho_3 \in \{\mathsf{int}, \mathsf{bool}\}$. So the meaning of a functional constant k can be given by a set $\mathbf{mean}(\mathsf{k})$ of pairs, i.e. if $(K_1, K_2) \in \mathbf{mean}(\mathsf{k})$ then $\mathsf{k}K_1$ evaluates to K_2. For example $(\mathsf{True}, \mathsf{False}) \in \mathbf{mean}(\mathsf{not})$ and $(\mathsf{Pair}(1,3), 4) \in \mathbf{mean}(+)$.

$$(\text{CAN})\ \frac{K \in \mathbf{V_T}}{K \Downarrow K} \quad (\text{FIX})\ \frac{M[x := \mathsf{fix}\, x.M] \Downarrow K}{\mathsf{fix}\, x.M \Downarrow K} \quad (\text{APP})\ \frac{M \Downarrow \lambda x.P \quad P[x := N] \Downarrow K}{MN \Downarrow K}$$

$$(\mathcal{K}\text{APP}_1)\ \frac{M \Downarrow \mathsf{k} \quad N \Downarrow K_1}{MN \Downarrow K_2}\ (K_1, K_2) \in \mathbf{mean}(\mathsf{k})$$

$$(\mathcal{K}\text{APP}_2)\ \frac{M \Downarrow \mathsf{k} \quad N \Downarrow \langle N_1, N_2\rangle \quad N_1 \Downarrow K_1 \quad N_2 \Downarrow K_2}{MN \Downarrow K_3}\ (\mathsf{Pair}(K_1, K_2), K_3) \in \mathbf{mean}(\mathsf{k})$$

$$\frac{(\text{CASE}_{1 \le i \le m}) \qquad P \Downarrow \mathsf{C}_i(P_1, \ldots, P_{a_i}) \quad Q_i[x_{i,1} := P_1, \ldots, x_{i,a_i} := P_{a_i}] \Downarrow K}{\mathsf{case}\, P\, \mathsf{of}\, \{\mathsf{C}_1(x_{1,1}, \ldots, x_{1,a_1})\ \mathsf{to}\ Q_1 \,\|\, \cdots \,\|\, \mathsf{C}_m(x_{m,1}, \ldots, x_{m,a_m})\ \mathsf{to}\ Q_m\} \Downarrow K}$$

Fig. 3. "Natural semantics" evaluation rules

Definition 3 (Evaluation relation). *Let $M \in \Lambda^{\mathsf{c}}_{\mathbf{T}}$. We write $M \Downarrow K$, and say that M evaluates to K, if this statement is derivable by the rules in Fig. 3.*

Let $M \Downarrow$, to be read "M is convergent", mean that, for some K, $M \Downarrow K$, and let $M \Uparrow$, to be read "M is divergent", mean that, for no K, $M \Downarrow K$. For every type $\rho \in \mathbf{T}$, let $\bot_\rho = \mathsf{fix}\, x^\rho.x$. It is easy to check that $\bot_\rho \Uparrow$, i.e. $\bot_\rho$ is the "typical" divergent computation of type ρ.

Following [13] we introduce the *ground contextual equivalence* on PCFD terms, which is the congruence on terms induced by the contextual preorder that compares the termination behaviour of programs just at the ground type int. This amounts to assuming that *complete PCFD programs* are closed terms of type int, and that the only observable behaviour of a complete program P is its divergence or convergence to some integer number. Let $(\mathcal{C}[\ ]^\rho)^\sigma$ denote a typed context of type σ with a hole of type ρ in it.

Definition 4 (Ground contextual equivalence). *Let M and N be terms of type ρ. Define $M \preceq^{\mathrm{gnd}}_{\mathrm{obs}} N$ whenever, for all closed contexts $(\mathcal{C}[\ ]^\rho)^{\mathsf{int}}$, if $\mathcal{C}[M]$ and $\mathcal{C}[N]$ are closed terms, then $\mathcal{C}[M] \Downarrow$ implies $\mathcal{C}[N] \Downarrow$. The relation $\preceq^{\mathrm{gnd}}_{\mathrm{obs}}$ is the* ground contextual preorder *and the equivalence induced by $\preceq^{\mathrm{gnd}}_{\mathrm{obs}}$, denoted by $\simeq^{\mathrm{gnd}}_{\mathrm{obs}}$, is the* ground observational equivalence.

The *closed term model* $\mathcal{M}^{\mathrm{gnd}}$ of PCFD is defined by interpreting each type ρ as the set of the equivalence classes of the relation $\simeq^{\mathrm{gnd}}_{\mathrm{obs}}$ on the closed terms of type ρ in $\Lambda^{\mathsf{c}}_{\mathbf{T}}$. Let $\mathbf{I}(\rho)$ denote the interpretation of type ρ in this model, and let $[M]$ denote the equivalence class of the closed term M. An *environment* is a mapping $e : \mathcal{V} \to \bigcup_{\rho \in \mathbf{T}} \mathbf{I}(\rho)$ which respects types, i.e. a mapping such that, for all x^ρ, $e(x^\rho) \in \mathbf{I}(\rho)$. The interpretation of a term M in an environment e is defined in a standard way by: $[\![M]\!]_e = [M[x_1 := N_1, \ldots, x_n := N_n]]$, where $\{x_1, \ldots, x_n\} = \mathrm{FV}(M)$ and $[N_i] = e(x_i)$ $(1 \le i \le n)$.

3 Partial equivalence relations as program properties

The language of program properties $\mathbf{L}$ (introduced in Section 4 as a language of non-standard types over $\mathbf{T}$) which is at the basis of the program analysis and transformation techniques proposed in this paper, will be interpreted as a subset of the *partial equivalence relations*[4] over the interpretation, $\mathbf{I}(\rho)$, of the types $\rho \in \mathbf{T}$ in the closed term model $\mathcal{M}^{\mathrm{gnd}}$. Let "p.e.r. over a type ρ" mean "p.e.r. over $\mathbf{I}(\rho)$". The following definition formally explains what is meant by "a term P of type ρ satisfies the property (p.e.r.) $\mathcal{R}$ over ρ".

Definition 5 (P satisfies $\mathcal{R}$). *Let $\mathcal{R}, \mathcal{R}_1, \ldots, \mathcal{R}_n$ $(n \geq 0)$ be p.e.r. over types $\rho, \rho_1, \ldots, \rho_n$, respectively. We say that a term P of type ρ with free variables $x_1^{\rho_1} \ldots, x_n^{\rho_n}$ satisfies the property $\mathcal{R}$ under the assumptions $\mathcal{R}_i$ for $x_i^{\rho_i}$ $(1 \leq i \leq n)$ if, for all the environments e and e′ such that $(e(x_i^{\rho_i}), e'(x_i^{\rho_i})) \in \mathcal{R}_i$, we have that $([\![M]\!]_e, [\![M]\!]_{e'}) \in \mathcal{R}$.*

For every type $\rho \in \mathbf{T}$, the diagonal p.e.r. over $\mathbf{I}(\rho)$, $\Delta^{\mathsf{int}} = \{([M],[M]) \mid [M] \in \mathbf{I}(\mathsf{int})\}$, which equates each element of $\mathbf{I}(\rho)$ with only itself, can be seen as the property satisfied by any term P of type ρ whose value (under some assumption on the free variables of the term) matters (it *can be used*). Note that any closed term M of type ρ satisfies this property. The trivial p.e.r. over $\mathbf{I}(\rho)$, $\Omega^{\rho} = \{([M],[N]) \mid [M],[N] \in \mathbf{I}(\rho)\}$, which equates all the elements of $\mathbf{I}(\rho)$, can be seen as the "true" property (satisfied by every term of type ρ) giving *no information* about the use of the term.
Given a p.e.r. $\mathcal{R}_1$ over ρ_1 and a p.e.r. $\mathcal{R}_2$ over ρ_2, let $\mathcal{R}_1 \twoheadrightarrow \mathcal{R}_2$ be the p.e.r. over $\rho_1 \to \rho_2$ defined as: $\mathcal{R}_1 \twoheadrightarrow \mathcal{R}_2 = \{([F],[G]) \mid \forall([M],[N]) \in \mathcal{R}_1.([FM],[GN]) \in \mathcal{R}_2\}$. The intuition behind this definition is that $\mathcal{R}_1 \twoheadrightarrow \mathcal{R}_2$ is the property of the programs F such that, for every program M having the property $\mathcal{R}_1$, the program FM has the property $\mathcal{R}_2$. For instance $\Omega^{\mathsf{int}} \to \Delta^{\mathsf{int}}$ is the property (satisfied by all the closed terms of type $\mathsf{int} \to \mathsf{int}$ which represent, necessarily constant, functions which do not use their argument) which says that the application of the function can be used without using the argument.

The set-theoretic inclusion between p.e.r.s over a type ρ represents a logical implication between properties, i.e., if $\mathcal{R}_1 \subseteq \mathcal{R}_2$ and a program P has the property $\mathcal{R}_1$, then P has also the property $\mathcal{R}_2$.

4 Evaluation types

In this section we introduce a language of non-standard types over $\mathbf{T}$, the language of the *evaluation types* (*e-types* for short), which is the basis for the program analyses and transformations technique proposed in this paper.

Let ϕ range over e-types and ϕ^{ρ} range over e-types with *underlying type* ρ (i.e. expressing properties of terms of type ρ). In the following we will often omit the superscript ρ when it is either not relevant or clear from the context. There are two e-types denoting properties for terms of type int: δ^{int}, which is the property of the terms of type int such that their value *can be used*, and ω^{int}, which is the "true" property, satisfied by every term of type int.

[4] A *partial equivalence relation* (*p.e.r.* for short) over a set A is a symmetric and transitive binary relation over A.

$\psi_0 = \omega^{\mathsf{list(int)}}$ $\qquad$ $\psi_1 = \mathsf{data}\, X.\mathsf{Nil}^{\partial} \parallel \mathsf{Cons}^{\partial}(\mathsf{int}, X)$

$\psi_2 = \mathsf{data}\, X.\mathsf{Nil}^{\partial} \parallel \mathsf{Cons}^{\ell}(\delta^{\mathsf{int}}, X)$ $\qquad$ $\psi_3 = \mathsf{data}\, X.\mathsf{Nil}^{\partial} \parallel \mathsf{Cons}^{\ell}(\omega^{\mathsf{int}}, X)$

$\psi_4 = \mathsf{data}\, X.\mathsf{Nil}^{\partial} \parallel \mathsf{Cons}^{\ell}(\delta^{\mathsf{int}}, \omega^{\mathsf{list(int)}})$ $\qquad$ $\psi_5 = \mathsf{data}\, X.\mathsf{Nil}^{\partial} \parallel \mathsf{Cons}^{\ell}(\omega^{\mathsf{int}}, \omega^{\mathsf{list(int)}})$

plus the 5 e-types (say $\psi_1', \ldots, \psi_5'$) obtained from $\psi_1, \ldots, \psi_5$ by replacing Nil^{∂} with Nil^{ℓ}

Fig. 4. The 11 e-types for terms of type list(int)

Others e-types are built following the standard type construction. Given two e-types ϕ_1 and ϕ_2 the e-type $\phi_1 \to \phi_2$ says that the application of the function to every argument of e-type ϕ_1 has e-type ϕ_2. Given a datatype $\mu = \mathsf{data}\, X.\mathsf{C}_1(\diamondsuit_{1,1}, \ldots, \diamondsuit_{1,a_1}) \parallel \cdots \parallel \mathsf{C}_m(\diamondsuit_{m,1}, \ldots, \diamondsuit_{m,a_m})$ the e-type ω^{μ} is the "true" property, satisfied by every term of type μ. There also e-types that provide information about which constructors of the datatype may be used: two *constructor annotations*, ∂ and ℓ (ranged over by u, v), are introduced to represent the fact that a datatype constructor *is ∂ead* (i.e. it *is not used*) or *is ℓive* (i.e. it *may be used*), respectively. In particular the syntax of the e-types ϕ^{μ} different from ω^{μ} is as follows: $\phi^{\mu} = \mathsf{data}\, X.\mathsf{C}_1{}^{\mathrm{u}_1}(\spadesuit_{1,1}, \ldots, \spadesuit_{1,a_1}) \parallel \cdots \parallel \mathsf{C}_m{}^{\mathrm{u}_m}(\spadesuit_{m,1}, \ldots, \spadesuit_{m,a_m})$, where for all $i \in \{1, \ldots, m\}$ and for all $j \in \{1, \ldots, a_i\}$

$$\spadesuit_{i,j} = \begin{cases} \text{either } X \text{ or } \omega^{\mu}, & \text{if } \mathrm{u}_i = \ell \text{ and } \diamondsuit_{i,j} = X \\ \text{an e-type } \phi^{\diamondsuit_{i,j}}, & \text{if } \mathrm{u}_i = \ell \text{ and } \diamondsuit_{i,j} \in \mathbf{T} \\ \diamondsuit_{i,j}, & \text{if } \mathrm{u}_i = \partial \ . \end{cases}$$

If a term Q of type μ has the e-type ϕ^{μ} above, then the constructors having annotation ∂ *are not used* while those having annotation ℓ *may be used* (with arguments having the specified e-types). For instance there are 11 e-types (listed in Fig. 4) for the type list(int). The "typical" terms having e-types ψ_1, ψ_1', and ψ_5 are $\mathsf{fix}\, x^{\mathsf{list(int)}}.x$, Nil, and $\mathsf{Cons}(N, M)$ (for any $N : \mathsf{int}$ and $M : \mathsf{list(int)}$), respectively. $\mathsf{Cons}(N, M)$ is also the "typical" term of e-type ψ_4. Moreover $\mathsf{fix}\, x^{\mathsf{list(int)}}.x$ has every e-type of list(int), while Nil has none of the e-types $\psi_1, \ldots, \psi_5$, and $\mathsf{Cons}(N, M)$ has neither e-type ψ_1 nor ψ_1'. Both the e-types ψ_2 and ψ_3 specify that the constructor Nil is not used while the second component of the constructor Cons (which may be used) has (recursively) the same e-type, so the "typical" terms having such e-types are of the form $\mathsf{Cons}(N_1, \mathsf{Cons}(N_2, \cdots \mathsf{Cons}(N_p, \mathsf{fix}\, x^{\mathsf{list(int)}}.x) \cdots))$, for some $N_1, \ldots, N_p$ $(p \geq 0)$.

We remark that the e-type syntax (see Definition 6 below) has been designed in such a way that (syntactically) different e-types denote (w.r.t. the semantics in Definition 7) different p.e.r.s. This observation justifies some choices that might appear quite arbitrary, e.g. not having e-types of the form δ^{ρ} where $\rho \neq \mathsf{int}$, and of the form $\phi \to \omega^{\sigma}$ (for any e-type ϕ and type σ). Each e-type ϕ^{ρ} is interpreted as a p.e.r., $[\![\rho]\!]$, over $\mathbf{I}(\rho)$ (see Section 3).

Definition 6 (Evaluation types). *The language of the e-types,* $\mathbf{L}$, *is defined by:* $\mathbf{L} = \cup_{\rho \in \mathbf{T}} \mathbf{L}(\rho)$, *where the sets* $\mathbf{L}(\rho)$ *are defined by the rules in Fig. 5.*

Definition 7 (Semantics of e-types). *The semantic function* $[\![\cdot]\!]$ *which maps e-types* ϕ^{ρ} *to p.e.r.s over* $\mathbf{I}(\rho)$ *is defined by the clauses in Fig. 6.*

$$(\delta)\ \delta^{\mathsf{int}} \in \mathbf{L}(\mathsf{int}) \qquad (\omega)\ \omega^{\rho} \in \mathbf{L}(\rho) \qquad (\rightarrow)\ \frac{\phi \in \mathbf{L}(\rho) \quad \psi \in \mathbf{L}(\sigma)}{\phi \rightarrow \psi \in \mathbf{L}(\rho \rightarrow \sigma)}\ \psi \neq \omega^{\sigma}$$

$$(\mathsf{data})\ \frac{\forall i \in \{1,\ldots,m\}.\forall j \in \{1,\ldots,a_i\}.\spadesuit_{i,j} \in \begin{cases} \{X, \omega^{\mu}\}, & \text{if } \mathrm{u}_i = \ell \text{ and } \diamondsuit_{i,j} = X \\ \mathbf{L}(\diamondsuit_{i,j}), & \text{if } \mathrm{u}_i = \ell \text{ and } \diamondsuit_{i,j} \in \mathbf{T} \\ \{\diamondsuit_{i,j}\}, & \text{if } \mathrm{u}_i = \partial \end{cases}}{\mathsf{data}\, X.\mathsf{C}_1{}^{\mathrm{u}_1}(\spadesuit_{1,1},\ldots,\spadesuit_{1,a_1}) \parallel \cdots \parallel \mathsf{C}_m{}^{\mathrm{u}_m}(\spadesuit_{m,1},\ldots,\spadesuit_{m,a_m}), \in \mathbf{L}(\mu)}$$
$$\text{where } \mu = \mathsf{data}\, X.\mathsf{C}_1(\diamondsuit_{1,1},\ldots,\diamondsuit_{1,a_1}) \parallel \cdots \parallel \mathsf{C}_m(\diamondsuit_{m,1},\ldots,\diamondsuit_{m,a_m})$$

Fig. 5. Evaluation types

$$\begin{aligned} [\![\delta^{\mathsf{int}}]\!] &= \Delta^{\mathsf{int}}, \\ [\![\omega^{\rho}]\!] &= \Omega^{\rho} \\ [\![\phi \rightarrow \psi]\!] &= [\![\phi]\!] \twoheadrightarrow [\![\psi]\!], \quad \text{where } \psi \text{ is not an } \omega\text{-e-type} \\ [\![\phi^{\mu}]\!] &= \textstyle\bigcup_{p \geq 0} [\![\phi^{\mu}]\!]^{p}, \quad \text{where} \end{aligned}$$

$$\phi^{\mu} = \mathsf{data}\, X.\mathsf{C}_1{}^{\mathrm{u}_1}(\spadesuit_{1,1},\ldots,\spadesuit_{1,a_1}) \parallel \cdots \parallel \mathsf{C}_m{}^{\mathrm{u}_m}(\spadesuit_{m,1},\ldots,\spadesuit_{m,a_m})$$

$$\begin{aligned} [\![\phi^{\mu}]\!]^{0} &= \{([\mathsf{fix}\, x^{\mu}.x], [\mathsf{fix}\, x^{\mu}.x])\} \\ [\![\phi^{\mu}]\!]^{p+1} &= [\![\phi^{\mu}]\!]^{p} \cup \textstyle\bigcup_{i \in \{1 \leq i \leq m \text{ and } \mathrm{u}_i = \ell\}} \{([\mathsf{C}_i(P_{i,1},\ldots,P_{i,a_i})], [\mathsf{C}_i(Q_{i,1},\ldots,Q_{i,a_i})]) \mid \\ &\qquad \forall j \in \{1,\ldots,a_i\}.([P_{i,j}],[Q_{i,j}]) \in \begin{cases} [\![\phi^{\mu}]\!]^{p}, & \text{if } \spadesuit_{i,j} = X \\ [\![\spadesuit_{i,j}]\!], & \text{if } \spadesuit_{i,j} \in \mathbf{L} \end{cases} \} \end{aligned}$$

Fig. 6. Evaluation types semantics

According to the e-type semantics (Definition 7) an e-type ω^{ρ} denotes the trivial p.e.r. Ω^{ρ}. Others "special" e-types are the δ-e-types: a δ-e-type $\phi^{\rho} \in \mathbf{L}$ denotes the diagonal p.e.r. Δ^{ρ}.

Definition 8 (ω-e-types and δ-e-types). *Let the ω-e-types be the e-types in the set $\mathbf{L}_{\omega} = \{\omega^{\rho} \mid \rho \in \mathbf{T}\}$. The set of the δ-e-types ($\mathbf{L}_{\delta}$) is the subset of the e-types which do not contain subexpressions of the form ω^{ρ} (for some ρ) and do not contain the annotation ∂. For every type $\rho \in \mathbf{T}$, let $\delta(\rho)$ be the corresponding unique δ-e-type*[5].

Two other subclasses of e-types which are useful for program simplification are the ∂-e-datatypes that characterize datatypes in which no constructor is used (and which are therefore, according to Definition 7[6], ground observational equivalent to a divergent computation), and the ℓ1-e-datatypes that characterize

[5] It is immediate to see that for every type $\rho \in \mathbf{T}$ there is exactly one δ-e-type ϕ^{ρ}. For instance $\delta(\mathsf{list}(\mathsf{int}))$ is the e-type ψ_2' in Fig. 4.

[6] Since an ∂-e-datatype ϕ^{μ} denotes the singleton p.e.r. $[\![\phi^{\mu}]\!] = \{([\bot_{\mu}], [\bot_{\mu}])\}$.

$$(\text{Ref})\ \phi \le \phi \qquad (\omega)\ \phi^{\rho} \le \omega^{\rho} \qquad (\to)\ \frac{\phi_1 \le \phi_2 \quad \psi_1 \le \psi_2}{\phi_2 \to \psi_1 \le \phi_1 \to \psi_2}\ \psi_2 \notin \mathbf{L}_{\omega}$$

$$(\mathsf{data})\ \frac{\forall i \in \{h \mid 1 \le h \le m \text{ and } \mathrm{u}_h = \ell\}.\forall j \in \{1,\ldots,a_i\}.\spadesuit_{i,j} \preceq \clubsuit_{i,j}}{\phi^{\mu} \le \psi^{\mu}}$$

where $\phi^{\mu} = \mathsf{data}\, X.\mathsf{C}_1{}^{\mathrm{u}_1}(\spadesuit_{1,1},\ldots,\spadesuit_{1,a_1}) \mathbin{[\!]} \cdots \mathbin{[\!]} \mathsf{C}_m{}^{\mathrm{u}_m}(\spadesuit_{m,1},\ldots,\spadesuit_{m,a_m})$

$\psi^{\mu} = \mathsf{data}\, X.\mathsf{C}_1{}^{\mathrm{v}_1}(\clubsuit_{1,1},\ldots,\clubsuit_{1,a_1}) \mathbin{[\!]} \cdots \mathbin{[\!]} \mathsf{C}_m{}^{\mathrm{v}_m}(\clubsuit_{m,1},\ldots,\clubsuit_{m,a_m})$

$\forall i \in \{1,\ldots,m\}.(\mathrm{u}_i = \partial \text{ or } \mathrm{v}_i = \ell)$

$\spadesuit \preceq \clubsuit$ is short for: $\begin{cases} \clubsuit \in \{X, \omega^{\mu}\}, & \text{if } \spadesuit = X \\ \clubsuit \in \mathbf{L} \text{ and } \spadesuit \le \clubsuit, & \text{if } \spadesuit \in \mathbf{L} \end{cases}$

Fig. 7. Entailment rules for e-types (system $\le$)

datatypes in which only one constructor is used and its arguments are not used (i.e. they characterize the minimum-information-code[7]).

Definition 9 (∂-e-datatypes and $\ell 1$-e-datatypes). *The set of the ∂-e-datatypes, $\mathbf{L}_{\partial}$, is the set of the e-types of the form:* $\mathsf{data}\, X.\mathsf{C}_1{}^{\partial}(\spadesuit_{1,1},\ldots,\spadesuit_{1,a_1}) \mathbin{[\!]} \cdots \mathbin{[\!]} \mathsf{C}_m{}^{\partial}(\spadesuit_{m,1},\ldots,\spadesuit_{m,a_m})$, *where $m \ge 0$. For every datatype μ, let $\partial(\mu)$ be the corresponding unique ∂-e-datatype*[8].

The set of the $\ell 1$-e-datatypes, $\mathbf{L}_{\ell 1}$, is the set of the e-types of the form: $\mathsf{data}\, X.\mathsf{C}_1{}^{\mathrm{u}_1}(\spadesuit_{1,1},\ldots,\spadesuit_{1,a_1}) \mathbin{[\!]} \cdots \mathbin{[\!]} \mathsf{C}_m{}^{\mathrm{u}_m}(\spadesuit_{m,1},\ldots,\spadesuit_{m,a_m})$, *where $m \ge 1$ and there exists $i \in \{1,\ldots,m\}$ such that $\mathrm{u}_i = \ell$, $\spadesuit_{i,1},\ldots,\spadesuit_{i,a_i} \in \mathbf{L}_{\omega}$, and $j \ne i$ implies $\mathrm{u}_j = \partial$. For every datatype μ with $m \ge 1$ data-constructors, let $\ell 1_{\mathsf{C}}(\mu)$ be the corresponding unique $\ell 1$-e-datatype in which the data-constructor C is the ℓive one*[9].

We conclude this section by introducing an entailment relation between e-types, $\le$. This relation models the set-theoretic inclusion between the interpretation of e-types, and so it represents the logical implication between properties.

Definition 10 (Entailment relation $\le$). *Let $\phi, \psi \in \mathbf{L}$. We write $\phi \le \psi$ to mean that $\phi \le \psi$ is derivable by the rules in Fig. 7. By $\cong$ we denote the equivalence relation induced by $\le$.*

Note that $\le$ is reflexive and transitive. Moreover, for any type $\rho \in \mathbf{T}$, $(\mathbf{L}(\rho), \le)$ is a complete lattice with top ω^{ρ} and bottom b^{ρ} (inductively defined by: $b^{\mathsf{int}} = \delta^{\mathsf{int}}$, $b^{\sigma \to \tau} = \omega^{\sigma} \to b^{\tau}$, and $b^{\mu} = \partial(\mu)$). For instance the lattice $(\mathbf{L}(\mathsf{list}(\mathsf{int})), \le)$ (whose elements are the 11 e-types listed in Fig. 4) is showed in Fig. 8.

Theorem 1 (Soundness $\le$). *$\phi \le \psi$ implies $[\![\phi]\!] \subseteq [\![\psi]\!]$.*

[7] Since an $\ell 1$-e-datatype ϕ^{μ} denotes the p.e.r. (with exactly two classes): $[\![\phi^{\mu}]\!] = \{([\bot_{\mu}],[\bot_{\mu}])\} \cup \{([\mathsf{C}(M_1,\ldots,M_a)],[\mathsf{C}(N_1,\ldots,N_a)]) \mid \mathsf{C}(M_1,\ldots,M_a), \mathsf{C}(N_1,\ldots,N_a) \in \mathbf{I}(\mu)\}$, where C is the unique ℓive constructor of μ.

[8] For instance $\partial(\mathsf{list}(\mathsf{int}))$ is the e-type ψ_1 in Fig. 4.

[9] For instance $\ell 1_{\mathsf{Nil}}(\mathsf{list}(\mathsf{int}))$ and $\ell 1_{\mathsf{Cons}}(\mathsf{list}(\mathsf{int}))$ are the e-types ψ_1' and ψ_5 in Fig. 4.

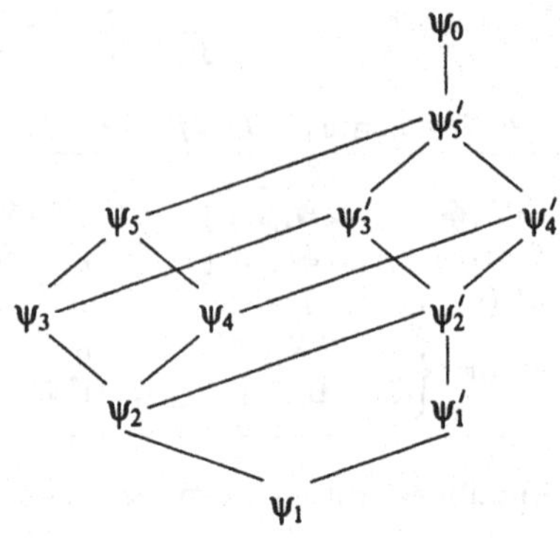

Fig. 8. The lattice $(\mathbf{L}(\mathsf{list}(\mathsf{int})), \leq)$

5 Detecting and removing useless-code

In this section we first introduce an e-type assignment system for detecting dead-code and minimum-information-code in PCFD programs, then we present simplification mappings for removing from a program the useless-code that can be detected with the e-type assignment system.

5.1 An e-type assignment system for detecting useless-code

If x^ρ is a term variable of type ρ, an assumption for x^ρ is an expression of the shape $x^\rho : \phi^\rho$, or $x : \phi^\rho$ for short. A basis is a set Σ of e-types assumptions for term variables. E-types are assigned to PCFD terms by a set of inference rules for judgments of the form $\Sigma \vdash_{\mathbf{L}} M^\Phi$ where M^Φ is a *decorated* term, i.e., it has written in it (some of) the e-types assigned to its subterms. Such a decorated term can then be processed by a transformation procedure (like those described in Section 5.2) that simplifies programs according to the information supplied by the e-types. For any e-type $\phi \in \mathbf{L}(\rho)$ let $\epsilon(\phi)$ denote the underlying type, ρ, of ϕ, and for any decorated term M^Φ, define $\epsilon(M^\Phi)$ as the term obtained from M^Φ by erasing all the e-type decorations. For each constant k a finite non-empty subset of e-types, $\mathbf{L}(\mathsf{k})$, is specified: for all integers n, $\mathbf{L}(\mathsf{n}) = \{\delta^{\mathsf{int}}\}$; for any binary operator $\odot^{\mathsf{pair}(\mathsf{int},\mathsf{int})\to\mathsf{int}}$, $\mathbf{L}(\odot) = \{\delta(\mathsf{pair}(\mathsf{int},\mathsf{int}) \to \delta^{\mathsf{int}}\}$; the constants involving the boolean datatype have more than one (non ω-) e-type, see Fig. 9. For instance the 4 e-types associated to not say that: if the result can be used then also the argument can be used, if in the argument the constructor True is ℓive and the constructor False is ∂ead then in the result False is ℓive and True is ∂ead (and vice versa), and if the argument diverges then also the result diverges.

Definition 11 (E-type assignment system $\vdash_{\mathbf{L}}$). *A $\vdash_{\mathbf{L}}$-typing statement is an expression $\Sigma \vdash_{\mathbf{L}} M^\Phi$ where Σ is a basis containing an assumption for each*

Let $\mathtt{t}^{\mathsf{bool}} = \mathsf{data}X.\mathsf{True}^{\ell}\|\mathsf{False}^{\partial}$, $\mathtt{f}^{\mathsf{bool}} = \mathsf{data}X.\mathsf{True}^{\partial}\|\mathsf{False}^{\ell}$ and, for any $\phi, \psi \in \mathbf{L}$, $\mathsf{pair}(\phi,\psi) = \mathsf{data}\ X.\mathsf{Pair}^{\ell}(\phi,\psi)$. Then (for any binary operator $\odot^{\mathsf{pair(int,int)}\to\mathsf{bool}}$):
$\mathbf{L}(\odot) = \{\delta(\mathsf{pair(int,int)}) \to \delta(\mathsf{bool}),\ \partial(\mathsf{pair(int,int)}) \to \partial(\mathsf{bool})\}$.
$\mathbf{L}(\mathsf{not}) = \{\delta(\mathsf{bool}) \to \delta(\mathsf{bool}),\ \mathtt{t}^{\mathsf{bool}} \to \mathtt{f}^{\mathsf{bool}},\ \mathtt{f}^{\mathsf{bool}} \to \mathtt{t}^{\mathsf{bool}},\ \partial(\mathsf{bool}) \to \partial(\mathsf{bool})\}$
$\mathbf{L}(\mathsf{and}) = \{\mathsf{pair}(\mathtt{t}^{\mathsf{bool}},\mathtt{t}^{\mathsf{bool}}) \to \mathtt{t}^{\mathsf{bool}},\ \mathsf{pair}(\phi_1,\mathtt{f}^{\mathsf{bool}}) \to \mathtt{f}^{\mathsf{bool}},\ \mathsf{pair}(\mathtt{f}^{\mathsf{bool}},\phi_2) \to \mathtt{f}^{\mathsf{bool}},$
$\partial(\mathsf{pair(bool,bool)}) \to \partial(\mathsf{bool}),\ \mathsf{pair}(\partial(\mathsf{bool}),\psi) \to \partial(\mathsf{bool}),\ \mathsf{pair}(\psi,\partial(\mathsf{bool})) \to \partial(\mathsf{bool}),$
$\mathsf{pair}(\phi_1,\phi_2) \to \delta(\mathsf{bool})\ \mid \phi_1,\phi_2 \in \{\mathtt{t}^{\mathsf{bool}},\mathtt{f}^{\mathsf{bool}}\}$ and $\psi \in \mathbf{L}(\mathsf{bool})\}$
$\mathbf{L}(\mathsf{or}) = \{\mathsf{pair}(\mathtt{f}^{\mathsf{bool}},\mathtt{f}^{\mathsf{bool}}) \to \mathtt{f}^{\mathsf{bool}},\ \mathsf{pair}(\phi_1,\mathtt{t}^{\mathsf{bool}}) \to \mathtt{t}^{\mathsf{bool}},\ \mathsf{pair}(\mathtt{t}^{\mathsf{bool}},\phi_2) \to \mathtt{t}^{\mathsf{bool}},$
$\partial(\mathsf{pair(bool,bool)}) \to \partial(\mathsf{bool}),\ \mathsf{pair}(\partial(\mathsf{bool}),\psi) \to \partial(\mathsf{bool}),\ \mathsf{pair}(\psi,\partial(\mathsf{bool})) \to \partial(\mathsf{bool}),$
$\mathsf{pair}(\phi_1,\phi_2) \to \delta(\mathsf{bool})\ \mid \phi_1,\phi_2 \in \{\mathtt{t}^{\mathsf{bool}},\mathtt{f}^{\mathsf{bool}}\}$ and $\psi \in \mathbf{L}(\mathsf{bool})\}$

Fig. 9. E-types of the PCFD constants

free variable of M, and M^{ϕ} is a decorated version of M. $\Sigma, x : \psi$ denotes the basis $\Sigma \cup \{x : \psi\}$ where it is assumed that x does not appear in Σ. By $\Sigma \vdash_{\mathbf{L}} M^{\phi}$ we mean that $\Sigma \vdash M^{\phi}$ can be derived by the rules in Fig. 10.

To state the soundness of the e-type assignment system $\vdash_{\mathbf{L}}$ w.r.t. Definitions 7 and 5 we introduce the following definition.

Definition 12. *Two environments e_1, e_2 are Σ-related if, for all $x^{\psi} \in \Sigma$, $(e_1(x), e_2(x)) \in [\![\psi]\!]$. Let $\Sigma \vdash_{\mathbf{L}} M^{\phi}$ and $\Sigma \vdash_{\mathbf{L}} N^{\phi}$. We write $\epsilon(M^{\phi}) \sim_{\phi}^{\Sigma} \epsilon(N^{\phi})$ to mean that for all e_1, e_2, if e_1, e_2 are Σ-related, then $([\![\epsilon(M^{\phi})]\!]_{e_1}, [\![\epsilon(N^{\phi})]\!]_{e_2}) \in [\![\phi]\!]$.*

Theorem 2 (Soundness of $\vdash_{\mathbf{L}}$). *Let $\Sigma \vdash_{\mathbf{L}} M^{\phi}$. Then $\epsilon(M^{\phi}) \sim_{\phi}^{\Sigma} \epsilon(M^{\phi})$.*

5.2 Useless-code elimination

In this section we first introduce a dead-code elimination mapping $\mathbf{O}$ that takes a $\vdash_{\mathbf{L}}$-decorated term M^{ϕ} and returns a simplified version of it in which the dead-code shown by the e-type decorations has been replaced by "dummy variables". Then we show how the simplification mapping $\mathbf{O}$ can be extended to a mapping $\mathbf{O}'$ which removes also the minumum-information-code.

For each type ρ, let d^{ρ}, d_1^{ρ}, d_2^{ρ}, ..., be *dummy variables* of type ρ. We remark that dummy variables are not present in the original programs: they are introduced by the dead-code elimination mapping $\mathbf{O}$ as place-holders for the dead-code removed. So in the following we assume that all the occurrences of dummy variables in a program are free (i.e. there are no bound dummy variables) and distinct (i.e. each dummy variable occurs at most once in a program). For every term M, let $\mathrm{DV}(M)$ be the set of the dummy variables in M. Let $\Lambda_{\mathbf{T}}^{\vdash_{\mathbf{L}}}$ be the set of $\vdash_{\mathbf{L}}$-decorated PCFD terms, i.e., $\Lambda_{\mathbf{T}}^{\vdash_{\mathbf{L}}} = \{M^{\phi} \mid \Sigma \vdash_{\mathbf{L}} M^{\phi}$ for some e-type ϕ and basis $\Sigma\}$.

Definition 13 (Simplification mapping O). *The function $\mathbf{O} : \Lambda_{\mathbf{T}}^{\vdash_{\mathbf{L}}} \to \Lambda_{\mathbf{T}}^{\vdash_{\mathbf{L}}}$ is defined by the clauses in Fig. 11, where the occurrences of "d" in the second and in the last row denote fresh dummy variables of the proper type. If Σ is a basis then define $\mathbf{O}(\Sigma) = \{x : \chi \mid x : \chi \in \Sigma$ and $\chi \notin \mathbf{L}_{\omega}\}$.*

$$(\omega)\ \frac{\Sigma \supseteq \mathrm{FV}(M) \quad \vdash_{\mathbf{T}} M : \rho}{\Sigma \vdash M^{\omega^{\rho}}} \qquad (\mathrm{Fix})\ \frac{\Sigma, x : \phi \vdash M^{\psi}}{\Sigma \vdash (\mathsf{fix}\, x^{\phi}.M)^{\psi}}\ \psi \notin \mathbf{L}_{\omega} \text{ and } \phi \in \{\psi, \omega^{\epsilon(\psi)}\}$$

$$(\mathrm{Var})\ \frac{\phi \leq \psi}{\Sigma, x : \phi \vdash x^{\psi}}\ \psi \notin \mathbf{L}_{\omega} \qquad (\mathrm{Con})\ \Sigma \vdash \mathsf{k}^{\phi}\ ,\ \phi \in \mathbf{L}(\mathsf{k})$$

$$(\rightarrow \mathrm{I})\ \frac{\Sigma, x : \phi \vdash M^{\psi}}{\Sigma \vdash (\lambda x.M)^{\phi \rightarrow \psi}}\ \psi \notin \mathbf{L}_{\omega} \qquad (\rightarrow \mathrm{E})\ \frac{\Sigma \vdash M^{\phi \rightarrow \psi} \quad \Sigma \vdash N^{\phi}}{\Sigma \vdash (MN^{\phi})^{\psi}}\ \psi \notin \mathbf{L}_{\omega}$$

$$(\mathrm{Data}_{1 \leq i \leq m})\ \frac{\Sigma \vdash P_1{}^{\psi_1} \quad \cdots \quad \Sigma \vdash P_{a_i}{}^{\psi_{a_i}}}{\Sigma \vdash (\mathsf{C}_i(P_1, \ldots, P_{a_i}))^{\phi^{\mu}}}$$

where $\phi^{\mu} = \mathsf{data}\, X.\mathsf{C}_1{}^{\mathsf{u}_1}(\spadesuit_{1,1}, \ldots, \spadesuit_{1,a_1}) \,\|\, \cdots \,\|\, \mathsf{C}_m{}^{\mathsf{u}_m}(\spadesuit_{m,1}, \ldots, \spadesuit_{m,a_m})$, $\mathsf{u}_i = \ell$,
and $\forall j \in \{1, \ldots, a_i\}$, $\psi_j = \begin{cases} \phi^{\mu}, & \text{if } \spadesuit_{i,j} = X \\ \spadesuit_{i,j}, & \text{if } \spadesuit_{i,j} \in \mathbf{L} \end{cases}$

$$(\mathrm{Case})\ \frac{\Sigma \vdash P^{\phi^{\mu}} \qquad \forall i \in \{1, \ldots, m\}. \vdash_{\mathbf{T}} Q_i : \tau \qquad \forall i \in \{h \mid 1 \leq h \leq m \text{ and } \mathsf{u}_i = \ell\}. \Sigma \cup \{x_{i,1} : \psi_{i,1}, \ldots, x_{i,a_i} : \psi_{i,a_i}\} \vdash Q_i{}^{\chi^{\tau}}}{\Sigma \vdash (\mathsf{case}\, P^{\phi^{\mu}} \mathsf{ of } \{\mathsf{cc}_1 \,\|\, \cdots \,\|\, \mathsf{cc}_m\})^{\chi^{\tau}}}$$

where $\chi^{\tau} \notin \mathbf{L}_{\omega}$, $m \geq 1$,
$\phi^{\mu} = \mathsf{data}\, X.\mathsf{C}_1{}^{\mathsf{u}_1}(\spadesuit_{1,1}, \ldots, \spadesuit_{1,a_1}) \,\|\, \cdots \,\|\, \mathsf{C}_m{}^{\mathsf{u}_m}(\spadesuit_{m,1}, \ldots, \spadesuit_{m,a_m})$,
$\forall i \in \{h \mid 1 \leq h \leq m \text{ and } \mathsf{u}_i = \ell\}. \forall j \in \{1, \ldots, a_i\}. \psi_{i,j} = \begin{cases} \phi^{\mu}, & \text{if } \spadesuit_{i,j} = X \\ \spadesuit_{i,j}, & \text{if } \spadesuit_{i,j} \in \mathbf{L} \end{cases}$,
and $\forall i \in \{1, \ldots, m\}. \mathsf{cc}_i = \begin{cases} \mathsf{C}_i(x_{i,1}, \ldots, x_{i,a_i}) \text{ to } Q_i, & \text{if } \mathsf{u}_i = \ell \\ \mathsf{C}_i(x_{i,1}, \ldots, x_{i,a_i}) \text{ to } \epsilon(Q_i^{\chi^{\tau}}), & \text{if } \mathsf{u}_i = \partial \end{cases}$

Fig. 10. Rules for e-type assignment (system $\vdash_{\mathbf{L}}$)

We have immediately that if $\Sigma \vdash_{\mathbf{L}} M^{\phi}$ then $\mathbf{O}(\Sigma) \subseteq \Sigma$ and $\Sigma' \vdash_{\mathbf{L}} \mathbf{O}(M^{\phi})$, where $\Sigma' = \mathbf{O}(\Sigma) \cup \{\mathsf{d}^{\sigma} : \omega^{\sigma} \mid \mathsf{d}^{\sigma} \in \mathrm{DV}(\epsilon(\mathbf{O}(M^{\phi})))\}$. Moreover we have that the simplification mapping is correct w.r.t. the e-type semantics, i.e., if $\Sigma \vdash_{\mathbf{L}} M^{\phi}$ then $\epsilon(M^{\phi}) \sim_{\phi}^{\Sigma'} \epsilon(\mathbf{O}(M^{\phi}))$ (where $\Sigma' = \Sigma \cup \{\mathsf{d}^{\sigma} : \omega^{\sigma} \mid \mathsf{d}^{\sigma} \in \mathrm{DV}(\epsilon(\mathbf{O}(M^{\phi})))\}$).

In order to use the simplification mapping $\mathbf{O}$ to simplify terms while preserving their meaning (w.r.t. $\mathcal{M}^{\mathrm{gnd}}$) we identify a subset of $\vdash_{\mathbf{L}}$-typings (that we call *faithful*) for which the $\sim_{\phi}^{\Sigma}$ relation implies the $\simeq_{\mathrm{obs}}^{\mathrm{gnd}}$ relation[10].

Definition 14 (Faithful $\vdash_{\mathbf{L}}$-typing). *$\Sigma \vdash_{\mathbf{L}} M^{\phi}$ is a* faithful *$\vdash_{\mathbf{L}}$-type assignment statement if $\phi \in \mathbf{L}_{\delta}$, and for all $x : \psi \in \Sigma$, $\psi \in \mathbf{L}_{\delta} \cup \mathbf{L}_{\omega}$.*

The proof that the simplifications performed by the mapping $\mathbf{O}$ on faithfully decorated terms preserve $\simeq_{\mathrm{obs}}^{\mathrm{gnd}}$ relies on the fact that, for all faithful $\vdash_{\mathbf{L}}$-typings $\Sigma \vdash_{\mathbf{L}} M^{\phi}$ and $\Sigma \vdash_{\mathbf{L}} N^{\phi}$, $\epsilon(M^{\phi}) \sim_{\phi}^{\Sigma} \epsilon(N^{\phi})$ implies $\epsilon(M^{\phi}) \simeq_{\mathrm{obs}}^{\mathrm{gnd}} \epsilon(N^{\phi})$.

[10] A faithful $\vdash_{\mathbf{L}}$-typing simply says that the term can be used (it has e-type $\in \mathbf{L}_{\delta}$) and that any of its free variables is either used (it has e-type $\in \mathbf{L}_{\delta}$) or not used at all (it has e-type $\in \mathbf{L}_{\omega}$).

$\mathbf{O}(M^{\psi}) = (\mathsf{o}(M,\psi))^{\psi}$, where

1. $\mathsf{o}(M,\psi) = \mathsf{d}$, if $\psi \in \mathbf{L}_\omega$
2. $\mathsf{o}(M,\psi) = \mathsf{fix}\, x^{\psi}.x$, if $\psi \in \mathbf{L}_\partial$
3. Otherwise:

$$\mathsf{o}(\mathsf{k},\psi) = \mathsf{k} \qquad \mathsf{o}(x,\psi) = x \qquad \mathsf{o}(MN^{\phi},\psi) = \mathsf{o}(M,\phi \to \psi)(\mathsf{o}(N,\phi))^{\phi}$$

$$\mathsf{o}(\lambda x.M,\psi_1 \to \psi_2) = \lambda x.\mathsf{o}(M,\psi_2) \qquad \mathsf{o}(\mathsf{fix}\, x^{\phi}.M,\psi) = \begin{cases} \mathsf{o}(M,\psi) & \text{if } \phi \in \mathbf{L}_\omega \\ \mathsf{fix}\, x^{\psi}.\mathsf{o}(M,\psi) & \text{if } \phi = \psi \end{cases}$$

$$\mathsf{o}(\mathsf{C}_i(P_1,\ldots,P_{a_i}),\phi^{\mu}) = \mathsf{C}_i(\mathsf{o}(P_1,\psi_1),\ldots,\mathsf{o}(P_{a_i},\psi_{a_i}))$$
$$\text{where } \phi^{\mu} = \mathsf{data}\, X.\mathbf{dc}_1^{\mathrm{u}_1} \parallel \cdots \parallel \mathbf{dc}_m^{\mathrm{u}_m},\ \mathbf{dc}_i^{\mathrm{u}_i} = \mathsf{C}_i^{\ell}(\spadesuit_{i,1},\ldots,\spadesuit_{i,a_i}),$$
$$\text{and } \forall j \in \{1,\ldots,a_i\}.\psi_j = \begin{cases} \phi^{\mu} & \text{if } \spadesuit_{i,j} = X \\ \spadesuit_{i,j} & \text{if } \spadesuit_{i,j} \in \mathbf{L} \end{cases}$$

$$\mathsf{o}(\mathsf{case}\, P^{\phi^{\mu}}\, \mathsf{of}\, \{\mathbf{cc}_1 \parallel \cdots \parallel \mathbf{cc}_m\},\chi) = \mathsf{fix}\, x^{\chi}.x \quad \text{if } \phi^{\mu} \in \mathbf{L}_\partial$$

$$\mathsf{o}(\mathsf{case}\, P^{\phi^{\mu}}\, \mathsf{of}\, \{\mathbf{cc}_1 \parallel \cdots \parallel \mathbf{cc}_m\},\chi) = \mathsf{case}\, \mathsf{o}(P,\phi^{\mu})^{\phi^{\mu}}\, \mathsf{of}\, \{\mathbf{cc}'_1 \parallel \cdots \parallel \mathbf{cc}'_m\}$$
$$\text{where } \phi^{\mu} = \mathsf{data}\, X.\mathbf{dc}_1^{\mathrm{u}_1} \parallel \cdots \parallel \mathbf{dc}_m^{\mathrm{u}_m} \notin \mathbf{L}_\partial$$
$$\forall i \in \{1,\ldots,m\}.\mathbf{dc}_i^{\mathrm{u}_i} = \mathsf{C}_i^{\ell}(\spadesuit_{i,1},\ldots,\spadesuit_{i,a_i})$$
$$\mathbf{cc}_i = \mathsf{C}_i(x_{i,1},\ldots,x_{i,a_i})\ \mathsf{to}\ Q_i$$
$$\mathbf{cc}'_i = \begin{cases} \mathsf{C}_i(x_{i,1},\ldots,x_{i,a_i})\ \mathsf{to}\ \mathsf{o}(Q_i,\chi) & \text{if } \mathrm{u}_i = \ell \\ \mathsf{C}_i(x_{i,1},\ldots,x_{i,a_i})\ \mathsf{to}\ \mathsf{d} & \text{if } \mathrm{u}_i = \partial \end{cases}$$

Fig. 11. Dead-code simplification mapping **O**

Theorem 3 (O on faithful $\vdash_{\mathbf{L}}$-typings preserves $\simeq^{\mathrm{gnd}}_{\mathrm{obs}}$). *Let $\Sigma \vdash_{\mathbf{L}} M^{\phi}$ be a faithful $\vdash_{\mathbf{L}}$-typing. Then $\epsilon(M^{\phi}) \simeq^{\mathrm{gnd}}_{\mathrm{obs}} \epsilon(\mathbf{O}(M^{\phi}))$.*

We now introduce the improved mapping, $\mathbf{O}'$, which performs more simplifications than **O**. In general $\mathbf{O}'$ does not preserve the meaning (w.r.t. $\mathcal{M}^{\mathrm{gnd}}$) of terms since it may give simplified terms which are strictly observationally greater than the original ones (see Example 2).

Definition 15 (Simplification mapping $\mathbf{O}'$). *The function $\mathbf{O}' : \Lambda_{\mathrm{T}}^{\vdash_{\mathrm{L}}} \to \Lambda_{\mathrm{T}}^{\vdash_{\mathrm{L}}}$ is defined exactly as the function **O** (see Definition 13) with the exception that we add to the last clause for the mapping $\mathsf{o}(\cdot,\cdot)$ in Fig. 11 the condition $\phi^{\mu} \notin \mathbf{L}_{\ell 1}$, and we add also the following two clauses for simplifying the minimum-information-code.*
Between the clauses 2 and 3:
2′. $\mathsf{o}(M,\psi) = \mathsf{C}_i(\mathsf{d}^{\rho_1},\ldots,\mathsf{d}^{\rho_{a_i}})$,
if ψ is the $\ell 1_{\mathsf{C}_i}$-datatype $\mathsf{data}\, X.\mathbf{dc}_1^{\partial} \parallel \cdots \parallel \mathsf{C}_i^{\ell}(\omega^{\rho_1},\ldots,\omega^{\rho_{a_i}}) \parallel \cdots \parallel \mathbf{dc}_m^{\partial}$.
As last clause:
$\mathsf{o}(\mathsf{case}\, P^{\phi^{\mu}}\, \mathsf{of}\, \{\mathbf{cc}_1 \parallel \cdots \parallel \mathbf{cc}_m\},\chi) = \mathsf{o}(Q_i,\chi)$,
where ϕ^{μ} is the $\ell 1_{\mathsf{C}_i}$-datatype $\mathsf{data}\, X.\mathbf{dc}_1^{\partial} \parallel \cdots \parallel \mathsf{C}_i^{\ell}(\omega^{\rho_1},\ldots,\omega^{\rho_{a_i}}) \parallel \cdots \parallel \mathbf{dc}_m^{\partial}$
and $\mathbf{cc}_i = \mathsf{C}_i(x_{i,1},\ldots,x_{i,a_i})\ \mathsf{to}\ Q_i$.

Theorem 4 (O′ on faithful $\vdash_{\mathbf{L}}$-typings). *Let $\Sigma \vdash_{\mathbf{L}} M^{\phi}$ be a faithful $\vdash_{\mathbf{L}}$-typing. Then $\epsilon(M^{\phi}) \preceq^{\mathrm{gnd}}_{\mathrm{obs}} \epsilon(\mathbf{O}'(M^{\phi}))$.*

Note that when dealing with terminating programs (like programs extracted from proofs) or with programs that are trusted to terminate (as well written pure functional programs should be) we have that the simplified programs are (or are trusted to be) ground observationally equivalent to the original ones.

Using the same technique of [6] (see also [7] Chapter 7) we can prove that for every PCFD term M there is a faithful $\vdash_{\mathbf{L}}$-typing (see Definition 14) showing all the useless-code that can be proved by system $\vdash_{\mathbf{L}}$. In this way we can also provide a complete (w.r.t. the system $\vdash_{\mathbf{L}}$) useless-code detection algorithm. Note that, since useless-code elimination may "rise" new useless-code (see Example 2), we have to apply the detection algorithm and the simplification mapping $\mathbf{O}'$ repeatedly, until no new useless-code is discovered.

6 Related work

The class of useless-code characterized in this paper strictly includes the useless-code characterized by the technique presented by Berardi an Boerio in [3]. The extra power is due to the use of subtyping (not used in [3]). For a discussion about the power gained by adding the e-type entailment relation we refer to [2] and [5] Chapter 4 (where a dead-code analysis with type entailment for a simply typed λ-calculus is presented), see also [6] and [7] Chapter 7 (where a constraint-based inference algorithm for the analysis of [2] is presented). We remark that in presence of datatypes the advantage of having the e-type entailment relation is even greater. Take for instance the term $P = (\lambda z^{\mu}.\cdots(\mathsf{case}\,B\,\mathsf{of}\,\{\ \mathsf{True\ to}\ z \ \mid\ \mathsf{False\ to}\ \mathsf{C}_i(\cdots)\ \})\cdots)Q$, where the case-expression is not useless-code. Suppose that we can assign to Q an e-type in which the constructor C_i is ∂ead. If we try to assign an e-type to P without using entailment, we are forced to assign to every (non-dead) occurrence of z in P (and also to Q) an e-type ϕ^{μ}_1 in which C_i is ℓive[11]. By using entailment, instead, we can assign such an e-type ϕ^{μ}_1 to the occurrence of z in the True branch of the case-expression, but we can assign e-types $\phi^{\mu}, \phi^{\mu}_2, \phi^{\mu}_3, \ldots$ in which C_i is ∂ead to Q and to the other occurrences of z in P, respectively (provided that $\phi^{\mu} \le \phi^{\mu}_1$, $\phi^{\mu} \le \phi^{\mu}_2, \ldots$).

Besides the use of the e-type entailment relation, the main differences between our approach and that of [3] are in the programming language considered and in the algorithm that finds the useless-code in a given term. The language considered in [3] is strongly normalizing and it can be seen as the language obtained from PCFD by removing the constructor fix and adding, for every type $\tau \in \mathbf{T}$, an operator it_{τ} for primitive recursion over datatypes and from datatypes to every type.

The algorithm described in [3] is a kind of "data flow" algorithm that analyzes a term by implicitly building a directed graph which represents the input-output

[11] Since ℓive branches of a case-expression must have the same (non-ω) e-type.

relation between the subterms of a given term. The principal advantage of our constraint-based algorithm (not presented here, for lack of space) is that (as shown in [6] and [7] Chapter 7) it is compositional while that of [3] is not.

Acknowledgements. I would thank Stefano Berardi, Mario Coppo and Paola Giannini for valuable remarks and suggestions, Jean-Christophe Filliâtre, for stimulating discussions on the topic of program extraction, and Jorge Pinto, for comments on a preliminary version of this paper. During the elaboration of this paper I was visiting the Laboratoire d'Informatique de l'École Polytechnique (LIX). I would like to thank my host Radhia Cousot and the whole LIX for the ideal working conditions they provided. I also thank the referees for many useful comments.

References

1. S. Berardi. Pruning Simply Typed Lambda Terms. *Journal of Logic and Computation*, 6(5):663–681, 1996.
2. S. Berardi and L. Boerio. Using Subtyping in Program Optimization. In *TLCA'95*, LNCS 902. Springer–Verlag, 1995.
3. S. Berardi and L. Boerio. Minimum Information Code in a Pure Functional Language with Data Types. In *TLCA'97*, LNCS 1210. Springer–Verlag, 1997.
4. S. K. Biswas. A Demand-Driven Set-Based Analysis. In *POPL'97*, pages 372–385. ACM, 1997.
5. L. Boerio. *Optimizing Programs Extracted from Proofs*. PhD thesis, Università di Torino, 1995.
6. M. Coppo, F. Damiani, and P. Giannini. Refinement Types for Program Analysis. In *SAS'96*, LNCS 1145, pages 143–158. Springer–Verlag, 1996.
7. F. Damiani. *Non-standard type inference for functional programs*. PhD thesis, Università di Torino, February 1998.
8. F. Damiani and F. Prost. Detecting and Removing Dead Code using Rank 2 Intersection. In *TYPES'96*, LNCS 1512. Springer–Verlag, 1998.
9. B. Barras et al. *The Coq Proof Assistant Reference Manual Version 6.2*. INRIA-Rocquencourt-CNRS-ENS Lyon, may 1998.
10. A. D. Gordon. Bisimilarity as a Theory of Functional Programs. Mini-Course. Technical Report NS-95-3 BRICS Notes Series, Computer Science Department of Aarhus University, 1995.
11. G. Kahn. Natural semantics. In K. Fuchi and M. Nivat, editors, *Programming Of Future Generation Computer*. Elsevier Sciences B.V. (North-Holland), 1988.
12. C. Paulin-Mohring. Extracting F_ω's Programs from Proofs in the Calculus of Constructions. In *POPL'89*. ACM, 1989.
13. A. M. Pitts. Operationally-based theories of program equivalence. In A. M. Pitts and P. Dybjer, editors, *Semantics and Logics of Computation*, pages 241–298. Cambridge University Press, 1997.
14. G. D. Plotkin. LCF considered as a programming language. *Theoretical Computer Science*, 5(3):223–255, 1977.
15. G. D. Plotkin. A structural approach to operational semantics. Technical Report DAIMI FN-19, Aarhus University, 1981.
16. Y. Takayama. Extraction of Redundancy-free Programs from Constructive Natural Deduction Proofs. *Journal of Symbolic Computation*, 12:29–69, 1991.

Every Unsolvable λ Term has a Decoration

René David

Laboratoire de Mathématiques. Université de Savoie F 73376 Le Bourget du Lac.
david@univ-savoie.fr

Abstract. I give a proof of the conjecture stated in [2] by R.Kerth : Every unsolvable λ term has a decoration.

1 Introduction

In this paper I give a proof of the conjecture stated in [2] by R. Kerth : Every unsolvable λ term has a decoration.

Let t be unsolvable. Denote by t_k the term obtained from t after k many steps of head reduction and by $(d \overrightarrow{u})$ the term d applied to the sequence $\overrightarrow{u}$ of arguments. If t reduces to t', say that a subterm d' of t' is a descendent (cf. definition 9) of a subterm d of t if it is a "copy" of d.

A sequence $(d_k)_{k \in N}$ of λ terms is a decoration for (the computation of) t if there is a strictly increasing function f from N to N such that for every k :

1. $t_{f(k)} = \overrightarrow{\lambda}(d_k \overrightarrow{u_k})$ for some finite (non empty) sequence $\overrightarrow{u_k}$ of λ terms.
2. d_k is solvable and d_{k+1} is a descendent of some element of $\overrightarrow{u_k}$.

Comments, notations and examples

1. The definition of a decoration given above is exactly the one of [2] but, in fact, the hypothesis "d_k is solvable" is useless since it is a consequence of the other hypothesis (cf the corollary 2)
2. Let $\delta = \lambda x\ (x\ x)$, $I = \lambda x\ x, B = \lambda b\ \lambda f\ (f\ (b\ b\ f))$ and $Y = (B\ B)$. Y is the Turing fixed point operator.
3. Let $t = (\delta\ \delta)$. Then the constant sequence (δ) is a decoration for t since t reduces by head reduction to $t' = (\delta\ \delta)$ and the first δ in t' is a descendent of the second δ in t.
4. Let $t = (B\ B\ I)$. Then the constant sequence (B) is a decoration for t since t reduces to itself (in 3 steps) and the first occurrence of B in this reduct is a descendent of the second occurence of B in t.
5. Let $w_1 = \lambda xyz\ (z\ x\ y)$, $w_2 = \lambda xyz\ (y\ (x\ (z\ x))), R = (w_1\ I\ w_2)$ and $w_3 = (w_2\ R)$. Then,
   - $t = (w_2\ R\ I\ w_2) \twoheadrightarrow (R\ w_3)$ (in 4 steps)
   - $(R\ w_3) \twoheadrightarrow (w_3\ I\ w_2) = t'$ (in 3 steps)
   - $(w_3\ I\ w_2) \twoheadrightarrow (w_2\ R\ I\ w_2) = t$ (in 7 steps)

It is easy to check that w_2, w_3 and R are solvable and that the descendent condition is satisfied. Thus the sequence $[w_2,\ R,\ w_3,\ w_2,\ R,\ w_3,\ w_2, ...]$ is a decoration for t. Note that t' is equal to t but t is written as w_2 applied to 3 arguments whereas t' is written as w_3 applied to 2 arguments and thus the R in t' is not seen as an argument of the head term.

6. Other examples can be found in [1].

The motivation (see [2]) of this conjecture is the following : A model of λ calculus is said to be sensible if all the unsolvable terms are equal in this model. It is not easy, in general, to check whether a given model of λ calculus is sensible or not. In [1] , [3] R Kerth built an uncountable number of graph models with different equational theories but he was unable to prove they were sensible, because the usual argument of reducibility did not work in his models. He was able to show that his models had no critical sequences (a semantical notion he introduced) and he showed that a graph model without critical sequences is sensible ... if his conjecture is true.

Thus, the constructions in [1] , [3] and the present paper show that there are uncountably many sensible distinct equational theories of continuous models (and similarly for the stable and strongly stable semantics).

Acknowledgements Rainer Kerth has read very carefully the first versions of this paper and suggested many improvements. Thanks, Rainer.

2 The idea of the proof

R. Kerth defines a decoration only for the head reduction of unsolvable terms, i.e. terms whose Böhm tree is $\perp$. I define below a decoration for the computation (by left reduction) of *any* branch of a term t. A branch in t is either an infinite branch of its Böhm tree or a finite one finishing with $\perp$, i.e. a branch in t corresponds to an infinite computation. I prove a more general result (The computation of any branch in any λ term admits a decoration. cf. Theorem 1) but this general notion of decoration is necessary for the proof of even the restricted case. The idea of the proof is the following.

1) Let a be a branch of t and b be a branch of a subterm u of t. I say that b is (t, a) useful if, intuitively (see the definition 10) the computation of the branch a of t "uses" all the nodes of addresses $b \upharpoonright i$ $(i < lg(b))$ of the Böhm tree of u. I first show that (cf the proposition 5) if a branch b of u is (t, a) useful and there is a decoration for (u, b), then there is a decoration for (t, a). This is the reason for which it is necessary to extend the notion of decoration to solvable terms. The decoration of an unsolvable term t may "come from" a decoration of a solvable subterm u of t.

2) Let $t = (u\ r_1\ ...\ r_n)$ and a be a branch in t. Say that a is created by the application of u to $r_1\ ...\ r_n$ if neither in u nor in any r_i there is a branch that is (t, a) useful. I also show (this is the key point of the proof, see the proposition 6) that if the branch a in $t = (u\ r_1\ ...\ r_n)$ is created by the application of u to $r_1\ ...\ r_n$, then t reduces to some $t' = \overrightarrow{\lambda}(r_i\ s_1\ ...\ s_m)$ for some $s_1\ ...\ s_m$ and

- the occurrence of r_i in t' is a descendent of the one in t.
- the branch a in t' still is created by the application of r_i to $s_1 \ldots s_m$.

Actually the proposition 6 is a bit more complicated because we have to deal with possible substitutions of the free variables.

3) The theorem 1 is then proved by induction on the complexity of t. If t is in head normal form the result follows immediately from the induction hypothesis. Otherwise $t = \overrightarrow{\lambda}\ (u\ r_1 \ldots r_p)$ for some $p \geq 1$. If the branch a is not created by the application of u to $r_1 \ldots r_n$, i.e. either in u or in some r_i there is a branch that is (t, a) useful, the result follows from the induction hypothesis and the first point above. Otherwise, we get a decoration by using repeatedly the second point above.

3 Definitions

Definition 1. *1. Let A be the set of finite or infinite lists of elements of* $N^* = N - \{0\}$. *A finite list is called an address.*

2. Let a, a' be in A. a ≤ a' means that a is an initial segment of a'. For i <lg(a), a ↾ i denotes the restriction of a to its first i elements.

3. The list a with i added at the beginning (resp at the end) will be denoted by [i :: a] (resp [a :: i]). The empty list is denoted by nil.

To be able to prove results on substitutions I need some extension of Λ. This is closely related (and a bit more general) to the directed λ calculus introduced in [4].

Definition 2. *1. Λ denotes the set of λ terms.*

2. The set Λ' of terms is defined by the following grammar :
$\Lambda' = V \mid \bot \mid c(a, \sigma) \mid \ \lambda x\ \Lambda' \mid (\Lambda'\ \Lambda')$
where
   (a) V is the set of variables
   (b) a substitution is a function from V to Λ' that is the identity except for a finite set (called its domain) of variables.
   (c) for every address a and every substitution, $c(a, \sigma)$ is a constant.

3. A Böhm function is a partial function $f : A \rightsquigarrow \{\bot\} \cup \{(E, x, p) \,/\, E \subset V, E$ finite, $x \in V, p \in N\}$ which satisfies :
   (a) f(nil) is defined.
   (b) f([a :: i]) is defined iff f(a) = (E, x, p) and $i \leq p$.
   (c) If f(a) = (E, x, p), f(a') = (E', x', p') and a ≠ a' then $E \cap E' = \emptyset$.

Notations, conventions and comments

- I adopt the Barendregt convention that variables are always named in such a way that there is no undesired capture and no confusion between different names.
- $\overrightarrow{\lambda}$ denotes a sequence (possibly empty) of abstractions and $(t\ \overrightarrow{r})$ represents the term t applied to a sequence (possibly empty) of arguments.

- $c(a, \sigma)$ represents the subterm (at the address a, in the environment given by σ) of the Böhm tree of some term u that will be substituted later on.
- A Böhm function codes a Böhm tree in the following way : $f(a) = (\{x_1, ..., x_k\}, x, p)$ (resp $\bot$) means that the node at the address a in the Böhm tree coded by f is $\lambda x_1 ... \lambda x_k \ (x \ t_1 ... t_p)$ for some terms $t_1, ..., t_p$ (resp $\bot$).

Definition 3. *Let σ, σ' be substitutions and t be in Λ'.*

1. *The free variables of t are defined by the usual rules and*
   - *$\bot$ has no free variables*
   - *x is a free variable of $c(a, \sigma)$ iff x is a free variable of $\sigma(y)$ for some y in the domain of σ.*
2. *The substitution $\sigma(t)$ is defined by the usual rules and*
   - *$\sigma(c(a, \tau)) = c(a, \sigma \circ \tau)$ for every τ and a.*
   - *$\sigma(\bot) = \bot$.*

Lemma 1. *Every term in Λ' can be uniquely written as $\overrightarrow{\lambda} \ (R \ \overrightarrow{r})$ where R is either a variable or $\bot$ or $(\lambda x \ u \ v)$ or $c(a, \sigma)$.*

Proof. By induction on the term.

Definition 4. *1. Let $t = \overrightarrow{\lambda} \ (R \ r_1 \ ... \ r_q)$ be in Λ' and f be a Böhm function. One step of f-reduction of t is defined as follows :*
   - *If $R = x$ then t is in f-head normal form and t has no f-reduct.*
   - *If $R = \bot$*
     - *If $t = \bot$ then t is in f-head normal form and t has no f-reduct.*
     - *otherwise, the f-reduct of t is $\bot$.*
   - *If $R = (\lambda x \ u \ v)$ then the f-reduct of t is $\overrightarrow{\lambda} \ (\sigma(u) \ r_1 \ ... \ r_q)$ where $\sigma(x) = v$.*
   - *If $R = c(a, \sigma)$*
     - *If $f(a) = (\{x_1, \ ..., x_k\}, \ x, \ p)$ then the f-reduct of t is $\overrightarrow{\lambda} \ \lambda x_{j+1} \ ... \ \lambda x_k \ (\sigma'(x) \ c([a :: 1], \sigma') \ ... \ c([a :: p], \sigma') \ r_{j+1} \ ... \ r_q)$ where j =Min(k, q), $\sigma' = \tau \circ \sigma$ and τ is defined by $\tau(x_i) = r_i$ for $1 \leq i \leq j$.*
     - *If $f(a) = \bot$, then the f-reduct of t is $\bot$.*
     - *If $f(a)$ is not defined the f-reduct of t is not defined.*

2. *$t \to_f t'$ (resp $t \twoheadrightarrow_f t'$) means that t' is the f-reduct of t (resp t' is obtained from t by some, possibly zero, steps of f-reductions).*

Comments and conventions

- An example of f-reduction is given after the definition 10.
- If t is in Λ the f-reduction is the ordinary head reduction (f is never used and thus can be anything).
- If t is in Λ' and f "represents" the term u (see the definition 8) the f-reduction "corresponds" to the (ordinary) head reduction of t' where

- t' is the term t where the constants $c(a, \sigma)$ have been replaced by the subterm of the Böhm tree of u at the address a in the environment σ.
- "corresponds" means that the reduction is the same except that the part of the computation of t' that "comes from" the computation of the node at the address a in the Böhm tree of u has been forgotten and is given by the "oracle" f.

– I allow $f(a)$ to be undefined in the definition of the f-reduction of t because I made no restrictions in the definition of Λ'. However the typical situation where the f-reduction is used is the following. Let $t = (u\ \vec{r})$ be in Λ, f "represents" u and $t' = (c(nil, Id)\ \vec{r})$. In this case the f-reduction will clearly always be defined.
– Similarly, if t "comes from" a λ term, since I only do head reductions the composition $\sigma' = \tau \circ \sigma$ (in the case $R = c(a, \sigma)$) in fact is a concatenation of substitutions (cf the definition 12 and the lemma 8) but I must allow also composition when I know nothing on t.
– When t is in Λ, I will not write the symbol f. For example I will write $t \twoheadrightarrow t'$ instead of $t \twoheadrightarrow_f t'$ and similarly for all the definitions in this section. For example $hnf(t)$ instead of $hnf(f,t)$ in the next definition.
– The letters $a, b, c, ...$ are reserved for elements of A, the letters $f, g, ...$ for Böhm functions and the letters $r, s, t, ...$ for terms in Λ'. This will avoid possible confusions.

Definition 5. *$hnf(f,t)$ (the f-head normal form of t) is defined by*

1. – *If some step of the f-reduction of t is undefined, then $hnf(f,t)$ is not defined.*
   – *If t $\twoheadrightarrow_f t'$ for some term t' in f-head normal form and t' $\neq \bot$, then $hnf(f,t) = t'$. In this case t is said to be f-solvable.*
   – *If the f-reduction of t does not terminate or if t $\twoheadrightarrow_f \bot$, then $hnf(f,t) = \bot$. In this case t is said to be f-unsolvable.*

Definition 6. *Let a be an address, t be in Λ' and f be a Böhm function*

1. *a is f-accessible in t is defined by*
   – *nil is f-accessible in t*
   – *$[i :: l]$ is f-accessible in t iff hnf(f, t) $= \vec{\lambda}(x\ t_1 ... t_n), 1 \leq i \leq n$ and l is f-accessible in t_i*
2. *Let a be f-accessible in t. $hnf(f,t,a)$ is defined by*
   – *$hnf(f,t,nil) = hnf(f,t)$.*
   – *$hnf(f,t,[i :: l]) = hnf(f,t_i,l)$ where $hnf(f,t) = \vec{\lambda}(x\ t_1 ... t_n)$*
3. *Let a be f-accessible in t. $adr(f,t,a)$ is defined by*
   – *$adr(f,t,nil) = t$.*
   – *$adr(f,t,[i :: l]) = adr(f,t_i,l)$ where $hnf(f,t) = \vec{\lambda}(x\ t_1 ... t_n)$*

Comments In the following t is assumed to be in Λ.

- a is accessible in t iff the Böhm tree of t (denoted by $BT(t)$) has a node at the address a.
- $hnf(t, a)$ is the λ term we get at the address a when the computation of the node at this address in $BT(t)$ is *terminated.*
- $adr(t, a)$ is the λ term we get at the *beginning* of the computation of the node at this address in $BT(t)$.

Definition 7. *Let a be in A, t be in Λ' and f be a Böhm function.*

1. *a is an f-branch in t iff*
   - *$\forall i < lg(a)$ $a \restriction i$ is f-accessible in t.*
   - *if a is finite, then $hnf(f, t, a) = \perp$*
2. *Assume a is an f- branch in t and k be in N. $Res(f, t, a, k)$ and $Br(f, t, a, k)$ are defined by*
   - *$Res(f, t, a, 0) = t$ and $Br(f, t, a, 0) = a$*
   - *If $Res(f, t, a, k)$ is not an f-head normal form then $Res(f, t, a, k+1) =$ the f- reduct of $Res(f, t, a, k)$ and $Br(f, t, a, k+1) = Br(f, t, a, k)$*
   - *If $Res(f, t, a, k) = \overrightarrow{\lambda}(x\ t_1 ... t_n)$ and $a = [i :: l]$ then $Res(f, t, a, k+1) = t_i$ and $Br(f, t, a, k+1) = l$*
   - *Otherwise $Res(f, t, a, k)$ and $Br(f, t, a, k)$ are undefined.*
3. *$t \twoheadrightarrow_{f,a} t'$ means that $t' = Res(f, t, a, k)$ for some k.*

Comments and examples In the following t is assumed to be in Λ.

1. $Res(t, a, k)$ is the term we get after k many steps in the computation of the branch a of $BT(t)$.
2. If $t' = Res(t, a, k)$ then $a' = Br(t, a, k)$ is the branch of t' that has to be computed to finish the computation of the branch a of t. Thus, if $t \twoheadrightarrow_a t'$ and $t' \twoheadrightarrow_{a'} t''$ then $t \twoheadrightarrow_a t''$.
3. Let t be in Λ. If t is unsolvable, then nil is the only accessible address (and the only branch) in t.
4. Let $t = (I\ \ \lambda x\ (x\ (\delta\ \delta)))$. Then $hnf(t, nil) = \lambda x\ (x\ (\delta\ \delta))$, $adr(t, [1]) = (\delta\ \delta)$ and $hnf(t, [1]) = \perp$. The only branch of t is $[1]$.
5. $hnf(Y, nil) = \lambda f\ \ (f\ (B\ B\ f))$. $hnf(Y, [1, 1, ..., 1]) = (f\ (B\ B\ f))$. The only branch of Y is $1^\infty = [1,\ 1,\ ...]$.
6. Let $w = \lambda xyz\ (z\ (y\ (x\ x\ y))\ z)$ and $t = (w\ w)$.
   - $hnf(t, nil) = \lambda yz\ (z\ (y\ (w\ w\ y))\ z)$,
   - $hnf(t, [1]) = (y\ (w\ w\ y))$,
   - $hnf(t, [2]) = z$,
   - $hnf(t, [1, 1]) = \lambda z_1(\ z_1\ (y\ (w\ w\ y))\ z_1)$
   - a is accessible in t iff $a = [1, 1, ..., 1]$ or $a = [1, 1, ..., 1, 2]$. The only branch of t is 1^∞.

Definition 8. *Let u be in Λ' and g be a Böhm function.*

1. *$\psi(g, u)$ is the Böhm function f defined as follows*
   - *$f(a)$ is defined iff a is g-accessible in u.*
   - *$f(a) = (\{x_1, \ldots, x_k\}, x, p)$ iff $hnf(g, u, a) = \lambda x_1 \ldots \lambda x_k\ (x\ t_1 \ldots t_p)$ for some terms $t_1, \ldots, t_p$*
   - *$f(a) = \bot$ iff $hnf(g, u, a) = \bot$.*
2. *Let t be in Λ'. $t[g, u]$ is the term obtained by replacing in t the occurrences of $c(a, \sigma)$ by $\sigma(adr(g, u, a))$ for every a and σ.*

Comment and example

- Most of the time the previous definition will be used with u in Λ and thus $t[g, u]$ also is in Λ and g is useless. In this case the function ψ describes the nodes of $BT(u)$. Remember (cf. the conventions after the definition 4) that, in this case, we "forget" the argument g i.e. we write $\psi(u)$ and $t[u]$. However the more general definition is necessary to prove that (see the proposition 2) "to be useful" is a transitive notion.
- Let $f = \psi(Y)$. Since $Y \twoheadrightarrow \lambda x\ (x\ (x\ (x \ldots$ we have $f(nil) = (\{x\}, x, 1)$ and $f([1, 1, 1 \ldots, 1]) = (\emptyset, x, 1)$

Definition 9. *Let t be in Λ'.*

1. *The notion of subterm of t is defined as usual, with the following additional rule. u is a (strict) subterm of $c(a, \sigma)$ if u is a subterm of $\sigma(x)$ for some x.*
2. *Let f be a Böhm function, b be f-accessible in t and $t \twoheadrightarrow_{f,b} t'$.*
   - *A subterm u' of t' is a residue of a subterm u of t if it is a "copy by β-reduction" of u where, possibly, the free variables have been substituted. u' is a descendent of u if it is a residue of u and the free variables have not been substituted.*
   - *The subterm $u' = c(a', \sigma')$ of t' is an immediate successor of the subterm $u = c(a, \sigma)$ of t if*

     $t \twoheadrightarrow_{f,b} t_1 = \overrightarrow{\lambda}(c(a, \tau)\ \overrightarrow{\tau}) \rightarrow_f t_2 = \overrightarrow{\lambda}(\tau'(x)\ c([a :: 1], \tau') \ldots c([a :: p], \tau')\ \overrightarrow{r'}) \twoheadrightarrow_{f,b} t'$

     u' is a residue of some element of the sequence $c([a :: 1], \tau') \ldots c([a :: p], \tau')$ in t_2

     the occurrence of $c(a, \tau)$ in t_1 is a residue of u.
3. *The successor relation (between terms as $c(a, \sigma)$) is the transitive closure of the immediate successor relation.*

Remark A more "formal" definition of these notions (that are intuitively very clear) is rather tedious. For more details see [2]. It is clear that the notion of descendent given above is exactly the one in [2]. In particular, if $t = (d\ \overrightarrow{u}) \twoheadrightarrow_a (d'\ \overrightarrow{u'})$ and d' is a residue of some element of the sequence $\overrightarrow{u}$ then it is also a descendent of this element.

Definition 10. *Let t, u be in Λ and assume that $t = D(\sigma(u))$ for some context D and some substitution σ. Let $t' = D(c(nil, \sigma))$ and $f = \psi(u)$. Let a be a branch in t.*

1. *Let b be an address accessible in u. b is (t, a) useful if, for some k, $\vec{v}$ and σ, $Res(f, t', a, k) = \vec{\lambda}\ (c(b, \sigma)\ \vec{v})$.*
2. *Let b be a branch in u. b is (t, a) useful if there is a sequence $< k_i, \sigma_i, \vec{v_i} >_{i<lg(b)}$ such that, for every i, $Res(f, t', a, k_i) = \vec{\lambda}\ (c(b \restriction i, \sigma_i)\ \vec{v_i})$; moreover the occurrence of $c(b \restriction i+1, \sigma_{i+1})$ in $Res(f, t', a, k_{i+1})$ is an immediate successor of the occurrence of $c(b \restriction i, \sigma_i)$ in $Res(f, t', a, k_i)$.*

Remarks and examples

- A context is a λ term (not a λ' term !) with some holes. As usual, in a substitution in a context some variables may be captured.
- It will be shown (see the proposition 1) that, with the notations of the previous definition, a is an f-branch in t' and thus the definition makes sense.
- Most often, either σ is the identity (i.e. u is a subterm of t) or D is an applicative context (i.e. $t = (\sigma(u)\ \vec{r})$) but it is not always the case (see the proposition 6) and I thus need this general definition. In fact both cases are essentially the same since it is not difficult to prove the following fact.
  Let $t = D(u)$ for some context D and a be a branch in t. Assume that the address nil in u is (t, a) useful, then $t \twoheadrightarrow_a \vec{\lambda}(\sigma(u)\ \vec{r})$ for some σ which is the identity except on the free variables of u that are captured by the context D.
- Let $t = (Y\ I)$. t is unsolvable and thus nil is a branch in t. 1^∞ is a branch in Y. It is easy to check that 1^∞ is (t, nil) useful.
- Note that a term t may have many subterms each of them has a branch that is (t, a) useful. For example, let $t = (Y_1\ F)\ (Y_2\ F)$ where $Y_1 = Y_2 = Y$ and $F = \lambda f \lambda g\ (g\ f)$. The following reduction shows that the branch 1^∞ in Y_1 (and similarly for Y_2) is (t, nil) useful.
  Let $f = \psi(Y)$ and $t' = (c(nil, Id)\ F\ (Y\ F))$. Remember that $f(nil) = (\{x\}, x, 1)$ and $f([1, 1, ..., 1]) = (\emptyset, x, 1)$. The f-reduction of t' is given by (where $\sigma(x) = F$) : $t' \to (F\ c([1], \sigma)\ (Y\ F)) \twoheadrightarrow (Y\ F\ c([1], \sigma)) \twoheadrightarrow$
  $(F\ (Y\ F)\ c([1], \sigma)) \twoheadrightarrow (c([1], \sigma)\ (Y\ F)) \to (F\ c([1, 1], \sigma)\ (Y\ F)) \twoheadrightarrow\ ...$
- Also note that, for an infinite branch b, being (t, a) useful is stronger that simply asking that for every i, $b \restriction i$ is (t, a) useful. Let $t = (Y_1\ H\ Y_2\ 0)$ where $Y_1 = Y_2 = Y$, $H = \lambda fnp\ (u\ n\ p\ (f\ n\ (s\ p)))$, $u = \lambda npa\ (n\ F\ (p\ F\ \lambda x\ a))$, $F = \lambda xy\ (y\ x)$, $0 = \lambda xy\ y$ and $s = \lambda nfx\ (f\ (n\ f\ x))$. For every k, the address 1^k is (t, nil) useful both in Y_2 and Y_1. The branch 1^∞ of Y_1 is (t, nil) useful but the branch 1^∞ of Y_2 is not. The reason is the following : u is a term (given by Maurey) such that $(u\ n\ p\ a) \to a$ for every Church integers $n \geq p$. Since Y may be seen as an "infinite" Church integer, $(u\ Y\ k\ a) \to a$ for every k and this computation "uses" the address $1^k$of Y. It follows that, letting $G = (Y_1\ H)$, $t = (G\ Y_2\ 0) \to (G\ Y_2\ 1) \to (G\ Y_2\ 2) \to \dots$. It is easy to

see that, in this computation, the node at the address 1^{k+1} of Y_1 that is used for the reduction $(G\, Y_2\, k) \to (G\, Y_2\, k+1)$ satisfies the descendent condition whereas, since the occurrence of Y_2 in $(G\, Y_2\, k+1)$ is a "new" one, the node at the address 1^{k+1} of Y_2 that is used in this reduction does not satisfy the condition.

Definition 11. *Let t be in Λ, a be a branch of t and (d_n) a sequence of λ terms. (d_n) is a decoration for (t, a) if there is a strictly increasing sequence (k_n) of integers and a sequence $(\overrightarrow{r}_n)$ such that for every $n \geq 0$*

1. *$Res(t, a, k_n) = \overrightarrow{\lambda}(d_n\, \overrightarrow{r_n})$*
2. *d_{n+1} is the descendent of an element of $\overrightarrow{r_n}$*
3. *d_n is solvable.*

Theorem 1. *Let t be in Λ and a be a branch in t. Then (t, a) has a decoration.*

Corollary 1. *Every unsolvable λ term has a decoration in the sense of [2].*

4 Proof of the theorem

4.1 Some lemmas on the f-reduction and usefulness

In this section I prove essentially two things : The notion of computation and the notion of usefulness are "transitive". Moreover in both cases the notion of descendence is preserved by this transitivity.

The first one (mainly the lemma 7) means that a computation (by left reduction) can be "partitioned" in the following way : Let u be a subterm of t. Get t' by replacing in t the subterm u by its Böhm tree. The computation of a branch a of t is the same as the computation of the branch a of t' where, when a node of $BT(u)$ appears in head position, the computation of this node is "inserted". There is a (non essential) technical difficulty showed in the following example : Assume $u \twoheadrightarrow \lambda x\; u_1 \twoheadrightarrow \lambda x\; (x\; v)$ then $(u\; r) \twoheadrightarrow (\lambda x\; u_1\; r) \to u_1[x := r] \twoheadrightarrow (r\; v[x := r])$ and the order is not exactly the same as $(u\, r) \twoheadrightarrow (\lambda x\, u_1\, r) \twoheadrightarrow (\lambda x\, (x\, v)\, r) \twoheadrightarrow (r\, v[x := r])$. This is why we have to use big steps of head reduction.

The second one is given by the proposition 2.

Lemma 2. *Let t, t' be in Λ', f be a Böhm function and a be f-accessible in t. Assume $t \twoheadrightarrow_{f,a} t'$. Then, for some $a' \leq a$, $t \twoheadrightarrow_{f,a} adr(f, t, a') \twoheadrightarrow_f t'$.*

Proof. Immediate from the definition.

Lemma 3. *Let v, v' be in Λ' and f be a Böhm function. Assume that $v \twoheadrightarrow_f v'$.*

1. *Let σ be a substitution. Then $\sigma(v) \twoheadrightarrow_f \sigma(v')$.*
2. *Let $\overrightarrow{r}$ be a sequence of terms and assume v' does not begin with λ. Then $(v\; \overrightarrow{r}) \twoheadrightarrow_f (v'\; \overrightarrow{r})$*
Moreover in both cases the length of the f-reduction remains the same.

Proof. Note that the more general case, where v' begins with λ, is treated in the lemma 5. The proof is by induction on the length of the reduction and case analysis. Use the fact that $\sigma(u[x := v]) = \sigma(u)[x := \sigma(v)]$.

Lemma 4. *Let t be in Λ' and f be a Böhm function such that t is f-unsolvable.*

1. *Let σ be a substitution. Then $\sigma(t)$ is f-unsolvable.*
2. *Let $\overrightarrow{r}$ be a sequence of terms. Then $(t\ \overrightarrow{r})$ is f-unsolvable. Moreover $(t\ \overrightarrow{r})$ has no reduct of the form $\overrightarrow{\lambda}(r_i\ \overrightarrow{v})$ where r_i is a descendent of an element of $\overrightarrow{r}$.*

Proof. 1. This follows immediately from the lemma 3.
2. If t does not reduce to a term beginning with λ this follows immediately from the lemma 3. Otherwise let $\overrightarrow{r} = (r_1 \ldots r_n)$ and t' be the least step where λ appears. Then (by the lemma 3) $(t\overrightarrow{r}\ ) \twoheadrightarrow_f (t'\ \overrightarrow{r}) = (\lambda x\ t_1\ \overrightarrow{r}) \twoheadrightarrow_f (\sigma(t_1)\ r_2 \ldots r_n)$ where $\sigma(x) = r_1$. The result follows by the lemma 3 and by repeating, if necessary, the same argument.

Corollary 2. *Let t be in Λ, a be a branch of t, (d_n) be a sequence of λ terms, (k_n) be a strictly increasing sequence of integers and $(\overrightarrow{r}_n)$ be a sequence of finite sequences of λ terms. Assume that for every $n \geq 0$*

1. *$Res(t, a, k_n) = \overrightarrow{\lambda}(d_n\ \overrightarrow{r_n})$*
2. *d_{n+1} is the descendent of an element of $\overrightarrow{r_n}$*

Then (d_n) is a decoration for (t, a).

Proof. The fact that d_n is solvable follows immediately from the lemma 4.

Lemma 5. *Let v, r_1, ..., r_p be in Λ', σ be a substitution and f be a Böhm function. Assume that $v \twoheadrightarrow_f \lambda x_1 \ldots \lambda x_k\ (u\ \overrightarrow{t})$. Then $(\sigma(v)\ r_1 \ldots r_p) \twoheadrightarrow_f \lambda x_{j+1} \ldots \lambda x_k\ (\sigma'(u)\ \overrightarrow{\sigma'(t)}\ r_{j+1} \ldots r_p)$ where $j = Min(k, p)$, $\sigma' = \tau \circ \sigma$ and τ is given by $\tau(x_i) = r_i$ for $1 \leq i \leq j$.*

Proof. By induction on k. The case $k = 0$ is given by the lemma 3. Assume $k \geq 1$. Look at the least step in the reduction $v \twoheadrightarrow_f v'$ where v' begins with λ, say $v' = \lambda x_1\ v_1$. Then, we have the following sequence of f-reductions : $(\sigma(v)\ r_1 \ldots r_p) \twoheadrightarrow_f (\lambda x_1\ \sigma(v_1)\ r_1 \ldots r_p) \rightarrow_f (\sigma_1(v_1)\ r_2 \ldots r_p) \twoheadrightarrow_f \lambda x_{j+1} \ldots \lambda x_k\ (\sigma'(u)\ \overrightarrow{\sigma'(t)}\ r_{j+1} \ldots r_p)$ where $\sigma_1 = \tau \circ \sigma$ and τ is given by : $\tau(x_1) = r_1$. The first $\twoheadrightarrow_f$ is given by the lemma 3 and the last $\twoheadrightarrow_f$ is given by the induction hypothesis.

Lemma 6. *Let t, u be in Λ', g be a Böhm function and $f = \psi(g, u)$. Assume $t = \overrightarrow{\lambda}\ (R\ r_1 \ldots r_p)$ and t' is the f-reduct of t. Then*

1. *if $R = x$, then $t[g, u]$ is in g-head normal form.*
2. *if $R = (\lambda x\ v\ w)$ or $\perp$, then the g-reduct of $t[g, u]$ is $t'[g, u]$.*
3. *if $R = c(a, \sigma)$*

- *If f(a)* $= \bot$, *then t[g, u] is not g-solvable.*
- *If f(a)* $= (\{x_1, ..., x_k\}, x, q)$ *then* $t[g, u] \twoheadrightarrow_g t'[g, u]$.

Proof. (1) and (2) are clear. (3.1) follows from the lemma 4 and (3.2) follows from the lemma 5.

Lemma 7. *Let t, u be in* Λ', *g be a Böhm function,* $f = \psi(g, u)$ *and a be f-accessible in t. Assume* $t \twoheadrightarrow_{f,a} t' = \overrightarrow{\lambda}(R\ \overrightarrow{s})$ *and R = either x or* $(\lambda x\ v\ w)$ *or* $c(b, \sigma)$ *and* $f(b) \neq \bot$. *Then,* $t[g, u] \twoheadrightarrow_{g,a} t'[g, u]$. *Moreover, let d' be a subterm of t' that is a residue (resp a descendent) of a subterm d of t. Then d'[g, u] is a residue (resp a descendent) of the corresponding subterm d[g, u].*

Proof. By induction on the length of the reduction of t. For $a = nil$ this follows from the lemma 6. If $a = [i :: b]$, then $t \twoheadrightarrow_f \overrightarrow{\lambda}(x\ t_1\ ...\ t_n)$. By the lemma 6, $t[g, u] \twoheadrightarrow_g \overrightarrow{\lambda}(x\ t_1[g, u]\ ...\ t_n[g, u]) \twoheadrightarrow_{g,a} t_i[g, u]$ and the result follows easily by induction on the length of a.

Proposition 1. *Let t, u be in* Λ', *g be a Böhm function and* $f = \psi(g, u)$. *Let a be in A. Then a is an f-branch in t iff a is a g-branch in t[g, u].*

Proof. It follows immediately from the lemma 6 that t has an f-head normal form iff $t[g, u]$ has a g-head normal form. Moreover if $hnf(f, t, nil) = \lambda x_1 ... \lambda x_k\ (x\ t_1 ... t_p)$ then $hnf(g, t[g, u], nil) = \lambda x_1 ... \lambda x_k\ (x\ t_1[g, u]\ ...\ t_p[g, u])$. The result follows easily.

Definition 12. *Let* σ, σ' *be substitutions.* $\tau = \sigma \oplus \sigma'$ *if for every variable x*

- *if* $\sigma(x) \neq x$ *then* $\tau(x) = \sigma(x)$ *and* $\sigma'(x) = x$
- *if* $\sigma'(x) \neq x$ *then* $\tau(x) = \sigma'(x)$ *and* $\sigma(x) = x$
- *otherwise* $\tau(x) = x$

Definition 13. *Let u be in* Λ. *Define, for a accessible in u, FV(u, a) by :*

- $FV(u, nil) = \emptyset$
- $FV(u, [a :: i]) = Fv(u, a) \cup \{x_1\ ...\ x_k\}$ *where* $hnf(u, a) = \lambda x_1\ ...\ x_k\ (x\ \overrightarrow{r})$

Lemma 8. 1. *Let* $t = (\sigma(u)\ \overrightarrow{r})$ *be in* Λ, $t' = (c(nil, \sigma)\ \overrightarrow{r})$, *b be accessible in t,* $f = \psi(u)$, $t' \twoheadrightarrow_{f,b} t''$ *and* $c(a, \tau)$ *be a subterm of t". Then* $\tau = \sigma \oplus \sigma'$ *for some* σ' *whose domain is included in FV(u,a). Moreover, for every variable y in the domain of* τ, *for every a' >a and every x in FV(u, a') - FV(u, a), x is not free in* $\tau(y)$.
2. *Similarly for t = D(*$\sigma(u)$*) with* $\tau = \sigma \oplus \sigma'' \oplus \sigma'$ *where the domain of* σ'' *is included in the set of variables captured by the context D.*
3. *Moreover if* $c(a', \tau')$ *is a descendent of* $c(a, \tau)$ *then* $\tau' = \tau \oplus \mu$ *for some* μ *whose domain is included in FV(u, a') - FV(u, a)*

Proof. This comes immediately from the fact that we are doing head reduction (and of course the renaming rule to avoid capture). More precisely, this is proved by induction on the length of the reduction $t' \twoheadrightarrow_{f,b} t''$ by a simple case analysis.

Lemma 9. *Let t = $(\sigma(u)\ \overrightarrow{r})$ be in Λ, b be a branch in t and f = $\psi(u)$. Let t' = $(c(nil, \sigma)\ \overrightarrow{r})$.*

1. *Assume t' $\twoheadrightarrow_{f,b} \overrightarrow{\lambda}\ (c(a,\tau)\ \overrightarrow{s})$ and u $\twoheadrightarrow_a adr(u,a) \twoheadrightarrow \lambda x_1 \dots \lambda x_k\ (d\ \overrightarrow{v}) \twoheadrightarrow \lambda x_1 \dots \lambda x_k \dots \lambda x_{k+k'}\ (d'\ \overrightarrow{v'})$ and d' is the descendent of an element of $\overrightarrow{v}$. Then t $\twoheadrightarrow_b \overrightarrow{\lambda}(\mu(d)\ \mu(\overrightarrow{v})\ \overrightarrow{w}) \twoheadrightarrow_b \overrightarrow{\lambda'}\ (\mu'(d')\ \mu'(\overrightarrow{v'})\ \overrightarrow{w'})$ and $\mu'(d') = \mu(d')$ is a descendent of the corresponding element of $\mu(\overrightarrow{v})$.*
2. *Similarly assume that :*
   - *t' $\twoheadrightarrow_{f,b} \overrightarrow{\lambda}\ (c(a,\tau)\ \overrightarrow{s}) \twoheadrightarrow_{f,b} \overrightarrow{\lambda'}\ (c(a',\tau')\ \overrightarrow{s'})$ for some $a < a'$ and $c(a',\tau')$ is a successor of $c(a,\tau)$.*
   - *u $\twoheadrightarrow_{a'} adr(u,a) \twoheadrightarrow \overrightarrow{\lambda}(d\ \overrightarrow{v}) \twoheadrightarrow_{a'} adr(u,a') \twoheadrightarrow \overrightarrow{\lambda}(d'\ \overrightarrow{v'})$ and d' is the descendent of an element of $\overrightarrow{v}$.*

Then t $\twoheadrightarrow_b \overrightarrow{\lambda}(\mu(d)\ \mu(\overrightarrow{v})\ \overrightarrow{w}) \twoheadrightarrow_b \overrightarrow{\lambda}(\mu'(d')\ \mu'(\overrightarrow{v'})\ \overrightarrow{w'})$ and $\mu'(d') = \mu(d')$ is a descendent of the corresponding element of $\mu(\overrightarrow{v})$.

Proof. 1. By the lemma 8, $\tau = \sigma \oplus \sigma_1$. By the lemma 7, $t \twoheadrightarrow_b \overrightarrow{\lambda}(\tau(adr(u,a))\ \overrightarrow{s[u]})$ and, by the lemma 5, $\overrightarrow{\lambda}(\tau(adr(u,a))\ \overrightarrow{s[u]}) \twoheadrightarrow \overrightarrow{\lambda}(\mu(d)\ \mu(\overrightarrow{v})\ \overrightarrow{w}) \twoheadrightarrow \overrightarrow{\lambda'}(\mu'(d')\mu'(\overrightarrow{v'})\ \overrightarrow{w'})$ where $\mu = \sigma' \circ \tau$ (resp $\mu' = \sigma" \circ \tau$) and the domain of σ' (resp $\sigma"$) is included in $\{x_1 \dots x_k\}$ (resp $\{x_1 \dots x_{k+k'}\}$). By the lemma 8, $\mu = \tau \oplus \sigma'$ and $\mu' = \tau \oplus \sigma"$. Since d' is the descendent of an element of $\overrightarrow{v}$ the variables $x_{k+1} \dots x_{k+k'}$ do not appear in d' and $\mu(d') = \mu'(d')$.
2. Similarly $t \twoheadrightarrow_b \overrightarrow{\lambda}(\mu(d)\ \mu(\overrightarrow{v})\ \overrightarrow{w}) \twoheadrightarrow_b \overrightarrow{\lambda}(\mu'(d')\ \mu'(\overrightarrow{v'})\ \overrightarrow{w'})$ where $\mu = \tau \oplus \sigma'$, $\mu' = \mu \oplus \sigma"$and the domain of $\sigma"$ is included in $FV(u,a') - FV(u,a)$. Since d' is the descendent of an element of $\overrightarrow{v}$, d' has no free variables in $FV(u,a') - FV(u,a)$ and thus $\mu'(d') = \mu(d')$.

Proposition 2. *Let t, u, v be in Λ, a (resp b, c) be a branch in t (resp in u, v). Assume that b is (t, a) useful and c is (u, b) useful. Then c is (t, a) useful.*

Proof. Let $t = D(\sigma(u))$, $u = E(\tau(v))$. Let $t' = D(c(nil,\sigma)), u' = E(c(nil,\tau))$. Let $F = D(\sigma(E))$. Then $t = F(\sigma \circ \tau(v))$. Let $t" = F(c(nil, \sigma \circ \tau))$. I only prove $t" \twoheadrightarrow_{g,a} \overrightarrow{\lambda}(c(c \upharpoonright j, \tau_j)\ \overrightarrow{r_j})$ for every $j < lg(c)$, where $g = \psi(v)$. I should prove a bit more, namely that the corresponding $c(c \upharpoonright j, \tau_j)$ are in the immediate successor relation (see the definition 10). This is rather tedious to write but this follows immediately from the proof.

Let $f = \psi(u)$ and $d = c \upharpoonright j$. Since c is (u,b) useful, $u' \twoheadrightarrow_{g,b} \overrightarrow{\lambda}(c(d,\tau')\ \overrightarrow{r})$. Thus, by the lemma 2, $u' \twoheadrightarrow_{g,b} adr(g,u',b') \twoheadrightarrow_g \overrightarrow{\lambda}(c(d,\tau')\ \overrightarrow{r})$ for some $b' \leq b$. Since b is (t,a) useful, $t' \twoheadrightarrow_{f,a} \overrightarrow{\lambda}(c(b',\sigma')\ \overrightarrow{s})$. Clearly $t" = t'[g,u']$. Thus, by the lemmas 7 and 5 , $t" \twoheadrightarrow_{g,b} \overrightarrow{\lambda}(\sigma'(adr(u',b')\ \overrightarrow{s}) \twoheadrightarrow_{g,b} \overrightarrow{\lambda}(c(d,\tau")\ \overrightarrow{r'})$.

Proposition 3. *Let t = $(\sigma(u)\ \overrightarrow{r})$ be in Λ and b be a branch in t. Let a be a branch in u that is (t, b) useful. Assume that Res(u, a, k) = $\overrightarrow{\lambda}(u_1\ \overrightarrow{v_1})$. Then,*

- *For some j and some* τ, $Res(t, b, j) = \overrightarrow{\lambda}(\tau(u_1)\ \tau(\overrightarrow{v_1})\ \overrightarrow{w})$.
- *Let c be a branch in* u_1 *that is* $(Res(u, a, k), Br(u, a, k))$ *useful. Then c is* $(Res(t, b, j), Br(t, b, j))$ *useful.*

Proof. By the lemma 2, $u \twoheadrightarrow_a adr(u, a_1) \twoheadrightarrow \overrightarrow{\lambda}(u_1\ \overrightarrow{v_1}) = u'$. Let $t' = (c(nil, \sigma), \overrightarrow{r})$ and $f = \psi(u)$. Since a is (t, b) useful $t' \twoheadrightarrow_{f,b} \overrightarrow{\lambda}(c(a_1, \sigma_1)\ \overrightarrow{s})$. Thus $t \twoheadrightarrow_b \overrightarrow{\lambda}(\sigma_1(adr(u, a_1))\ \overrightarrow{s}) \twoheadrightarrow_b \overrightarrow{\lambda}(\tau(u_1)\ \tau(\overrightarrow{v_1})\ \overrightarrow{w}) = Res(t, b, j) = t"$. Let $a' = Br(u, a, k)$ and $b" = Br(t, b, j)$. Since a is (t, b) useful, it is clear that a' is $(t", b")$ useful and since c is (u', a') useful, by the proposition 2, c is $(t", b")$ useful.

4.2 The key results

The propositions 5 and 6 give the key points mentioned in the section 2. Intuitively the proposition 6 gives the next step of the decoration and the proposition 7 is the technical result that allows to iterate the construction.

Proposition 4. *Let u be in* Λ. *Assume that u is unsolvable and* (d_k) *is a decoration for* (u, nil).

1. *Let* σ *be a substitution. Then* $(\sigma(d_k))$ *is a decoration for* $(\sigma(u), nil)$.
2. *Let* $t = (u\ \overrightarrow{r})$. *Then there is a sequence* (σ_k) *of substitutions such that* $(\sigma_k(d_k))$ *is a decoration for* (t, nil).

Proof. The first case is trivial since, by the lemma 3, if $u \twoheadrightarrow u'$ then $\sigma(u) \twoheadrightarrow \sigma(u')$. For the second case let p be the length of $\overrightarrow{r}$. If $p = 0$, this is trivial. Assume $p \geq 1$. If, for every k, $Res(u, nil, k)$ does not begin with λ the result follows from the lemma 3. Otherwise, let k be the least integer such that $Res(u, nil, k) = \lambda x\ u'$. Since (d_k) is a decoration for (u, nil), let (k_n) be the sequence such that $Res(u, nil, k_n) = \overrightarrow{\lambda}(d_n\ \overrightarrow{v_n})$.

Assume first that $k_0 > k$. Then (by the lemma 3) $(u \overrightarrow{r}) \twoheadrightarrow (\lambda x\ u'\ \overrightarrow{r}) \twoheadrightarrow (\sigma(u')\ r_2 .. r_p)$ where $\sigma(x) = r_1$. Repeating the same argument with $(\sigma(u')\ r_2\ r_p)$ yields the result.

Assume that $k_0 \leq k$. Let n_0 be the largest integer such that $k_{n_0} \leq k$. Then (by the lemma 3) for $n \leq n_0$ $Res(t, nil, k_n) = (d_n\ \overrightarrow{v}_n\ \overrightarrow{r})$. $Res(t, nil, k_{n_0}) \twoheadrightarrow (\lambda x\ u'\ \overrightarrow{r}) \twoheadrightarrow (\sigma(u')\ r_2 \ldots r_p)$ where $\sigma(x) = r_1$. Since $(d_n)_{n>n_0}$ is a decoration for (u', nil), $(\sigma(d_n))_{n>n_0}$ is a decoration for $(\sigma(u'), nil)$. Since d_{n_0+1} is a descendent of an element of v_{n_0}, x is not free in d_{n_0+1}. Repeating the same argument with $((\sigma(u')\ r_2 \ldots r_p)\ , nil)$ yields the result.

Proposition 5. *Let t, u be in* Λ *and b (resp a) be a branch in t (resp u). Assume a is* (t, b) *useful and let* (d_k) *be a decoration for* (u, a). *Then there is a sequence* (σ_k) *of substitutions such that* $(\sigma_k(d_k))$ *is a decoration for* (t, b).

Proof. - If a is infinite, the sequence (σ_k) is easily constructed by using the lemma 9.

- If a is finite the sequence (σ_k) is easily constructed by using the lemma 9 for the finite part of the branch and the proposition 4 for its last node.

Proposition 6. *Let $t = (u\ r_1\ ...\ r_n)$ be a λ term and a be a branch in t. Assume there is no branch neither in u nor in any r_i that is (t, a) useful. Then there is $< i,\ k,\ u_1,\ \nu >$ such that, letting $t' = Res(t,\ a,\ k)$ and $a' = Br(t,\ a,\ k)$:*

- *$t' = \overrightarrow{\lambda}\ (\nu(u_1)\ \overrightarrow{v})$ for some $\overrightarrow{v}$,*
- *$u_1 = (r_i\ s_1\ ...\ s_m)$ and $\nu(r_i) = r_i$ is a descendent of its occurrence in t.*
- *For $1 \leq j \leq m$, s_j has no branch that is (t', a') useful*
- *u_1 has a branch that is (t', a') useful.*

Comments The intuition of the proof is the following : Since there is no useful branch in u the set of useful nodes in $BT(u)$ is (by König's lemma) finite. Assume, for example, that $t = (\lambda x \lambda y\ (x\ s_1\ s_2)\ r_1\ r_2)$. Then $t \twoheadrightarrow (r_1\ s_1'\ s_2')$. If there is no useful branch neither in s_1' nor in s_2' we are done. Otherwise there is such a useful branch in, say, s_1'. Thus $t \twoheadrightarrow \overrightarrow{\lambda}(s_1'\ \overrightarrow{w})$ for some $\overrightarrow{w}$. By the lemmas of the section 4.1 it is mainly enough to prove the result for s_1'. But $t' = (\lambda x \lambda y\ s_1\ r_1\ r_2) \twoheadrightarrow s_1'$ and the cardinality of the set of useful nodes of t' is smaller than the one of t. We get the result by repeating the previous argument.

Before giving the proof I give an example of the difficult case (the case 2.b in the proof). This is the example 4.3.6 in [1]. Let $w = \lambda xyz\ (y\ (x\ (z\ x)))$, $R = \lambda z\ (z\ I\ w)$ and $t = (w\ R\ I\ w)$. t is unsolvable. w, R, I are normal and so they do not have a branch that is (t, nil) useful. $t \twoheadrightarrow (I\ (R\ (w\ R))) \twoheadrightarrow (R\ (w\ R))$. We cannot choose the step $(I\ (R\ (w\ R)))$ and the argument I as the first element of the decoration for t since the unsolvability is already created (and "used") in $(R\ (w\ R))$. We will choose the next step $(R\ (w\ R))$ and the argument R because, at this step, the unsolvability is not yet created since R and $(w\ R)$ are solvable. Thus, here, the solution is : $k = 4, u_1 = (R\ (w\ R)), i = 1, \nu = Id$ and $\overrightarrow{v}$ is empty.

Proof. Let $E = \{b\ /\ b$ is an address accessible in u, that is (t, a) useful$\}$. Note that for b in $E, hnf(u, b) \neq \perp$ because otherwise b would be a branch in u that is (t, a) useful.

I define a procedure to construct the desired $< i, k, u_1, \nu >$ and a branch in u. This procedure halts (and I thus get the result) because otherwise this means we always are in the case (1) below and this procedure has constructed an infinite branch in u that is (t, a) useful and this is a contradiction. Note that I cannot use the fact that E is finite (and prove the result by induction on the cardinality of E). Intuitively this is actually the argument used but we cannot formalize it in this way. If E is infinite, by König's lemma, there is an infinite branch b such that for every i, $b \upharpoonright i \in E$ but (see the example after the definition 10) this does not imply that b is (t, a) useful.

nil clearly is in E. Let $hnf(u, nil) = \lambda x_1\ ...\ x_k\ (x\ w_1\ ...\ w_p)$, $j_0 = Min(k, n)$ and σ is given by $\sigma(x_j) = r_j$ for $j \leq j_0$. It is clear that $j_0 \geq 1$ because otherwise t reduces to $\overrightarrow{\lambda}(x\ \overrightarrow{w}\ \overrightarrow{r})$ and then u or some r_i would have a branch that is (t, a) useful.

1) Assume first that $x \notin \{x_1\ ...\ x_k\}$. Then $t \twoheadrightarrow \lambda x_{j_0+1}\ ...\ x_k\ (x\ \ \sigma(w_1)\ ...\ \sigma(w_p)$ $r_{j_0+1}\ ...\ r_n)$ and thus $a \neq nil$. Let $a = [i :: l]$. If $i > p$, there is a branch in r_i

that is (t, a) useful and this contradicts the hypothesis. Thus $i \leq p$. Let $u' = \lambda x_1 \ldots x_{j_0}\ w_i$. Then $t \twoheadrightarrow_a \sigma(w_i)$ and $(u'\ \ r_1 \ldots r_n) \twoheadrightarrow \sigma(w_i)$. The first node of the branch constructed by the procedure is i. Repeat the procedure (to get the other nodes) with $(u'\ \ r_1 \ldots r_n)$.
2) Assume that $x = x_i$. Then $t \twoheadrightarrow \lambda x_{j_0+1} .. x_k\ (r_i\ \ \sigma(w_1) .. \sigma(w_p)\ r_{j_0+1} .. r_n)$.

a) Assume first that for $1 \leq q \leq p$, $\sigma(w_q)$ has no branch that is (t, a) useful. Then $< i, j_0, u_1, Id >$ where $u_1 = (r_i\ \ \sigma(w_1) \ldots \sigma(w_p)\ r_{j_0} \ldots r_n)$ clearly satisfies the conclusion of the proposition.

b) Assume that, for some $1 \leq q \leq p$, $\sigma(w_q)$ has a branch that is (t, a) useful.

Claim

There is b in E and $j \leq j_0$ such that $hnf(u, b) = \overrightarrow{\lambda}(x_j\ \ s_1\ \ldots\ s_l)$ and $\sigma(hnf(u, b))$ has a branch that is (t, a) useful but no $\sigma(s_m)$ has such a branch.
Proof

Note that $adr(u, [q]) = w_q$. By the hypothesis, $[q]$ is in E. Let $hnf(u, [q]) = \overrightarrow{\lambda}(y\ s_1 \ldots s_l)$. If $y = x_j$ and no $\sigma(s_m)$ has a branch that is (t, a) useful, $b = [q]$ satisfies the conclusion of the claim. Otherwise some $\sigma(s_m)$ has a branch that is (t, a) useful. (*Proof :* If $y = x_j$ this is clear. If $y \notin \{x_1 \ldots x_k\}$, $\sigma(hnf(u, [q])) = \overrightarrow{\lambda}(y\ \sigma(s_1) \ldots \sigma(s_l))$ and this is again clear since a branch in $\sigma(hnf(u, [q]))$ is a branch in some $\sigma(s_m)$). We may repeat the argument with $b = [q :: m]$. If the claim fails we get in this way an infinite branch in u that is (t, a) useful. (Q.E.D. of the claim)

Let (b, j) be given by the claim. Let $t' = (c(nil, Id)\ r_1 \ldots r_n)$ and $f = \psi(u)$. $t' \twoheadrightarrow_{f,a} \overrightarrow{\lambda}(c(b, \tau)\ \overrightarrow{w})$ for some $\tau = \sigma \oplus \sigma'$ and thus $t \twoheadrightarrow_a \overrightarrow{\lambda}(\tau(adr(u, b))\ \overrightarrow{w})$. By the lemmas 5 and 8, there is a substitution τ' such that $\overrightarrow{\lambda}(\tau(adr(u, b))\ \overrightarrow{w}) \twoheadrightarrow \overrightarrow{\lambda}(\mu(x_j)\ \ \mu(\overrightarrow{s})\ \ \overrightarrow{v}) = Res(t, a, k)$ where $\mu = \tau \oplus \tau' = \sigma \oplus \sigma' \oplus \tau'$. Then, $< j, k, u_1, \sigma' \oplus \tau' >$ satisfies the conclusion of the proposition, where $u_1 = (r_j\ \overrightarrow{\sigma(s)}) = \sigma((x_j\ s_1 \ldots\ \ s_l))$.

Proposition 7. *Let $(d_n)_{n \geq 0}$ (resp. $(\overrightarrow{u_n})_{n \geq 0}$, $(\overrightarrow{v_n})_{n \geq 1}$, resp. $(a_n)_{n \geq 0}$, resp. $(\sigma_n)_{n \geq 1}$) be a sequence of λ terms (resp. be sequences of finite sequences of λ terms, resp. be a sequence of elements of A, resp. be a sequence of substitution). Assume that for every $n \geq 0$*

- *$t_n = (d_n\ \overrightarrow{u_n})$ and a_n is a branch in t_n.*
- *For some k_n, $Res(t_n, a_n, k_n) = \overrightarrow{\lambda_n}(\sigma_{n+1}(t_{n+1})\ \ \overrightarrow{v}_{n+1})$ and a_{n+1} is $(Res(t_n, a_n, k_n), Br(t_n, a_n, k_n))$ useful.*
- *d_{n+1} is the descendent of an element of the sequence $\overrightarrow{u_n}$*
- *$\sigma_{n+1}(d_{n+1}) = d_{n+1}$.*

  Then, there is an increasing sequence (τ_n) of substitutions such that the sequence $(\tau_n(d_n))$ is a decoration for $(t_0,\ a_0)$.

Proof. I construct (by induction on n) a sequence $< j_n, r_n, b_n, \tau_n >$ such that : $r_0 = t_0, j_0 = 0, \tau_0 = Id, b_0 = a_0$ and, for $n \geq 1$, $r_n = Res(r_0, b_0, j_n) = \overrightarrow{\lambda}(\tau_n(t_n)\ \overrightarrow{w_n}), b_n = Br(r_0, b_0, j_n), \tau_n(d_n) = \tau_{n-1}(d_n)$ and a_n is (r_n, b_n) useful. It is clear that the sequence (τ_n) satisfies the conclusion.

$t_n \twoheadrightarrow_{a_n} \overrightarrow{\lambda_n}(\sigma_{n+1}(t_{n+1})\ \overrightarrow{v_{n+1}})$. Since a_n is (r_n, b_n) useful and by the proposition 3, $r_n \twoheadrightarrow_{b_n} r'_n = \overrightarrow{\lambda}(\overrightarrow{\lambda}_n\ (\tau_n(\sigma_{n+1}(t_{n+1}))\ \tau_n(\overrightarrow{v_{n+1}}))\ \overrightarrow{w_n})$ for some τ_n and $\overrightarrow{w_n}$.

Clearly $r'_n \twoheadrightarrow \overrightarrow{\lambda'}\ (\tau_{n+1}(t_{n+1})\ \overrightarrow{w_{n+1}}) = Res(r_0, a_0, j_{n+1})$ for some $\overrightarrow{w_{n+1}}$ where $\tau_{n+1} = \tau_n \circ \sigma_{n+1} \oplus \mu_n$ and the domain of μ_n is included in the variables in $\overrightarrow{\lambda}_n$. Since d_{n+1} is the descendent of an element of $\overrightarrow{u_n}$, d_{n+1} is not affected by μ_n . Since, by the hypothesis, $\sigma_{n+1}(d_{n+1}) = d_{n+1}$, we have $\tau_{n+1}(d_{n+1}) = \tau_n(d_{n+1})$. Finally, again by the proposition 3, a_{n+1} is $(r_{n+1},\ b_{n+1})$ useful.

4.3 End of the proof of the theorem

Let t be a λ term and a be branch in t. The existence of a decoration is proved by induction on the complexity of t.

- If $t = \lambda x\ u$ or $t = (x \overrightarrow{r})$ the result follows immediately from the induction hypothesis.
- If $t = (u\ r_1 ...\ r_n)$ and there is, either in u or in some r_i, a branch that is (t, a) useful. For example, say b is such a branch in u. By the induction hypothesis there is a decoration of (u, b) and by the proposition 5 there is a decoration for (t, a).
- Otherwise $t = (u\ r_1 ...\ r_n)$ and there is no branch neither in u nor in any r_i that is (t, a) useful. Let $a_0 = a, d_0 = u$, $\overrightarrow{u_0} = r_1\ ...\ r_n, t_0 = (d_0\ \overrightarrow{u_0})$ and $\overrightarrow{v_0}$ be the empty sequence. By the proposition 6 there is $< i, k_0, t_1, \sigma >$ such that, letting $t' = Res(t_0, a_0, k_0)$ and $a' = Br(t_0, a_0, k_0)$:
  - $t' = \overrightarrow{\lambda}\ (\sigma(t_1)\ \ \overrightarrow{v_1})$, $t_1 = (r_i\ s_1\ ...\ s_m), \sigma(r_i) = r_i$ for some terms $s_1\ ...\ s_m\ \overrightarrow{v_1}$ and some substitution σ.
  - For $1 \leq j \leq m$, s_j has no branch that is (t', a') useful
  - t_1 has a branch a_1 that is (t', a') useful.

Let $d_1 = r_i$ and $\overrightarrow{u_1} = s_1\ ...\ s_m$. No s_j has a branch that is (t_1, a_1) useful since, otherwise, by the proposition 2 such a branch would be (t', a') useful. We may again use the proposition 6 with t_1 and the branch a_1. By repeating the same argument we get sequences satisfying the hypothesis of the proposition 7 and thus a decoration for t.

References

1. R. Kerth *"Isomorphisme et équivalence équationnelle entre modèles du λ calcul"* Ph.D. thesis Université Paris 7, 1995.
2. R. Kerth *"The interpretation of Unsolvable λ Terms in Models of Untyped λ Calculus"*. To appear in the JSL
3. R. Kerth *"On the Construction of Stable Models of Untyped λ Calculus"*. To appear in TCS
4. R. David & K. Nour *"Storage operators and directed λ calculus"*. JSL 60, n°4, 1054-1086, 1995.

Game Semantics for Untyped $\lambda\beta\eta$-Calculus*

Pietro Di Gianantonio, Gianluca Franco, and Furio Honsell

Dipartimento di Matematica e Informatica
Università di Udine, Italy.
{digianantonio,gfranco,honsell}@dimi.uniud.it

Abstract. We study extensional models of the untyped lambda calculus in the setting of game semantics. In particular, we show that, somewhat unexpectedly and contrary to what happens in ordinary categories of domains, all reflexive objects in the category of games $\mathcal{G}$, introduced by Abramsky, Jagadeesan and Malacaria, induce the same λ-theory. This is $\mathcal{H}^*$, the maximal theory induced already by the classical CPO model D_∞, introduced by Scott in 1969. This results indicates that the current notion of game carries a very specific bias towards *head reduction*.

Introduction

λ-theories are congruences over λ-terms, which extend pure β-conversion. Their interest lies in the fact that they correspond to the possible *operational* (*observational*) semantics of λ-calculus. Although researchers have mainly focused on only three such operational semantics, namely those given by head reduction, head lazy reduction or call-by-value reduction, the class of λ-theories is, in effect, unfathomly rich, see *e.g.* [6, 12, 11, 7] for interesting examples of this complexity. Brute force, purely syntactical techniques are usually extremely difficult to use in the study of λ-theories. Therefore, since the seminal work of Dana Scott on D_∞ in 1969 [16], semantical tools have been extensively investigated.

A large number of mathematical models for λ-calculus, arising from syntax-free constructions, have been introduced, since then, in various categories of domains (see e.g. [17, 8, 6, 10, 12, 5, 7]). And a rich host of *different* λ-theories now have a "fully abstract" syntax-free model, *i.e.* a model which induces precisely those identities which hold in the given theory. However, the denotational semantics supported by these models do not match all the possible operational semantics of λ-calculus.

For example, in most existing categories of domains, λ-models have too many functions, and hence many interesting λ-theories, such as those arising from observing termination under some natural sequential reduction strategies (see e.g. [11]), do not have fully abstract models [12, 5]. An example of such a strategy is the one which tries to reduce a term to a closed term. In the case of CPOs, the sequentiality embedded in these strategies clashes with the existence of Scott-continuous "parallel" functions. While, in the case of coherent spaces, and stable

* This work was partially supported by ESPRIT WG 21900-TYPES, MURST-97 "Tecniche formali ..." and TMR Network n. ERBFMRXCT980170-LINEAR

functions, the presence of so called "parasitic" functions, prevents other kinds of identities deriving from monotonicity.

In this paper we explore the methodology for giving denotational semantics based on games, recently introduced by Abramsky, Jagadeesan, Malacaria, and Hyland, Ong (see [3,14]). This methodology has been extremely successful in modeling sequential languages [3,15]. It should be reasonable to expect, therefore, that one could obtain fully abstract game models, at least for those λ-theories mentioned above, which escape domain models. Of course, the very fact that game semantics faithfully captures *sequentiality*, should suggest also that even game semantics is not rich enough to provide fully abstract models for *all* λ-theories. It is possible to show, in fact, that there are λ-theories where, say, the behavior of an unsolvable term, *i.e.* a term with no head normal form, is that of a "parallel function", which checks if at least one of its arguments evaluates to a fixed term.

Somewhat surprisingly, however, it turns out that *all* reflexive objects, *i.e.* extensional λ-models, in the standard category of games of [3], determine λ-models which have the *same* theory. This is the *well known* maximal λ-theory $\mathcal{H}^*$ [6], already induced by Scott's D_∞. We recall that, if M, N are closed λ-terms (*i.e.* $M, N \in \Lambda^0$), and *HNF* denotes the set of λ-terms which have a *head normal form*, then $M =_{\mathcal{H}^*} N$ if and only if

$$\forall C[\,]\,.\, C[M], C[N] \in \Lambda^0 \Longrightarrow (C[M] \in HNF \iff C[N] \in HNF)$$

Alternatively, this is the theory where two terms are equal if we cannot observe that head reduction terminates when one is placed in a given context, but does not terminate when the other is.

More specifically, in this paper we show that all reflexive objects in the Cartesian closed category of games $K_!(\mathcal{G})$ [3] determine λ-models which are isomorphic to models which can be constructed as special *non-initial* colimits in a category $\mathcal{G}^e$ of games and "embeddings", which mimics the traditional Scott's construction in CPOs and embedding-projection pairs. By extending the methodology of approximants originally introduced in [18,13,12] for the continuous case, to the setting of the game semantics, we study the fine structure of these models.

The paper [9] is a companion to the present one. Finitary logical descriptions of game models, in the spirit of [8,1], are introduced. The case of one of the models introduced in this paper is discussed in detail.

One can elaborate in various ways on the main result of this paper. In any case, we think that it shows that existing game semantics is more rigid than CPO semantics, which can model a very rich collection of λ-theories. Since the current notion of game appears to carry a very strong bias towards *head reduction*, a new notion of game seems to be necessary to model λ-theories different from $\mathcal{H}^*$.

The present paper is organized as follows. In section 1, we introduce the categories of games that we shall utilize, namely $\mathcal{G}$ and $K_!(\mathcal{G})$. In Section 2 we discuss initial and non-initial solutions of recursive game equations. In Section 3 we introduce the special class of extensional λ-models $\mathcal{D}^*$, and we prove that all reflexive objects in $K_!(\mathcal{G})$ determine models belonging to $\mathcal{D}^*$. In Section 4 we study the fine structure of the models in $\mathcal{D}^*$ and prove that such models induce the theory $\mathcal{H}^*$. In Section 5 we give some concrete examples of extensional game λ-models, including the model arising from applying Scott's trick [17] to the game setting. Final remarks and directions for future work appear in section 6.

We assume the reader familiar with the basic notions and definitions of λ-calculus, see e.g. [6]. For the benefit of a reader coming from the λ-calculus community, this paper is self-contained as far as the theory of games, however the reader can refer to [2–4,14] for more details on this topic.

The authors are grateful to Fabio Alessi, Samson Abramsky, and Marina Lenisa for useful discussions.

1 Categories of games

In this section, we introduce two categories of games. Both are introduced by Abramsky, Jadgadeesan and Malacaria in 1993 [3]. Notice however that for our purposes the machinery of "questions and answers" *i.e.* the bracketing condition, seems unnecessary. One can safely, and more simply, focus only on the full and faithful sub-category of this category consisting of all those games all whose moves are labeled as questions.

We begin by giving the basic definitions.

Definition 1 (Games). *A* game *has two participants: the* Player *and the* Opponent. *A game A is a quadruple* $(M_A, \lambda_A, P_A, \approx_A)$ *where:*

- M_A *is the set of* moves *of the game.*
- $\lambda_A : M_A \to \{O, P\} \times \{Q, A\}$ *is the* labeling *function: it tells us if a move is taken by the Opponent or by the Player, and if it is a Question or an Answer. We can decompose* λ_A *into* $\lambda_A^{OP} : M_A \to \{O, P\}$ *and* $\lambda_A^{QA} : M_A \to \{Q, A\}$ *and put* $\lambda_A = \langle \lambda_A^{OP}, \lambda_A^{QA} \rangle$. *We denote by* $^-$ *the function which exchanges Player and Opponent, i.e.* $\overline{O} = P$ *and* $\overline{P} = O$. *We also denote with* $\overline{\lambda_A^{OP}}$ *the function defined by* $\overline{\lambda_A^{OP}}(a) = \overline{\lambda_A^{OP}(a)}$. *Finally, we denote with* $\overline{\lambda_A}$ *the function* $\langle \overline{\lambda_A^{OP}}, \lambda_A^{QA} \rangle$.
- P_A *is a non-empty and prefix-closed subset of the set* $M_A^{\circledast}$ *(which will be written as* $P_A \subseteq^{nepref} M_A^{\circledast}$*), where* $M_A^{\circledast}$ *is the set of all sequences of moves which satisfy the following conditions:*
  - $s = at \Rightarrow \lambda_A(a) = OQ$
  - $(\forall i : 1 \leq i \leq |s|)[\lambda_A^{OP}(s_{i+1}) = \overline{\lambda_A^{OP}(s_i)}]$
  - $(\forall\, t \sqsubseteq s)[|t \upharpoonright M_A^A| \leq |t \upharpoonright M_A^Q|]$

  where M_A^A *and* M_A^Q *denote the subsets of game moves labeled respectively as Answers and as Questions,* $s \upharpoonright M$ *denotes the set of moves of* M *which appear in* s *and* $\sqsubseteq$ *is the substring relation.* P_A *is called the set of* positions *of the game A.*
- $\approx_A$ *is an equivalence relation on* P_A *which satisfies the following properties:*
  - $s \approx_A s' \Rightarrow |s| = |s'|$
  - $sa \approx_A s'a' \Rightarrow s \approx_A s'$
  - $s \approx_A s' \;\wedge\; sa \in P_A \Rightarrow (\exists a')[sa \approx_A s'a']$

In the above s, s', t and t' range over sequences of moves, while a, a', b and b' range over moves. The empty sequence is written ϵ.

Definition 2 (Strategies).
A strategy for the Player in a game A is a non-empty set $\sigma \subseteq P_A^{even}$ of positions of even length such that $\overline{\sigma} = \sigma \cup dom(\sigma)$ is prefix-closed, where $dom(\sigma) = \{t \in P_A^{odd} \mid (\exists a)[ta \in \sigma]\}$, and P_A^{odd} and P_A^{even} denote the sets of positions of odd and even length respectively.

In this paper we shall consider only *history-free* strategies, *i.e.* strategies which depend only on the *last* move by the Opponent.

Definition 3 (History-free strategies).
A strategy σ for a game A is history-free if it satisfies the following properties:
- $sab, tac \in \sigma \Rightarrow b = c$
- $sab, t \in \sigma, ta \in P_A \Rightarrow tab \in \sigma$.

Definition 4. *Let σ, τ be strategies for a game A, we write $\sigma \approx \tau$ if and only if:*
- $sab \in \sigma, s'a'b' \in \tau, sa \approx_A s'a' \Rightarrow sab \approx_A s'a'b'$
- $s \in \sigma, s' \in \tau, sa \approx_A s'a' \Rightarrow (\exists b)[sab \in \sigma]$ *iff* $(\exists b')[s'a'b' \in \tau]$.

The above relation on strategies is not an equivalence relation since it might lack reflexivity. If σ is a strategy for a game A such that $\sigma \approx \sigma$, we write $\sigma : A$.

Definition 5 (Tensor product).
Given games A and B the tensor product $A \otimes B$ is the game defined as follows:

- $M_{A\otimes B} = M_A + M_B$
- $\lambda_{A\otimes B} = [\lambda_A, \lambda_B]$
- $P_{A\otimes B} \subseteq M_{A\otimes B}^{\circledast}$ *is the set of positions, s, which satisfy the following:*
  i) the projections on each component (written as $s \upharpoonright A$ or $s \upharpoonright B$) are positions for the games A and B respectively;
  ii) every answer in s must be in the same component game as the corresponding question.
- $s \approx_{A\otimes B} s' \Longleftrightarrow s \upharpoonright A \approx_A s' \upharpoonright A, s \upharpoonright B \approx_B s' \upharpoonright B, (\forall i)[s_i \in M_A \Leftrightarrow s'_i \in M_A]$

Here $+$ denotes disjoint union of sets, that is $A + B = \{in_l(a) \mid a \in A\} \cup \{in_r(b) \mid b \in B\}$, and $[-,-]$ is the usual (unique) decomposition of a function defined on disjoint unions.

Definition 6 (Unit). *The unit element for the tensor product is given by the empty game $I = (\varnothing, \varnothing, \{\epsilon\}, \{(\epsilon, \epsilon)\})$.*

Definition 7 (Linear implication). *Given games A and B the compound game $A \multimap B$ is defined as follows:*

- $M_{A\multimap B} = M_A + M_B$
- $\lambda_{A\multimap B} = [\overline{\lambda_A}, \lambda_B]$

- $P_{A\otimes B} \subseteq M^{\circledast}_{A\otimes B}$ *is the set of positions, s, which satisfy:*
  i) the projections on each component are positions for the games A and B respectively;
  ii) every answer in s must be in the same component game as the corresponding question.
- $s \approx_{A\multimap B} s' \iff s \restriction A \approx_A s' \restriction A, s \restriction B \approx_B s' \restriction B, (\forall i)[s_i \in M_A \Leftrightarrow s'_i \in M_A]$

It is easy to see that in the "tensor game" only the Opponent can switch component, while in the "linear implication game" only the Player can switch.

Definition 8 (Exponential). *Given a game A the game !A is defined by:*

- $M_{!A} = \omega \times M_A = \sum_{i\in\omega} M_A$
- $\lambda_{!A}(\langle i, a\rangle) = \lambda_A(a)$
- $P_{!A} \subseteq M^{\circledast}_{!A}$ *is the set of positions, s, which satisfy the following conditions:*
  i) $(\forall i \in \omega)[s \restriction A_i \in P_{A_i}]$;
  ii) every answer in s is in the same index as the corresponding question.
- $s \approx_{!A} s' \iff \exists$ *a permutation of indexes* $\alpha \in S(\omega)$ *such that:*
  - $\pi_1^*(s) = \alpha^*(\pi_1^*(s'))$
  - $(\forall i \in \omega)[\pi_2^*(s \restriction \alpha(i)) \approx \pi_2^*(s \restriction i)]$

  where π_1 *and* π_2 *are the projections of* $\omega \times M_A$ *and* $s \restriction i$ *is an abbreviation of* $s \restriction A_i$.

One can easily see that the following definition is well posed and that the objects introduced in Definitions 5, 6 provide indeed a categorical tensor product and its unit.

Definition 9 (The category of games $\mathcal{G}$).
The category $\mathcal{G}$ has as objects games and as morphisms, between games A and B, the equivalence classes, for the relation $\approx_{A\multimap B}$, *of history-free strategies* $\sigma : A \multimap B$. *We denote the equivalence class of* σ *by* $[\sigma]$.

The identity for each game A is given by the (equivalence class) of the copycat *strategy* $id_A = \{s \in P_{A'\multimap A''} \mid s \restriction A' = s \restriction A''\}$ *where the superscripts are introduced to distinguish between the two different occurrences of the game A.*

Composition is given by the extension on equivalence classes of the following composition of strategies. Given strategies $\sigma : A \multimap B$ *and* $\tau : B \multimap C$, $\tau \circ \sigma : A \multimap C$ *is defined by*
$\tau\circ\sigma = \{s \restriction (A, C) \mid s \in (M_A + M_B + M_C)^* \wedge s \restriction (A, B) \in \overline{\sigma}, s \restriction (B, C) \in \overline{\tau}\}^{even}$

Throughout this paper, without loss of generality, we shall restrict ourselves to "irredundant" games, *i.e.* to games such that every move appears in at least one position. Any redundant game is in fact categorically isomorphic to an irredundant one.

One can easily see that the constructions introduced in Definitions 5, 7 and 8 can be made to be functorial. Thus the category $\mathcal{G}$ is a monoidal closed category [3], which however is not Cartesian closed.

Definition 10 (A Cartesian closed category of games). *The category $K_!(\mathcal{G})$ is the category obtained by taking the co-Kleisli category over $\mathcal{G}$ over the co-monad $(!, \mathtt{der}, \delta)$ [3], where for each game A the strategies $\mathtt{der}_A : !A \multimap A$ and $\delta_A : !A \multimap !!A$ are defined as follows:*
- $\mathtt{der}_A = [\{s \in P_{!A \multimap A} \mid s \upharpoonright (!A)_0 = s \upharpoonright A\}]$
- $\delta_A = [\{s \in P_{!A \multimap !!A} \mid s \upharpoonright (!A)_{p(i,j)} = s \upharpoonright (!(!A)_i)_j\}]$

where $p : \mathbb{N} \times \mathbb{N} \to \mathbb{N}$ is a pairing function.

The category $K_!(\mathcal{G})$ has as objects games and as morphisms between games A and B the equivalence classes of history-free strategies for the game $!A \multimap B$. Moreover it is Cartesian closed.

Definition 11 (Cartesian product).
Given games A and B the Cartesian product $A\&B$ is the game defined as follows:
- $M_{A\&B} = M_A + M_B$
- $\lambda_{A\&B} = [\lambda_A, \lambda_B]$
- $P_{A\&B} = P_A + P_B$
- $\approx_{A\&B} = \approx_A + \approx_B$.

1.1 Order-enrichment

Following [3] we can enrich each homset of $\mathcal{G}$ with a partial order structure:

Definition 12. *Given a game A, and strategies $\sigma : A$ and $\tau : A$ we write $\sigma \lesssim \tau$ iff*
$(\forall s, s', a, b, a')[sab \in \sigma \wedge \ s' \in \tau \wedge \ sa \approx s'a' \implies \exists b'.(s'a'b' \in \tau \ \wedge \ sab \approx s'a'b')]$,
and we define $[\sigma] \sqsubseteq_A [\tau] \iff \sigma \lesssim \tau$.

Given a game A let $\hat{A}$ be the set of equivalence classes of history-free strategies for A. $\sqsubseteq_A$ is a partial order over $\hat{A}$, whose least element is $[\{\epsilon\}]$. Notice that $\hat{\mathfrak{A}} \simeq \mathcal{G}(I, \mathfrak{A})$.

We now prove that this partial order is not complete. This answers a question raised in [3] page 21.

Definition 13 (Game $\mathfrak{N}$). *The game $\mathfrak{N}$ is defined as follows:*
- $M_{\mathfrak{N}} = \{q, !\} \cup \{n, \overline{n} \mid n \in \mathbb{N}\}$
- $\lambda_{\mathfrak{N}}(q) = \lambda_{\mathfrak{N}}(\overline{n}) = OQ \quad and \quad \lambda_{\mathfrak{N}}(!) = \lambda_{\mathfrak{N}}(n) = PQ$
- $P_{\mathfrak{N}} = \{qn(\overline{n-1})(n-1)(\overline{n-2})\ldots\overline{0}0q!q!q!\ldots \mid n \in \mathbb{N}\}^{nepref}$
- $s \approx_{\mathfrak{N}} t \Leftrightarrow |s| = |t|$.

Theorem 1. *$(\hat{\mathfrak{N}}, \sqsubseteq_{\mathfrak{N}})$ is not a complete partial order.*

Proof. *Consider the following strategies indexed by $n \geq 1$:*
$\sigma_n = \{qn(\overline{n-1})(n-1)(\overline{n-2})\ldots 1\}^{nepref}$.
It is easy to check that $\sigma_n \lesssim \sigma_m$ for $n \leq m$. The chain $[\sigma_0], [\sigma_1], \ldots, [\sigma_n], \ldots$ has no lub, since there is no infinite history-free strategy in $\mathfrak{N}$. □

Corollary 1. *The categories $\mathcal{G}$ and $K_!(\mathcal{G})$ are not cpo-enriched categories under the order relation on morphisms of Definition 12.*

2 Solution of recursive games equations

The categories of games $\mathcal{G}$ and $K_!(\mathcal{G})$ allow for the existence of *recursive* objects, *i.e.* objects that are fixed points of particular functors. In this section we analyze and elaborate the method proposed by Abramsky and McCusker in [4], for defining recursive games. In a well-founded setting, this method allows to define only *initial* fixed points of functors. However in order to model non-trivially $\lambda\beta\eta$-calculus, it is well known that we need to define models which arise from *non*-initial fixed points. To this end we have to change the functor altogether and use some form of encoding or, equivalently, generalize the method of [4] and consider games "up to" isomorphisms, or consider *non-well-founded* sets. In this section we shall explore the first two alternatives.

2.1 Initial fixed points

We start by discussing briefly the method of Abramsky and McCusker [4] in a well-founded setting. This method follows the pattern used for building initial fixed points in the context of information systems. First a complete partial order $\trianglelefteq$ on games is introduced.

Definition 14. *Let A, B be games, A is a* sub-game *of B ($A \trianglelefteq B$) iff*
- $M_A \subseteq M_B$;
- $\lambda_A = \lambda_B \restriction M_A$;
- $P_A = P_B \cap M_A^{\circledast}$;
- $s \approx_A s'$ *iff* $s \approx_B s'$ *and* $s \in P_A$.

One can easily see that the sub-game relation defines a complete partial order on games. Hence a functor F which is continuous with respect to $\trianglelefteq$ has a (minimal) fixed point $D = F(D)$ given by $\bigsqcup_{\trianglelefteq} F^n(I)$. Notice that we have indeed an identity between D and $F(D)$.

In domain theory, non-initial fixed points for a functor F are usually obtained by carrying out the above construction starting from some object A, different from the initial one (*i.e.* I in this case), such that $A \trianglelefteq F(A)$. However one can prove that for functors F obtained from constant functors by composition of the basic functors $\&$, $\otimes$, $\multimap$, $(\ )_\perp$, and $!$, and for every game A, whose moves are well-founded sets, if $A \trianglelefteq F(A)$ then $\exists n \in \mathbb{N}$ s.t. $A \trianglelefteq F^n(I)$. Hence *only* initial fixed points can be obtained using this technique in well-founded Set Theory.

As remarked earlier, even if no non-trivial model of $\lambda\beta\eta$-calculus can be obtained applying this technique directly to the functor $!D \multimap D$, nevertheless using Scott's trick (see [17]) we can still define models of $\lambda\beta\eta$. What we need is a non-trivial game which satisfies the equivalence $D \simeq D\&D$. To see this consider the initial fixed point, E, of the functor $F(X) = X \to D$ in a general Cartesian closed category. This is clearly non trivial. One can easily see that the following chain of equivalences holds $E = E \to D = (E \to D) \to D \simeq (E \to (D \times D)) \to D \simeq ((E \to D) \times (E \to D)) \to D = ((E \to D) \times E) \to D \simeq (E \to D) \to (E \to D) = E \to E$. We shall present this model in Section 5.

2.2 Non-initial fixed points

In order to obtain a non-initial fixed point of a functor, without having to deal with the subtleties of non-well-founded sets, or with indirect encodings, we present a generalization of the method proposed in [4], "up to isomorphism".

The basic idea is to obtain a fixed point of a functor F as a limit of a chain of approximations $D_0, D_1, D_2, \ldots$ where, not necessarily $D_n \trianglelefteq D_{n+1}$, but only a weaker relation between D_n and D_{n+1} holds. We simply ask that each D_n is isomorphic to a sub-game B of D_{n+1}. In order to formalize our construction we need to introduce a new category $\mathcal{G}^e$. A similar category was introduced also in [2] for other purposes.

Definition 15. *Given games A and B an* embedding $f : A \rightarrowtail B$ *is a total injective function $f : M_A \to M_B$ such that:*
- $\lambda_A = \lambda_B \circ f$
- $f^*(P_A) = P_B \cap (f^*(M_A))^{\circledast}$
- $s \approx_A s'$ *iff* $f^*(s) \approx_B f^*(s')$

In the above we have used the notation f^* to denote the natural extension of f both to sequences and sets of sequences.

Definition 16. *The category of games $\mathcal{G}^e$ has as objects games and as morphisms embeddings.*

Proposition 1. *The category $\mathcal{G}^e$ is ω-cocomplete.*

Proof. *Given an ω-chain $\langle D_n, f_n\rangle$ with $f_n : D_n \rightarrowtail D_{n+1}$ its colimit is $\langle D_\infty, \mu_n\rangle$ where D_∞ is the game:*
- $M_{D_\infty} = (\bigcup_{n\in\omega} M_{D_n})/_{\equiv}$, *where $\equiv$ is the least equivalence relation such that* $\forall n \in \mathbb{N}\ \forall a \in D_n\ \forall b \in D_{n+1}.\ f_n(a) = b \Rightarrow a \equiv b.$
- $\lambda_{D_\infty}([a]_\equiv) = \lambda_{D_n}(a)$ *if* $a \in D_n$
- $P_{D_\infty} = \bigcup_{n\in\omega}\{[a_1]_\equiv [a_2]_\equiv \ldots [a_p]_\equiv \mid a_1 a_2 \ldots a_p \in P_{D_n}\}$
- $\approx_{D_\infty} = \bigcup_{n\in\omega} = \{([a_1]_\equiv [a_2]_\equiv \ldots [a_p]_\equiv, [a'_1]_\equiv [a'_2]_\equiv \ldots [a'_p]_\equiv) \mid (a_1 a_2 \ldots a_p, a'_1 a'_2 \ldots a'_p) \in \approx_{D_n}\}$.

The colimit functions $\mu_n : D_n \rightarrowtail D_\infty$ are defined by $\mu_n(a) = [a]_\equiv$. □

Each embedding $f : A \rightarrowtail B$ in $\mathcal{G}^e$ induces two morphisms $f^+ : A \multimap B$ and $f_- : B \multimap A$ in $\mathcal{G}$ defined as follows.

Definition 17. *Given an embedding $f : A \rightarrowtail B$, put:*

$$f^+ = \{t \in P_{A\multimap B} \mid t \in s_f\}$$
$$f_- = \{t' \in P_{B\multimap A} \mid t' \in s_f\}$$

where s_f is the least set satisfying:

$$s_f = \{t\, a\, f(a) \mid t \in s_f,\ a \in M_A\} \cup \{t' f(a)\, a \mid t' \in s_f,\ a \in M_A\} \cup \{\epsilon\}.$$

One can easily see that $(g \circ f)^+ = g^+ \circ f^+$ and $(g \circ f)_- = f_- \circ g_-$. The category $\mathcal{G}^e$ is indeed isomorphic to a sub-category of $\mathcal{G}$ and to a sub-category of $\mathcal{G}^{op}$. Now, using the well-known machinery, we can obtain fixed points of any continuous functor F in $\mathcal{G}^e$.

Theorem 2. *Given a game D and an embedding $f : D \rightarrowtail F(D)$, let $\langle D_\infty, \mu_n \rangle_{n\in\omega}$ be the colimit of the chain $\langle (F)^n(D), (F)^n(f) \rangle_{n\in\omega}$. Then, the game D_∞ is the fixed point of the functor F. The isomorphic embeddings $\varphi : D_\infty \rightarrowtail F(D_\infty)$ and $\psi : F(D_\infty) \rightarrowtail D_\infty$ are given by $\varphi = \bigsqcup_{n\in\omega} F(\mu_n) \circ \mu_n^{-1}$ and $\psi = \bigsqcup_{n\in\omega} \mu_n \circ F(\mu_n)^{-1}$, where the lubs are taken in the category of* partial *embeddings.*

Proposition 2. *Given a game D and an embedding $f : D \rightarrowtail F(D)$ let $\langle D_\infty, \mu_n \rangle_{n\in\omega}$ be the fixed point of the functor F. For each $n \in \mathbb{N}$ let $p_n : D_\infty \multimap D_\infty = (\mu_n)^+ \circ (\mu_n)_-$. Then for each game A and for each strategy $\sigma : A \multimap D_\infty$, we have that $p_n \circ \sigma = \{s \mid s \in (\sigma \cap (in_l(M_A) \cup in_r(\mu_n(M_{F^n(D)})))^{\circledast})\}$. moreover for each $n \in \mathbb{N}$:*
- $p_n \sqsubseteq p_{n+1}$
- $\bigsqcup_{n\in\omega} p_n = id$
- $p_n \circ p_m = p_{min\{m,n\}}$.

Using the above machinery, given an endofunctor F in $\mathcal{G}$ (either variant or covariant), one can obtain a fixed point of F provided there exists a covariant continuous functor F^e in $\mathcal{G}^e$, which coincides with F on objects.

One can easily see that this is the case for constant functors, the functors &, $\otimes$, $\multimap$, $!$, $(\ )_\perp$ and their compositions.

3 Extensional λ-models in $K_!(\mathcal{G})$

As it is well known, a model for $\lambda\beta\eta$-calculus is a pair $\mathsf{D} = \langle D, f \rangle$, where D is an *extensional reflexive object* in a Cartesian closed category, *i.e.* an object D such that D isomorphic to $D \to D$, and $f : D \to [D \to D]$ is an isomorphism. Two models $\mathsf{D} = \langle D, f \rangle$, $\mathsf{D}' = \langle D', f' \rangle$ are isomorphic if there exists an isomorphism $g : D \to D'$ such that $f' \circ g = [g^{-1} \to g] \circ f$. This implies that the two "applicative structures" are the same, *i.e.* for each $\sigma, \tau : A \to D$ we have that $g \circ ev \circ \langle f \circ \sigma, \tau \rangle = ev \circ \langle f' \circ (g \circ \sigma), g \circ \tau \rangle$.

In this section, using the techniques outlined in Section 2, we define a subclass, $\mathcal{D}^*$, of extensional models in $K_!(\mathcal{G})$, and prove the crucial result, namely, that each extensional model in $K_!(\mathcal{G})$ is isomorphic to a model in $\mathcal{D}^*$. In Section 4 we will prove that all models in $\mathcal{D}^*$ induce the λ-theory $\mathcal{H}^*$.

The endofunctor *Fun* on the category $\mathcal{G}^e$ is defined by putting:
- $Fun(D) = [D \to D] = (!D \multimap D)$
- $Fun(f) = [!f, f]$, for $f : A \rightarrowtail B$, where $!f(\langle i, a \rangle) = \langle i, f(a) \rangle$.

One can easily see that *Fun* is continuous.

Definition 18. *Let $\mathcal{D}^*$ be the class of λ-models $\mathsf{D} = \langle D, f \rangle$ where D is the limit of a chain generated by iterating the functor Fun on an initial game D_0, using an initial embedding $f^* : D_0 \rightarrowtail Fun(D_0)$, such that for each $m \in M_{D_0}$, $f^*(m) = in_r(m')$ for some $m' \in M_{D_0}$. And where the isomorphism $f : D \to Fun(D)$ in $K_!(\mathcal{G})$ is $\varphi^+ \circ \mathbf{der}_D$, where φ is the isomorphic embedding given by the colimit construction.*

Isomorphisms in $K_!(\mathcal{G})$ can be reduced to isomorphisms in $\mathcal{G}^e$:

Proposition 3. *For each isomorphism $\sigma : A \to B$ in $K_!(\mathcal{G})$, there exists an isomorphic strategy $\sigma' : A \multimap B$ such that $\sigma = \sigma' \circ \mathtt{der}_A$. And, for each isomorphic strategy $\sigma : A \multimap B$, there exists an isomorphic embedding $f_\sigma : A \rightarrowtail B$ such that $\sigma = (f_\sigma)^+$.*

Proof. *The proof of the first part is straightforward. In order to show the second part, recall from [3] that an history-free strategy $\sigma : A \multimap B$ can be described as a map g_σ from Opponent's moves to Player's moves in the game $A \multimap B$. If σ is an isomorphism, with inverse σ^{-1} then g_σ maps each Player's move of A in a Player's move of B, and each Opponent's move in B to a Opponent's move in A. In fact, suppose by contradiction, that an Opponent's move $b \in M_B$ is such that $g_\sigma(b) = b'$ is a Player's move in B, then $g_{\sigma^{-1}\circ\sigma}(b) = b'$ and therefore $\sigma^{-1} \circ \sigma$ is not the copy-cat strategy. By a similar argument one can prove that g_σ and $g_{\sigma^{-1}}$ are one the inverse of the other. The function f_σ is then defined by*

$$f_\sigma(a) = \begin{cases} g_{\sigma(a)} & \text{if } \lambda_B^{OP}(a) = O \\ g_{\sigma^{-1}}(a) & \text{if } \lambda_A^{OP}(a) = P \end{cases}$$

By an analysis similar to the one above, and using the bracketing condition it is possible to prove that f_σ preserves the labeling and that $f_\sigma^*(P_A) = P_B$. □

In order to establish the main result of this section, we need a new definition, and prove a technical lemma.

Definition 19. *Given a game A and a move $a \in M_A$, the rank of a, $r(a)$, is the smallest integer n such that there exists a sequence of moves $a_1, \ldots, a_n$ such that $a_1, \ldots, a_n, a \in P_A$.*

Lemma 1. *For each game A, for each embedding $f : A \rightarrowtail \mathit{Fun}(A)$ and for each move $a \in M_A$, if $f(a) = in_l(\langle n, a'\rangle)$ then $r(a') < r(a)$; if $f(a) = in_r(a')$ then $r(a') \leq r(a)$.*

Proof. *Let s_a be a minimal position with end point a. The projection of $f^*(s_a)$ on the left component must still be a position in $P_{!A}$. Its length is strictly smaller than that of s_a, since the initial move of s_a has to be mapped onto a move on the right component.* □

Theorem 3. *Each extensional model in $K_!(\mathcal{G})$ is isomorphic to a model in $\mathcal{D}^*$.*

Proof. *Let $\langle D, \sigma\rangle$ be an extensional model in $K_!(\mathcal{G})$. Then, by Proposition 3 there exists an isomorphic embedding $f : D \rightarrowtail \mathit{Fun}(D)$, such that $\sigma = f^+ \circ \mathtt{der}_A$. Let M_{D_0} be the largest subset of M_D such that $\forall d \in M_{D_0}\ \exists d' \in M_{D_0}$ such that $f(d) = in_r(d')$. Alternatively, with a slight abuse of notation, we can define $M_{D_0} = \{d \in D \mid \forall n \in \mathbb{N}\ .\ ({in_r}^{-1} \circ f)^n(d) \text{ is defined}\}$.*

It is immediate to verify that the quadruple $D_0 = (M_{D_0}, \lambda_D \restriction M_{D_0}, P_D \cap M_{D_0}^{\circledast}, \approx_D \cap (M_{D_0} \times M_{D_0})\ )$ is indeed a sub-game of D. By the construction of D_0, it follows that $f_0 = f_{|D_0}$ is an embedding from D_0 to $\mathit{Fun}(D_0)$.

Let D^* be the limit of the ω-chain $\langle \mathit{Fun}^n(D_0), \mathit{Fun}^n(f_0)\rangle_{n\in\mathbb{N}}$, and let $f^* : D^* \rightarrowtail \mathit{Fun}(D^*)$ be the isomorphic embedding induced by the limit construction.

We will prove that there exists an isomorphic embedding $f' : D^* \rightarrowtail D$ such that $f \circ f' = Fun(f') \circ f^*$.

The isomorphism f' is defined as follows: given $d \in Fun^n(D_0)$ $f'([d]_\equiv) = f_{0,n}^{-1}(d)$, where $f_{0,n} : D \rightarrowtail Fun^n(D)$ is the isomorphism $Fun^{n-1}(f) \circ \ldots \circ Fun(f) \circ f$. Since, for each $n \in \mathbb{N}$, $M_{Fun^n(D_0)} \subseteq M_{Fun^n(D)}$, f' is a well defined function from M_{D^*} to M_D. Moreover it is not difficult to verify that f' is an embedding.

We need to prove that f' is surjective. This can be done by induction on the rank of the moves in D. Formally, we will prove that for each move $d \in M_D$ there exists a move d' in M_{D^*} such that $d = f'(d')$.

- *Basic step. This follows from the fact that all initial moves (i.e. moves of rank 0) are in M_{D_0}.*
- *Induction step. Let $d \in M_D$ be a move of rank $n + 1$, two possible cases arise. Either $d \in D_0$, and therefore $d = f'([d]_\equiv)$, or there exist $p, i \in \mathbb{N}$ and $d' \in M_D$ such that $\forall m \leq p, ({in_r}^{-1} \circ f)^m(d)$ is defined and $f(({in_r}^{-1} \circ f)^p(d)) = in_l(\langle i, d' \rangle)$. By Lemma 1 the rank of d' is less than $n + 1$ and, hence, by induction hypothesis, there exists $k \in \mathbb{N}$ and $d'' \in Fun^k(D_0)$ such that $d' = f'([d'']_\equiv)$. Let $d''' = {in_r}^p \circ in_l(\langle i, d'' \rangle) \in Fun^{k+p}(D_0)$, it is not difficult to verify that $d = f'([d''']_\equiv)$.*

Moreover, it is straightforward to verify that $f \circ f' = Fun(f') \circ f^*$, and from the fact that: $[(f^+ \circ \mathbf{der}_D)^{-1} \to (f^+ \circ \mathbf{der}_D)] = Fun(f)^+ \circ \mathbf{der}_{D \to D}$, the theorem follows straightforwardly. □

4 The fine structure of models in $\mathcal{D}^*$

In order to analyze the equational theories induced by the models in $\mathcal{D}^*$, we establish an *Approximation Theorem*, in the style of [18, 12]. Using this result we will be able to characterize the meaning of a term in the model as the lub of the set of the meanings of the syntactical *approximants* of the term.

To our knowledge this is the first time such a theorem is proved for models in "non-concrete" categories such as game models.

As usual it is convenient to consider $\Lambda(\Omega)$, an extension of λ-calculus with a constant to denote divergence, and its indexed version $\Lambda(\Omega)^{\mathbb{N}}$.

Definition 20. 1. *The set of $\lambda\Omega$-terms, $\Lambda(\Omega)(\ni M)$ is defined from a set of variables $Var(\ni x)$ as follows: $M ::= x \mid MM \mid \lambda x.M \mid \Omega$.*
2. *The set of (possibly)* indexed terms *$\Lambda(\Omega)^{\mathbb{N}}(\ni M)$ is the superset of $\Lambda(\Omega)$ defined as follows: $M ::= x \mid MM \mid \lambda x.M \mid \Omega \mid M^n$.*
3. *A term is* truly indexed *if it is of the shape M^n. A term is* completely indexed *if all its subterms of the shape constant, variable, abstraction, and application are immediate subterms of truly indexed terms.*

The intended meaning of an indexed term M^n is the n-th projection of the interpretation of the term M. Hence we give:

Definition 21. *Let $D = \langle D, \varphi \rangle$ be in $\mathcal{D}^*$. The interpretation of a term $M \in \Lambda(\Omega)^{\mathbb{N}}$ (whose free variables are among the list $\Delta = \{x_1, \ldots, x_n\}$) in the model is the strategy $[\![M]\!]^{D}_{\Delta} : !(\overbrace{D \,\&\ldots\&\, D}^{|\Delta|}) \multimap D$ defined inductively as follows:*
- $[\![x_i]\!]^{D}_{\Delta} = \pi^{\Delta}_{i}$;
- $[\![MN]\!]^{D}_{\Delta} = ev \circ \langle (\varphi \circ [\![M]\!]^{D}_{\Delta}) \ , [\![N]\!]^{D}_{\Delta} \rangle$;
- $[\![\lambda x.M]\!]^{D}_{\Delta} = \psi \circ \Lambda(\ [\![M]\!]^{D}_{\Delta,x})$;
- $[\![M^n]\!]^{D}_{\Delta} = p_n \circ [\![M]\!]^{D}_{\Delta}$;
- $[\![\Omega]\!]^{D}_{\Delta} = \sigma_\epsilon$;

where π^{Δ}_{i} are the canonical projection morphisms, ev and Λ denote "evaluation" and "abstraction" in the Cartesian closed category $K_!(\mathcal{G})$, $\sigma_\epsilon = [\{\epsilon\}]$, $\psi = \varphi^{-1}$ and the p_n are the strategies defined in Proposition 2.

Given strategies σ, τ with codomain D, we use the abbreviation $\sigma \cdot \tau$ to denote the strategy $ev \circ \langle (\varphi \circ \sigma), \ \tau \rangle$, and we will denote with $(D)^n$ the game $\overbrace{D \,\&\ldots\&\, D}^{n}$.

The main result of this section is Theorem 6. In order to establish it we need several preliminary results.

Lemma 2. *For each model $D = \langle D^*, \varphi \rangle$ in $\mathcal{D}^*$, for each game A and pair of strategies $\sigma, \tau : !A \multimap D^*$, we have:*
- $(p_0 \circ \sigma) \cdot \tau = (p_0 \circ \sigma) \cdot \sigma_\epsilon = p_0 \circ (\sigma \cdot \sigma_\epsilon)$
- $(p_{n+1} \circ \sigma) \cdot \tau \sqsubseteq p_{n+1} \circ (\sigma \cdot (p_n \circ \tau)) \ \forall n \in \mathbb{N}$.

Notice that in the statement of Lemma 2.2, we have not taken equality but only inequality. This is done in order to be able to deal simultaneously not only with models in $\mathcal{D}^*$, but also, in Section 5, with models obtained using the trick of Scott outlined in Section 2.

The following Lemmata and Definitions follow closely the pattern of [18, 12], and they amount essentially to the game theoretic version of the corresponding "continuous result".

Definition 22. *The erasing function $\mathcal{R} : \Lambda(\Omega)^{\mathbb{N}} \to \Lambda(\Omega)$ is inductively defined as follows: $\mathcal{R}(x) = x$; $\mathcal{R}(\Omega) = \Omega$; $\mathcal{R}(PQ) = \mathcal{R}(P)\mathcal{R}(Q)$, $\mathcal{R}(\lambda x.P) = \lambda x.\mathcal{R}(P)$, $\mathcal{R}(M^n) = \mathcal{R}(M)$.*

Lemma 3. *For each model $D = \langle D^*, \varphi \rangle$ in $\mathcal{D}^*$, for each term $M \in \Lambda(\Omega)$ whose free variables are in Δ, given a finite strategy $\sigma : !(D^*)^{|\Delta|} \multimap D^*$ s.t. $\sigma \sqsubseteq [\![M]\!]^{D}_{\Delta}$ there exists a natural number n s.t. $\sigma \sqsubseteq [\![M^n]\!]^{D}_{\Delta}$.*

Lemma 4. *For each model $D = \langle D^*, \varphi \rangle$ in $\mathcal{D}^*$, for each term $M \in \Lambda(\Omega)$ whose free variables are in Δ, given a finite strategy $\sigma : !(D^*)^{|\Delta|} \multimap D^*$ s.t. $\sigma \sqsubseteq [\![M]\!]^{D}_{\Delta}$ there exists a completely indexed term $Q \in \Lambda(\Omega)^{\mathbb{N}}$ such that $\mathcal{R}(Q) = M$ and $\sigma \sqsubseteq [\![Q]\!]^{D}_{\Delta}$.*

Lemma 5. *Let $\sigma : A$, then $\sigma = \bigsqcup\{\tau : A \mid \tau$ finite and $\tau \sqsubseteq \sigma\}$.*

Proposition 4. *For each model* $D = \langle D^*, \varphi \rangle$ *in* $\mathcal{D}^*$, *for each term* $M \in \Lambda$, $[\![M]\!]^D_\Delta = \bigsqcup\{[\![Q]\!]^D_\Delta \mid Q$ *is a completely indexed term s.t.* $\mathcal{R}(Q) = M\}$.

Definition 23. *The following reduction rules are definable on* $\Lambda(\Omega)$:

$$(\Omega_1)\ \lambda x.\Omega \to \Omega \qquad\qquad (\Omega_2)\ \Omega M \to \Omega.$$

The following reductions are definable on completely indexed terms of $\Lambda(\Omega)^{\mathrm{N}}$:

$$(\Omega_1^n)\ \lambda x.\Omega^n \to \Omega^0$$
$$(\Omega_2^n)\ \Omega^n M \to \Omega^0$$
$$(\beta_I)\ ((\lambda x.P^n)^{m+1} Q^p)^h \to (P[x/Q^a])^b$$

where $b = min\{n, m+1, h\}, a = min\{m, p\}$

$$(\beta_0)\ ((\lambda x.P)^0 Q)^h \to (P[x/\Omega])^0$$
$$(\beta_{i,j})\ (M^i)^j \to M^{min\{i,j\}}.$$

Notice again that the above definition of the (β_I) indexed reduction rule and the statement of the following Theorem are not formulated as in [18], but are relaxed so as to take care of the model D^{N} (see Section 5).

Theorem 4 (Validity of indexed reduction). *For each model* $D = \langle D^*, \varphi \rangle$ *in* $\mathcal{D}^*$, *the rules* $(\Omega_1^n), (\Omega_2^n), (\beta_I), (\beta_0)$ *and* $(\beta_{i,j})$ *are valid in the following sense: let* $P, Q \in \Lambda(\Omega)^{\mathrm{N}}$ *then:* $(P \twoheadrightarrow_{\Omega_1^n \Omega_2^n \beta_0 \beta_I \beta_{i,j}} Q) \Longrightarrow [\![P]\!]^D_\Delta \sqsubseteq [\![Q]\!]^D_\Delta$.

Lemma 6. *A completely indexed term* Q *is* $\Omega_1^n \Omega_2^n \beta_0 \beta_I \beta_{i,j}$*-normalizing.*

Lemma 7. *For each model* $D = \langle D^*, \varphi \rangle$ *in* $\mathcal{D}^*$, *for each term* $M \in \Lambda$, $[\![M]\!]^D_\Delta = \bigsqcup\{[\![N]\!]^D_\Delta \mid \exists Q$ *completely indexed term such that* $\mathcal{R}(Q) = M$ *and* N *is the* $\Omega_1^n \Omega_2^n \beta_0 \beta_I \beta_{i,j}$*-normal form of* $Q\}$.

Definition 24. *The* direct approximant *of a* λ*-term* $M \in \Lambda$ *is a normal form* $A \in \Lambda(\Omega)$ *obtained from* M *by replacing each redex in* M *by* Ω, *and performing all the* $\Omega_1^n \Omega_2^n$*-reductions.*

Definition 25. *The set of* approximants *of* M *is the set* $\mathcal{A}(M) = \{A \mid \exists M'$, $M \twoheadrightarrow_{\beta\eta} M'$ *and* A *is the direct approximant of* $M'\}$.

Theorem 5 (Approximation theorem). *For each model* $D = \langle D^*, \varphi \rangle$ *in* $\mathcal{D}^*$, *for each term* $M \in \Lambda$, $[\![M]\!]^D_\Delta = \bigsqcup\{[\![A]\!]^D_\Delta \mid A \in \mathcal{A}(M)\}$.

Theorem 6. *For each model* $D = \langle D^*, \varphi \rangle$ *in* $\mathcal{D}^*$, $Th(D) = \mathcal{H}^*$.

Proof. (sketch) Using theorems 4 and 5, the standard argument for the continuous case (see e.g. [6] Sec. 19.2) can be mimicked in the game setting. □

5 Examples of game models for $\lambda\beta\eta$-calculus

We introduce four extensional λ-models in $K_!(\mathcal{G})$. The first three are defined using Theorem 2. The first two belong to $\mathcal{D}^*$, while the third does not. The fourth is the model obtained by Scott's trick as outlined in Section 2.

Definition 26. *1. Let $D_0^\circ = (\{*\}, \lambda(*) = OQ, \{\epsilon, *\}, id)$ and define $f_\circ : D_0^\circ \rightarrowtail (!D_0^\circ \multimap D_0^\circ)$ by $f_\circ(*) = in_r(*)$;*
2. let $D_0^{*\circ} = (\{*, \circ\}, (\lambda(*) = OQ, \lambda(\circ) = PQ), \{\epsilon, *, *\circ\}, id)$ and define $f_{*\circ} : D_0^{*\circ} \rightarrowtail (!D_0^{*\circ} \multimap D_0^{*\circ})$ by $f_{*\circ}(*) = in_r(*)$ and $f_{*\circ}(\circ) = in_r(\circ)$;
3. let $D_0^{**} = (\{*, \circ\}, (\lambda(*) = OQ, \lambda(\circ) = PQ), \{\epsilon, *, *\circ\}, id)$ and define $f_{**} : D_0^{**} \to (!D_0^{**} \multimap D_0^{**})$ by $f_{**}(*) = in_r(*)$ and $f_{**}(\circ) = in_l(\langle 0, *\rangle)$.

Definition 27. *The models $D_\infty^\circ, D_\infty^{*\circ}, D_\infty^{**}$ are determined by the limits of the chains generated by iterating the functor Fun on the embeddings $f_*, f_{*\circ}, f_{**}$, and by the corresponding injection φ respectively.*

Definition 28. *Let $A_{\mathbb{N}} = (\mathbb{N}, \lambda n.OQ, \{\epsilon\} \cup \mathbb{N}, id)$. The model $D^{\mathbb{N}}$ is the one naturally induced by the least fixed point of the functor $F(D) = {!D} \multimap A_{\mathbb{N}}$ where the following chain holds for every $n \in \mathbb{N}$: $D_{n+1}^{\mathbb{N}} \simeq {!D_n^{\mathbb{N}}} \multimap A_{\mathbb{N}} \simeq {!D_n^{\mathbb{N}}} \multimap (A_{\mathbb{N}} \,\&\, A_{\mathbb{N}}) \simeq ({!D_n^{\mathbb{N}}} \multimap A_{\mathbb{N}}) \,\&\, ({!D_n^{\mathbb{N}}} \multimap A_{\mathbb{N}}) \simeq D_{n+1}^{\mathbb{N}} \,\&\, D_{n+1}^{\mathbb{N}}$ and hence, $D_{n+1}^{\mathbb{N}} \simeq {!D_n^{\mathbb{N}}} \multimap A_{\mathbb{N}} \simeq {!(D_n^{\mathbb{N}} \,\&\, D_n^{\mathbb{N}})} \multimap A_{\mathbb{N}} \simeq {!D_n^{\mathbb{N}}} \multimap ({!D_n^{\mathbb{N}}} \multimap A_{\mathbb{N}}) \simeq {!D_n^{\mathbb{N}}} \multimap D_{n+1}^{\mathbb{N}}$.*

One can easily see that $D^{\mathbb{N}}$ is a λ-model since any bijection $p : \mathbb{N} + \mathbb{N} \to \mathbb{N}$, induces an isomorphism between $A_{\mathbb{N}}$ and $A_{\mathbb{N}} \,\&\, A_{\mathbb{N}}$.

6 Conclusions and Final Remarks

In this paper we have shown that all extensional λ-models in the category $K_!(\mathcal{G})$ of [3] induce the same λ-theory, this is the well-known theory $\mathcal{H}^*$. It is natural to conjecture, therefore, that also there is only one non-extensional *sensible λ-theory* which can be modeled using games. We recall that a sensible λ-theory is a theory where all unsolvable terms are equated. This would be the theory $\mathcal{B}$ of Böhm trees, and would be the theory of any reflexive object in $K_!(\mathcal{G})$, $\langle D, f\rangle$, for which f maps the undefined strategy on $!D \multimap D$ on the undefined strategy on D, but it is not an isomorphism.

Our results clearly indicate that existing game models are *even* more rigid than continuous models. But is this really a "surprise", or a "bad surprise"? Definitely there must be some intrinsic feature of games, as they are currently defined, that is intimately related with head reduction. Probably it is not the fact that we have considered only "history-free" strategies, more likely it has to do with the "strict" protocol of alternation of moves between Opponent and Player. We feel however that when the appropriate constraint will be relaxed, the perspicuous analytic power of *games* will become applicable also to other reduction strategies, besides head reduction.

We end this paper with two technical remarks. In the game models of λ-calculus that we have introduced it is not necessary to take an extensional quotient at the end in order to get a "fully abstract" model, as is done in the typed case [3] or in the lazy case. The essential ingredient in the proof of Theorem 6 is Lemma 2. The same argument used there implies also that CPO models obtained using "Scott's trick" as presented in [17] induce the theory $\mathcal{H}^*$.

References

1. S. Abramsky. Domain theory in logical form. In *Annals of Pure and Applied Logic*, volume 51, pages 1–77, 1991.
2. S. Abramsky and R. Jagadeesan. Games and full completeness for multiplicative linear logic. *Journal of Symbolic Logic*, 59(2):543–574, June 1994.
3. S. Abramsky, R. Jagadeesan, and P. Malacaria. Full abstraction for PCF. Ftp-available at `http://www.dcs.ed.ac.uk/home/samson`, 1994.
4. S. Abramsky and G. McCusker. Games for recursive types. In *Theory and Formal Methods of Computing 1994*. Imperial College Press, October 1995.
5. S. Abramsky and C.H.L. Ong. Full abstraction in the lazy λ-calculus. *Information and Computation*, 105:159–267, 1993.
6. H. Barendregt. *The Lambda Calculus: Its Syntax and Semantics*. North-Holland, Amsterdam, 1984. revised edition.
7. C. Berline. From Computation to Foundations via Functions and Application: the λ-calculus and its Webbed Models. To appear.
8. M. Coppo, M. Dezani-Ciancaglini, F. Honsell, and G.Longo. Extended type structures and filter λ-models. In G.Lolli G.Longo and A.Marcja, editors, *Logic Colloquium '82*. Elsevier Science Publishers, 1984.
9. P. Di Gianantonio and G. Franco. A type assignment system for the game semantics. In *ICTCS Proceedings, Prato*, 1998.
10. J.Y. Girard. The system F of variable types, fifteen years later. *Theoretical Computer Science*, 45:159–192, 1986.
11. F. Honsell and M. Lenisa. Final semantics for untyped λ-calculus. In *LNCS*, volume 902, pages 249–265. Springer-Verlag, 1995.
12. F. Honsell and S. Ronchi Della Rocca. An Approximation Theorem for Topological Lambda Models and the Topological Incompleteness of Lambda Calculus. *Journal of Computer and System Sciences*, 45:49–75, 1992.
13. J.M.E. Hyland. A syntatic characterization of the equality in some models of the λ-calculus. *Journal of London Mathematical Society*, 12(2):361–370, 1976.
14. J.M.H. Hyland and C.H.L. Ong. On full abstraction for PCF:I, II, III. ftp-available at `theory.doc.ic.ac.uk` in directory `papers/Ong`.
15. G. A. McCusker. *Games and full abstraction for a functional language with recursive types*. PhD thesis, Imperial College, 1996.
16. D. Scott. Continuous lattices. In *Toposes, Algebraic Geometry and Logic - Lecture Notes in Mathematics*, volume 274. Springer-Verlag, Berlin, New York, 1972.
17. D. Scott and C. Gunter. Semantic domains. In *Handbook of Theorical Computer Science*. North Holland, 1990.
18. C. P. Wadsworth. The Relation between computational and Denotational Properties for Scott's D_∞-models of the λ-calculus. *SIAM*, 5(3):488–521, 1976.

A Finite Axiomatization of Inductive-Recursive Definitions

Peter Dybjer[1] and Anton Setzer[2]

[1] Department of Mathematics and Computing Science, Chalmers University of Technology.
peterd@cs.chalmers.se
[2] Department of Mathematics, Uppsala University.
setzer@math.uu.se

Abstract. Induction-recursion is a schema which formalizes the principles for introducing new sets in Martin-Löf's type theory. It states that we may inductively define a set while simultaneously defining a function from this set into an arbitrary type by structural recursion. This extends the notion of an inductively defined set substantially and allows us to introduce universes and higher order universes (but not a Mahlo universe). In this article we give a finite axiomatization of inductive-recursive definitions. We prove consistency by constructing a set-theoretic model which makes use of one Mahlo cardinal.

1 Introduction

In this article we present an elegant, uniform method for introducing large sets in type theory. We draw on experience from proof theory, category theory, and set theory to formulate a compact, completely formal theory of inductive-recursive definitions, and to prove its consistency.

Induction-recursion is a schema for introducing new sets in type theory developed by Dybjer [18]. All the usual sets in Martin-Löf's type theory and practically all sets (data types), which are defined in analogy with it, are instances of this schema. Applications of induction-recursion include not only a variety of type-theoretic analogues of large cardinals (inaccessible cardinals, hyper-inaccessible cardinals, etc) but also various powerful notions needed for the type-theoretic formalization of metamathematics (such as reducibility predicates and logical relations for dependent types). Induction-recursion can also provide novel ways to formalize simple concepts such as the set of lists with distinct elements [18].

The original presentation of induction-recursion was as an external schema [18]. In this article we internalize this concept. The new theory has a special type of codes for inductive-recursive definitions. New sets defined by induction-recursion are introduced by deriving codes in this type. Therefore we achieve full precision of the concept of an inductive-recursive definition. The meta-theory becomes easier, as will be demonstrated by building a full function space model.

Ordinary dependent type theory with generalized inductive definitions (that is, Martin-Löf's type theory without universes) has a natural full function space interpretation in classical set theory [5,20]. As shown by our construction of a set-theoretic model the step from inductive to inductive-recursive definitions in type theory is roughly analogous to moving from ordinary ZF set theory to ZF set theory with a Mahlo cardinal. The proof-theoretic strength of type theory increases accordingly when inductive-recursive definitions are added. The consistency of the theory is shown without assuming the positivity restriction on parameters needed for Dybjer's original realizability model of inductive-recursive definitions [18].

The new theory explains that induction-recursion can be viewed as a very general reflection principle: given finitely many (possibly infinitary) operations on a type D, we can construct by simultaneous induction-recursion a universe U with decoding function $\mathrm{T} : \mathrm{U} \to D$, which reflects each of the D-operations. This reflection principle can be expressed formally by a diagram which extends the initial algebra diagram used for categorical semantics of inductively defined sets. The resulting theory has been implemented in the *Half* system, a proof assistant for Martin-Löf's type theory developed by Coquand and Synek, see Cederquist [15].

Plan of the paper. In Section 2 we present Martin-Löf's Logical Framework. In Section 3 we recall how to use initial algebras for giving categorical semantics of inductive types in the simply typed lambda calculus. In Section 4 we discuss the step from induction to induction-recursion and how we need to modify the notion of an endofunctor Φ and of an initial Φ-algebra in order to capture the formal rules for induction-recursion. We then show how to give a finite axiomatization of inductive-recursive definitions by introducing a type of codes for such modified endofunctors. In Section 5 we show how to recover some well-known set constructors by giving appropriate codes. In Section 6 we build a set-theoretic model. In Section 7 we mention some related work.

2 An Extension of the Logical Framework

The Logical Framework (see [21]) has the following forms of judgements: Γ context, and $A : \mathrm{type}$, $A = B : \mathrm{type}$, $a : A$, $a = b : A$, depending on contexts Γ (written as $\Gamma \Rightarrow A : \mathrm{type}$, etc.). We have set : type and if $A : \mathrm{set}$, then $A : \mathrm{type}$. The collection of types is closed under the formation of dependent function types written as $(x : A) \to B$, with elements formed by abstraction $(x : A)a$, application written in the form $a(b)$ and which has the η-rule. Types are also closed under the formation of dependent products written as $(x : A) \times B$, with elements $\langle a, b \rangle$, projections π_0 and π_1 and again the η-rule (surjective pairing). There is also the type 1, with unique element $\langle\rangle : 1$ and η-rule expressing, that if $a : 1$, then $a = \langle\rangle : 1$.

We will add a level between set and type, which we call stype for small types: stype : type. (The reason for the need for stype is discussed in [18].) If a : set then a : stype. Moreover, stype is also closed under dependent function types, dependent products and includes the one-element type. However, set itself will not be in stype.

Finally, in order to make it possible to code all constructors into one (see the remark on page 132), we add the set $\mathbb{B}$ of booleans with elements tt for true and ff for false and as elimination rule case distinction if a then b else $c : D$ for $a : \mathbb{B}$, D : type and $b, c : D$.

We also use some abbreviations, such as omitting the type in an abstraction, that is, writing $(x)a$ instead of $(x : A)a$, and writing repeated application as $a(b_1, \ldots, b_n)$ instead of $a(b_1) \cdots (b_n)$ and repeated abstraction as $(x_1 : A_1, \ldots, x_n : A_n)a$ instead of $(x_1 : A_1) \cdots (x_n : A_n)a$.

3 Inductive Types as Initial Algebras

Let us first consider the question of how to formalize inductive types in the setting of the simply typed λ-calculus. We shall consider *generalized inductive definitions* of types given by a finite number of constructors

$$\mathrm{intro}_i : \Phi_i(\mathrm{U}) \to \mathrm{U} \ ,$$

where Φ_i are *strictly positive* in the following restricted sense:

- The constant functor $\Phi(D) = 1$ is strictly positive. This is the base case corresponding to an introduction rule with no premises.
- If Ψ is strictly positive and A is an stype, then $\Phi(D) = A \times \Psi(D)$ is strictly positive. This corresponds to the addition of a *non-inductive*[1] premise.
- If Ψ is strictly positive and A is an stype, then $\Phi(D) = (A \to D) \times \Psi(D)$ is strictly positive. This corresponds to the addition of an *inductive* premise, where A corresponds to the hypotheses of this premise in a generalized inductive definition (and when $A = 1$ we have the special case of an *ordinary inductive definition*).

Note that all occurrences of U in $\Phi(\mathrm{U})$ are strictly positive in the standard sense that U does not occur to the left of an arrow in $\Phi(\mathrm{U})$.

Assume $\Phi_1, \ldots, \Phi_n$ are strictly positive functors, and let $\boldsymbol{\Phi} := (\Phi_1, \ldots, \Phi_n)$. Then the inductive type generated by $\boldsymbol{\Phi}$ can be captured categorically as an initial $\boldsymbol{\Phi}$-algebra, that is, a sequence of arrows $(i = 1, \ldots, n)$

$$\Phi_i(\mathrm{U}) \xrightarrow{\mathrm{intro}_i} \mathrm{U}$$

such that for any other $\boldsymbol{\Phi}$-algebra

$$\Phi_i(D) \xrightarrow{d_i} D$$

[1] In [18] the terminology "non-recursive premise" was used, but "non-inductive premise" seems better in connection with induction-recursion, since it primarily has to do with the inductively defined set and not with the recursively defined function. Similarly we will use "inductive premise" instead of "recursive premise".

there is a unique arrow $\mathrm{T} : \mathrm{U} \longrightarrow D$, such that the following diagrams commute

$$\begin{array}{ccc} \Phi_i(\mathrm{U}) & \xrightarrow{\mathrm{intro}_i} & \mathrm{U} \\ {\scriptstyle \Phi_i(\mathrm{T})}\downarrow & & \downarrow{\scriptstyle \mathrm{T}} \\ \Phi_i(D) & \xrightarrow{d_i} & D \end{array}$$

4 Inductive-Recursive Definitions

4.1 From Inductive to Inductive-Recursive Definitions

In the presence of dependent types more inductive definitions become possible. Let us look at some examples:

The set $\Sigma(A, B)$ has one constructor $\mathrm{p} : (x : A) \rightarrow (y : B(x)) \rightarrow \Sigma(A, B)$. It has two non-inductive arguments, where the type $B(x)$ of the second argument depends on the first premise $x : A$.

The well-ordering set $\mathrm{W}(A, B)$ has one constructor $\sup : (x : A) \rightarrow (y : B(x) \rightarrow \mathrm{W}(A, B)) \rightarrow \mathrm{W}(A, B)$. It has a first non-inductive argument x and a second $B(x)$-indexed inductive argument y. So the second argument depends on the first non-inductive argument.

Both are examples of inductive definitions (no simultaneously defined function participates in the definition yet). For this case later premises can only depend on earlier non-inductive premises, but not on earlier inductive premises. We cannot make use of inductive premises, because they only give information about the set we are currently defining.

To capture inductive definitions of sets in the presence of dependent types [20, 1], we thus only need to change the notion of a strictly positive functor Φ above by replacing the non-inductive case by:

- If A is an stype, and Ψ_x is a strictly positive functor depending on $x : A$, then $\Phi(D) = (x : A) \times \Psi_x(D)$ is strictly positive.

We shall now replace the sequence of functors $(\Phi_1, \ldots, \Phi_n)$ by a single functor by defining $\Phi(D) := (x : \mathrm{N}_n) \times \Phi_x(D)$. In order to make this possible we need the existence of finite sets with n elements N_n. An easy observation shows that $\mathbb{B}$ and the empty set N_0 suffice. (It will however be possible to define N_0, see section 5).

In the case of inductive-recursive definitions however, a later premise may also depend on an earlier inductive premise. We consider the key example, the ordinary first universe U à la Tarski [3], which is defined inductively, while simultaneously defining the decoding function $\mathrm{T} : \mathrm{U} \rightarrow \mathrm{set}$ recursively. Consider one of its constructors, $\widehat{\Sigma} : (x : \mathrm{U}) \rightarrow (y : \mathrm{T}(x) \rightarrow \mathrm{U}) \rightarrow \mathrm{U}$ with the defining equality $\mathrm{T}(\widehat{\Sigma}(a, b)) = \Sigma(\mathrm{T}(a), \mathrm{T} \circ b) : \mathrm{set}$. Here we have two inductive premises: $x : \mathrm{U}$ (implicitly indexed by the one-element type 1) and $y : \mathrm{U}$ indexed by $\mathrm{T}(x)$. The second argument depends on the first inductive argument via T.

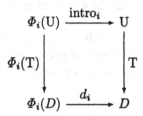

Is U inductively generated by a strictly positive functor Φ as was the case for inductively defined sets? If this is the case, Φ must depend on the recursively defined function T as well: we need something like $\Phi : (U : \mathrm{set}) \to (T : U \to \mathrm{set}) \to \mathrm{set}$ defined by $\Phi(U,T) = (x : U) \times (T(x) \to U)$!

In general, induction-recursion allows that a simultaneously defined function $\mathrm{T} : \mathrm{U} \to D$ for an arbitrary fixed type D may participate in the inductive generation of the set U.

- The modified non-inductive case thus becomes: if A is an stype, and Ψ_x is a strictly positive functor depending on $x : A$, then $\Phi(U,T) = (x : A) \times \Psi_x(U,T)$ is strictly positive.
- The modified inductive case becomes: if A is an stype, and Ψ_g is a strictly positive functor depending on $g : A \to D$, then $\Phi(U,T) = (f : A \to U) \times \Psi_{T \circ f}(U,T)$ is strictly positive.

We see that the Φ which generates U (as defined above), is isomorphic to the following strictly positive functor: $\Phi(U,T) = (f : 1 \to U) \times (T(f(\langle\rangle)) \to U)$ [2].

Furthermore, T is defined by $\mathrm{T}(\widehat{\Sigma}(a,b)) = \Sigma(\mathrm{T}(a), \mathrm{T} \circ b)$, i.e. $\mathrm{T}(\widehat{\Sigma}(a,b)) = d(\mathrm{T}(a), \mathrm{T} \circ b)$ with $d : (A : \mathrm{set}) \to (A \to \mathrm{set}) \to \mathrm{set}$ and $d(A,B) := \Sigma(A,B)$.

In general, we need an additional component Φ^{Arg} which specifies the domain of d. (Note that this domain only depends on D and not on U and T!). Finally, we need a third component $\Phi^{\mathrm{map}}(U,T) : \Phi^{\mathrm{arg}}(U,T) \to \Phi^{\mathrm{Arg}}$ and then we can draw a diagram

$$
\begin{array}{ccc}
\Phi^{\mathrm{arg}}(\mathrm{U},\mathrm{T}) & \xrightarrow{\ \mathrm{intro}\ } & \mathrm{U} \\
{\scriptstyle \Phi^{\mathrm{map}}(\mathrm{U},\mathrm{T})}\downarrow & & \downarrow{\scriptstyle \mathrm{T}} \\
\Phi^{\mathrm{Arg}} & \xrightarrow{\ d\ } & D
\end{array}
$$

which summarizes the inductive definition of U and recursive definition of T. Think of D as a type of "semantic" objects and of $d : \Phi^{\mathrm{Arg}} \to D$ as a (possibly infinitary) "semantic" operation with Φ^{Arg} as the domain (or generalized arity) of d. U is a universe of codes for objects in D and $\mathrm{T} : \mathrm{U} \to D$ is the decoding function. The constructor intro is the syntactic reflection of $d : \Phi^{\mathrm{Arg}} \to D$.[3]

Note the similarity between the above diagram for induction-recursion and the ordinary diagram for an initial algebra of an endofunctor which was displayed in Section 2! The key difference is that here Φ is no longer a functor in the ordinary sense, but consists of three components: Φ^{Arg}, Φ^{arg}, and Φ^{map}. These will be axiomatized below.

---

2 As this example shows the term "strictly positive" may no longer be wholly appropriate, since the T-argument now can appear negatively. Allen [12] used the alternative term "half-positive" for this reason. U always appears strictly positively however.

3 Recall that the term "universe à la Tarski" was chosen by Martin-Löf [3] because of the similarity between the definition of T (for the ordinary first universe) and Tarski's definition of truth.

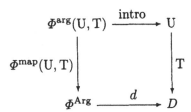

4.2 A Finite Axiomatization

We shall now give the formal rules for the inductive-recursive definition of a set U and a function $\mathrm{T} : \mathrm{U} \to D$. Such a definition is always parameterized with respect to the target type D of T, since a particular inductive-recursive definition generates a universe for a finite number of D-operations.

The main step is to introduce a new type SP_D, the objects of which are representatives of strictly positive "functors" Φ as above:

$$\frac{D : \text{type}}{\mathrm{SP}_D : \text{type}}$$

(There is also a rule, which lets us infer that $\mathrm{SP}_D = \mathrm{SP}_{D'}$ if $D = D'$, but we will omit all such equality preservation rules.)

SP_D has three associated operations corresponding to Φ^{Arg}, Φ^{arg}, and Φ^{map} in the informal exposition above:

$$\frac{D : \text{type} \qquad \phi : \mathrm{SP}_D}{\mathrm{Arg}_{D,\phi} : \text{type}}$$

$$\frac{\phi : \mathrm{SP}_D \qquad U : \text{set} \qquad T : U \to D}{\mathrm{arg}_\phi(U, T) : \text{stype}}$$

$$\frac{\phi : \mathrm{SP}_D \qquad U : \text{set} \qquad T : U \to D}{\mathrm{map}_\phi(U, T) : (\mathrm{arg}_\phi(U, T)) \to \mathrm{Arg}_{D,\phi}}$$

To simplify notation we have suppressed the parameter (the "global" premise) D : type and the argument D for the second and third operation[4]. It should be emphasized that arg_ϕ and map_ϕ are only abbreviations of the proper formal expressions $\mathrm{arg}_{D,\phi}$ and $\mathrm{map}_{D,\phi}$. Similarly, we will suppress the D in some later operations as well.

With this new notation the diagram for the inductive-recursive definition of U and T becomes:

$$\begin{array}{ccc}
\mathrm{arg}_\phi(\mathrm{U}_{\phi,d}, \mathrm{T}_{\phi,d}) & \xrightarrow{\mathrm{intro}_{\phi,d}} & \mathrm{U}_{\phi,d} \\
\downarrow {\scriptstyle \mathrm{map}_\phi(\mathrm{U}_{\phi,d}, \mathrm{T}_{\phi,d})} & & \downarrow {\scriptstyle \mathrm{T}_{\phi,d}} \\
\mathrm{Arg}_{D,\phi} & \xrightarrow{\quad d \quad} & D
\end{array}$$

[4] In Arg we have not suppressed it, since the equality rules for it will make use of D.

We have the following introduction rules for SP_D (again with D suppressed):

$$\mathrm{nil} : \mathrm{SP}_D$$

$$\frac{A \text{ stype} \qquad \phi : A \to \mathrm{SP}_D}{\mathrm{nonind}(A, \phi) : \mathrm{SP}_D}$$

$$\frac{A \text{ stype} \qquad \phi : (A \to D) \to \mathrm{SP}_D}{\mathrm{ind}(A, \phi) : \mathrm{SP}_D}$$

$$\begin{aligned}
\mathrm{Arg}_{D,\mathrm{nil}} &= 1 \\
\mathrm{Arg}_{D,\mathrm{nonind}(A,\phi)} &= (x : A) \times \mathrm{Arg}_{D,\phi(x)} \\
\mathrm{Arg}_{D,\mathrm{ind}(A,\phi)} &= (f : A \to D) \times \mathrm{Arg}_{D,\phi(f)}
\end{aligned}$$

$$\begin{aligned}
\mathrm{arg}_{\mathrm{nil}}(U, T) &= 1 \\
\mathrm{arg}_{\mathrm{nonind}(A,\phi)}(U, T) &= (x : A) \times (\mathrm{arg}_{\phi(x)}(U, T)) \\
\mathrm{arg}_{\mathrm{ind}(A,\phi)}(U, T) &= (f : A \to U) \times (\mathrm{arg}_{\phi(T \circ f)}(U, T))
\end{aligned}$$

$$\begin{aligned}
\mathrm{map}_{\mathrm{nil}}(U, T, \langle\rangle) &= \langle\rangle \\
\mathrm{map}_{\mathrm{nonind}(A,\phi)}(U, T, \langle a, \gamma\rangle) &= \langle a, \mathrm{map}_{\phi(a)}(U, T, \gamma)\rangle \\
\mathrm{map}_{\mathrm{ind}(A,\phi)}(U, T, \langle f, \gamma\rangle) &= \langle T \circ f, \mathrm{map}_{\phi(T \circ f)}(U, T, \gamma)\rangle
\end{aligned}$$

We are now ready to give the formal rules for U and T. These rules have the common premises D : type, $\phi : \mathrm{SP}_D$ and $d : \mathrm{Arg}_{D,\phi} \to D$ which will be omitted.

Formation rules:

$$\mathrm{U}_{\phi,d} : \mathrm{set}$$

$$\mathrm{T}_{\phi,d} : \mathrm{U}_{\phi,d} \to D$$

Introduction rule:

$$\frac{a : \mathrm{arg}_\phi(\mathrm{U}_{\phi,d}, \mathrm{T}_{\phi,d})}{\mathrm{intro}_{\phi,d}(a) : \mathrm{U}_{\phi,d}}$$

Equality rule:

$$\frac{a : \mathrm{arg}_\phi(\mathrm{U}_{\phi,d}, \mathrm{T}_{\phi,d})}{\mathrm{T}_{\phi,d}(\mathrm{intro}_{\phi,d}(a)) = d(\mathrm{map}_\phi(\mathrm{U}_{\phi,d}, \mathrm{T}_{\phi,d}, a))}$$

Moreover, structural recursion on U into a type D', that is, the analogue of universe elimination, is expressed by the following diagram (we omit the indices ϕ, d of U, T, intro, R, write $D'[t]$ for the substitution of some fixed variable in D' by t and, when used as an argument, D' instead of $(x)D'[x]$; assume in the following $x : \mathrm{U}_{\phi,d} \Rightarrow D'[x]$: type as a global premise)

$$\begin{array}{ccc}
\mathrm{arg}_\phi(\mathrm{U}, \mathrm{T}) & \xrightarrow{\quad \mathrm{intro} \quad} & \mathrm{U} \\
\Big\downarrow {\scriptstyle \langle \mathrm{id}, \mathrm{mapIH}_{\phi,\mathrm{U},\mathrm{T},D'}(\mathrm{R}_{D'}(e))\rangle} & & \Big\downarrow {\scriptstyle \langle \mathrm{id}, \mathrm{R}_{D'}(e)\rangle} \\
(\gamma : \mathrm{arg}_\phi(\mathrm{U}, \mathrm{T})) \times \mathrm{IH}_{\phi,\mathrm{U},\mathrm{T},D'}(\gamma) & \xrightarrow{\langle \mathrm{intro} \circ \pi_0, e\rangle} & (x : \mathrm{U}) \times D'[x]
\end{array}$$

where we have used the operation IH which generates the induction hypothesis and the operation mapIH which generates the recursive call:

$$\frac{U : \mathrm{set} \qquad T : (x : U) \to D \qquad x : U \Rightarrow D'[x] : \mathrm{type} \qquad \gamma : \mathrm{arg}_\phi(U, T)}{\mathrm{IH}_{\phi,U,T,D'}(\gamma) : \mathrm{type}}$$

$$\frac{U : \mathrm{set} \qquad T : (x : U) \to D \qquad x : U \Rightarrow D'[x] : \mathrm{type} \qquad R : (x : U) \to D'[x]}{\mathrm{mapIH}_{\phi,U,T,D'}(R) : (x : \mathrm{arg}_\phi(U, T)) \to \mathrm{IH}_{\phi,U,T,D'}(x)}$$

$$\begin{aligned} \mathrm{IH}_{\mathrm{nil},U,T,D'}(\langle\rangle) &= 1 \\ \mathrm{IH}_{\mathrm{nonind}(A,\phi),U,T,D'}(\langle a, \gamma\rangle) &= \mathrm{IH}_{\phi(a),U,T,D'}(\gamma) \\ \mathrm{IH}_{\mathrm{ind}(A,\phi),U,T,D'}(\langle f, \gamma\rangle) &= ((y : A) \to D'[f(y)]) \times (\mathrm{IH}_{\phi(T\circ f),U,T,D'}(\gamma)) \end{aligned}$$

$$\begin{aligned} \mathrm{mapIH}_{\mathrm{nil},U,T,D'}(R, \langle\rangle) &= \langle\rangle \\ \mathrm{mapIH}_{\mathrm{nonind}(A,\phi),U,T,D'}(R, \langle a, \gamma\rangle) &= \mathrm{mapIH}_{\phi(a),U,T,D'}(R, \gamma) \\ \mathrm{mapIH}_{\mathrm{ind}(A,\phi),U,T,D'}(R, \langle f, \gamma\rangle) &= \langle R \circ f, \mathrm{mapIH}_{\phi(T\circ f),U,T,D'}(R, \gamma)\rangle \end{aligned}$$

Elimination rule (universe elimination):

$$\frac{e : (\gamma : \mathrm{arg}_\phi(\mathrm{U}_{\phi,d}, \mathrm{T}_{\phi,d})) \to (\mathrm{IH}_{\phi,\mathrm{U}_{\phi,d},\mathrm{T}_{\phi,d},D'}(\gamma)) \to (D'[\mathrm{intro}_{\phi,d}(\gamma)])}{\mathrm{R}_{\phi,d,D'}(e) : (a : \mathrm{U}_{\phi,d}) \to D'[a]}$$

Equality rule (universe elimination, premises omitted):

$$\mathrm{R}_{\phi,d,D'}(e, \mathrm{intro}_{\phi,d}(\gamma)) = e(\gamma, \mathrm{mapIH}_{\phi,\mathrm{U}_{\phi,d},\mathrm{T}_{\phi,d},D'}(\mathrm{R}_{\phi,d,D'}(e), \gamma))$$

5 Examples

We shall show how to find ϕ : SP_D for some well-known set constructors. (Compare the informal discussion in Section 4.1.) We will write intro instead of $\mathrm{intro}_{\phi,d}$. Let in the first examples $D := 1$ and $d := (x : C)\langle\rangle$ for some suitable type C, since this is how we obtain inductive definitions as degenerate cases of inductive-recursive definitions.

Σ-sets. Let

$$\phi_{A,B} := \mathrm{nonind}(A, (x)\mathrm{nonind}(B(x), (y)\mathrm{nil}))$$

in the context $A : \mathrm{set}, B : A \to \mathrm{set}$. It follows that $\Sigma(A, B) := \mathrm{U}_{\phi_{A,B},d} : \mathrm{set}$. This set has the constructor $\mathrm{intro} : ((x : A) \times (B(x) \times 1)) \to \Sigma(A, B)$. If we define $\mathrm{p} := (A, B, x, y)\mathrm{intro}(\langle x, \langle y, \langle\rangle\rangle\rangle)$, then $\mathrm{p} : (A : \mathrm{set}, B : A \to \mathrm{set}, x : A, y : B(x)) \to \Sigma(A, B)$ and one can easily derive the ordinary elimination rules as if p were the constructor of Σ. Note that this illustrates that we get dependencies on parameters (in the sense of Dybjer [1,18]) like A, B for free.

Natural numbers. Let

$$\begin{aligned}
&\phi := \mathrm{nonind}(\mathbb{B}, (x)\ \mathrm{if}\ x\ \mathrm{then\ nil\ else\ ind}(1, (y)\mathrm{nil}))\ ,\\
&\mathrm{N} := \mathrm{U}_{\phi,d}\ ,\\
&0 := \mathrm{intro}(\langle \mathrm{tt}, \langle\rangle\rangle) : \mathrm{N}\ ,\\
&\mathrm{S} := (n)\mathrm{intro}(\langle \mathrm{ff}, \langle (y)n, \langle\rangle\rangle\rangle) : \mathrm{N} \to \mathrm{N}\ .
\end{aligned}$$

Although this definition is like the definition of N by the equation $\mathrm{N} = 1 + (1 \to \mathrm{N}) \times 1$, because of the η-rule on 1 this is equivalent to the ordinary definition of N. The usual elimination rules for N can be derived.

The empty set. Let

$$\phi := \mathrm{ind}(1, (x)\mathrm{nil})\ ,$$

and define $\mathrm{N}_0 := \mathrm{U}_{\phi,d}$. Then we can show the elimination rule for the empty set N_0. Note that this corresponds to the definition of N_0 by having one constructor intro : $\mathrm{N}_0 \to \mathrm{N}_0$. We can define now $\mathrm{N}_0' := \mathrm{U}_{\mathrm{nonind}(\mathrm{N}_0,(x)\mathrm{nil}),d}$, which can be regarded as the empty set with no constructors. However, one might prefer to add the set N_0 like the set $\mathbb{B}$ as a basic set.

Well-orderings. Let

$$\phi_{A,B} := \mathrm{nonind}(A, (x)\mathrm{ind}(B(x), (y)\mathrm{nil}))\ ,$$

in the context $A : \mathrm{set}, B : (x : A) \to \mathrm{set}$, and define $\mathrm{W}(A, B) := \mathrm{U}_{\phi_{A,B},d} : \mathrm{set}$ with the constructor intro : $((x : A) \times ((B(x) \to \mathrm{W}(A, B)) \times 1)) \to \mathrm{W}(A, B)$. As before we can define the ordinary constructor sup with its elimination rules.

A universe closed under N and Σ. Let $D := \mathrm{set}$,

$$\phi := \mathrm{nonind}(\mathbb{B}, (x)\ \mathrm{if}\ x\ \mathrm{then\ nil\ else\ ind}(1, (f)\mathrm{ind}(f(\langle\rangle), (y)\mathrm{nil})))\ .$$

Hence $\mathrm{Arg}_{D,\phi} = (x : \mathbb{B}) \times \mathrm{E}(x)$, with

$$\begin{aligned}
&\mathrm{E}(\mathrm{tt}) = 1\ ,\\
&\mathrm{E}(\mathrm{ff}) = (x : 1 \to \mathrm{set}) \times (f : x(\langle\rangle) \to \mathrm{set}) \times 1\ .
\end{aligned}$$

Moreover, let $\mathrm{d} : \mathrm{Arg}_{D,\phi} \to \mathrm{set}$ be defined such that

$$\begin{aligned}
\mathrm{d}(\langle \mathrm{tt}, \langle\rangle\rangle) &= \mathrm{N}\ ,\\
\mathrm{d}(\langle \mathrm{ff}, \langle A, \langle B, \langle\rangle\rangle\rangle\rangle) &= \Sigma(A(\langle\rangle), (y)B(y))\ ,
\end{aligned}$$

using the elimination rules for $\mathbb{B}$ and product. Define $\mathrm{U}' := \mathrm{U}_{\phi,d}$, $\mathrm{T}' := \mathrm{T}_{\phi,d}$, and

$$\begin{aligned}
&\widehat{\mathrm{N}} := \mathrm{intro}(\langle \mathrm{tt}, \langle\rangle\rangle) : \mathrm{U}'\ ,\\
&\widehat{\Sigma} := (a, b)\mathrm{intro}(\langle \mathrm{ff}, \langle (x)a, \langle b, \langle\rangle\rangle\rangle\rangle) : (a : \mathrm{U}', b : \mathrm{T}'(a) \to \mathrm{U}') \to \mathrm{U}'\ .
\end{aligned}$$

$\widehat{\mathrm{N}}$ and $\widehat{\Sigma}$ are essentially the two constructors of the universe U', T' and we have $\mathrm{T}'(\widehat{\mathrm{N}}) = \mathrm{N}$, $\mathrm{T}'(\widehat{\Sigma}(a, b)) = \Sigma(\mathrm{T}'(a), \mathrm{T}' \circ b)$.

Lists with distinct elements. Assume A : set and $\# : (A \times A) \to A$, where $\#$ is an (infix) apartness relation on A. In [18] the set Dlist of lists with elements which are distinct with respect to the relation $\#$ is defined inductively together with the recursively defined relation (family of sets) $\mathrm{Fresh} : \mathrm{Dlist} \to A \to \mathrm{set}$, where $\mathrm{Fresh}(l, a)$ expresses that a is distinct from all elements in l. (If we wish to make the dependence on the parameters A and $\#$ explicit, we may write $\mathrm{Dlist}(A, \#)$ and $\mathrm{Fresh}(A, \#)$.) Dlist has the constructors[5]

$$\begin{aligned} \mathrm{empty} &: \mathrm{Dlist} \ , \\ \mathrm{cons} &: (a : A, u : \mathrm{Dlist}, \mathrm{Fresh}(u, a)) \to \mathrm{Dlist} \ , \end{aligned}$$

and $\mathrm{Fresh}(l, a)$ is defined such that

$$\begin{aligned} \mathrm{Fresh}(\mathrm{empty}) &= (b)1 \ , \\ \mathrm{Fresh}(\mathrm{cons}(a, u, p)) &= (b)((b\#a) \wedge \mathrm{Fresh}(u, a)) \ . \end{aligned}$$

Then $\mathrm{Dlist} = \mathrm{U}_{\phi_{A,\#}, \mathrm{d}_{A,\#}}$, $\mathrm{Fresh} = \mathrm{T}_{\phi_{A,\#}, \mathrm{d}_{A,\#}}$, where $D := A \to \mathrm{set}$,

$$\begin{aligned} \phi_{A,\#} &:= \mathrm{nonind}(\mathbb{B}, (x) \text{ if } x \text{ then nil} \\ &\qquad \text{else } \mathrm{nonind}(A, (a)\mathrm{ind}(1, (u)\mathrm{nonind}(u(\langle\rangle, a), \mathrm{nil})))) \ , \\ \mathrm{d}_{A,\#}(\langle \mathrm{tt}, \langle\rangle\rangle) &= (b)1 \ , \\ \mathrm{d}_{A,\#}(\langle \mathrm{ff}, \langle a, \langle u, \langle p, \langle\rangle\rangle\rangle\rangle\rangle) &= (b)((b\#a) \wedge u(\langle\rangle, a)) \ . \end{aligned}$$

The above examples show that we can derive all inductive-recursive sets in a form, which is close to the way we would ordinarily like to write them down. We must for example write the arguments in list notation and, if we have a non-indexed inductive argument, write it as an argument depending on the type 1. In an implementation of the calculus one could of course easily avoid this administrative overhead.

6 Set-Theoretic Model

6.1 Interpretation of Expressions

The idea behind the model is simple: interpret all constructions in set theory in the obvious way! In particular, each type is interpreted as a set, equal types are interpreted as equal sets, $a : A$ is interpreted as $a \in A$, and $a = b : A$ is interpreted as a and b are equal elements of A. Moreover, $A \to B$ is interpreted as the set of all functions from A to B in the set-theoretic sense, and $(x : A) \to B$ as the set-theoretic cartesian product $\Pi_{x \in A} B$, etc.

The inductively defined type SP_D of codes for strictly positive operators is interpreted as an inductively defined set in the set-theoretic sense, that is, as a set generated by iterating a monotone operator up to a fixed point. Similarly, the inductive-recursively defined set U and function $\mathrm{T} : \mathrm{U} \to D$, are also interpreted by iterating a monotone operator up to a fixed point.

[5] We have here renamed the constructor nil in [18] to empty.

In order to ensure that a fixed point indeed can be reached we postulate the existence of one Mahlo cardinal in addition to the ordinary axioms of ZF set theory.[6] We also need the the axiom of choice to deal with cardinals, and for simplicity we assume the generalized continuum hypothesis.[7]

Note, that a cardinal κ is inaccessible, iff it is regular and $\aleph_\kappa = \kappa$, where $\aleph_\alpha$ enumerates the infinite cardinals. An inaccessible cardinal κ is a Mahlo cardinal, iff every normal function $f : \kappa \to \kappa$ has a regular fixed point. (A normal function f is a (strictly) monotone function, which is continuous at limit ordinals λ, i.e. $f(\lambda) = \sup_{\alpha<\lambda} f(\alpha)$.) The standard model of our extension of ZF is $\mathrm{V}_{\mathrm{M}^+}$, where M^+ is the first inaccessible above M, however all types will be interpreted as elements of V_Λ, where Λ is the first (non-regular) fixed point of $\lambda\alpha.\aleph_\alpha$ above M.

We will develop the semantics following the approach in [20]. Let $\lambda_0 := \aleph_{\mathrm{M}+1}$, $\lambda_{n+1} := \aleph_{\lambda_n}$, and $\Lambda := \sup_{n\in\omega} \lambda_n$.

If $a, a_1, \ldots, a_n, c$ are sets, and b is a function with domain a, let

$$\begin{aligned} \Pi_{x\in a} b(x) &:= \{f \mid f \text{ function } \wedge \operatorname{dom}(f) = a \wedge \forall x \in a. f(x) \in b(x)\} \ , \\ \lambda x \in a.b(x) &:= \{\langle x, b(x)\rangle \mid x \in a\} \ , \\ \Sigma_{x\in a} b(x) &:= \{\langle c, d\rangle \mid c \in a \wedge d \in b(c)\} \ , \\ a_0 + \cdots + a_n &:= \Sigma_{i\in\{0,\ldots,n\}} a_i \ (\text{if } n \geq 1) \ , \\ (a \to c) &:= \Pi_{x\in a} c \ . \end{aligned}$$

Moreover, $(a)_i := a_i$, if $a = \langle a_0, \ldots, a_i\rangle$ and undefined otherwise.

Whenever we introduce sets A^α indexed by ordinals α, let in the following

$$A^{<\alpha} := \bigcup_{\beta<\alpha} A^\beta \ .$$

We shall use the set V_Λ as the set-theoretic universe for our interpretation. All types and objects of types will thus be interpreted as elements of V_Λ [8]. Terms which depend on free variables will be interpreted relative to an assignment ρ, that is, a function, which maps a finite set of variables to elements of V_Λ. In the following ρ (possibly with indices or accents) will always be an assignment. If $a \in \mathrm{V}_\Lambda$, then ρ^a_x is the assignment with $\operatorname{dom}(\rho^a_x) := \operatorname{dom}(\rho) \cup \{x\}$, such that

$$\rho^a_x(y) := \begin{cases} a & \text{if } x = y, \\ \rho(y) & \text{otherwise.} \end{cases}$$

Let terms be the set of expressions which possibly occur as elements of a type or as types: So variables are terms and if $a, b, a_1, \ldots, a_n$ are terms, x is a variable,

[6] In Sect. 7 ("Constructive versions of the model") we will discuss how to replace these strong set theoretic requirements by far weaker ones.

[7] Without the generalized continuum hypothesis one has to replace Mahlo and inaccessible by strongly Mahlo and strongly inaccessible, respectively, and $\aleph_\alpha$ by $\beth_\alpha$.

[8] We here use a notion of model which only requires all *derivable* types to be interpreted as elements of V_Λ. Note however that V_Λ is not closed under the formation of dependent function types. If we wish to satisfy this requirement we can either reinterpret type as the class of all sets or as V_I for some inaccessible cardinal $\mathrm{I} > \mathrm{M}$.

and C is an n-ary constructor (including set, stype and constructors like $\langle\cdot,\cdot\rangle$, π_0 but excluding type) of the system, then $(x : a) \to b$, $(x : a)b$, $(x : a) \times b$ and $C(a_1, \dots, a_n)$ are terms.

For terms t and assignments ρ we will determine, whether its interpretation t^*_ρ is defined, and if it is defined, the value of t^*_ρ. This will be done in such a way that for every term t and every $n \in \omega$ there exists an $m \in \omega$ such that, if $\mathrm{rng}(\rho) \subseteq \mathrm{V}_{\lambda_n}$, $t^*_\rho \in \mathrm{V}_{\lambda_m}$. For closed terms, t^*_ρ will not depend on ρ and we therefore omit the subscript ρ. We will use $\simeq$ for partial equality in the usual sense, and also let $t^*_\rho :\simeq s$ mean that the interpretation of t under assignment ρ is defined to be s, provided s is defined, and is undefined otherwise. We extend this definition further by defining

$$\mathrm{type}^* := V_\Lambda \ .$$

The interpretation of terms is given by

$$\begin{array}{ll}
x^*_\rho :\simeq \rho(x) \ , & \mathrm{set}^* :\simeq \mathrm{stype}^* :\simeq \mathrm{V_M} \ , \\
((x : A) \to B)^*_\rho :\simeq \Pi_{y \in A^*_\rho} B^*_{\rho^y_x} \ , & ((x : A)a)^*_\rho :\simeq \lambda y \in A^*_\rho . a^*_{\rho^y_x} \ , \\
(a(b))^*_\rho :\simeq a^*_\rho(b^*_\rho) \ , & ((x : A) \times B)^*_\rho :\simeq \Sigma_{y \in A^*_\rho} B^*_{\rho^y_x} \ , \\
\langle a, b\rangle^*_\rho :\simeq \langle a^*_\rho, b^*_\rho\rangle \ , & (\pi_0(a))^*_\rho :\simeq (a)_0 \ , \\
(\pi_1(a))^*_\rho :\simeq (a)_1 \ , & 1^* :\simeq 1 \ , \\
\langle\rangle^* :\simeq 0 \ , & \mathbb{B}^* :\simeq \{0, 1\} \ , \\
\mathrm{tt}^* :\simeq 0 \ , & \mathrm{ff}^* :\simeq 1 \ ,
\end{array}$$

$$(\text{if } a \text{ then } b \text{ else } c)^*_\rho :\simeq \begin{cases} b^*_\rho & \text{if } a^*_\rho = 0 \ , \\ c^*_\rho & \text{if } a^*_\rho = 1 \ , \\ \text{undefined} & \text{otherwise} \ . \end{cases}$$

To interpret terms with constructors SP, nonind, ind, Arg, …, we first define SP^*, nonind^*, ind^*, Arg^*, arg^*, map^*, IH^*, mapIH^*, U^*, T^* and interpret

$$\begin{aligned}
(\mathrm{SP}_D)^*_\rho &:\simeq \mathrm{SP}^*(D^*_\rho) \ , \\
(\mathrm{nonind}(a, b))^*_\rho &:\simeq \mathrm{nonind}^*(a^*_\rho, b^*_\rho) \ , \\
(\mathrm{Arg}_{D,\phi})^*_\rho &:\simeq \mathrm{Arg}^*(D^*_\rho, \phi^*_\rho) \ , \\
(\mathrm{map}_{D,\phi}(U, T))^*_\rho &:\simeq \lambda x \in \mathrm{arg}^*(D^*_\rho, \phi^*_\rho, U^*_\rho, T^*_\rho) . \mathrm{map}^*(D^*_\rho, \phi^*_\rho, U^*_\rho, T^*_\rho, x) \ , \\
&\text{etc.}
\end{aligned}$$

$\mathrm{SP}^*(D)$ is defined for $D \in \mathrm{type}^*$ as the least set such that

$$\mathrm{SP}^*(D) = 1 + \Sigma_{a \in \mathrm{set}^*}(a \to \mathrm{SP}^*(D)) + \Sigma_{a \in \mathrm{set}^*}((a \to D) \to \mathrm{SP}^*(D)) \ ,$$

which we get by iterating the appropriate operator κ times, if for all $a \in \mathrm{set}^*$ the cardinality of a and of $a \to D$ is less than κ. If $D \in \mathrm{V}_{\lambda_n}$, therefore $\mathrm{SP}^*(D) \in \mathrm{V}_{\lambda_{n+1}}$.

$$\mathrm{nil}^* :\simeq \langle 0, 0\rangle, \ \mathrm{nonind}^*(a, b) :\simeq \langle 1, \langle a, b\rangle\rangle, \ \mathrm{ind}^*(a, b) :\simeq \langle 2, \langle a, b\rangle\rangle \ .$$

$\mathrm{Arg}^*(D, \phi)$ is defined, if $\phi \in \mathrm{SP}^*(D)$, and then defined in accordance with the equations for Arg, that is,

$$\begin{aligned}
\mathrm{Arg}^*(D, \mathrm{nil}^*) &:\simeq 1 \ , \\
\mathrm{Arg}^*(D, \mathrm{nonind}^*(A, \phi)) &:\simeq \Sigma_{x \in A} \mathrm{Arg}^*(D, \phi(x)) \ , \\
\mathrm{Arg}^*(D, \mathrm{ind}^*(A, \phi)) &:\simeq \Sigma_{f \in (A \to D)} \mathrm{Arg}^*(D, \phi(f)) \ .
\end{aligned}$$

Similarly, we define $\mathrm{arg}^*(D, \phi, U, T)$, $\mathrm{map}^*(D, \phi, U, T, a)$, and for $D' \in (U \to \mathrm{type}^*)$, $\mathrm{IH}^*(D, \phi, U, T, D', \gamma)$, $\mathrm{mapIH}^*(D, \phi, U, T, D', R, a)$.

$$\begin{aligned}
\mathrm{U}^*(D, \phi, d) &:\simeq \mathrm{U}^{\mathrm{M}}(D, \phi, d) \ , \\
\mathrm{T}^*(D, \phi, d) &:\simeq \lambda x \in \mathrm{U}^{\mathrm{M}}(D, \phi, d). \mathrm{T}^{\mathrm{M}}(D, \phi, d, x) \ ,
\end{aligned}$$

where $\mathrm{U}^\alpha(D, \phi, d)$ and $\mathrm{T}^\alpha(D, \phi, d)$ or shorter U^α and T^α are simultaneously defined by recursion on α as

$$\begin{aligned}
\mathrm{U}^\alpha &:\simeq \mathrm{arg}^*(D, \phi, \mathrm{U}^{<\alpha}, \mathrm{T}^{<\alpha}) \ , \\
\mathrm{T}^\alpha(a) &:\simeq d(\mathrm{map}^*(D, \phi, \mathrm{U}^{<\alpha}, \mathrm{T}^{<\alpha}, a)) \ ,
\end{aligned}$$

$$\begin{aligned}
\mathrm{intro}(a)^*_\rho &:\simeq a^*_\rho \ , \\
\mathrm{R}^*(D, \phi, d, D', e, a) &:\simeq \mathrm{R}^{\mathrm{M}}(D, \phi, d, D', e, a), \text{ where} \\
\mathrm{R}^\alpha(D, \phi, d, D', e, a) &:\simeq e(a, \mathrm{mapIH}^*(D, \phi, \mathrm{U}^*(D, \phi, d), \mathrm{T}^*(D, \phi, d) \ , \\
&\qquad D', \mathrm{R}^{<\alpha}(D, \phi, d, D', e), a)) \ .
\end{aligned}$$

Contexts will be interpreted as sets of assignments:

$$\emptyset^* :\simeq \emptyset \ , \quad (\Gamma, x : A)^* :\simeq \{\rho^a_x \mid \rho \in \Gamma^*_\rho \wedge a \in A^*_\rho\} \ .$$

6.2 Soundness of the Rules

Theorem 1. (Soundness theorem)

(a) *If* $\vdash \Gamma$ context, *then* Γ^* *is defined.*
(b) *If* $\vdash \Gamma \Rightarrow A : E$, *where* $E \equiv \mathrm{type}$ *or* E *is a term, then* Γ^* *is defined,* $\forall \rho \in \Gamma^*. A^*_\rho \in E^*_\rho$, *and if* $E \not\equiv \mathrm{type}$, $\forall \rho \in \Gamma^*. E^*_\rho \in \mathrm{type}^*$.
(c) *If* $\vdash \Gamma \Rightarrow A = B : E$, *where* $E \equiv \mathrm{type}$ *or* E *is a term, then* Γ^* *is defined,* $\forall \rho \in \Gamma^*(A^*_\rho \in E^*_\rho \wedge B^*_\rho = A^*_\rho)$, *and if* $E \not\equiv \mathrm{type}$, $\forall \rho \in \Gamma^*. E^*_\rho \in \mathrm{type}^*$.
(d) $\not\vdash a : \mathrm{N}_0$, *where* N_0 *is the empty set, for any of the possibilities mentioned in Section 5.*

The proof of the Soundness theorem is more or less routine, except for the verification that U : set. In order to prove this we will need some lemmata.

First we need to verify that U^α is increasing with α and that for $\alpha < \beta$ T^α and T^β coincide on U^α. In order to prove this we need to verify that $\mathrm{arg}^*(D, \phi, U, T)$ and $\mathrm{map}^*(D, \phi, U, T)$ are monotone in U, T, as expressed by the following lemma:

Lemma 1. *Assume* $D \in \mathrm{type}^*$, $\phi \in \mathrm{SP}^*(D)$, $U \subseteq U' \in \mathrm{set}^*$, $T' : U' \to D$, $T = T' \restriction U$. *Then*

(a) $\mathrm{arg}^*(D,\phi,U,T) \subseteq \mathrm{arg}^*(D,\phi,U',T')$ *and*
(b) $\mathrm{map}^*(D,\phi,U',T') \restriction \mathrm{arg}^*(D,\phi,U,T) = \mathrm{map}^*(D,\phi,U,T)$.

We want to show that there is a $\kappa < \mathrm{M}$ such that $\mathrm{U}^{<\kappa} = \mathrm{U}^{\kappa}$. This is the case if κ is a limit ordinal such that $\mathrm{arg}^*(D,\phi,U,T)$ is κ-continuous in U and T, that is,

$$\mathrm{arg}^*(D,\phi,\mathrm{U}^{<\kappa},\mathrm{T}^{<\kappa}) = \bigcup_{\alpha<\kappa} \mathrm{arg}^*(D,\phi,\mathrm{U}^{\alpha},\mathrm{T}^{\alpha}) \ . \tag{1}$$

To obtain this we need that all index sets, which start an inductive argument, have cardinality less than κ. The set $\mathrm{Aux}(D,\phi,U,T) \in \mathrm{set}^*$, where $D \in \mathrm{type}^*$, $\phi \in \mathrm{SP}^*(D)$, $U \in \mathrm{set}^*$, $T \in U \to D$, collects all possible such index sets. It is defined by induction on ϕ:

$$\begin{aligned}
\mathrm{Aux}(D,\mathrm{nil}^*,U,T) &:= 1 \ , \\
\mathrm{Aux}(D,\mathrm{nonind}^*(A,\phi),U,T) &:= \Pi_{x\in A}\mathrm{Aux}(D,\phi(x),U,T) \ , \\
\mathrm{Aux}(D,\mathrm{ind}^*(A,\phi),U,T) &:= A + \Pi_{f\in(A\to U)}\mathrm{Aux}(D,\phi(T\circ f),U,T) \ .
\end{aligned}$$

Lemma 2. *Assume* $D \in \mathrm{type}^*$, $\phi \in \mathrm{SP}^*(D)$. *Let* κ *be inaccessible and let for* $\alpha < \kappa$ $U^\alpha \in \mathrm{set}^*$, $T^\alpha : U^\alpha \to D$ *such that for* $\alpha < \beta$, $U^\alpha \subseteq U^\beta$, $T^\alpha = T^\beta \restriction U^\alpha$. *Assume also for some* $\alpha_0 < \kappa$ *and for all* $\alpha_0 \leq \alpha < \kappa$

$$\mathrm{Aux}(D,\phi,U^\alpha,T^\alpha) \in V_\kappa \ . \tag{2}$$

Then $\mathrm{arg}^*(D,\phi,U,T)$ *is* κ*-continuous in* U *and* T, *that is, (1) holds.*

Proof: "$\supseteq$" follows by Lemma 1b.

"$\subseteq$" follows by induction on ϕ. We treat only the main case $\phi = \mathrm{ind}^*(A,\gamma)$. Assume $a \in \mathrm{arg}^*(D,\phi,U^{<\kappa},T^{<\kappa})$, and show $a \in \mathrm{arg}^*(D,\phi,U^\alpha,T^\alpha)$ for some $\alpha < \kappa$. We know $a = \langle f,c\rangle$ for some $f : A \to U^{<\kappa}$, $c \in \mathrm{arg}^*(D,\gamma(T^{<\kappa} \circ f),U^{<\kappa},T^{<\kappa})$. By (2) it follows $A \in V_\kappa$, and by the inaccessibility of κ there exists a $\beta < \kappa$ such that $f : A \to U^{<\beta}$, especially $f : A \to U^\beta$. W.l.o.g. $\alpha_0 \leq \beta$. For $\beta \leq \alpha < \kappa$ it follows $\mathrm{Aux}(D,\gamma(T^\alpha \circ f),U^\alpha,T^\alpha) \in V_\kappa$ and therefore by induction hypothesis there exists a β' such that $c \in \mathrm{arg}^*(D,\gamma(T^{\beta'} \circ f),U^{\beta'},T^{\beta'})$. With $\alpha := \max\{\beta,\beta'\}$ follows the assertion. □

Lemma 3. *Assume* $\phi \in \mathrm{SP}^*(D)$, $s \in \mathrm{Arg}^*(D,\phi) \to D$. *Abbreviate* $\mathrm{U}^\alpha := \mathrm{U}^\alpha(D,\phi,d)$, $\mathrm{T}^\alpha := \mathrm{T}^\alpha(D,\phi,d)$ *and note that* $\mathrm{U}^*(D,\phi,d) = \mathrm{U}^{\mathrm{M}}$, $\mathrm{T}^*(D,\phi,d) = \mathrm{T}^{\mathrm{M}}$.

(a) $\mathrm{T}^\alpha : \mathrm{U}^\alpha \to D$, *and if* $\alpha < \mathrm{M}$, $\mathrm{U}^\alpha \in \mathrm{V}_{\mathrm{M}}$.
(b) If $\alpha < \beta$ *then* $\mathrm{U}^\alpha \subseteq \mathrm{U}^\beta$ *and* $\mathrm{T}^\beta \restriction \mathrm{U}^\alpha = \mathrm{T}^\alpha$.
(c) There exists $\kappa < \mathrm{M}$ *such that* $\mathrm{U}^\alpha = \mathrm{U}^\kappa$ *(and therefore* $\mathrm{T}^\alpha = \mathrm{T}^\kappa$*) for all* $\alpha > \kappa$.
(d) $\mathrm{U}^{\mathrm{M}} \in \mathrm{V}_{\mathrm{M}}$, $\mathrm{arg}^*(D,\phi,\mathrm{U}^{\mathrm{M}},\mathrm{T}^{\mathrm{M}}) \subseteq \mathrm{U}^{\mathrm{M}}$.

Proof:

(a) Easy induction on α.

(b) Induction on α, β, by using Lemma 1(b).

(c) Define $f : \mathrm{Ord} \to \mathrm{Ord}$ by transfinite recursion:

$$f(\beta) = \min\{\alpha \mid \forall \beta' < \beta(f(\beta') < \alpha) \wedge \\ \forall \beta' < \mathrm{M}(\mathrm{U}^{\beta'} \subseteq \mathrm{V}_\beta \to \mathrm{U}^{\beta'+1} \cup \mathrm{Aux}(D, \phi, \mathrm{U}^{\beta'}, \mathrm{T}^{\beta'}) \subseteq \mathrm{V}_\alpha\}$$

$f : \mathrm{M} \to \mathrm{M}$ follows immediately by M being inaccessible, since

$$\{\mathrm{U}^{\beta'} \mid \beta' < \mathrm{M} \wedge \mathrm{U}^{\beta'} \subseteq \mathrm{V}_\beta\} \in \mathrm{V}_{\beta+1} \subseteq \mathrm{V}_M \ .$$

Let for $\alpha < \mathrm{M}$ $\theta(\alpha) := f^\alpha(0)$. By the regularity of M we have $\theta : \mathrm{M} \to \mathrm{M}$. Since f is increasing, θ is normal. Hence, since M is Mahlo, θ has an inaccessible fixed point $\kappa < \mathrm{M}$.

Therefore $f : \kappa \to \kappa$: Assume $\alpha < \kappa$. κ is a limit ordinal, therefore $\alpha < \theta(\beta)$ for some $\beta < \kappa$, $f(\alpha) < f(\theta(\beta)) = \theta(\beta+1) < \theta(\kappa) = \kappa$. By induction on α, using the regularity of κ, for $\alpha < \kappa$ $\mathrm{U}^\alpha \in \mathrm{V}_\kappa$, $\mathrm{Aux}(D, \phi, \mathrm{U}^\alpha, \mathrm{T}^\alpha) \in V_\kappa$, and therefore by Lemma 2

$$\begin{aligned} \mathrm{U}^\kappa &= \arg^*(D, \phi, \mathrm{U}^{<\kappa}, \mathrm{T}^{<\kappa}) \\ &= \bigcup_{\alpha<\kappa} \arg^*(D, \phi, \mathrm{U}^\alpha, \mathrm{T}^\alpha) \\ &= \bigcup_{\alpha<\kappa} \mathrm{U}^{\alpha+1} = \mathrm{U}^{<\kappa}. \end{aligned}$$

By induction on α for all $\alpha \geq \kappa$ $\mathrm{U}^\alpha = \mathrm{U}^{<\kappa} = \mathrm{U}^\kappa$.

(d) $\mathrm{U}^\mathrm{M} = \mathrm{U}^\kappa \in \mathrm{V}_\mathrm{M}$, $\arg^*(D, \phi, \mathrm{U}^\mathrm{M}, \mathrm{T}^\mathrm{M}) = \arg^*(D, \phi, \mathrm{U}^\kappa, \mathrm{T}^\kappa) \subseteq \mathrm{U}^{\kappa+1} \subseteq \mathrm{U}^\mathrm{M}$. □

7 Related and Future Work

Universes in type theory. The first example of an inductive-recursive definition in type theory was Martin-Löf's universe à la Tarski [3]. [9] Then Palmgren [22] defined external and internal universe hierarchies and also a super universe. Rathjen, Griffor, and Palmgren [23] defined quantifier universes and Palmgren [2] defined higher order universe hierarchies. All these constructions use induction-recursion, whereas Setzer [10] defined a Mahlo universe, which goes beyond it.

Inductive definitions in type theory. Previous work on formalization of inductive definitions in Martin-Löf's type theory has mainly used external schemata in the style of Martin-Löf's intuitionistic theory of iterated inductive definitions in predicate logic [17]. See for example Backhouse [19], Dybjer [1], and Paulin [16]. A schema for inductive-recursive definitions was introduced by Dybjer [18].

[9] There are earlier examples of informal inductive-recursive definitions, for example, Martin-Löf's simultaneous definition of the notions of computable type and term [4] from 1972. However, the explicitly inductive-recursive nature of type-theoretic universes was only brought out when they were formulated à la Tarski rather than à la Russell.

Categorical semantics of inductive types and of universes. The categorical semantics of inductively defined dependent types has been discussed for example by Coquand and Paulin [14] and Mendler [11]. The latter article also discusses categorical semantics of universes in type theory. In a future article we plan to extend Mendler's work, by giving categorical semantics of inductive-recursive definitions in terms of initial algebras on endofunctors in slice categories. We will also show how such semantics suggest an alternative finite axiomatization of inductive-recursive definitions.

Set-theoretic semantics of type theory. It is well-known that Martin-Löf's type theory has a "naive" full function-space model, see for example the introduction in Troelstra [7]. Dybjer [20] gives a full function space model of Martin-Löf's type theory with an external schema for inductive definitions. Aczel's recent article [5] contains further information about set-theoretic interpretations of type theory.

Large cardinals in set theory. Induction-recursion gives quite a general approach to type-theoretic analogues of large cardinals in set theory. See for example Drake [9] for an introduction to large cardinals. Induction-recursion gives rise to analogues of for example inaccessible, hyper-inaccessible cardinals, and more generally Mahlo's π-numbers [23], but does not justify the definition of a *set*, which is an analogue of a Mahlo cardinal. However, the *type* of sets has closure properties similar to those of a Mahlo cardinal.

Constructive versions of the model. The current model requires much more proof theoretic power than is actually needed: the strength of the type theory considered is very weak relative to ZF, even without any addition of large cardinals. Aczel [5] shows that the set theoretic models interpret as well the principle of excluded middle of type theory, an enormous strengthening of the type theory. In order to get a model in a theory which has the same strength, Aczel modifies the model and replaces ZF by constructive set theory CZF. One can as well define a model in a theory of the same strength by giving a realizability interpretation in Kripke-Platek set theory extended by a recursive Mahlo ordinal and ω admissibles above, extending [24, 25, 6]. Both models require some extra work, which exceeds the space available in this article.[10]

Proof-theoretic strength of type theory. It should be easy to develop a term model of the theory in KPM^+ used in [6] for the interpretation of Mahlo type theory. Such a model, which will make use of a (countable) *recursive Mahlo* ordinal and ω admissibles above it only, would show that the strength of the current type theory is at most as big as the Mahlo universe. On the other hand, set can be seen as being almost a Mahlo-universe, since we have induction over arbitrary types. What is missing to get the full strength is the possibility of

[10] The interpretation in the extension of Kripke-Platek set theory will be presented in an extended version of this article.

having the W-type on top of the universe. In [13] together with [26], [25], [8] it was shown that in case of one universe such a restriction reduces the strength from $|\mathrm{KPI}^+|$ to $|\mathrm{KPI}|$ and with a similar argument for the lower bound as in [13] it is very likely that using the Mahlo-feature of set we have a lower bound $|\mathrm{KPM}|$. Therefore it seems that the strength of our theory lies in the interval $[|\mathrm{KPM}|, |\mathrm{KPM}^+|]$.

Inductive-recursive definitions seem to cover what is by many (but not all) researchers considered at the moment as predicative type theory. Even if some extensions are not covered by our calculus, it seems unlikely that such extensions will get beyond the strength of the Mahlo universe. This indicates that Mahloness is a natural boundary in the world of predicativity, which can only be crossed by adding principles such as the existence of the Mahlo universe as a set. The second author regards such principles as predicatively justifiable.

Inductive-recursive definition of indexed families. The external schema by Dybjer [18] considers the more general case of the simultaneous inductive-recursive definition of a set-indexed family of sets and functions. The present finite axiomatization can be extended to this case too, but we postpone the presentation of this to a future article.

References

1. P. Dybjer. Inductive families. *Formal Aspects of Computing*, 6:440–465, 1994.
2. E. Palmgren. On universes in type theory. To appear in: G. Sambin, and J. Smith, editors: *Twenty-Five Years of Constructive Type Theory*.
3. P. Martin-Löf. *Intuitionistic Type Theory*. Bibliopolis, 1984.
4. P. Martin-Löf. An intuitionistic theory of types. In G. Sambin and J. Smith, editors, *Twenty-Five Years of Constructive Type Theory*. Oxford University Press, 1998. To appear. Reprinted version of an unpublished report from 1972.
5. P. Aczel. On relating type theories and set theories. Submitted to *TYPES' 98*, LNCS, Springer-Verlag.
6. A. Setzer. A model for a type theory with Mahlo universe. Draft, 1996.
7. A. S. Troelstra. On the syntax of Martin-Löf's type theories. *Theoretical Computer Science*, 51:1–26, 1987.
8. A. Setzer. Well-ordering proofs for Martin-Löf type theory. *Annals of Pure and Applied Logic*, 92:113 – 159, 1998.
9. F. R. Drake. *Set Theory - an Introduction to Large Cardinals*. North Holland, 1974.
10. A. Setzer. Extending Martin-Löf Type Theory by one Mahlo-universe. To appear in *Archive for Mathematical Logic*.
11. P. F. Mendler. Predicative type universes and primitive recursion. In *Proceedings Sixth Annual Synposium on Logic in Computer Science*. IEEE Computer Society Press, 1991.
12. S. Allen. *A Non-Type-Theoretic Semantics for Type-Theoretic Language*. PhD thesis, Department of Computer Science, Cornell University, 1987.
13. E. Griffor and M. Rathjen. The strength of some Martin-Löf type theories. *Archive for Mathematical Logic*, 33:347 – 385, 1994.

14. T. Coquand and C. Paulin. Inductively defined types, preliminary version. In *LNCS 417, COLOG '88, International Conference on Computer Logic*. Springer-Verlag, 1990.
15. J. Cederquist. *Pointfree Approach to Constructive Analysis in Type Theory*. PhD thesis, Department of Computing Science, Chalmers University of Technology and University of Göteborg, 1997.
16. C. Paulin-Mohring. Inductive definitions in the system Coq - rules and properties. In *Proceedings Typed λ-Calculus and Applications*, pages 328–245. Springer-Verlag, LNCS, March 1993.
17. P. Martin-Löf. Hauptsatz for the intuitionistic theory of iterated inductive definitions. In J. E. Fenstad, editor, *Proceedings of the Second Scandinavian Logic Symposium*, pages 179–216. North-Holland, 1971.
18. P. Dybjer. A general formulation of simultaneous inductive-recursive definitions in type theory. To appear in *Journal of Symbolic Logic*.
19. R. Backhouse. On the meaning and construction of the rules in Martin-Löf's theory of types. In A. Avron, B. Harper, F. Honsell, I. Mason, and G. Plotkin, editors, *Proceedings of the Workshop on General Logic, Edinburgh, February 1987*. Laboratory for Foundations of Computer Science, Department of Computer Science, University of Edinburgh, 1988. ECS-LFCS-88-52.
20. P. Dybjer. Inductive sets and families in Martin-Löf's type theory and their set-theoretic semantics. In G. Huet and G. Plotkin, editors, *Logical Frameworks*, pages 280–306. Cambridge University Press, 1991.
21. B. Nordström, K. Petersson, and J. Smith. *Programming in Martin-Löf's Type Theory: an Introduction*. Oxford University Press, 1990.
22. E. Palmgren. *On Fixed Point Operators, Inductive Definitions and Universes in Martin-Löf's Type Theory*. PhD thesis, Uppsala University, 1991.
23. M. Rathjen, E. R. Griffor, and E. Palmgren. Inaccessibility in constructive set theory and type theory. *Annals of Pure and Applied Logic*, 94:181 - 200, 1998.
24. A. Setzer. *Proof theoretical strength of Martin-Löf Type Theory with W-type and one universe.* PhD thesis, Fakultät für Mathematik der Ludwig-Maximilians-Universität München, 1993.
25. A. Setzer. An upper bound for the proof theoretical strength of Martin-Löf Type Theory with W-type and one Universe. Draft, 1996.
26. A. Setzer. *Proof theoretical strength of Martin-Löf Type Theory with W-type and one universe.* PhD thesis, Universität München, 1993.

Lambda Definability with Sums via Grothendieck Logical Relations

Marcelo Fiore[1] and Alex Simpson[2]

[1] COGS, University of Sussex
<marcelo@cogs.susx.ac.uk>
[2] LFCS, Division of Informatics, University of Edinburgh
<Alex.Simpson@dcs.ed.ac.uk>

Abstract. We introduce a notion of *Grothendieck logical relation* and use it to characterise the definability of morphisms in *stable* bicartesian closed categories by terms of the simply-typed lambda calculus with finite products and finite sums. Our techniques are based on concepts from topos theory, however our exposition is elementary.

Introduction

The use of logical relations as a tool for characterising the λ-definable elements in a model of the simply-typed λ-calculus originated in the work of Plotkin [10], who obtained such a characterisation of the definable elements in the full type hierarchy using a notion of *Kripke logical relation*. Subsequently, the more general notion of a *Kripke logical relation of varying arity* was developed by Jung and Tiuryn, and shown to characterise the definable elements in any Henkin model [4]. Although not emphasised in [4], relations of varying arity are powerful enough to characterise *relative definability* with respect to any given set of elements considered as constants. The full generality of the approach is demonstrated in Alimohamed [1], where such relations are used to characterise relative definability in an arbitrary cartesian closed category.

In general, results about the pure simply-typed λ-calculus extend easily to analogous results for systems containing finite product types. This is not the case for finite coproduct (sum) types. Although the equational theory of bicartesian closed categories provides a basic formal system, the syntactic techniques used to study systems without coproducts fall over in their presence. Two fundamental properties of this equational theory, decidability (Ghani [3]) and its completeness relative to the equalities valid in the category, **Set**, of sets (Dougherty and Subrahmanyam [2]), were established only recently. It is apparently still an open question whether the finite model property holds for this theory (although it is inconceivable that it does not). Also, both the above results have been proved only for nonempty sums (i.e. with the empty type omitted).

In this paper, we extend the logical relations characterization of relative definability to the simply-typed λ-calculus with products and sums (including the empty type). As might be expected, this requires some development of the theory of logical relations. It turns out that what is needed is a natural generalization of Kripke logical relations of varying arity, in which the base poset (or, more generally, category) for the relation is endowed with a *Grothendieck topology* [6]. Using such *Grothendieck logical relations*, we characterise relative definability in any bicartesian closed category in which the finite coproducts are stable (as is the case in **Set**). We do not know if the characterisation extends also to the non stable case.

From the categorical point of view our results are best explained in terms of *glueing* [12, 1]. However, for this conference version of the paper, we keep our exposition elementary, in the hope that it will be accessible to most type theorists with some background in categorical semantics.

It should be said that the research in this paper originated as part of a strategy conceived by the authors for attacking the full abstraction problem for call-by-value FPC (which includes finite sums). Kripke logical relations of varying arity had already been used to obtain full abstraction for PCF by O'Hearn and Riecke [8]. The extension of these results to FPC seemed to us to require an additional analysis of both partiality and sums. This line of research was never fully pursued because similar full abstraction results for FPC were soon obtained by Riecke and Sandholm [11]. However, their treatment of coproducts is somewhat *ad hoc* (although one does get the feeling that a Grothendieck topology is at work behind the scenes). We believe that it would be very worthwhile to integrate our more conceptual approach to coproducts into the full abstraction picture.

It seems likely that the notion of Grothendieck logical relation will have other applications. For example, the lengthy and heavily syntactic proof of equational completeness relative to **Set** in [2], has hints of Grothendieck toplologies within it. It is plausible that Grothendieck logical relations will lead to simpler and more general such completeness proofs.

1 Simply typed lambda calculus with sums

The language we work with is a simply-typed λ-calculus with additional types for finite products and sums. In this section we describe the syntax of the language, and its interpretation in any bicartesian closed category.

Syntax. We use $\mathtt{T}, \ldots$ to range over a set T of *base types*, and $\tau, \ldots$ to range over types which are specified by the grammar below.

$$\tau ::= \mathtt{T} \mid \tau_1 \to \tau_2 \mid \times^{(n)}(\tau_1, \ldots, \tau_n) \mid +^{(n)}(\tau_1, \ldots, \tau_n) \qquad n \in \mathbb{N}$$

We write $\mathtt{1}$ and $\mathtt{0}$ for $\times^{(0)}()$ and $+^{(0)}()$ respectively. We use n-ary products and sums as primitive to emphasize that all our definitions for the zero-ary cases are just the natural instances of the general n-ary scheme. This is of particular interest in the case of the empty type $\mathtt{0}$, which is generally thought of as troublesome, and often omitted from consideration altogether [3, 2].

We use $\mathtt{x}, \ldots$ to range over a countably infinite set of variables. A *(type) environment* is a finite sequence $\mathtt{x}_1 : \tau_1, \ldots, \mathtt{x}_n : \tau_n$ where all the variables are distinct. We use $\Gamma, \ldots$ to range over environments. We write $\langle\rangle$ for the empty sequence in general, and the empty environment in particular.

Terms are specified according to a *T-signature*, Σ, which is a set of pairs of the form $(c : \tau)$ assigning types τ to *constants* c, such that each constant symbol in Σ is assigned only one type. The terms are generated by the rules in Fig. 1. For notational convenience, we will always omit the superscripts from the injections $\mathtt{in}_i^{\tau_1,\ldots,\tau_n}(t)$. As usual we consider terms as identified up to α-equivalence.

For the remainder of the paper we consider a fixed (though arbitrary) set of base types T and signature Σ.

Semantics. For the purpose of this paper, a *bicartesian closed category* is a category with finite coproducts, finite products and exponentials (we do not assume finite limits). Let $\mathcal{S}$ be bicartesian closed with chosen structure (0, +, 1, $\times$, $\Rightarrow$) (here we are distinguishing initial object, binary coproduct, terminal object, binary product and exponential). We define canonical finite coproducts by $\coprod^{(0)} \stackrel{\text{def}}{=} 0$ and $\coprod^{(n+1)}(A_1, \ldots, A_n, A_{n+1}) \stackrel{\text{def}}{=} \coprod^{(n)}(A_1, \ldots, A_n) + A_{n+1}$. Canonical finite products $\prod^{(n)}(A_1, \ldots, A_n)$ are defined similarly. We use standard notation for injections, projections, the universal maps, and the "evaluation" map and "Currying" operation associated with the closed structure.

A *T-interpretation in* $\mathcal{S}$ is a function from T to objects of $\mathcal{S}$. Under a T-interpretation $\mathcal{I}$ every type τ is interpreted as an object $[\![\tau]\!]_{\mathcal{I}}$ in the obvious way. The interpretation of types extends to environments by the usual definition:

$$[\![\mathtt{x}_1 : \tau_1, \ldots, \mathtt{x}_n : \tau_n]\!]_{\mathcal{I}} \stackrel{\text{def}}{=} \prod^{(n)}([\![\tau_1]\!]_{\mathcal{I}}, \ldots, [\![\tau_n]\!]_{\mathcal{I}})$$

A *(T, Σ)-interpretation $\mathcal{I}$ in $\mathcal{S}$* is a pair $(\mathcal{I}_T, \mathcal{I}_\Sigma)$ where $\mathcal{I}_T$ is a T-interpretation, and $\mathcal{I}_\Sigma$ is a function mapping each constant $(c : \tau) \in \Sigma$ to a global element $\mathcal{I}_\Sigma(c) : 1 \to [\![\tau]\!]$ in $\mathcal{S}$. Under a (T, Σ)-interpretation every term $\Gamma \vdash t : \tau$ is interpreted as a generalised element $[\![\Gamma \vdash t : \tau]\!]_{\mathcal{I}} : [\![\Gamma]\!] \to [\![\tau]\!]$ in $\mathcal{S}$ by:

$$\begin{aligned}
&[\![\mathtt{x}_1 : \tau_1, \ldots, \mathtt{x}_n : \tau_n \vdash \mathtt{x}_i : \tau_i]\!] \stackrel{\text{def}}{=} \pi_i \\
&[\![\Gamma \vdash c : \tau]\!] \stackrel{\text{def}}{=} \mathcal{I}_\Sigma(c) \circ \langle\rangle \\
&[\![\Gamma \vdash \lambda\mathtt{x}. : \tau_1.t : \tau_1 \to \tau_2]\!] \stackrel{\text{def}}{=} \lambda[\![\Gamma, \mathtt{x} : \tau_1 \vdash t : \tau_2]\!] \\
&[\![\Gamma \vdash t(t_1) : \tau_2]\!] \stackrel{\text{def}}{=} \mathsf{ev} \circ \langle [\![\Gamma \vdash t : \tau_1 \to \tau_2]\!], [\![\Gamma \vdash t_1 : \tau_1]\!] \rangle \\
&[\![\Gamma \vdash \langle t_1, \ldots, t_n\rangle : \times^{(n)}(\tau_1, \ldots, \tau_n)]\!] \stackrel{\text{def}}{=} \langle [\![\Gamma \vdash t_1 : \tau_1]\!], \ldots, [\![\Gamma \vdash t_n : \tau_n]\!] \rangle \\
&[\![\Gamma \vdash \mathtt{proj}_i(t) : \tau_i]\!] \stackrel{\text{def}}{=} \pi_i \circ [\![\Gamma \vdash t : \times^{(n)}(\tau_1, \ldots, \tau_n)]\!] \\
&[\![\Gamma \vdash \mathtt{in}_i(t) : +^{(n)}(\tau_1, \ldots, \tau_n)]\!] \stackrel{\text{def}}{=} \amalg_i \circ [\![\Gamma \vdash t : \tau_i]\!] \\
&[\![\Gamma \vdash \mathtt{case}\ t\ \mathtt{of}\ [\mathtt{in}_1(\mathtt{x}_1).t_1, \ldots, \mathtt{in}_n(\mathtt{x}_n).t_n] : \tau]\!] \stackrel{\text{def}}{=} \\
&\qquad [[\![\Gamma, \mathtt{x}_1 : \tau_1 \vdash t_1 : \tau]\!], \ldots, [\![\Gamma, \mathtt{x}_n : \tau_n \vdash t_n : \tau]\!]] \circ \\
&\qquad\qquad \delta^{(n)} \circ \langle \mathrm{id}_{[\![\Gamma]\!]}, [\![\Gamma \vdash t : +^{(n)}(\tau_1, \ldots, \tau_n)]\!] \rangle
\end{aligned}$$

where $\delta^{(n)} : C \times (\coprod^{(n)}(A_1, \ldots, A_n)) \to \coprod^{(n)}(C \times A_1, \ldots, C \times A_n))$ is the distributivity isomorphism.

$$\frac{}{\mathbf{x}_1 : \tau_1, \ldots, \mathbf{x}_n : \tau_n \vdash \mathbf{x}_i : \tau_i}\ 1 \leq i \leq n \qquad \frac{}{\Gamma \vdash c : \tau}\ (c : \tau) \in \Sigma$$

$$\frac{\Gamma, \mathbf{x} : \tau_1 \vdash t : \tau_2}{\Gamma \vdash \lambda \mathbf{x} : \tau_1 . t : \tau_1 \to \tau_2} \qquad \frac{\Gamma \vdash t : \tau_1 \to \tau_2 \quad \Gamma \vdash t_1 : \tau_1}{\Gamma \vdash t(t_1) : \tau_2}$$

$$\frac{\Gamma \vdash t_1 : \tau_1 \quad \ldots \quad \Gamma \vdash t_n : \tau_n}{\Gamma \vdash \langle t_1, \ldots, t_n \rangle : \times^{(n)}(\tau_1, \ldots, \tau_n)} \qquad \frac{\Gamma \vdash t : \times^{(n)}(\tau_1, \ldots, \tau_n)}{\Gamma \vdash \mathbf{proj}_i(t) : \tau_i}\ 1 \leq i \leq n$$

$$\frac{\Gamma \vdash t : \tau_i}{\Gamma \vdash \mathbf{in}_i^{\tau_1, \ldots, \tau_n}(t) : +^{(n)}(\tau_1, \ldots, \tau_n)}\ 1 \leq i \leq n$$

$$\frac{\Gamma \vdash t : +^{(n)}(\tau_1, \ldots, \tau_n) \qquad \Gamma, \mathbf{x}_i : \tau_i \vdash t_i : \tau \quad 1 \leq i \leq n}{\Gamma \vdash \mathbf{case}\ t\ \mathbf{of}\ [\mathbf{in}_1(\mathbf{x}_1).t_1, \ldots, \mathbf{in}_n(\mathbf{x}_n).t_n] : \tau}$$

Fig. 1. Term syntax

$$\frac{}{\Gamma \mid \Xi \vdash t = t : \tau} \qquad \frac{\Gamma \mid \Xi \vdash t = t' : \tau}{\Gamma \mid \Xi \vdash t' = t : \tau}$$

$$\frac{\Gamma \mid \Xi \vdash t_1 = t_2 : \tau \quad \Gamma \mid \Xi \vdash t_2 = t_3 : \tau}{\Gamma \mid \Xi \vdash t_1 = t_3 : \tau}$$

$$\frac{}{\mathbf{x}_1 : \tau_1, \ldots, \mathbf{x}_n : \tau_n \mid t_1 =_{\tau_1'} t_1', \ldots, t_n =_{\tau_n'} t_n' \vdash t_i = t_i' : \tau_i'}\ 1 \leq i \leq n$$

$$\frac{\Gamma \mid \Xi \vdash t_1 = t_1' : \tau_1}{\Gamma \mid \Xi \vdash t(t_1) = t(t_1') : \tau_2} \qquad \frac{\Gamma, \mathbf{x} : \tau_1 \mid \Xi, \mathbf{x} =_{\tau_1} \mathbf{x} \vdash t = t' : \tau_2}{\Gamma \mid \Xi \vdash \lambda \mathbf{x} : \tau_1 . t = \lambda \mathbf{x} : \tau_1 . t' : \tau_1 \to \tau_2}$$

$$\frac{}{\Gamma \mid \Xi \vdash (\lambda \mathbf{x} : \tau_1 . t)(t') = t[t'/\mathbf{x}] : \tau_2} \qquad \frac{}{\Gamma \mid \Xi \vdash t = \lambda \mathbf{x} : \tau_1 . t(\mathbf{x}) : \tau_1 \to \tau_2}\ \mathbf{x} \notin FV(t)$$

$$\frac{}{\Gamma \mid \Xi \vdash \mathbf{proj}_i \langle t_1, \ldots, t_n \rangle = t_i : \tau_i}\ 1 \leq i \leq n$$

$$\frac{}{\Gamma \mid \Xi \vdash t = \langle \mathbf{proj}_1(t), \ldots, \mathbf{proj}_n(t) \rangle : \times^{(n)}(\tau_1, \ldots, \tau_n)}$$

$$\frac{}{\Gamma \mid \Xi \vdash \mathbf{case}\ \mathbf{in}_i(t)\ \mathbf{of}\ [\mathbf{in}_1(\mathbf{x}_1).t_1, \ldots, \mathbf{in}_n(\mathbf{x}_n).t_n] = t_i[t/x_i] : \tau}\ 1 \leq i \leq n$$

$$\frac{\Gamma, \mathbf{x}_i : \tau_i \mid \Xi, \mathbf{in}_i(\mathbf{x}_i) = t \vdash t_i = t' : \tau \qquad 1 \leq i \leq n}{\Gamma \mid \Xi \vdash \mathbf{case}\ t\ \mathbf{of}\ [\mathbf{in}_1(\mathbf{x}_1).t_1, \ldots, \mathbf{in}_n(\mathbf{x}_n).t_n] = t' : \tau}$$

Fig. 2. Equational rules

2 Stable coproducts

To obtain our characterisation of definability, we shall be interested in bicartesian closed categories which enjoy the additional property that coproducts are stable.

Definition 1 (Stable coproducts). In an arbitrary category, a coproduct $\{A_i \to A\}_{i \in I}$ is said to be *stable* if, for every arrow $X \to A$ and $i \in I$, there is a pullback square

$$\begin{array}{ccc} X_i & \longrightarrow & X \\ \downarrow & & \downarrow \\ A_i & \longrightarrow & A \end{array}$$

and the family $\{X_i \to X\}_{i \in I}$ is also a coproduct.

Note that, the stability of the empty coproduct amounts to the strictness of initial objects, which holds in any cartesian closed category [5, Proposition 8.3].

We call a bicartesian closed category *stable* if it has stable finite coproducts (for which it suffices that binary coproducts are stable). Any elementary topos provides an example of a stable bicartesian closed category, and so does any Heyting algebra (note that the latter example shows that stable coproducts need not be disjoint).

We next present a sound formal system for deriving equalities between terms, which is naturally interpreted in stable bicartesian closed categories. The formal system is essentially equivalent to the system WBCT of [2], which was introduced as a critical tool in their proof of the completeness of the equational theory of bicartesian closed categories relative to the valid equations in **Set**. The fact that this system has a natural interpretation in any stable bicartesian closed category has not been observed before.

The proof system is based on a notion of *constrained (type) environment* implementing equational assumptions about terms of sum type.

Definition 2 (Constrained environment). The *constrained environments* $\Gamma \mid \Xi$, consisting of an environment Γ subject to *constraints* Ξ, are defined inductively by the following rules.

$$\frac{}{\langle\rangle \mid \langle\rangle} \qquad\qquad \frac{\Gamma \mid \Xi}{\Gamma, \mathbf{x} : \tau \mid \Xi, \mathbf{x} =_\tau \mathbf{x}} \ \mathbf{x} \notin \Gamma$$

$$\frac{\Gamma \mid \Xi \qquad \Gamma \vdash t : +^{(n)}(\tau_1, \ldots, \tau_n)}{\Gamma, \mathbf{x} : \tau_i \mid \Xi, \ \mathbf{in}_i(\mathbf{x}) =_{+^{(n)}(\tau_1, \ldots, \tau_n)} t} \ \mathbf{x} \notin \Gamma, \quad 1 \le i \le n$$

The equational rules manipulate judgements of the form $\Gamma \mid \Xi \vdash t = t' : \tau$ where both $\Gamma \vdash t : \tau$ and $\Gamma \vdash t' : \tau$ are terms. The rules are given in Fig. 2. They are to be understood as applying only when all the premises and conclusions are genuine (well-typed) terms as specified above.

Henceforth in this section, let $\mathcal{S}$ be a stable bicartesian closed category with chosen structure. (In addition to the chosen bicartesian closed structure, described earlier, we assume a choice of pullbacks for coproduct morphisms. It is not necessary to assume any coherence conditions for these!) Let $\mathcal{I}$ be an interpretation in $\mathcal{S}$. We interpret constrained environments $\Gamma \mid \Xi$ as monos $[\![\Gamma \mid \Xi]\!] \rightarrowtail [\![\Gamma]\!]$. The definition is by structural induction as follows.

- $([\![\langle\rangle \mid \langle\rangle]\!] \rightarrowtail [\![\langle\rangle]\!]) \stackrel{\text{def}}{=} \mathrm{id}_1$.
- $([\![\Gamma, \mathtt{x} : \tau \mid \Xi, \mathtt{x} =_\tau \mathtt{x}]\!] \rightarrowtail [\![\Gamma]\!] \times [\![\tau]\!]) \stackrel{\text{def}}{=} ([\![\Gamma \mid \Xi]\!] \rightarrowtail [\![\Gamma]\!]) \times \mathrm{id}_{[\![\tau]\!]}$.
- $[\![\Gamma, \mathtt{x}_i : \tau_i \mid \Xi, \mathtt{in}_i(\mathtt{x}_i) =_{+^{(n)}(\tau_1,\ldots,\tau_n)} t]\!] \rightarrowtail [\![\Gamma]\!] \times [\![\tau_i]\!]$ is the pairing $\langle m \circ p_i, q_i \rangle$ arising from the following pullback square.

$$\begin{array}{ccc}
[\![\Gamma, \mathtt{x}_i : \tau_i \mid \Xi, \mathtt{in}_i(\mathtt{x}_i) = t]\!] & \xrightarrow{p_i} & [\![\Gamma \mid \Xi]\!] \\
 & & \big\downarrow m \\
\big\downarrow q_i & & [\![\Gamma]\!] \\
 & & \big\downarrow [\![\Gamma \vdash t : +^{(n)}(\tau_1,\ldots,\tau_n)]\!] \\
[\![\tau_i]\!] & \xrightarrow[\amalg_i]{} & \coprod^{(n)}([\![\tau_1]\!], \ldots, [\![\tau_n]\!])
\end{array} \tag{1}$$

Note that, by stability, the family

$$\{p_i : [\![\Gamma, \mathtt{x}_i : \tau_i \mid \Xi, \mathtt{in}_i(\mathtt{x}_i) = t]\!] \to [\![\Gamma \mid \Xi]\!]\}_{1 \le i \le n}$$

from (1) is a coproduct. Observe also that, by definition, for a constrained environment $\Gamma \mid \Xi$ of the form $\mathtt{x}_1 : \tau_1, \ldots, \mathtt{x}_n : \tau_n \mid \mathtt{x}_1 =_{\tau_1} \mathtt{x}_1, \ldots, \mathtt{x}_n =_{\tau_n} \mathtt{x}_n$, we have that $([\![\Gamma \mid \Xi]\!] \rightarrowtail [\![\Gamma]\!]) = \mathrm{id}_{[\![\Gamma]\!]}$. Thus the interpretation of constrained environments extends that of environments. Furthermore, for any $\Gamma \mid \Xi$ of the form $(\mathtt{x}_1 : \tau_1, \ldots, \mathtt{x}_n : \tau_n \mid t_1 =_{\tau_1'} t_1', \ldots, t_n =_{\tau_n'} t_n')$, we have an equaliser diagram

$$[\![\Gamma \mid \Xi]\!] \rightarrowtail [\![\Gamma]\!] \;\underset{\langle [\![\Gamma \vdash t_i' : \tau_i']\!] \rangle_{i=1,n}}{\overset{\langle [\![\Gamma \vdash t_i : \tau_i']\!] \rangle_{i=1,n}}{\rightrightarrows}}\; \prod^{(n)}([\![\tau_1']\!], \ldots, [\![\tau_n']\!]) \tag{2}$$

Proposition 1 (Soundness). *If $\Gamma \mid \Xi \vdash t = t' : \tau$ is derivable then*

$$([\![\Gamma \mid \Xi]\!] \rightarrowtail [\![\Gamma]\!] \xrightarrow{[\![\Gamma \vdash t : \tau]\!]} [\![\tau]\!]) = ([\![\Gamma \mid \Xi]\!] \rightarrowtail [\![\Gamma]\!] \xrightarrow{[\![\Gamma \vdash t' : \tau]\!]} [\![\tau]\!]) \quad .$$

The proof is the usual straightforward induction on the structure of derivations, using the facts observed above.

It would be interesting to obtain a completeness converse to Proposition 1. We do not know if such a result holds, although weaker versions can be obtained by not insisting that all exponentials exist in $\mathcal{S}$. Also, following [2, Theorem 5.3], one can show that the proof system is sound and complete for deriving the equalities between terms in *unconstrained* environments that are valid in an arbitrary bicartesian closed category. These issues will be discussed further in the full version of this paper.

3 Grothendieck logical relations

For each object A of the semantic category $\mathcal{S}$ we define the notion of a (categorical) *Kripke relation* of varying arity over A. The idea is that the arity of the relation varies over a category $\mathbb{W}$ (of *worlds*), as specified by a functor $a : \mathbb{W} \to \mathcal{S}$ (that associates *arities* to worlds). For each object w of $\mathbb{W}$, the object $a(w)$ is considered as an arity in the natural internal sense that $a(w)$-tuples of A are given by morphisms $x : a(w) \to A$ in $\mathcal{S}$. The action of the arity functor a on morphisms allows such a tuple x of arity $a(w)$ to be reinterpreted along any change of world $\psi : v \to w$ in $\mathbb{W}$ to obtain the $a(v)$-tuple $x \circ a(\psi)$. For notational convenience, we write $x \cdot \psi$ for $x \circ a(\psi)$ when a is clear from the context.

Definition 3 (Kripke relation). Given a small category $\mathbb{W}$ and a functor $a : \mathbb{W} \to \mathcal{S}$, a $\mathbb{W}$-*Kripke relation* R *of arity* a over an object A of $\mathcal{S}$ is a family $\{R(w) \subseteq \mathcal{S}(a(w), A)\}_{w \in |\mathbb{W}|}$ satisfying

(Monotonicity) For every $\psi : w \to v$ in $\mathbb{W}$ and every $x : a(v) \to A$ in $\mathcal{S}$, if $x \in R(v)$ then $x \cdot \psi \in R(w)$.

The notion of Kripke relation has a natural formulation in the language of presheaves. Writing $\widehat{\mathbb{W}}$ for the category of presheaves $[\mathbb{W}^{\mathrm{op}}, \mathbf{Set}]$, any arity functor $a : \mathbb{W} \to \mathcal{S}$ induces a *hom functor* $a* : \mathcal{S} \to \widehat{\mathbb{W}}$ given by $(a * A)(\_) \stackrel{\mathrm{def}}{=} \mathcal{S}(a\_, A) : \mathbb{W}^{\mathrm{op}} \to \mathbf{Set}$. A Kripke relation of arity a over $A \in \mathcal{S}$ is just a subpresheaf $R \subseteq a * A$ in $\widehat{\mathbb{W}}$. So, a Kripke relation of arity a is a *unary* relation on $a * A$ in the internal logic of the presheaf topos $\widehat{\mathbb{W}}$.

Our generalisation of Kripke relation allows us to impose additional structure on the category of worlds in the form of a *Grothendieck topology.* A Grothendieck topology is a collection of *covers*, which are families of morphisms with the same codomain, subject to axioms on the collection. A cover $\{\varphi_i : w_i \to w\}_{i \in I}$ of w specifies that information about w can be recovered "locally" by piecing together relevant information about each of the w_i along φ_i. The formal definition of a Grothendieck topology specifies the properties that the collection of covers must satisfy in order for such local determination to behave properly.

Definition 4 (Basis for a topology). A *(basis for a Grothendieck) topology* K on a category $\mathbb{W}$ consists of a family of *(basic) covers* $K(w) \subseteq \bigcup_{v \in \mathbb{W}} \mathbb{W}(v, w)$ for each object w in $\mathbb{W}$, satisfying:

(Identity) The singleton family $\{\mathrm{id}_w\} \in K(w)$.
(Stability) For every family $\{\varphi_i\}_{i \in I} \in K(w)$ and morphism $\psi : v \to w$ there exists a family $\{\gamma_j\}_{j \in J} \in K(v)$ such that, for each $\gamma_j \in K(v)$, there exists $\varphi_i \in K(w)$ such that $\psi \circ \gamma_j$ factors through φ_i.
(Transitivity) If $\{\varphi_i : w_i \to w\}_{i \in I} \in K(w)$ and $\{\gamma_{ij}\}_{j \in J_i} \in K(w_i)$ for every $i \in I$ then the family $\{\varphi_i \circ \gamma_{ij}\}_{i \in I,\, j \in J_i} \in K(w)$.

A small category together with a Grothendieck topology is called a *site*.

Example 1. In any category the *trivial topology*, I, consists only of the singleton families $\{\mathrm{id}\}$.

Example 2. In a category with stable finite coproducts, the *finite coproduct topology* is given by

$$\{\{\varphi_i : w_i \to w\}_{1 \le i \le n} \mid n \ge 0 \text{ and } \{\varphi_i : w_i \to w\}_{1 \le i \le n} \text{ is a coproduct}\}.$$

The stability of coproducts ensures that the stability axiom for a Grothendieck topology is satisfied. Note that the empty family covers an object if and only if the object is (necessarily strict) initial.

In order to generalise the notion of Kripke relation to take into account a Grothendieck topology, we add an extra condition establishing that the relation is determined locally in the sense discussed above.

Definition 5 (Grothendieck relation). Given a site $(\mathbb{W}, K)$ and a functor $a : \mathbb{W} \to \mathcal{S}$, a $(\mathbb{W}, K)$-*Grothendieck relation of arity* a *over* $A \in \mathcal{S}$ is a $\mathbb{W}$-Kripke relation $\{R(w) \subseteq \mathcal{S}(a(w), A)\}_{w \in |\mathbb{W}|}$ that further satisfies:

(Local character) For every cover $\{\varphi_i : w_i \to w\}_{i \in I} \in K(w)$ and for all maps $x : a(w) \to A$ in $\mathcal{S}$, if $x \cdot \varphi_i \in R(w_i)$ for all $i \in I$ then $x \in R(w)$.

In the case of the trivial topology, the local character property is vacuous and so any Kripke relation is a Grothendieck relation.

It is instructive to reformulate the notion of a Grothendieck relation in terms of standard concepts from sheaf theory. For notational convenience, given a presheaf P in $\widehat{\mathbb{W}}$, for any $\psi : v \to w$ in $\mathbb{W}$ and $x \in P(w)$ we write $x \cdot \psi$ for the element $P(\psi)(x) \in P(v)$. (This generalises our previous notation for presheaves $a * A$ to arbitrary presheaves.)

Definition 6 (Closed subpresheaf). Given a site $(\mathbb{W}, K)$ and a presheaf P in $\widehat{\mathbb{W}}$, a subpresheaf $R \subseteq P$ is said to be K-*closed* if, for every cover $\{\varphi_i : w_i \to w\}_{i \in I} \in K(w)$ and for all $x \in P(w)$ if $x \cdot \varphi_i \in R(w_i)$ for all $i \in I$ then $x \in R(w)$.

Hence, a Grothendieck relation R of arity a over A is precisely a K-closed subpresheaf $R \subseteq a * A$.

There is another, less elementary, characterisation of Grothendieck relations. Writing $\underline{\mathrm{Sh}}(\mathbb{W}, K)$ for the full subcategory of $\widehat{\mathbb{W}}$ whose objects are *sheaves* (for K) [6], it is well-known (see [6, III.5 and V.3] for example) that the embedding $\underline{\mathrm{Sh}}(\mathbb{W}, K) \hookrightarrow \widehat{\mathbb{W}}$ has a (left-exact) left adjoint, the associated sheaf functor $\mathbf{a} : \widehat{\mathbb{W}} \to \underline{\mathrm{Sh}}(\mathbb{W}, K)$. For every presheaf P, the closed subpresheaves of P are in natural bijective correspondence with the subsheaves of $\mathbf{a}(P)$ [6]. Thus, a Grothendieck relation of arity a over A is just a subsheaf of $\mathbf{a}(a * A)$ in $\underline{\mathrm{Sh}}(\mathbb{W}, K)$. In particular, when the presheaf $a * A$ is already a sheaf for K, a Grothendieck relation over A is just a subsheaf of $a * A$. However, we shall *not* assume in general that $a * A$ is a sheaf.

We define a *category* of Grothendieck relations over $\mathcal{S}$ whose morphisms are given by those morphisms of $\mathcal{S}$ that preserve the relations.

Definition 7. Given a site $(\mathbb{W}, K)$ and an arity functor $a : \mathbb{W} \to \mathcal{S}$:

1. $\underline{\mathrm{G}}(\mathbb{W}, K, a)$ is the category with
   objects: given by pairs (A, R) consisting of an object $A \in \mathcal{S}$ and a $(\mathbb{W}, K)$-Grothendieck relation R of arity a over A,
   arrows $(A, R) \to (B, S)$: given by arrows $f : A \to B$ in $\mathcal{S}$ such that,
   $$\text{for all } x : a(w) \to A,\ x \in R(w) \text{ implies } f \circ x \in S(w) \quad , \tag{3}$$
   identity and composition: as in $\mathcal{S}$.
2. We write $U : \underline{\mathrm{G}}(\mathbb{W}, K, a) \to \mathcal{S}$ for the forgetful functor mapping (A, R) to A.

Proposition 2. *For $\mathcal{S}$ bicartesian closed, the category $\underline{\mathrm{G}}(\mathbb{W}, K, a)$ is bicartesian closed and the forgetful functor $U : \underline{\mathrm{G}}(\mathbb{W}, K, a) \to \mathcal{S}$ is faithful, and preserves and creates the bicartesian closed structure.*

Proof. Finite coproducts: $\coprod_n (A_n, R_n) = (\coprod_n A_n, \bigvee_n R_n)$ where $(a(w) \xrightarrow{x} \coprod_n A_n) \in (\bigvee_n R_n)(w)$ iff$_{\mathsf{def}}$ there exists a cover $\{\varphi_i : w_i \to w\}_{i \in I} \in K(w)$ such that for all $i \in I$, there exist n_i with $1 \le n_i \le n$ and $(a(w_i) \xrightarrow{x_i} A_{n_i}) \in R_{n_i}(w_i)$ such that $x \cdot \varphi_i = \amalg_{n_i} \circ x_i : a(w_i) \to \coprod_n A_n$.

Finite products: $\prod_n (A_n, R_n) = (\prod_n A_n, \bigwedge_n R_n)$ where $(a(w) \xrightarrow{x} \prod_n A_n) \in (\prod_n R_n)(w)$ iff$_{\mathsf{def}}$ for all n, $(a(w) \xrightarrow{x} \prod_n A_n \xrightarrow{\pi_n} A_n) \in R_n(w)$.

Exponentials: $(A, R) \Rightarrow (B, S) = (A \Rightarrow B, S^R)$ where $(a(w) \xrightarrow{f} (A \Rightarrow B)) \in S^R(w)$ iff$_{\mathsf{def}}$ for all $\psi : v \to w$ and all $(a(v) \xrightarrow{x} A) \in R(v)$, we have
$$(a(v) \xrightarrow{\langle f \cdot \psi,\, x \rangle} (A \Rightarrow B) \times A \xrightarrow{\mathsf{ev}} B) \in S(v). \quad \square$$

Although straightforward, the proposition above is the categorical analogue of the *fundamental lemma of logical relations* [7], which states that any syntactically definable morphism in $\mathcal{S}$ automatically preserves relations. To formulate this result explicitly, we require further definitions.

Definition 8. Given a site $(\mathbb{W}, K)$, an arity functor $a : \mathbb{W} \to \mathcal{S}$ and a Grothendieck relation R of arity a over $A \in \mathcal{S}$, we say that a global element $x : 1 \to A$ in $\mathcal{S}$ *satisfies* R if, for all $w \in |\mathbb{W}|$, it holds that $(a(w) \to 1 \xrightarrow{x} A) \in R(w)$.

Definition 9 (Grothendieck logical relation). Let $\mathcal{I}$ be a (T, Σ)-interpretation in a bicartesian closed category $\mathcal{S}$. A *Grothendieck logical relation for Σ under $\mathcal{I}$* is given by: a site $(\mathbb{W}, K)$; an arity functor $a : \mathbb{W} \to \mathcal{S}$; and, a family $\{R_{\mathrm{T}}\}_{\mathrm{T} \in T}$ such that:

1. each R_{T} is a Grothendieck relation of arity a over $\mathcal{I}_T(\mathrm{T})$, and
2. for all $(c : \tau) \in \Sigma$, it holds that $\mathcal{I}_\Sigma(c)$ satisfies R_τ, where we write R_τ (R_Γ) for the Grothendieck relation on $[\![\tau]\!]$ ($[\![\Gamma]\!]$) determined by the bicartesian closed structure on $\underline{\mathrm{G}}(\mathbb{W}, K, a)$ according to the structure of τ (Γ).

Lemma 1 (Fundamental Lemma of GLRs). *Let $\mathcal{S}$ be a bicartesian closed category and let $\mathcal{I}$ be a $(\mathcal{T}, \Sigma)$-interpretation in $\mathcal{S}$. For any Grothendieck logical relation $((\mathbb{W}, K), a, \{R_{\mathrm{T}}\}_{\mathrm{T}\in\mathcal{T}})$ for Σ under $\mathcal{I}$, the following two equivalent statements hold.*

1. *For every term $\Gamma \vdash t : \tau$, the interpretation $[\![\Gamma \vdash t : \tau]\!]$ is an arrow $([\![\Gamma]\!], R_\Gamma) \to ([\![\tau]\!], R_\tau)$ in $\underline{\mathrm{G}}(\mathbb{W}, K, a)$.*
2. *For every term $\vdash t : \tau$, the global element $[\![\vdash t : \tau]\!] : 1 \to [\![\tau]\!]$ satisfies R_τ.*

Our motivation for generalising Kripke relations to Grothendieck relation is to obtain the converse: any global element of $\mathcal{S}$ that satisfies all Grothendieck logical relations is syntactically definable. At present we have such a result only in the special case that $\mathcal{S}$ is stable. This is the content of the theorem below, which is the principal result of the paper.

Theorem 1 (Definability). *Suppose $\mathcal{S}$ is a stable bicartesian closed category and $\mathcal{I}$ is a $(\mathcal{T}, \Sigma)$-interpretation in $\mathcal{S}$. Then there exists a Grothendieck logical relation $((\mathbb{W}, K), a, \{R_{\mathrm{T}}\}_{\mathrm{T}\in\mathcal{T}})$ for Σ under $\mathcal{I}$, such that every global element of $[\![\tau]\!]$ that satisfies R_τ is definable by a closed term of type τ.*

4 Proof of Definability

In this section we prove Theorem 1. Accordingly, suppose $\mathcal{S}$ is a stable bicartesian closed category (with chosen structure) and $\mathcal{I}$ is a $(\mathcal{T}, \Sigma)$-interpretation in $\mathcal{S}$. We construct a Grothendieck logical relation, satisfying the property of Theorem 1, based on a syntactic site $(\mathbf{W}, \mathsf{K})$ defined below. The construction has similarities with the syntactic sites used in recent approaches to obtaining intuitionistic completeness results for intuitionistic logic, see e.g. [9].

Definition 10 (Syntactic site).

1. The category $\mathbf{W}$ has
   <u>objects:</u> given by constrained environments as in Definition 2,
   <u>arrows</u> $\Gamma' \mid \Xi' \to \Gamma \mid \Xi$: given by *renamings* ($\stackrel{\text{def}}{=}$ monotone injections) $\rho : \mathrm{dom}(\Gamma) \to \mathrm{dom}(\Gamma')$, where $\mathrm{dom}(\mathbf{x}_1 : \tau_1, \ldots, \mathbf{x}_n : \tau_n) \stackrel{\text{def}}{=} (\mathbf{x}_1 \leq \cdots \leq \mathbf{x}_n)$, that preserve typing:
   $$\mathbf{x} : \tau \in \Gamma \Rightarrow \rho(\mathbf{x}) : \tau \in \Gamma' \quad ,$$
   and preserve constraints:
   $$t =_\tau t' \in \Xi \Rightarrow t[\rho] =_\tau t'[\rho] \in \Xi' \quad ,$$
   <u>identities</u> and <u>composition:</u> as for functions.

2. The covers in K are defined inductively by the following rules:

$$\{\mathrm{id}_{\mathrm{dom}(\Gamma)}\} \in \mathsf{K}(\Gamma \mid \Xi)$$

$$\frac{\{\rho_j\} \cup \{\rho : \Gamma' \mid \Xi' \to \Gamma \mid \Xi\} \in \mathsf{K}(\Gamma \mid \Xi) \qquad \Gamma \vdash t : +^{(n)}(\tau_1, \ldots, \tau_n)}{\{\rho_j\} \cup \{\rho \circ \iota_k : \Gamma'_k \mid \Xi'_k \to \Gamma \mid \Xi\}_{1 \le k \le n} \in \mathsf{K}(\Gamma \mid \Xi)}$$

where $\Gamma'_k \mid \Xi'_k = (\Gamma', \mathbf{x}'_k : \tau_k \mid \Xi', \mathbf{in}_k(\mathbf{x}'_k) = t)$ for any choice of fresh variables $\mathbf{x}'_1, \ldots, \mathbf{x}'_n$ and the renamings $\iota_k : \mathrm{dom}(\Gamma') \to \mathrm{dom}(\Gamma', \mathbf{x}'_k : \tau_k)$ are the inclusion functions.

It follows that any cover $\{\rho_j\}$ consists entirely of inclusion functions (which is why $\Gamma'_k \mid \Xi'_k$ can be defined using t rather than $t[\rho]$). Observe also that a constrained environment $\Gamma \mid \Xi$ is covered by the empty family if and only if there exists a term $\Gamma \vdash t : 0$.

The above definition provides, for every $\Gamma \vdash t : +^{(n)}(\tau_1, \ldots, \tau_n)$, *sub-basic* covers of the form

$$\{\ (\Gamma,\ \mathbf{x}_i : \tau_i \mid \Xi,\ \mathbf{in}_i(\mathbf{x}_i) = t) \longrightarrow (\Gamma \mid \Xi)\ \}_{1 \le i \le n}$$

keeping the morphisms as simple a possible whilst allowing the axioms of a Grothendieck topology to hold. For instance, the stability axiom holds because for any inclusion

$$\iota_i\ :\ (\Gamma,\ \mathbf{x}_i : \tau_i \mid \Xi,\ \mathbf{in}_i(\mathbf{x}_i) = t) \longrightarrow \Gamma \mid \Xi$$

(as present in the non-trivial covers) and any renaming $\rho : \Gamma' \mid \Xi' \to \Gamma \mid \Xi$, we have a commuting diagram:

$$\begin{array}{ccc}
(\Gamma',\ \mathbf{x}' : \tau_i \mid \Xi',\ \mathbf{in}_i(\mathbf{x}') = t[\rho]) & \xrightarrow{\ \rho[\mathbf{x}_i \mapsto \mathbf{x}']\ } & (\Gamma,\ \mathbf{x}_i : \tau_i \mid \Xi,\ \mathbf{in}_i(\mathbf{x}_i) = t) \\
\downarrow {\scriptstyle \iota'_i} & & \downarrow {\scriptstyle \iota_i} \\
(\Gamma' \mid \Xi') & \xrightarrow[\ \rho\ ]{} & (\Gamma \mid \Xi)
\end{array}$$

for any $\mathbf{x}'$ not in Γ'. Observe that the possibility of morphisms renaming variables is crucial here, as the variable $\mathbf{x}_i$ may already appear in the environment Γ'. Thus the stability of covers would not hold if we only allowed inclusions as morphisms in $\mathbf{W}$. Indeed, the category $\mathbf{W}$ is not a preorder.

Definition 11 (Standard arity functor). The *standard arity functor* $\boldsymbol{s} : \mathbf{W} \to \mathcal{S}$ sends any constrained environment $\Gamma \mid \Xi$ to its interpretation $[\![\Gamma \mid \Xi]\!]$, and any renaming $\rho : \Gamma' \mid \Xi' \to \Gamma \mid \Xi$ to the unique map $\boldsymbol{s}(\rho)$, given by the universal property of the equaliser $[\![\Gamma \mid \Xi]\!] \rightarrowtail [\![\Gamma]\!]$ of (2) in Section 2, such that the square below commutes.

$$\begin{array}{ccc}
[\![\Gamma' \mid \Xi']\!] & \rightarrowtail & [\![\Gamma']\!] \\
{\scriptstyle \boldsymbol{s}(\rho)} \downarrow & & \downarrow {\scriptstyle \langle \pi_{\rho \mathbf{x}} \rangle_{\mathbf{x} \in \Gamma}} \\
[\![\Gamma \mid \Xi]\!] & \rightarrowtail & [\![\Gamma]\!]
\end{array} \qquad (4)$$

For a cover $\{\iota_i : (\Gamma, \mathbf{x}_i : \tau_i \mid \Xi, \mathbf{in}_i(\mathbf{x}_i) = t) \to \Gamma \mid \Xi\}_{1 \le i \le n}$ in K it follows, from (1) and the stability of coproducts, that the family $\{s(\iota_i)\}_{1 \le i \le n}$ is a coproduct in $\mathcal{S}$. By induction, this property extends to arbitrary covers in K and hence we have the following consequence.

Proposition 3. *For every cover* $\{\, \rho_i : \Gamma_i \mid \Xi_i \to \Gamma \mid \Xi \,\}$, *the family* $\{\, s(\rho_i) : [\![\Gamma_i \mid \Xi_i]\!] \to [\![\Gamma \mid \Xi]\!] \,\}$ *is a coproduct.*

Corollary 1. *For all* $A \in |\, \mathcal{S} \,|$, *the presheaf* $s * A$ *in* $\widehat{\mathbf{W}}$ *is a sheaf for* K.

The key lemma for establishing the definability result follows.

Lemma 2. *For every cover* $\{\, \rho_i : \Gamma_i \mid \Xi_i \to \Gamma \mid \Xi \,\}$ *and every family of terms* $\{\, \Gamma_i \vdash t_i : \tau \,\}$ *there exists a term* $\Gamma \vdash t : \tau$ *such that*

1. $\Gamma_i \mid \Xi_i \vdash t_i = t : \tau$.
2. *If* $\Gamma \vdash t' : \tau$ *is such that* $\Gamma_i \mid \Xi_i \vdash t_i = t' : \tau$ *for all* i, *then* $\Gamma \mid \Xi \vdash t' = t : \tau$.
3. *The diagram below commutes for all* i

$$\begin{array}{ccc} [\![\Gamma_i \mid \Xi_i]\!] & \xrightarrow{\;s(\rho_i)\;} & [\![\Gamma \mid \Xi]\!] \\ \big\downarrow & & \big\downarrow{\scriptstyle x} \\ [\![\Gamma_i]\!] & \xrightarrow[{[\![\Gamma_i \vdash t_i : \tau]\!]}]{} & [\![\tau]\!] \end{array}$$

iff $x = ([\![\Gamma \mid \Xi]\!] \rightarrowtail [\![\Gamma]\!] \xrightarrow{[\![\Gamma \vdash t : \tau]\!]} [\![\tau]\!])$.

Proof. (1)–(2) To a derivation D of a cover $\{\, \rho_i : \Gamma_i \mid \Xi_i \to \Gamma \mid \Xi \,\}$ and terms $\{\, \Gamma_i \vdash t_i : \tau \,\}$ we associate a term $\Gamma \vdash \mathcal{T}(D, \{\Gamma_i \vdash t_i : \tau\}) : \tau$ by induction on the structure of the derivation as follows.

- $\mathcal{T}(\{\mathrm{id}_{\mathrm{dom}(\Gamma)}\}, \{\Gamma \vdash t : \tau\}) \stackrel{\mathrm{def}}{=} t$.
- For r the rule
$$\frac{\{\rho_j\}_{j \in J} \cup \{\rho\}}{\{\rho_j\}_{j \in J} \cup \{\rho \circ \iota_k\}_{1 \le k \le n}}$$
where $\iota_k : (\Gamma, \mathbf{x}_k : \tau_k \mid \Xi, \mathbf{in}_k(\mathbf{x}_k) = t) \to \Gamma \mid \Xi$, we set
$$\mathcal{T}(D.r, \{\Gamma_j \vdash t_j : \tau\}_{j \in J} \cup \{\Gamma, \mathbf{x}_k : \tau_k \vdash t_k : \tau\}_{1 \le k \le n})$$
$$\stackrel{\mathrm{def}}{=} \mathcal{T}(D, \{\Gamma_j \vdash t_j : \tau\}_{j \in J} \cup \{\Gamma \vdash \mathbf{case}\; t \;\mathbf{of}\; [\mathbf{in}_1(\mathbf{x}_1).t_1, \ldots, \mathbf{in}_n(\mathbf{x}_n).t_n] : \tau\}).$$

That the term $\mathcal{T}(D, \{\Gamma_i \vdash t_i : \tau\})$ has the desired properties can be shown by induction using the equational rules.

(3) By Proposition 3, because

$$\begin{aligned} &([\![\Gamma_i \mid \Xi_i]\!] \rightarrowtail [\![\Gamma_i]\!] \xrightarrow{[\![\Gamma_i \vdash t_i : \tau]\!]} [\![\tau]\!]) \\ &= ([\![\Gamma_i \mid \Xi_i]\!] \rightarrowtail [\![\Gamma_i]\!] \xrightarrow{[\![\Gamma_i \vdash t : \tau]\!]} [\![\tau]\!]) && \text{, by Proposition 1} \\ &= ([\![\Gamma_i \mid \Xi_i]\!] \rightarrowtail [\![\Gamma_i]\!] \xrightarrow{\langle \pi_{\rho_i(x)} \rangle_{x \in \Gamma}} [\![\Gamma]\!] \xrightarrow{[\![\Gamma \vdash t : \tau]\!]} [\![\tau]\!]) \\ &= ([\![\Gamma_i \mid \Xi_i]\!] \xrightarrow{s(\rho_i)} [\![\Gamma \mid \Xi]\!] \rightarrowtail [\![\Gamma]\!] \xrightarrow{[\![\Gamma \vdash t : \tau]\!]} [\![\tau]\!]) && \text{, by (4)} \end{aligned}$$

Proposition 4. *Let $\mathcal{S}$ be a stable bicartesian closed category (with chosen structure) and let $\mathcal{I}$ be a (T, Σ)-interpretation in $\mathcal{S}$. Then*

1. *for*

$$\mathcal{R}_{\mathrm{T}}(\Gamma \mid \Xi) \stackrel{\text{def}}{=} \{\, [\![\Gamma \mid \Xi]\!] \rightarrowtail [\![\Gamma]\!] \xrightarrow{[\![\Gamma \vdash t:\mathrm{T}]\!]} [\![\mathrm{T}]\!] \,\} \quad , \tag{5}$$

   $((\mathbf{W}, \mathsf{K}), \boldsymbol{s}, \{\mathcal{R}_{\mathrm{T}}\}_{\mathrm{T}\in T})$ *is a Grothendieck logical relation for Σ under $\mathcal{I}$;*
2. *for every type τ,*

$$\mathcal{R}_{\tau}(\Gamma \mid \Xi) = \{\, [\![\Gamma \mid \Xi]\!] \rightarrowtail [\![\Gamma]\!] \xrightarrow{[\![\Gamma \vdash t:\tau]\!]} [\![\tau]\!] \,\} \quad .$$

Proof. (1) Follows from (2) below.

(2) By induction on the structure of τ.

$\tau = \mathrm{T}$: By (5).

$\tau = \tau_1 \to \tau_2$:

($\supseteq$) Let $m = ([\![\Gamma \mid \Xi]\!] \rightarrowtail [\![\Gamma]\!])$ and $m' = ([\![\Gamma' \mid \Xi']\!] \rightarrowtail [\![\Gamma']\!])$.
For $\rho : \Gamma' \mid \Xi' \to \Gamma \mid \Xi$ and $x \in \mathcal{R}_{\tau}(\Gamma' \mid \Xi')$ we have, by induction, that $x = [\![\Gamma' \vdash t' : \tau_1]\!] \circ m'$ for some t'. Thus, to establish that $[\![\Gamma \vdash t : \tau_1 \to \tau_2]\!] \circ m$ is in $\mathcal{R}_{\tau_1 \to \tau_2}(\Gamma \mid \Xi)$ we need show that $\mathsf{ev} \circ \langle [\![\Gamma \vdash t : \tau_1 \to \tau_2]\!] \circ m \circ \boldsymbol{s}(\rho), [\![\Gamma' \vdash t' : \tau_1]\!] \circ m' \rangle$ is in $\mathcal{R}_{\tau_2}(\Gamma' \mid \Xi')$.
Using that $m \circ \boldsymbol{s}(\rho) = \langle \pi_{\rho\mathbf{x}} \rangle_{\mathbf{x}\in\Gamma} \circ m'$ and that $[\![\Gamma \vdash t : \tau_1 \to \tau_2]\!] \circ \langle \pi_{\rho\mathbf{x}} \rangle_{\mathbf{x}\in\Gamma} = [\![\Gamma' \vdash t[\rho] : \tau_1 \to \tau_2]\!]$ one sees that $\mathsf{ev} \circ \langle [\![\Gamma \vdash t : \tau_1 \to \tau_2]\!] \circ m \circ \boldsymbol{s}(\rho), [\![\Gamma' \vdash t' : \tau_1]\!] \circ m' \rangle = [\![\Gamma' \vdash t[\rho](t') : \tau_2]\!] \circ m'$ and, by induction, we are done.
($\subseteq$) Let

$$f \in \mathcal{R}_{\tau_1 \to \tau_2}(\Gamma \mid \Xi) \quad . \tag{6}$$

Recall that $([\![\Gamma, \mathbf{x} : \tau_1 \mid \Xi, \mathbf{x} =_{\tau_1} \mathbf{x}]\!] \rightarrowtail [\![\Gamma]\!] \times [\![\tau_1]\!]) = m \times \mathrm{id}_{[\![\tau]\!]}$ where $m = ([\![\Gamma \mid \Xi]\!] \rightarrowtail [\![\Gamma]\!])$. Thus, for $\iota : (\Gamma, \mathbf{x} : \tau_1 \mid \Xi, \mathbf{x} =_{\tau_1} \mathbf{x}) \to \Gamma \mid \Xi$ the inclusion, we have that $\boldsymbol{s}(\iota) = \pi_1 : [\![\Gamma \mid \Xi]\!] \times [\![\tau_1]\!] \to [\![\Gamma \mid \Xi]\!]$.
Since, by induction, $\pi_2 = [\![\Gamma, \mathbf{x} : \tau_1 \vdash \mathbf{x} : \tau_1]\!] \circ (m \times \mathrm{id}_{[\![\tau_1]\!]}) : [\![\Gamma \mid \Xi]\!] \times [\![\tau_1]\!] \to [\![\tau_1]\!]$ is in $\mathcal{R}_{\tau_1}(\Gamma, \mathbf{x} : \tau_1 \mid \Xi, \mathbf{x} =_{\tau_1} \mathbf{x})$ it follows from (6) that $\mathsf{ev} \circ \langle f \circ \pi_1, \pi_2 \rangle$ is in $\mathcal{R}_{\tau_2}(\Gamma, \mathbf{x} : \tau_1 \mid \Xi, \mathbf{x} =_{\tau_1} \mathbf{x})$. So, again by induction, $\mathsf{ev} \circ \langle f \circ \pi_1, \pi_2 \rangle = [\![\Gamma, \mathbf{x} : \tau_1 \vdash t : \tau_2]\!] \circ (m \times \mathrm{id}_{[\![\tau_1]\!]})$ for some t, and hence $f = [\![\Gamma \vdash \lambda \mathbf{x} : \tau_1 . t : \tau_1 \to \tau_2]\!] \circ m$.

$\tau = \times^{(n)}(\tau_1, \ldots, \tau_n)$:

($\supseteq$) Let $m = ([\![\Gamma \mid \Xi]\!] \rightarrowtail [\![\Gamma]\!])$.
By induction, for $1 \le i \le n$, $\pi_i \circ [\![\Gamma \vdash t : \times^{(n)}(\tau_1, \ldots, \tau_n)]\!] \circ m = [\![\Gamma \vdash \mathtt{proj}_i(t) : \tau_i]\!] \circ m$ is in $\mathcal{R}_{\tau_i}(\Gamma \mid \Xi)$. Thus, $[\![\Gamma \vdash t : \times^{(n)}(\tau_1, \ldots, \tau_n)]\!] \circ m$ is in $\mathcal{R}_{\times^{(n)}(\tau_1, \ldots, \tau_n)}(\Gamma \mid \Xi)$.
($\subseteq$) Let $x \in \mathcal{R}_{\times^{(n)}(\tau_1, \ldots, \tau_n)}(\Gamma \mid \Xi)$. Then, for $1 \le i \le n$, we have that $\pi_i \circ x \in \mathcal{R}_{\tau_i}(\Gamma \mid \Xi)$. By induction, $\pi_i \circ x = [\![\Gamma \vdash t_i : \tau_i]\!] \circ m$, where $m = ([\![\Gamma \mid \Xi]\!] \rightarrowtail [\![\Gamma]\!])$, for some t_i $(1 \le i \le n)$. Thus, $x = [\![\Gamma \vdash \langle t_1, \ldots, t_n \rangle : \times^{(n)}(\tau_1, \ldots, \tau_n)]\!] \circ m$.

$\tau = +^{(n)}(\tau_1, \ldots, \tau_n)$:

($\supseteq$) Let $m = ([\![\Gamma \mid \Xi]\!] \rightarrowtail [\![\Gamma]\!])$ and, for $\mathbf{x}_i \notin \Gamma$ $(1 \le i \le n)$, let $m_i = ([\![\Gamma, \mathbf{x}_i : \tau_i \mid \Xi, \mathtt{in}_i(\mathbf{x}_i) =_{+(\tau_1,\ldots,\tau_n)} t]\!] \rightarrowtail [\![\Gamma]\!] \times [\![\tau_1]\!])$.
By induction, we have that $\pi_2 \circ m_i = [\![\Gamma, \mathbf{x}_i : \tau_i \vdash \mathbf{x}_i : \tau_i]\!] \circ m_i$ is in $\mathcal{R}_{\tau_i}(\Gamma, \mathbf{x}_i : \tau_i \mid \Xi, \mathtt{in}_i(\mathbf{x}_i) =_{+^{(n)}(\tau_1,\ldots,\tau_n)} t)$ for all i.
Consider the cover

$$\{\, (\Gamma, \mathbf{x}_i : \tau_i \mid \Xi, \mathtt{in}_i(\mathbf{x}_i) =_{+^{(n)}(\tau_1,\ldots,\tau_n)} t) \xrightarrow{\iota_i} \Gamma \mid \Xi \,\}_{1 \le i \le n} \quad .$$

Then since, for $1 \le i \le n$, the diagram below commutes,

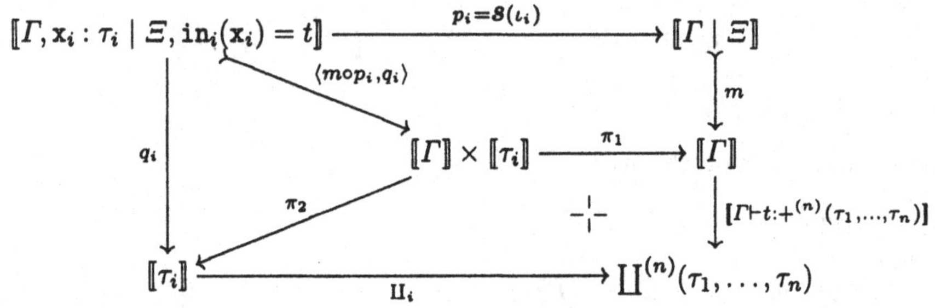

it follows that $[\![\Gamma \vdash t : +^{(n)}(\tau_1, \ldots, \tau_n)]\!] \circ m$ is in $\mathcal{R}_{+^{(n)}(\tau_1,\ldots,\tau_n)}(\Gamma \mid \Xi)$.
($\subseteq$) If $x \in \mathcal{R}_{+^{(n)}(\tau_1,\ldots,\tau_n)}(\Gamma \mid \Xi)$ then there exists a cover $\{\, \rho_i : \Gamma_i \mid \Xi_i \to \Gamma \mid \Xi \,\}$ such that for all i, using the induction hypothesis, there exist $\Gamma_i \vdash t_i : \tau_{n_i}$ with $1 \le n_i \le n$ such that for all i

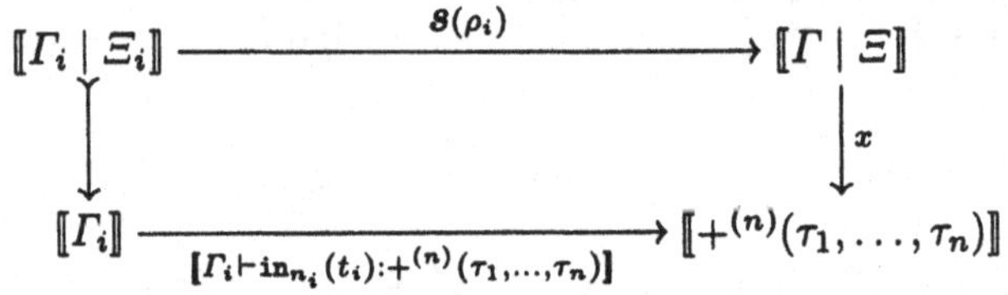

Hence, by Lemma 2, we are done.

$\square$

Corollary 2. *For the Grothendieck logical relation* $((\mathbf{W}, \mathsf{K}), s, \{\mathcal{R}_{\mathrm{T}}\}_{\mathrm{T} \in \mathcal{T}})$, *a global element of* $[\![\tau]\!]$ *in* S *satisfies* $\mathcal{R}_\tau$ *if and only if it is definable by a closed term of type* τ.

5 Further results

In the full version of this paper, we shall show that Theorem 1 can be strengthened by requiring that a "universal" site $(\mathbb{W}, K)$ can be found in which $\mathbb{W}$ is a partial order. This strengthening could be proved directly by making clumsy modifications to the construction of the syntactic site $(\mathbf{W}, \mathsf{K})$ given in Section 4. It is preferable, however, to derive the result by means of an elegant general construction. As in the well-known construction of the *Diaconescu cover* of a

Grothendieck topos [6, IX.9], any site $(\mathbb{W}, K)$ determines a related site $\underline{D}(\mathbb{W}, K)$ over a poset $\underline{D}(\mathbb{W})$ together with a surjective functor $d_{\mathbb{W}} : \underline{D}(\mathbb{W}) \to (\mathbb{W})$. We have proved that, for any arity functor $a : \mathbb{W} \to \mathcal{S}$ (for $\mathcal{S}$ bicartesian closed), there is an associated full and faithful bicartesian closed functor $\underline{G}(\mathbb{W}, K, a) \to \underline{G}(\underline{D}(\mathbb{W}), \underline{D}(K), a\, d_{\mathbb{W}})$. This means that our definability result for the syntactic site $(\mathbf{W}, \mathsf{K})$ yields the desired poset-based definability result for $\underline{D}(\mathbf{W}, \mathsf{K})$.

Other aspects of the paper also benefit from a more abstract categorical treatment. For example, the construction of the category $\underline{G}(\mathbb{W}, K, a)$ is an example of the *subscone* variant of glueing [1], in which the objects are restricted to K-closed monos (in $\widehat{\mathbb{W}}$). Essentially this amounts to glueing relative to a factorization system. The analysis of the structure on $\underline{G}(\mathbb{W}, K, a)$ can be performed entirely at this more general level.

Finally, it is also possible to give syntax-free account of definability. For any bicartesian closed functor $F : \mathbb{B} \to \mathcal{S}$ where $\mathbb{B}$ is small and $\mathcal{S}$ is stable, there exists a site $(\mathbb{W}, K)$ (with $\mathbb{W}$ a poset) and an arity functor $a : \mathbb{W} \to \mathcal{S}$ such that F factors as UG where $G : \mathbb{B} \to \underline{G}(\mathbb{W}, K, a)$ is a *full* bicartesian closed functor.

References

1. M. Alimohamed. A characterization of lambda definability in categorical models of implicit polymorphism. *Theoretical Computer Science*, 146:5–23, 1995.
2. D. Dougherty and R. Subrahmanyam. Equality between functionals in the presence of coproducts. Submitted to *Information and Computation*. An earlier version appeared in *Proceedings of 10th LICS*, pages 282–291, 1995.
3. N. Ghani. $\beta\eta$-equality for coproducts. In *Typed Lambda Calculi and Applications, Proceedings of TLCA '95*, pages 171–185. Springer LNCS 902, 1995.
4. A. Jung and J. Tiuryn. A new characterisation of lambda definability. In *Typed Lambda Calculi and Applications, Proceedings of TLCA '93*, pages 230–244. Springer LNCS 664, 1993.
5. J. Lambek and P. J. Scott. *Introduction to Higher Order Categorical Logic*. Number 7 in Cambridge studies in advanced mathematics. Cambridge University Press, 1986.
6. S. Mac Lane and I. Moerdijk. *Sheaves in Geometry and Logic: A First Introduction to Topos Theory*. Springer-Verlag, 1992.
7. J.C. Mitchell. Type systems for programming languages. In J. van Leeuwen, editor, *Handbook of Theoretical Computer Science*, volume II, pages 365 – 458. Elsevier Science Publishers, 1990.
8. P.W. O'Hearn and J.G. Riecke. Kripke logical relations and PCF. *Information and Computation*, 120:107–116, 1995.
9. E. Palmgren. Constructive sheaf semantics. *Mathematical Logic Quarterly*, 43:321 – 325, 1997.
10. G.D. Plotkin. Lambda-definability in the full type hierarchy. In J. P. Seldin and J. R. Hindley, editors, *To H. B. Curry: Essays on Combinatory Logic, Lambda Calculus and Formalism*. Academic Press, New York, 1980.
11. J.G. Riecke and A.B. Sandholm. A relational account of call-by-value sequentiality. In *Proceedings of 12th Annual Symposium on Logic in Computer Science*, pages 258–267, 1997.
12. G. Wraith. Artin glueing. *Journal of Pure and Applied Algebra*, 4:345–348, 1974.

Explicitly Typed $\lambda\mu$-Calculus for Polymorphism and Call-by-Value

Ken-etsu Fujita

Kyushu Institute of Technology, Iizuka 820-8502, Japan,
fujiken@dumbo.ai.kyutech.ac.jp

Abstract. We introduce an explicitly typed $\lambda\mu$-calculus of call-by-value as a short-hand for the 2nd order Church-style. Our motivation comes from the observation that in Curry-style polymorphic calculi, control operators such as `callcc` or μ-operators cannot, in general, treat the terms placed on the control operator's left. Following the continuation semantics, we also discuss the notion of values in classical system, and propose an extended form of values. It is shown that the CPS-translation is sound with respect to $\lambda 2$ (2nd order λ-calculus). Next, we provide an explicitly and an implicitly typed Damas-Milner systems with μ-operators. Finally, we give a brief comparison with standard ML plus `callcc`, and discuss a natural way to avoid the unsoundness of ML with `callcc`.

1 Introduction

On the basis of the Curry-Howard-De Bruijn isomorphism [19], proof reductions can be regarded as computational rules, and the algorithmic contents of proofs can be used to obtain correct programs that satisfy logical specifications. The computational meaning of proofs has been investigated in a wide range of fields, including not only intuitionistic logic but also classical logic and modal logic [21]. In the area of classical logic, there have been a number of noteworthy investigations including Griffin[12], Murthy[26], Parigot[30], Berardi&Barbanera[4], Rehof&Sørensen[35], de Groote[8] and Ong[28].

As far as we know, however, polymorphic call-by-value calculus is less studied from the viewpoint of classical logic. In this paper, we introduce an explicitly typed $\lambda\mu$-calculus of call-by-value as a short-hand for the 2nd order Church-style. Our motivation comes from the observation that in Curry-style polymorphic calculi, control operators such as `callcc` or μ-operators cannot, in general, treat the terms placed on the control operator's left. Following the continuation semantics, we also discuss the notion of values in classical system, and propose an extended form of values. It is shown that the CPS-translation is sound with respect to $\lambda 2$ (System F of Girard, Polymorphic calculus of Reynolds). We observe that the inverse of the soundness does not hold, and that adding $\bot$-reduction in Ong&Stewart [29] breaks down the soundness of the CPS-translation. As one of by-products, it can be obtained that the 2nd order call-by-value $\lambda\mu$-calculus has the strong normalization property. Next, we provide an explicitly and an implicitly typed Damas-Milner systems with μ-operators, and compare those from a viewpoint of polymorphic control operators under call-by-value. Finally, we give a brief comparison with standard ML plus `callcc`, and discuss a natural way to avoid the unsoundness of ML with `callcc` [14].

2 Curry-Style vs. Church-Style

With respect to the simply typed lambda calculus $\lambda^{\rightarrow}$, there is a forgetful map from $\lambda^{\rightarrow}$ à la Church to à la Curry, and conversely, well-typed terms in $\lambda^{\rightarrow}$-Curry can be lifted to well-typed terms in $\lambda^{\rightarrow}$-Church [3]. In the case of ML [25], there also exists implicitly typed and explicitly typed systems, and they are essentially equivalent [17]. Hence, the implicitly typed system serves as a short-hand for the explicitly typed system.

However, the equivalence between Curry-style and Church-style does not always hold for complex systems. Parigot [30] introduced $\lambda\mu$-calculus in Curry-style as 2nd order classical logic although $\lambda\mu$-calculus à la Church was also given [32]. An intrinsically classical reduction is called the structural reduction that is a kind of permutative proof reductions in Prawitz [34] or the so-called commutative cut. The $\lambda\mu$-calculus of Parigot is now known as a call-by-name system. If we construct a call-by-value $\lambda\mu$-calculus, then the Curry-style cannot work for a consistent system. In a call-by-value system of $\lambda\mu$, we can adopt a certain permutative reduction [30, 29], called the symmetric structural reduction, to manage the terms placed on the μ-operator's left. However, the symmetric structural proof reduction, in general, violates the subject reduction property in the Curry-style. Consider the following figures:

$$\dfrac{V : (\forall t.\sigma_1) \to \sigma_2 \qquad \dfrac{\dfrac{\begin{array}{c}\dfrac{M_1 : \sigma_1}{[\alpha]M_1 : \bot; \sigma_1^{\alpha}} \\ \vdots \\ M : \bot; \sigma_1^{\alpha}\end{array}}{\mu\alpha.M : \sigma_1}}{\mu\alpha.M : \forall t.\sigma_1}}{V(\mu\alpha.M) : \sigma_2} \qquad \triangleright \qquad \dfrac{\begin{array}{c}\dfrac{\dfrac{V : (\forall t.\sigma_1) \to \sigma_2 \quad \dfrac{M_1 : \sigma_1}{M_1 : \forall t.\sigma_1}\,(\forall I)^*}{VM_1 : \sigma_2}}{[\alpha](VM_1) : \bot; \sigma_2^{\alpha}} \\ \vdots \\ M[V \Rightarrow \alpha] : \bot; \sigma_2^{\alpha}\end{array}}{\mu\alpha.M[V \Rightarrow \alpha] : \sigma_2}$$

where $M[V \Rightarrow \alpha]$ denotes a term obtained by replacing each subterm of the form $[\alpha]N$ in M with $[\alpha](VN)$. Here, when M is in the form of $[\alpha](\lambda x_1 \cdots x_n.M')$ and the type σ_1 depends on type of some x_i $(1 \le i \le n)$, the eigenvariable condition of $(\forall I)^*$ is broken down. For instance,

$$\lambda x.(\lambda f.(\lambda x_1 x_2.x_2)(fx)(f(\lambda x.x)))\ (\mu\alpha.[\alpha](\lambda y.\mu\beta.[\alpha](\lambda v.y)))$$

has type $t \to t \to t$. But this term is reduced to $\lambda x.x$ by the use of the symmetric structural reduction. Let $P \equiv \lambda f.(\lambda x_1 x_2.x_2)(fx)(f(\lambda x.x))$ and $Q \equiv \mu\alpha.[\alpha](\lambda y.\mu\beta.[\alpha](\lambda v.y))$. Then similarly

$$\lambda g.(\lambda x.g(PQx))(\lambda x.g(PQx)) : (\forall t'.(t' \to t')) \to t \to t$$

is reduced to $\lambda g.(\lambda x.g(xx))(\lambda x.g(xx))$. On the other hand, the case $\mu\alpha.M$ of $\mu\alpha.[\alpha](\lambda v.\mu\beta.[\alpha](\lambda x.x))$ is a special case where the symmetric structural reduction is applicable even to polymorphic $\mu\alpha.M$, and then, for example,

$$\lambda x.((\lambda f.(\lambda x_1 x_2.x_2)(fx)(f(\lambda x.x)))\ (\mu\alpha.[\alpha](\lambda v.\mu\beta.[\alpha](\lambda x.x)))\ x) : t \to t$$

is reduced to $\lambda x.x$. This kind of phenomenon was first discovered by Harper & Lillibridge [14] as a counterexample for ML with `callcc`. From the viewpoint of classical proof reductions, the fatal defect can be explained such that in $\lambda\mu$-calculus à la Curry (2nd order classical logic), an application of the symmetric structural reduction, in general, breaks down the eigenvariable condition of polymorphic generalization, and then the terms placed on the polymorphic μ-operator's left cannot be managed by the symmetric structural reduction. In terms of explicit polymorphism, in other words, an evaluation under Λ-abstractions cannot be allowed without restricting $\Lambda t.M$ to $\Lambda t.V$ [15]. Even in the Damas-Milner style [6] (implicitly typed ML) plus control operators, a similar defect still happens under a ML-like call-by-value [15, 16].

To avoid such a problem in implicitly typed ML with control operators, one can adopt an η-like expansion for polymorphic control operators [11], such that

$$\texttt{let } f = \mu\alpha.M_1 \texttt{ in } M_2 \triangleright \texttt{let } f = \lambda x.\mu\alpha.M_1[\alpha \Leftarrow x] \texttt{ in } M_2,$$

where each subterm in the form of $[\alpha](\lambda y.w)$ in M_1 is replaced with $[\alpha](\lambda y.w)x$. Another natural way to avoid the problem in call-by-value $\lambda\mu$ is to take an explicitly typed system. In the above example, the term Q is a polymorphic term, and this type becomes $\forall t.(t \to t)$. Here, the explicitly typed term as a form of a value, $V \equiv \Lambda t.Q$ is used for β_v-reductions, such that

$$\lambda x.(\lambda f.(\lambda x_1 x_2.x_2)(ftx)(f(t \to t)(\lambda x.x)))\ V : t \to t \to t$$

is now reduced to $\lambda vx.x$. In the next section, under the call-by-value strategy we introduce an explicitly typed $\lambda\mu$-calculus especially for polymorphic terms, which is regarded as a short-hand for the complete Church-style. To obtain the results in this paper, it is enough to consider a system such that $\Lambda t.M$ is represented simply by ΛM such as lifting and $M\sigma$ by $M()$, and $(\Lambda M)()$ is reduced to M. A similar observation is given for let-polymorphism in Leroy [23]. The annotations Λ and () for polymorphic terms play a role of choosing an appropriate computation under call-by-value. However, from the viewpoint of logic, a call-by-value $\lambda\mu$-calculus with explicit polymorphism, called a domain-free system in Barthe&Sørensen [5], is considered here rather than such a simplified polymorphism using the annotations or implicit polymorphism by name [23].

On the other hand, Harper&Lillibridge [15] extensively studied explicit polymorphism and CPS-conversion for F_w with `callcc`. The call-by-value system $\lambda_V\mu$ introduced in section 3 can be regarded as a meaningful simplification of the 2nd order fragment of their system. Moreover, the Damas-Milner style $\lambda\mu_{ml}$ introduced in section 5 has no restriction for establishing the subject reduction and Meyer-Wand typing properties, as compared with those of [15, 16].

3 Explicitly Typed $\lambda_V\mu$-Calculus

Following the observation in the previous section, we introduce an explicitly typed $\lambda\mu$-calculus of call-by-value especially for polymorphic terms, called a domain-free system [5], which is regarded as a short-hand for the Church-style.

The types σ are defined from type variables t and a type constant $\bot$. We have a set of (λ-)variables $x, y, z, \cdots$, and a set of names (that will be called continuation variables later) $\alpha, \beta, \cdots$. The type assumptions are defined as usual, and Δ is used for a set of name-indexed types. The terms M are defined as variables, λ- or Λ-abstractions, applications, μ-abstractions, or named terms. From a logical viewpoint, the typing rule $(\bot E)$ for $\mu\alpha.M$ is regarded as a classical inference rule such that infer $\Gamma, \neg\Delta \vdash \mu\alpha.M : \sigma$ from $\Gamma, \neg\Delta, \alpha{:}\neg\sigma \vdash M : \bot$. The typing rule $(\bot I)$ for $[\alpha]M$ can be considered as a special case of $\bot$-introduction by the use of $(\to E)$. On the basis of the continuation semantics in the next section, a name can be interpreted as a continuation variable. In the rule $(\bot I)$, the continuation variable α appears only in the function-position, but not in the argument-position. Here, the negative assumption $\alpha : \neg\sigma$ corresponding to σ^α of $(\bot I)$ can be discharged only by $(\bot E)$. This style of proofs consisting of the special case of $\bot$-introduction is called a regular proof in Andou [1]. The notion of values is introduced below as an extended form; the class of values is closed under both value-substitutions induced by (β_v) and left and right context-replacements induced by $(\mu_{l,r})$, as defined later. The definition of the reduction rules is given below under call-by-value. In particular, the classical reductions $(\mu_{l,r,t})$ below can be explained as a logical permutative reduction in the sense of Prawitz [34] and Andou [1]. Here, in the reduction of $(\mu\alpha.M)N \rhd \mu\alpha.M[\alpha \Leftarrow N]$, since both type of $\mu\alpha.M$ and type of each subterm M' with the form $[\alpha]M'$ in M can be considered as members of the segments ending with the type of $\mu\alpha.M$, the application of $(\to E, \forall E)$ is shifted up to each occurrence M', and then $M[y \Leftarrow N]$ (each $[\alpha]M'$ is replaced with $[\alpha](M'N)$) is obtained. This reduction is also called a structural reduction in Parigot [30]. On the other hand, since a term of the form $\mu\alpha.M$ is not regarded as a value, $(\lambda x.M_1)(\mu\alpha.M_2)$ will not be a β-contractum, but will be a contractum of (μ_l) below, which can be considered as a symmetric structural reduction. $FV(M)$ stands for the set of free variables in M, and $FN(M)$ for the set of free names in M.

$\lambda_V\mu$:

Types $\sigma ::= t \mid \bot \mid \sigma \to \sigma \mid \forall t.\sigma$

Type Assumptions $\Gamma ::= \langle\,\rangle \mid x{:}\sigma, \Gamma$ $\qquad \Delta ::= \langle\,\rangle \mid \sigma^\alpha, \Delta$

Terms $M ::= x \mid \lambda x.M \mid MM \mid \Lambda t.M \mid M\sigma \mid \mu\alpha.M \mid [\alpha]M$

Type Assignment

$$\Gamma \vdash x : \Gamma(x); \Delta$$

$$\frac{\Gamma \vdash M_1 : \sigma_1 \to \sigma_2; \Delta \quad \Gamma \vdash M_2 : \sigma_1; \Delta}{\Gamma \vdash M_1 M_2 : \sigma_2; \Delta}\,(\to E) \qquad \frac{\Gamma, x{:}\sigma_1 \vdash M : \sigma_2; \Delta}{\Gamma \vdash \lambda x.M : \sigma_1 \to \sigma_2; \Delta}\,(\to I)$$

$$\frac{\Gamma \vdash M : \forall t.\sigma_1; \Delta}{\Gamma \vdash M\sigma_2 : \sigma_1[t := \sigma_2]; \Delta}\,(\forall E) \qquad \frac{\Gamma \vdash M : \sigma; \Delta}{\Gamma \vdash \Lambda t.M : \forall t.\sigma; \Delta}\,(\forall I)^*$$

$$\frac{\Gamma \vdash M : \sigma; \Delta}{\Gamma \vdash [\alpha]M : \bot; \Delta, \sigma^\alpha}\,(\bot I) \qquad \frac{\Gamma \vdash M : \bot; \Delta, \sigma^\alpha}{\Gamma \vdash \mu\alpha.M : \sigma; \Delta}\,(\bot E)$$

where $(\forall I)^*$ denotes the eigenvariable condition.

Values $V ::= x \mid \lambda x.M \mid \Lambda t.M \mid [\alpha]M$

Term reductions

(β_v) $(\lambda x.M)V \triangleright M[x := V]$; (η_v) $\lambda x.Vx \triangleright V$ if $x \notin FV(V)$;
(β_t) $(\Lambda t.M)\sigma \triangleright M[t := \sigma]$; (μ_t) $(\mu\alpha.M)\sigma \triangleright \mu\alpha.M[\alpha \Leftarrow \sigma]$;
(μ_r) $(\mu\alpha.M_1)M_2 \triangleright \mu\alpha.M_1[\alpha \Leftarrow M_2]$; (μ_l) $V(\mu\alpha.M) \triangleright \mu\alpha.M[V \Rightarrow \alpha]$;
(rn) $[\alpha](\mu\beta.V) \triangleright V[\beta := \alpha]$; $(\mu\text{-}\eta)$ $\mu\alpha.[\alpha]M \triangleright M$ if $\alpha \notin FN(M)$,
where the term $M[\alpha \Leftarrow N]$ denotes a term obtained by M replacing each subterm of the form $[\alpha]M'$ in M with $[\alpha](M'N)$. That is, the terms (context) placed on $\mu\alpha.M$'s right is replaced in an argument position of M' in $[\alpha]M'$. In turn, the term $M[V \Rightarrow \alpha]$ denotes a term obtained by M replacing each subterm of the form $[\alpha]M'$ in M with $[\alpha](VM')$.

Values are reduced to simpler values by (η_v), eta-reduction and (rn), renaming rules, and those rules are restricted to values, whose condition is necessary to establish a sound CPS-translation in section 4. We note that as observed in Ong&Stewart [29], there are closed normal forms which are not values, called canonical forms, e.g., $\mu\alpha.[\alpha](\lambda x.\mu\beta.[\alpha](\lambda v.x))$. Those terms can be reduced by (S_3) in [31] or ζ^{ext}_{fun} in [29], but in this case, $(\mu\alpha.M)(\mu\beta.N)$ is reduced in the two ways (not confluent). Note also that the failure of operational extensionality for μPCF^-_v is demonstrated in [29]. In fact, ζ^{ext}_{fun} becomes admissible under the eta-reduction and (μ_r). Here, however a term in the form of $\mu\alpha.M$ is not a value, and we have the value-restricted (η_v) rather than the eta-reduction itself.

We denote $\triangleright_\mu$ by the one-step reduction induced by $\triangleright$. We write $=_\mu$ for the reflexive, symmetric, transitive closure of $\triangleright_\mu$. The notations such as $\triangleright_\beta$, $\triangleright_{\beta\eta}$, $\triangleright^+_\beta$, $\triangleright^*_{\beta\eta}$, $=_{\beta\eta}$, etc. are defined as usual, and $\triangleright^i_\beta$ denotes i-step β-reductions $(i \geq 0)$.

Proposition 1 (Subject reduction property for $\lambda_V\mu$). *If we have $\Gamma \vdash M_1 : \sigma; \Delta$ and $M_1 \triangleright_\mu M_2$ in $\lambda_V\mu$, then $\Gamma \vdash M_2 : \sigma; \Delta$ in $\lambda_V\mu$.*

Proof. By induction on the derivation of $M_1 \triangleright_\mu M_2$. Note that in $\lambda_V\mu$, typing rules are uniquely determined depending on the shape of terms. □

The well-known type erasure M° is defined as follows:
$(x)^\circ = x$; $(\lambda x.M)^\circ = \lambda x.M^\circ$; $(M_1M_2)^\circ = M_1^\circ M_2^\circ$;
$(\Lambda t.M)^\circ = M^\circ$; $(M\sigma)^\circ = M^\circ$; $(\mu\alpha.M)^\circ = \mu\alpha.M^\circ$; $([\alpha]M)^\circ = [\alpha]M^\circ$.
Then it can be seen that the typing relation is preserved between $\lambda_V\mu$ and implicitly typed $\lambda\mu$:
(i) If we have $\Gamma \vdash M : \sigma; \Delta$ in $\lambda_V\mu$, then $\Gamma \vdash M^\circ : \sigma; \Delta$ in implicit $\lambda\mu$.
(ii) If we have $\Gamma \vdash M_1 : \sigma; \Delta$ in implicit $\lambda\mu$, then there exists M_2 such that $M_1 = M_2^\circ$ and $\Gamma \vdash M_2 : \sigma; \Delta$ in $\lambda_V\mu$.
The set of types inhabited by terms coincides between implicit $\lambda\mu$ and $\lambda_V\mu$. However, erasing type information makes much more reductions possible, such as η-reduction of the erasure in Mitchell [24], and the subject reduction property for M° is broken down, for example, a counterexample in section 2.

4 CPS-Translation for $\lambda_V\mu$-Calculus

To provide the CPS-translation, we define a simplified version of $\lambda 2$ à la Church as the intuitionistic fragment of $\lambda_V\mu$ (This system of $\lambda 2$ is the so-called domain-free system [5]). Here, besides λ-variables $x, y, z, \cdots$ used in λ-calculus as usual,

$\lambda 2$ has the distinguished variables $\alpha, \beta, \cdots$ called continuation variables. Reduction rules in $\lambda 2$ are also defined as usual under call-by-name. The term with the form $[\alpha]M$ (value) will be interpreted as $\lambda k.k(\overline{M}\alpha)$, where the representation of $\overline{M}\alpha$ is consumed by the continuation k, such as the case of λ-abstraction. The translation from $\lambda_V\mu$ to $\lambda 2$, with an auxiliary function Ψ for values, comes from Plotkin [33].

Definition 1 (CPS-translation). $\overline{x} = \lambda k.kx$; $\overline{\lambda x.M} = \lambda k.k(\lambda x.\overline{M})$;
$\overline{M_1M_2} = \lambda k.\overline{M_1}(\lambda m.\overline{M_2}(\lambda n.mnk))$; $\overline{\Lambda t.M} = \lambda k.k(\Lambda t.\overline{M})$;
$\overline{M\sigma} = \lambda k.\overline{M}(\lambda m.m\sigma^q k)$; $\overline{\mu\alpha.M} = \lambda\alpha.\overline{M}(\lambda x.x)$; $\overline{[\alpha]M} = \lambda k.k(\overline{M}\alpha)$.
$\Psi(x) = x$; $\Psi(\lambda x.M) = \lambda x.\overline{M}$; $\Psi(\Lambda t.M) = \Lambda t.\overline{M}$; $\Psi([\alpha]M) = \overline{M}\alpha$.
$t^q = t$; $(\sigma_1 \to \sigma_2)^q = \sigma_1^q \to \neg\neg\sigma_2^q$; $(\forall t.\sigma)^q = \forall t.\neg\neg\sigma^q$.

According to the continuation semantics of Meyer&Wand [27], our definition of the CPS-translation can be read as follows: If we have a variable x, then the value x is passed on to the continuation k. In the case of a λ- or Λ-abstraction, a certain function that will take two arguments is passed on to the continuation k. If we have a term with a continuation variable α, then a certain function with the argument α is passed on to the continuation k, where the variable α will be substituted by a continuation. Here, it would be natural that a value is regarded as the term that is mapped by Ψ to some term consumed by the continuation k, since the continuation is the context in which a term is evaluated and then to which the value is sent. Our notion of values as an extended form is derived following this observation.

Lemma 1. *Let $=$ denote the definitional equality of the CPS-translation.*
(i) For any term M where $k \notin FV(M)$, $\lambda k.\overline{M}k \rhd_\beta \overline{M}$.
(ii) For any value V, $\overline{V} = \lambda k.k\Psi(V)$.
(iii) For any term M, value V, and type σ, we have $\overline{M[x := V]} = \overline{M}[x := \Psi(V)]$ and $\overline{M[t := \sigma]} = \overline{M}[t := \sigma^q]$.

The above lemma can be proved by straightforward induction. On the basis of the CPS-translation, the left and right context-replacements $M[\alpha \Leftarrow M_1]$ and $M[V \Rightarrow \alpha]$ can be interpreted as the following substitutions for continuation variables, respectively.

Lemma 2. *Let M contain i free occurrences of $[\alpha]$ where $i \geq 0$. Then we have that $\overline{M[\alpha \Leftarrow M_1]} \rhd_\beta^i \overline{M}[\alpha := \lambda m.\overline{M_1}(\lambda n.mn\alpha)]$ and $\overline{M[\alpha \Leftarrow \sigma]} \rhd_\beta^i \overline{M}[\alpha := \lambda m.m\sigma^q\alpha]$.*

Proof. By induction on the structure of M. □

Lemma 3. *For any term M and value V, $\overline{M[V \Rightarrow \alpha]} \rhd_\beta^{3i} \overline{M}[\alpha := \lambda n.\Psi(V)n\alpha]$, where M contains i free occurrences of $[\alpha]$.*

Proof. By induction on the structure of M. □

Lemma 4. *If we have $M \rhd_\mu N$ in $\lambda_V\mu$, then $\overline{M} =_{\beta\eta} \overline{N}$ in $\lambda 2$.*

Proof. By induction on the derivation of $M \triangleright_\mu N$. □

Now, we have confirmed the soundness of the translation in the sense that equivalent $\lambda_V\mu$-terms are translated into equivalent $\lambda 2$-terms. This property essentially holds for untyped terms.

Proposition 2 (Soundness of the CPS-translation). *If we have $M =_\mu N$ in $\lambda_V\mu$, then then $\overline{M} =_{\beta\eta} \overline{N}$ in $\lambda 2$.*

The translation logically establishes the double negation translation of Kuroda. For a set of name-indexed formulae Δ, we define $(\sigma^\alpha, \Delta)^q$ as $\alpha : \neg\sigma^q, \Delta^q$.

Proposition 3. *If $\lambda_V\mu$ has $\Gamma \vdash M : \sigma; \Delta$, then $\lambda 2$ has $\Gamma^q, \Delta^q \vdash \overline{M} : \neg\neg\sigma^q$.*

Proof. By induction on the derivation. □

From the consistency of $\lambda 2$, it is derived that $\lambda_V\mu$ is consistent in the sense that there is no closed term M such that $\vdash M : \bot;$ in $\lambda_V\mu$.

With respect to Proposition 2, it is known that the implication is, in general, not reversible. The counterexample in [33] is not well-typed. Even though we consider well-typed $\lambda_V\mu$-terms, the completeness does not hold for $\lambda_V\mu$: If we have $M_1 \equiv (\lambda x.x)(xy)$ and $M_2 \equiv xy$ in $\lambda_V\mu$, then $\overline{M_1} =_{\beta\eta} xy =_{\beta\eta} \overline{M_2}$ in $\lambda 2$, but $M_1 \neq_\mu M_2$ in $\lambda_V\mu$. Note that in this counterexample, if one excluded η-reduction, then $\overline{M_1} \neq_\beta \overline{M_2}$. Following Hofmann [18], the rewriting rules of $\lambda_V\mu$ are weak from the viewpoint of the semantics, since `Ident`, $(\lambda x.x)M = M$ is necessary in this case.

According to Ong&Stewart [29], their call-by-value $\lambda\mu$-calculus has more reduction rules with the help of type annotation; $\bot$-reduction:

$$V^{\bot\to\sigma}M^\bot \triangleright \mu\beta^\sigma.M^\bot \quad \text{if } \sigma \not\equiv \bot.$$

Here, assume that we have $N_1 \equiv (\lambda x.x)(x([\alpha]y))$ and $N_2 \equiv x([\alpha]y)$, such that $x : \bot \to \sigma, y : \sigma \vdash N_i : \sigma; \sigma^\alpha$ $(i = 1, 2)$ where $\sigma \not\equiv \bot$ in $\lambda_V\mu$. Then N_1 and N_2 are reduced to $N_3 \equiv \mu\beta.[\alpha]y$ by the use of $\bot$-reduction. Now, we have $\overline{N_1} =_{\beta\eta} x(\alpha y) =_{\beta\eta} \overline{N_2}$ in $\lambda 2$, but $\overline{N_3} =_\beta \lambda\beta.\alpha y$ in $\lambda 2$. This example means that the soundness of the CPS-translation is broken down for $\lambda_V\mu$ with $\bot$-reduction, even in the absence of η-reduction. However, on the basis of the correspondence between μ-operator and Felleisen's $\mathcal{C}$-operator [9] such that $\mu\alpha.M = \mathcal{C}(\lambda\alpha.M)$ and $[\alpha]M = \alpha M$, one obtains that $x(\alpha y) =_\mathcal{C} (\lambda x.\mathcal{A}(x))(\alpha y) =_\mathcal{C} \mathcal{A}(\alpha y) =_\mathcal{C} \mathcal{C}(\lambda\beta.\alpha y)$ in the equational theory $\lambda_\mathcal{C}$ [18]. From the naive observation, Hofmann's categorical models for $\lambda_\mathcal{C}$ would also work for an equational version of call-by-value $\lambda\mu$-calculus.

Let $\triangleright_{\beta\eta r}$ be one-step $\triangleright_\mu$ consisting of (β_v), (β_t), (η_v), $(\mu\text{-}\eta)$, or (rn). Let $\triangleright_{st}$ be one-step $\triangleright_\mu$ consisting of (μ_l), (μ_r), or (μ_t). Following the proof of lemma 2, if $M_1 \triangleright_{\beta\eta r} M_2$, then $\overline{M_1} \triangleright^+_{\beta\eta} \overline{M_2}$. On the one hand, each $\triangleright_{st}$-step from M does not simply induce β-steps from $\overline{M}$, i.e., β-conversion may be used. To demonstrate the strong normalization for well-typed $\lambda_V\mu$-terms, it is enough to construct an infinite reduction path from $\overline{M}$ if M has an infinite reduction path. In the case of $\triangleright_{st}$, following lemmata 2 and 3, the CPS-translated terms without the β-conversion still have enough β-, η-redexes to construct an infinite reduction. For instance, in the case M_1 of $(V(\mu\alpha.M))N$, we have $M_1 \triangleright_{st} M_2 \triangleright_{st} M_3$, where

$M_2 \equiv (\mu\alpha.M[V \Rightarrow \alpha])N$ and $M_3 \equiv \mu\alpha.M[V \Rightarrow \alpha][\alpha \Leftarrow N]$. Here, $\overline{M_1}$ can be reduced as follows:

$\overline{M_1} \rhd_\beta^+ N_2 \equiv \lambda k.(\lambda\alpha.\overline{M}id\theta_1)(\lambda m.\overline{N}(\lambda n.mnk)) \rhd_\beta N_3 \equiv \lambda\alpha.\overline{M}id\theta_1\theta_2$,

where $id = \lambda x.x$, $\theta_1 = [\alpha := \lambda n.\Psi(V)n\alpha]$, and $\theta_2 = [\alpha := \lambda m.\overline{N}(\lambda n.mn\alpha)]$. We now have $\overline{M_2} \rhd_\beta^* N_2$ and $\overline{M_3} \rhd_\beta^* N_3$. Let $[N/\alpha]$ be either $[N \Rightarrow \alpha]$ or $[\alpha \Leftarrow N]$.

Lemma 5. *(i) If $M_1 \rhd_{st} M_2 \rhd_{st} M_3$, then $\overline{M_1} \rhd_\beta^+ N_2 \rhd_\beta^+ N_3$ for some λ2-terms N_2 and N_3 such that $\overline{M_2} \rhd_\beta^* N_2$ and $\overline{M_3} \rhd_\beta^* N_3$.*
(ii) Let $\alpha \notin FN(N)$. If $M_1[N/\alpha] \rhd_{\beta\eta r} M_2$, then $\overline{M_1}\theta_1 \rhd_{\beta\eta}^+ \overline{N_2}\theta_2$ for some $\lambda_V\mu$-term N_2 and substitutions θ_1 and θ_2 such that $\overline{M_1[N/\alpha]} \rhd_\beta^* \overline{M_1}\theta_1$ and $\overline{M_2} \rhd_\beta^* \overline{N_2}\theta_2$.

Proof. By induction on the derivations of $\rhd_{st}$ and $\rhd_{\beta\eta r}$. □

Lemma 6. *If there exits an infinite $\rhd_\mu$-reduction path from $\lambda_V\mu$-term M, then $\overline{M}$ also has an infinite $\rhd_{\beta\eta}$-reduction path.*

Proof. From Lemma 5 and the proof of Lemma 4. □

From Proposition 3, Lemma 6 and the fact that λ2 is strongly normalizing [5], the strong normalization property for $\lambda_V\mu$ can be obtained.

Proposition 4 (Strong Normalization Property for $\lambda_V\mu$). *Any well-typed $\lambda_V\mu$-term is strongly normalizable.*

It is observed [10] that the straightforward use of the Tait&Martin-Löf parallel reduction [37] could not work for proving the Church-Rosser property for λμ including renaming rule, contrary to the comments on Theorem 2.5 in [29]. Even though one defines parallel reduction $\gg$ as usual, we cannot establish that if $M_i \gg N_i$ $(i = 1, 2)$, then $M_1[\alpha \Leftarrow M_2] \gg N_1[\alpha \Leftarrow N_2]$; fact (iv) in the proof of Theorem 1 in [30].

Lemma 7 (Weak Church-Rosser Property for $\lambda_V\mu$). *If $M \rhd_\mu M_1$ and $M \rhd_\mu M_2$, then $M_1 \rhd_\mu^* N$ and $M_2 \rhd_\mu^* N$ for some N.*

From Proposition 4 and Lemma 7, we can obtain the Church-Rosser property using Newman's lemma [2].

Proposition 5 (Church-Rosser Theorem). *$\lambda_V\mu$ has the Church-Rosser property for well-typed terms.*

5 Damas-Milner Style with μ-Operators

There exist implicitly typed and explicitly typed ML, and with respect to the implicitly typed ML, there also exist two styles; the conventional ML [6] and the system ML* [22] in which assumption types are universal and derived types are monomorphic. Those two implicitly typed ML are essentially equivalent [22], and moreover, implicitly and explicitly typed ML are also equivalent [17]. First, we provide an explicitly typed ML with μ-operators, called $\lambda\mu_{eml}$, and define CPS-translation into λ2. Next, an implicitly typed ML* with μ-operators, called $\lambda\mu_{iml}$ is provided, and CPS-translation into ML is defined. Finally, we give a brief comparison between those.

5.1 $\lambda\mu_{eml}$: Explicit let-polymorphism by value

Following the observation in section 2, we introduce an explicitly typed ML with μ-operators especially for polymorphic terms, which is regarded as a short-hand for the completely explicitly typed ML plus μ-operators. We define the system $\lambda\mu_{eml}$ under call-by-value in the following. The types τ and the type schemes σ are defined as usual. A type assumption Γ is a finite set of declarations with the form $x:\sigma$, and Δ is a finite set of name-indexed types with the form τ^α.

$$\tau ::= t \mid \bot \mid \tau \to \tau \qquad \sigma ::= \tau \mid \forall t.\sigma \qquad \Gamma ::= \langle\,\rangle \mid x:\sigma, \Gamma \qquad \Delta ::= \langle\,\rangle \mid \tau^\alpha, \Delta$$

$$\Gamma \vdash x : \Gamma(x); \Delta$$

$$\frac{\Gamma \vdash M_1 : \tau_1 \to \tau_2; \Delta \quad \Gamma \vdash M_2 : \tau_1; \Delta}{\Gamma \vdash M_1 M_2 : \tau_2; \Delta}\ (\to E) \qquad \frac{\Gamma, x:\tau_1 \vdash M : \tau_2; \Delta}{\Gamma \vdash \lambda x.M : \tau_1 \to \tau_2; \Delta}\ (\to I)$$

$$\frac{\Gamma \vdash M : \forall t.\sigma; \Delta}{\Gamma \vdash M\tau : \sigma[t := \tau]; \Delta}\ (\text{Inst}) \qquad \frac{\Gamma \vdash M : \sigma; \Delta}{\Gamma \vdash \Lambda t.M : \forall t.\sigma; \Delta}\ (\text{Gen})^*$$

$$\frac{\Gamma \vdash M_1 : \sigma; \Delta \quad \Gamma, x:\sigma \vdash M_2 : \tau; \Delta}{\Gamma \vdash \mathtt{let}\ x = M_1\ \mathtt{in}\ M_2 : \tau; \Delta}\ (\text{Let})$$

$$\frac{\Gamma \vdash M : \tau; \Delta}{\Gamma \vdash [\alpha]M : \bot; \Delta, \tau^\alpha}\ (\bot I) \qquad \frac{\Gamma \vdash M : \bot; \Delta, \tau^\alpha}{\Gamma \vdash \mu\alpha.M : \tau; \Delta}\ (\bot E)$$

where (Gen)* denotes the eigenvariable condition.
Reduction rules:
(β_v) $(\lambda x.M)V \rhd M[x := V]$; (η_v) $\lambda x.Vx \rhd V$ if $x \notin FV(V)$;
(let) $\mathtt{let}\ x = V\ \mathtt{in}\ M \rhd M[x := V]$; (β_t) $(\Lambda t.M)\tau \rhd M[t := \tau]$;
$(let\text{-}\mu_e)$ $\mathtt{let}\ x = \mu\alpha.M_1\ \mathtt{in}\ M_2 \rhd \mu\alpha.M_1[\lambda x.M_2 \Rightarrow \alpha]$;
(μ_r) $(\mu\alpha.M_1)M_2 \rhd \mu\alpha.M_1[\alpha \Leftarrow M_2]$; (μ_l) $V(\mu\alpha.M) \rhd \mu\alpha.M[V \Rightarrow \alpha]$;
(rn) $[\alpha](\mu\beta.V) \rhd V[\beta := \alpha]$; $(\mu\text{-}\eta)$ $\mu\alpha.[\alpha]M \rhd M$ if $\alpha \notin FN(M)$,
where the notion of values is the same as that in section 3.

We denote $\rhd_{eml}$ by the above one-step reduction induced by $\rhd$. We write $=_{eml}$ for the reflexive, symmetric, transitive closure of $\rhd_{eml}$.

Since in the reduction $(let\text{-}\mu_e)$, the let-bound expression $\mu\alpha.M$ must have a monomorphic type, that is, $\mathtt{let}\ x = \mu\alpha.M_1\ \mathtt{in}\ M_2$ can be read as an abbreviation for $(\lambda x.M_2)(\mu\alpha.M_1)$ for well-typed terms, we have the subject reduction property for $\lambda\mu_{eml}$ without any restrictions.

Proposition 6 (Subject reduction property for $\lambda\mu_{eml}$). *If we have $\Gamma \vdash M_1 : \sigma; \Delta$ and $M_1 \rhd_{eml} M_2$ in $\lambda\mu_{eml}$, then $\Gamma \vdash M_2 : \sigma; \Delta$ in $\lambda\mu_{eml}$.*

Definition 2 (CPS-translation from $\lambda\mu_{eml}$ to $\lambda 2$). $\overline{x} = \lambda k.kx$;
$\overline{\lambda x.M} = \lambda k.k(\lambda x.\overline{M})$; $\overline{\Lambda t.M} = \lambda k.k(\Lambda t.\overline{M})$; $\overline{[\alpha]M} = \lambda k.k(\overline{M}\alpha)$;
$\overline{M_1 M_2} = \lambda k.\overline{M_1}(\lambda m.\overline{M_2}(\lambda n.mnk))$; $\overline{M\tau} = \lambda k.\overline{M}(\lambda m.m\tau^q k)$;
$\overline{\mathtt{let}\ x = M_1\ \mathtt{in}\ M_2} = \lambda k.\overline{M_1}(\lambda x.\overline{M_2}k)$; $\overline{\mu\alpha.M} = \lambda\alpha.\overline{M}(\lambda x.x)$.
$\Psi(x) = x$; $\Psi(\lambda x.M) = \lambda x.\overline{M}$; $\Psi(\Lambda t.M) = \Lambda t.\overline{M}$; $\Psi([\alpha]M) = \overline{M}\alpha$.
$\tau^q = \tau$ *where τ is atomic;* $(\tau_1 \to \tau_2)^q = \tau_1^q \to \neg\neg\tau_2^q$.

Lemma 8. *If $M \rhd_{eml} N$, then $\overline{M} =_{\beta\eta} \overline{N}$.*

Proof. By induction on the derivation of $M \rhd_{eml} N$. □

We have confirmed the soundness of the CPS-translation for untyped terms.

Proposition 7 (Soundness of the CPS-translation). *If we have $M =_{eml} N$ in $\lambda\mu_{eml}$, then $\overline{M} =_{\beta\eta} \overline{N}$ in $\lambda 2$.*

Without any restriction, the translation logically establishes Kuroda's translation. We define $(\tau^\alpha, \Delta)^q$ as $\alpha : \neg\tau^q, \Delta^q$ and $(\forall \boldsymbol{t}.\tau)^q$ as $\forall \boldsymbol{t}.\neg\neg\tau^q$.

Proposition 8. *If $\lambda\mu_{eml}$ has $\Gamma \vdash M : \sigma; \Delta$, then $\lambda 2$ has $\Gamma^q, \Delta^q \vdash \overline{M} : \neg\neg\sigma^q$.*

Similarly to Proposition 4, we also have the strong normalization property for well-typed $\lambda\mu_{eml}$-terms.

Proposition 9. *$\lambda\mu_{eml}$ has the strong normalization property for well-typed terms.*

5.2 $\lambda\mu_{iml}$: Implicit let-polymorphism by value

We introduce an implicitly typed ML* (see also [22]) with μ-operators.

$$\Gamma \vdash x : \tau; \Delta \quad \text{if } \tau \le \Gamma(x)$$

$$\frac{\Gamma \vdash M_1 : \tau_1 \to \tau_2; \Delta \quad \Gamma \vdash M_2 : \tau_1; \Delta}{\Gamma \vdash M_1 M_2 : \tau_2; \Delta} (\to E) \qquad \frac{\Gamma, x{:}\tau_1 \vdash M : \tau_2; \Delta}{\Gamma \vdash \lambda x.M : \tau_1 \to \tau_2; \Delta} (\to I)$$

$$\frac{\Gamma \vdash M_1 : \tau_1; \Delta \quad \Gamma, x{:}\forall \boldsymbol{t}.\tau_1 \vdash M_2 : \tau_2; \Delta}{\Gamma \vdash \texttt{let } x = M_1 \texttt{ in } M_2 : \tau_2; \Delta} (\text{let})^*$$

$$\frac{\Gamma \vdash M : \tau; \Delta}{\Gamma \vdash [\alpha]M : \bot; \Delta, \tau^\alpha} (\bot I) \qquad \frac{\Gamma \vdash M : \bot; \Delta, \tau^\alpha}{\Gamma \vdash \mu\alpha.M : \tau; \Delta} (\bot E)$$

where (let)* denotes the eigenvariable condition, and $\bot$ is a type constant.
Reduction rules:
(β_v) $(\lambda x.M)V \rhd M[x := V]$; (η_v) $\lambda x.Vx \rhd V$ if $x \notin FV(V)$;
(let) $\texttt{let } x = V \texttt{ in } M \rhd M[x := V]$;
$(let\text{-}\mu_i)$ $\texttt{let } x = \mu\alpha.M_1 \texttt{ in } M_2 \rhd \texttt{let } x = \lambda x.\mu\alpha.M_1[\alpha \Leftarrow x] \texttt{ in } M_2$
where M_1 contains a subterm in the form $[\alpha](\lambda y.w)$;
(μ_r) $(\mu\alpha.M_1)M_2 \rhd \mu\alpha.M_1[\alpha \Leftarrow M_2]$; (μ_l) $V(\mu\alpha.M) \rhd \mu\alpha.M[V \Rightarrow \alpha]$;
(rn) $[\alpha](\mu\beta.V) \rhd V[\beta := \alpha]$; $(\mu\text{-}\eta)$ $\mu\alpha.[\alpha]M \rhd M$ if $\alpha \notin FN(M)$,
where the notion of values is the same as that in section 3.

Note the similarity of ζ^{ext}_{fun} in [29] to $(let\text{-}\mu_i)$, but note also that in $\lambda\mu_{iml}$, ζ^{ext}_{fun} is applied only to the top level of the let-bound expression, whose result can be reduced by (let) under call-by-value. On the other hand, since a term in the form of $\mu\alpha.M$ is not a value, $\texttt{let } x = \mu\alpha.M_1 \texttt{ in } M_2$ cannot be reduced by (let) directly. In other words, the application of $(let\text{-}\mu_i)$ regards polymorphic $\mu\alpha.M$ as a value in the form of $\lambda x.\mu\alpha.M[\alpha \Leftarrow x]$ (polymorphic $\mu\alpha.M$ cannot cooperate with (μ_l); see below), and then $\texttt{let } f = \mu\alpha.M \texttt{ in } (\lambda x.N)f$ is not reduced by (μ_l) but by (β_v), and $\texttt{let } f = \mu\alpha.M \texttt{ in } f(\mu\beta.N)$ is not by (μ_r) but by (μ_l).

As in $\lambda\mu_{eml}$, we can establish the subject reduction property for $\lambda\mu_{iml}$. However, in $\lambda\mu_{iml}$ we cannot adopt (*let-*μ_e): $\mathtt{let}\ x = \mu\alpha.M\ \mathtt{in}\ N \triangleright \mu\alpha.M[\lambda x.N \Rightarrow \alpha]$, since (*let-*$\mu_e$) cannot be well-typed in $\lambda\mu_{iml}$. One may still consider (*let-*μ_e) in the β-reduced form:

(*let-*μ_e') $\mathtt{let}\ x = \mu\alpha.M\ \mathtt{in}\ N\ \triangleright\ \mu\alpha.M[[\alpha](N[x := w])/[\alpha]w]$,

where the term $M[[\alpha](N[x := w])/[\alpha]w]$ is a term obtained from M, replacing each subterm of the form $[\alpha]w$ with $[\alpha](N[x := w])$. In general, (*let-*μ_e') cannot represent a correct proof reduction either. To verify this, assume that we have the following proof figure for the left-hand side of (*let-*μ_e'), where $\mu\alpha.M$ is used polymorphically in N more than once:

$$
\dfrac{
\dfrac{\dfrac{\begin{array}{c}\Pi_1\\ P:\tau_1\end{array}}{[\alpha]P:\bot;\tau_1^{\alpha}}\\ \Pi_2\\ M:\bot;\tau_1^{\alpha}}{\mu\alpha.M:\tau_1}
\qquad
\begin{array}{c}\dfrac{[x:\forall t.\tau_1]}{x:\tau_1[t:=\tau_3]}\qquad\dfrac{[x:\forall t.\tau_1]}{x:\tau_1[t:=\tau_4]}\\ \Pi_3\\ N:\tau_2\end{array}
}{\mathtt{let}\ x=\mu\alpha.M\ \mathtt{in}\ N:\tau_2}\ (\mathrm{let})^*
$$

Then one obtains the following type assignment for the right-hand side:

$$
\dfrac{
\begin{array}{c}
\dfrac{\begin{array}{c}\Pi_1[t:=\tau_3]\circ S\qquad\qquad \Pi_1[t:=\tau_4]\circ S\\ P:\tau_1[t:=\tau_3]\circ S\qquad P:\tau_1[t:=\tau_4]\circ S\\ \Pi_3 S\\ N[x:=P]:\tau_2 S\end{array}}{[\alpha](N[x:=P]):\bot;(\tau_2 S)^{\alpha}}\\
\Pi_2[t:=\tau_3]\circ S\\
M[[\alpha](N[x:=w])/[\alpha]w]:\bot;(\tau_2 S)^{\alpha}
\end{array}
}{\mu\alpha.M[[\alpha](N[x:=w])/[\alpha]w]:\tau_2 S}
$$

Here, τ_3 and τ_4 must be unifiable under some substitution S, since the assumption whose type contains a free variable t in Π_1 may be discharged by $(\rightarrow I)$ in Π_2, and in this case those assumptions must be chancelled by the single application of $(\rightarrow I)$ after the reduction.

Following the above observation, we obtain that (*let-*μ_e') represents a correct proof reduction only if all types of x in N can be unified, where the merit of polymorphism is lost. It can also be observed that, in the above proof figure, if Π_2 contains no $(\rightarrow I)$ that discharges the type containing free t, then there is no need to unify each type of x in N, and (*let-*μ_e') becomes correct in this case. For example, in the case of $\mathtt{let}\ x = \mu\alpha.[\alpha]\lambda y.\mu\beta.[\alpha](\lambda v.y)\ \mathtt{in}\ N$, one has to unify each type of x in N. On the other hand, (*let-*μ_e') is a correct reduction for the case of $\mathtt{let}\ x = \mu\alpha.[\alpha]\lambda v.\mu\beta.[\alpha](\lambda y.y)\ \mathtt{in}\ N$. See also observation in section 2.

It would not be straightforward to give a CPS-translation to $\lambda\mu_{iml}$, since, as observed in the above, polymorphic let-expressions cannot be read as an abbreviation of λ-expressions, which can cooperate with (μ_l) under call-by-value. Hence, we start with separating the λ-variables x into two categories; monomorphic $x:\tau$ and polymorphic $X:\forall t.\tau$. We also consider a strict class of values, excluding a single occurrence X: $V ::= x \mid \lambda x.M \mid [\alpha]M$. To establish the CPS-translation, the call-by-value reduction rules are applied for the strict class. Then a call-by-value CPS-translation is given to monomorphic x, and a call-by-name CPS-translation is to polymorphic X. The translation from $\lambda\mu_{iml}$ (ML* plus μ-operator) to ML ($\lambda\mu_{eml}$ without μ-operators) is defined as follows:

Definition 3 (CPS-translation from $\lambda\mu_{iml}$ to ML). $\overline{x} = \lambda k.kx$;
$\overline{X} = \lambda k.Xk$; $\overline{\lambda x.M} = \lambda k.k(\lambda x.\overline{M})$; $\overline{[\alpha]M} = \lambda k.k(\overline{M}\alpha)$;
$\overline{M_1M_2} = \lambda k.\overline{M_1}(\lambda m.\overline{M_2}(\lambda n.mnk))$;
$\overline{\texttt{let } X{=}M_1 \texttt{ in } M_2} = \lambda k.(\texttt{let } X{=}\overline{M_1} \texttt{ in } (\overline{M_2}k))$; $\overline{\mu\alpha.M} = \lambda\alpha.\overline{M}(\lambda x.x)$.
$\Psi(x) = x$; $\Psi(\lambda x.M) = \lambda x.\overline{M}$; $\Psi([\alpha]M) = \overline{M}\alpha$.

The Meyer-Wand typing property (Kuroda's double negation) can be established for $\lambda\mu_{iml}$ without any restriction.

Proposition 10. *If $\lambda\mu_{iml}$ has $\Gamma \vdash M : \tau; \Delta$, then* ML *has $\Gamma^q, \Delta^q \vdash \overline{M} : \neg\neg\tau^q$.*

Now, $\lambda\mu_{iml}$ without (*let*-μ_i) has the soundness of the CPS-translation.

Lemma 9. *If $M \rhd_{iml} N$ in $\lambda\mu_{iml}$ without (let-μ_i), then $\overline{M} =_{\beta\eta} \overline{N}$ in* ML.

Proof. By induction on the derivation of $M \rhd_{iml} N$. □

Following the proof of lemma 9, not only (*let*) itself but also (*let*) without restricting to values (call-by-name) can be interpreted. This point would justify the 'by-name' semantics for let-expressions in Harper et al. [13, 16] and the implicit let-polymorphism by name in Leroy [23], which is quite similar to $\lambda\mu_{eml}$.

The type erasure M° from $\lambda\mu_{eml}$ to $\lambda\mu_{iml}$ is defined as that in section 3. Then the typing relations between $\lambda\mu_{eml}$ and $\lambda\mu_{iml}$ are equivalent as follows:
(i) If we have $\Gamma \vdash M_1 : \tau; \Delta$ in $\lambda\mu_{iml}$, then there exists M_2 such that $M_1 = M_2^\circ$ and $\Gamma \vdash M_2 : \tau; \Delta$ in $\lambda\mu_{eml}$.
(ii) If we have $\Gamma \vdash M : \forall t.\tau; \Delta$ in $\lambda\mu_{eml}$, then $\Gamma \vdash M^\circ : \tau; \Delta$ in $\lambda\mu_{iml}$.
However, computationally they are different with respect to (*let*) of $\lambda\mu_{eml}$ and (*let*-μ_i). We compare the two rules in the case the let-bound expression of the polymorphic control operators.

On the one hand, if we did not consider a reduction strategy, then there were two critical cases such that (1) $(\mu\alpha.M)(\mu\beta.N)$; and (2) $(\lambda x.N)(\mu\alpha.M)$. One can apply (μ_r) and (μ_l) in the case of (1), and (μ_l) and (β) in the case of (2).
(1) In $\lambda\mu_{eml}$, $\texttt{let } f{=}\Lambda t.\mu\alpha.M \texttt{ in } (f\tau)(\mu\beta.N)$ is reduced by (μ_r) after (*let*). In turn, $\texttt{let } f{=}\mu\alpha.M \texttt{ in } f(\mu\beta.N)$ can be reduced by (μ_l) in $\lambda\mu_{iml}$.
(2) In $\lambda\mu_{eml}$, $\texttt{let } f{=}\Lambda t.\mu\alpha.M \texttt{ in } (\lambda x.N)(f\tau)$ is reduced by (μ_l) after (*let*). On the other hand, $\texttt{let } f{=}\mu\alpha.M \texttt{ in } (\lambda x.N)f$ can be reduced by (β_v) in $\lambda\mu_{iml}$.
With respect to the critical cases, $\lambda\mu_{eml}$ (explicit let-polymorphism by value) and $\lambda\mu_{iml}$ (implicit let-polymorphism by value) choose different computations.

6 Comparison with Related Work and Concluding Remarks

We briefly compare $\lambda\mu_{ml}$ with ML [25] together with `callcc` [13]. In ML, the class of type variables is partitioned into two subclasses, i.e., the applicative and

the imperative type variables. The type of `callcc` is declared with imperative type variables to guarantee the soundness of the type inference. On the basis of the classification, the typing rule for let-expressions is given such that if the let-bound expression is not a value, then generalization is allowed only for applicative type variables; otherwise generalization is possible with no restriction. There is a simple translation from the ML-programs to the $\lambda\mu_{ml}$-terms, such that the two subclasses of type variables in ML are degenerated into a single class: $\lceil \texttt{callcc}(M) \rceil = \mu\alpha.[\alpha](\lceil M \rceil(\lambda x.[\alpha]x))$;
$\lceil \texttt{throw}\ M\ N \rceil = \mu\beta.\lceil M \rceil \lceil N \rceil$ where β is fresh.
However, according to Harper et al. [13], the following program:

$\texttt{let}\ f = \texttt{callcc}(\lambda k.\lambda x.\texttt{throw}\ k\ (\lambda v.x))\ \texttt{in}\ (\lambda x_1 x_2.x_2)(f\ \texttt{1})(f\ \texttt{true})$

is not typable in ML, since $\texttt{callcc}(\lambda k.\lambda x.\texttt{throw}\ k\ (\lambda v.x))$ with imperative type variables is not a value, and in the case of non-value expressions, polymorphism is allowed only for expressions with applicative type variables. If it were typable with `bool`, then this was reduced to `1` following the operational semantics. Under the translation $\lceil\ \rceil$ together with type annotation, in $\lambda\mu_{eml}$ we have

$\texttt{let}\ f = \Lambda t.\mu\alpha.[\alpha]\lambda x.\mu\beta.[\alpha](\lambda v.x)\ \texttt{in}\ (\lambda x_1 x_2.x_2)(f\ \texttt{int}\ \texttt{1})(f\ \texttt{bool}\ \texttt{true})$

with type `bool`, and this is now reduced to `true`, as in F_ω plus `callcc` under call-by-value, not under ML-like call-by-value [15]. In turn, the following term

$\texttt{let}\ f = \mu\alpha.[\alpha]\lambda x.\mu\beta.[\alpha](\lambda v.x)\ \texttt{in}\ (\lambda x_1 x_2.x_2)(f\ \texttt{1})(f\ \texttt{2})$

with type `int` is reduced to `1` by (μ_l). On the other hand, in $\lambda\mu_{iml}$ we have

$\texttt{let}\ f = \mu\alpha.[\alpha]\lambda x.\mu\beta.[\alpha](\lambda v.x)\ \texttt{in}\ (\lambda x_1 x_2.x_2)(f\ \texttt{1})(f\ \texttt{true})$

with type `bool`, and this is also reduced to `true`. $\lambda\mu_{ml}$ could overcome the counterexample of polymorphic `callcc` in ML, and moreover, the typing conditions for let-expressions could be deleted, which is observed in section 5. In particular, $\lambda\mu_{iml}$ is another candidate for implicit polymorphism by value, compared with implicit polymorphism by name in Leroy [23].

Ong&Stewart [29] extensively studied a call-by-value programming language based on a call-by-value variant of finitely typed $\lambda\mu$-calculus. There are some distinctions between Ong&Stewart and our finite type fragment; their reduction rules have type annotations like the complete Church-style, and, using the annotation, more reduction rules are defined than ours, which can give a stronger normal form. In addition, our notion of values is an extended one, which would be justified by observation based on the CPS-translation. Moreover, our renaming rule is applied for the extended values, and following the proof of lemma 4, this distinction is essential for the CPS-translation of renaming rule. Otherwise the reductions by renaming rule would not be simulated by β-reductions. On the other hand, in the equational theory λ_C of Hofmann [18], one obtains $\alpha(\mathcal{C}(\lambda\beta.M)) =_C M[\beta := \alpha]$ without restricting to values, which would be distinction between equational theory and rewriting theory.

We used the CPS-translation as a useful tool to show consistency and strong normalization of the system. With respect to Proposition 2 (soundness of CPS-translation); for call-by-name $\lambda\mu$, on the one hand, the completeness is obtained in de Groote [7], i.e., the call-by-name CPS-translation is injective. For a call-by-value system with Felleisen's control operators [9], on the other hand, the completeness is established with respect to categorical models [18], and moreover, this method is successfully applied to call-by-name $\lambda\mu$ [20]. We believe that our CPS-translation would be natural along the line of [33], and it is worth pursuing the detailed relation to such categorical models [20, 36].

Acknowledgements I am grateful to Susumu Hayashi, Yukiyoshi Kameyama, and the members of the Proof Animation Group for helpful discussions. I would also like to thank the referees for most helpful comments and suggestions.

References

1. Y.Andou: A Normalization-Procedure for the First Order Classical Natural Deduction with Full Logical Symbols. Tsukuba Journal of Mathematics 19 (1) pp.153–162, 1995.
2. H.P.Barendregt: *The Lambda Calculus, Its Syntax and Semantics* (revised edition), North-Holland, 1984.
3. H.P.Barendregt: Lambda Calculi with Types, Handbook of Logic in Computer Science Vol.II, Oxford University Press, pp.1–189, 1992.
4. F.Barbanera and S.Berardi: Extracting Constructive Context from Classical Logic via Control-like Reductions, Lecture Notes in Computer Science 664, pp.45–59, 1993.
5. G.Barthe and M.H.Sørensen: Domain-free Pure Type Systems, Lecture Notes in Computer Science 1234, pp.9–20, 1997.
6. L.Damas and R.Milner: Principal type-schemes for functional programs, *Proc. 9th Annual ACM Symposium on Principles of Programming Languages*, pp.207–212, 1982.
7. P.de Groote: A CPS-Translation for the $\lambda\mu$-Calculus, Lecture Notes in Computer Science 787, pp.85–99, 1994.
8. P.de Groote: A Simple Calculus of Exception Handling, Lecture Notes in Computer Science 902, pp.201–215, 1995.
9. M.Felleisen, D.P.Friedman, E.Kohlbecker, and B.Duba: Reasoning with Continuations, *Proc. Annual IEEE Symposium on Logic in Computer Science*, pp.131–141, 1986.
10. K.Fujita: Calculus of Classical Proofs I, Lecture Notes in Computer Science 1345, pp.321–335, 1997.
11. K.Fujita: Polymorphic Call-by-Value Calculus based on Classical Proofs, Lecture Notes in Artificial Intelligence 1476, pp.170–182, 1998.
12. T.G.Griffin: A Formulae-as-Types Notion of Control, *Proc. 17th Annual ACM Symposium on Principles of Programming Languages*, pp.47–58, 1990.
13. R.Harper, B.F.Duba, and D.MacQueen: Typing First-Class Continuations in ML, *J.Functional Programming*, 3 (4) pp.465–484, 1993.
14. R.Harper and M.Lillibridge: ML with callcc is unsound, *The Types Form*, 8, July, 1991.
15. R.Harper and M.Lillibridge: Explicit polymorphism and CPS conversion, *Proc. 20th Annual ACM Symposium on Principles of Programming Languages*, pp.206–219, 1993.
16. R.Harper and M.Lillibridge: Polymorphic type assignment and CPS conversion, *LISP and Symbolic Computation* 6, pp.361–380, 1993.
17. R.Harper and J.C.Mitchell: On The Type Structure of Standard ML, *ACM Transactions on Programming Languages and Systems*, Vol. 15, No.2, pp.210–252, 1993.
18. M.Hofmann: Sound and complete axiomatisations of call-by-value control operators, *Math.Struct. in Comp. Science* 5, pp.461–482, 1995.

19. W.Howard: *The Formulae-as-Types Notion of Constructions, To H.B.Curry: Essays on combinatory logic, lambda-calculus, and formalism*, Academic Press, pp.479–490, 1980.
20. M.Hofmann and T.Streicher: Continuation models are universal for $\lambda\mu$-calculus, *Proc. 12th Annual IEEE Symposium on Logic in Computer Science*, 1997.
21. S.Kobayashi: Monads as modality, *Theor.Comput.Sci.* 175, pp.29–74, 1997.
22. A.J.Kfoury, J.Tiuryn, and P.Urzyczyn: An Analysis of ML Typability, *Journal of the Association for Computing Machinery*, Vol.41, No.2, pp.368–398, 1994.
23. X.Leroy: Polymorphism by name for references and continuations, *Proc. 20th Annual ACM Symposium of Principles of Programming Languages*, pp.220–231, 1993.
24. J.C.Mitchell: Polymorphic Type Inference and Containment, *Information and Computation* 76, pp.211–249, 1988.
25. R.Milner: A Theory of Type Polymorphism in Programming, *Journal of Computer and System Sciences* 17, pp.348–375, 1978.
26. C.R.Murthy: An Evaluation Semantics for Classical Proofs, *Proc. 6th Annual IEEE Symposium on Logic in Computer Science*, pp.96–107, 1991.
27. A.Meyer and M.Wand: Continuation Semantics in Typed Lambda-Calculi, Lecture Notes in Computer Science 193, pp.219–224, 1985.
28. C.-H.L.Ong: A Semantic View of Classical Proofs: Type-Theoretic, Categorical, and Denotational Characterizations, *Linear Logic '96 Tokyo Meeting*, 1996.
29. C.-H.L.Ong and C.A.Stewart: A Curry-Howard Foundation for Functional Computation with Control, *Proc. 24th Annual ACM Symposium of Principles of Programming Languages*, 1997.
30. M.Parigot: $\lambda\mu$-Calculus: An Algorithmic Interpretation of Classical Natural Deduction, Lecture Notes in Computer Science 624, pp.190–201, 1992.
31. M.Parigot: Classical Proofs as Programs, Lecture Notes in Computer Science 713, pp.263-276, 1993.
32. M.Parigot: Proofs of Strong Normalization for Second Order Classical Natural Deduction, J.Symbolic Logic 62 (4), pp.1461–1479, 1997.
33. G.Plotkin: Call-by-Name, Call-by-Value and the λ-Calculus, *Theor.Comput.Sci.* 1, pp. 125–159, 1975.
34. D.Prawitz: Ideas and Results in Proof Theory, *Proc. 2nd Scandinavian Logic Symposium*, edited by N.E.Fenstad, North-Holland, pp.235–307, 1971.
35. N.J.Rehof and M.H.Sørensen: The λ_Δ-Calculus, Lecture Notes in Computer Science 789, pp.516–542, 1994.
36. T.Streicher and B.Reus: Continuation semantics: abstract machines and control operators, to appear in *J.Functional Programming*.
37. M.Takahashi: Parallel Reductions in λ-Calculus, *J.Symbolic Computation* 7, pp.113–123, 1989.

Soundness of the Logical Framework for Its Typed Operational Semantics

Extended Abstract

Healfdene Goguen*

Department of Computer Science, University of Edinburgh
The King's Buildings, Edinburgh, EH9 3JZ, United Kingdom
Fax: (+44) (131) 667-7209

Abstract. Typed operational semantics [4, 5] is a technique for describing the operational behavior of the terms of type theory. The combination of operational information and types provides a strong induction principle that allows an elegant and uniform treatment of the metatheory of type theory. In this paper, we adapt the new proof of strong normalization by Joachimski and Matthes [6] for the simply-typed λ-calculus to prove soundness of the Logical Framework for its typed operational semantics. This allows an elegant treatment of strong normalization, Church–Rosser, and subject reduction for $\beta\eta$-reduction for the Logical Framework. Along the way, we also give a cleaner presentation of typed operational semantics than has appeared elsewhere.

1 Introduction

Typed operational semantics [4, 5] is a technique for describing the operational behavior of the terms of type theory. Originally developed for Luo's type theory *UTT* [11], a system with dependent types, type universes, inductive types, and impredicative propositions, it has also been applied to modal logics [8] and higher-order subtyping [2].

A presentation similar to typed operational semantics was discovered independently by van Raamsdonk and Severi [18][1]. Their approach is to define an operational definition of strong normalization, by elaborating the weak-head normal forms and the one-step weak-head β-expansions with suitable premisses.

However, their system is limited by their adherence to capturing strong normalization. If the operational system is instead equipped with types then it can serve as the basis for developing the full metatheory of type theory, including strengthening, subject reduction, Church–Rosser and strong normalization. This is the basis of the technique developed for *UTT* [4], which gave a new proof of subject reduction for $\beta\eta$-reduction using the strength of the induction principle of typed operational semantics.

* Now at AT&T Labs, 180 Park Ave., Florham Park NJ 07932 USA.

[1] Loader [9] also developed his work using the same system, after reading [4].

It seems that these benefits of including types in the operational presentation have been ignored elsewhere because others have worked in the framework of simple types. Among the properties that are easy for simple types and more difficult for dependent types are:

- The well-formedness of types. In the simply-typed λ-calculus types are always well-formed. In systems with dependent types, there is an interdependency between the well-formed terms and the well-formed types.
- The inversion of typing judgements. The simply-typed λ-calculus can be formulated as a syntax-directed system, meaning that each term constructor has exactly one corresponding rule of inference. This cannot be done with dependent types because of judgemental equality, which can interfere at any point in the derivation of a judgement.
- The enumeration of variables. In the simply-typed λ-calculus, it can be assumed that there are an infinite number of variables at each type. In systems with dependent types, this cannot be achieved so easily[2], and so we need to formulate lemmas about the manipulation of the hypotheses in the context, such as Thinning and Strengthening.

For these reasons, the metatheory of dependent types is much more complex than that of simple types, and results such as subject reduction require an extensive background development.

Recently, Joachimski and Matthes [6] developed a simple and elegant proof of strong normalization for the simply-typed λ-calculus using the operational definition of strong normalization, replacing the Tait–Girard style proof using saturated sets or candidates of reducibility. The proof using the operational system follows by simultaneously showing the admissibility of substitution and application in the operational system, by complete induction on the type of the substituted variable or the domain of the application.

Our goal in the present paper is to show that this proof can be adapted to the typed operational semantics for the Logical Framework. This shows that the technique lifts successfully to dependent types, and serves to make explicit the information about types necessary in the proof that can be left implicit and informal for the simply-typed λ-calculus. Along the way, we also give a cleaner presentation of typed operational semantics than has appeared elsewhere for the Logical Framework, using ideas incorporated from recent work with Compagnoni on typed operational semantics for subtyping [2].

The final result of the paper, Corollary 2, is the equivalence of the usual typing rules of the Logical Framework and the typed operational semantics. This equivalence allows us to use an approach quite different to the traditional one for Pure Type Systems [1, 10, 17]. We develop all of the properties of the type theory in the typed operational semantics—the only induction on derivations of

[2] Adding a new variable to the context extends the possible types, because types can depend on that variable, and so infinite contexts require some kind of diagonalization. Pottinger's infinite contexts [10, 14] are one solution, but they are quite heavy technical machinery.

the Logical Framework is in the proof of soundness—and then use the equivalence to transfer these properties to the usual presentation.

We believe that the elegance of the development outlined in this paper justifies our approach. Furthermore, although the traditional development of the metatheory of type theory works well with non-normalizing type theories, our approach is more robust for strongly normalizing type theories. For example, as the current paper shows, the difference between the approach for systems with or without η-reduction is very small when using typed operational semantics, as opposed to other developments [3, 15, 19].

The operational understanding of type theory, through the operational definition of strong normalization or through typed operational semantics, seems to have been crucial to the discovery of the new proof technique. The same technique almost certainly works directly for strong normalization, without the intermediary operational definition, but verifying the details using traditional tools of λ-calculus such as residuals is likely to be difficult and tedious.

Our terminology for soundness and completeness of the typed operational semantics has been controversial. Our view is that the operational semantics defines a term model for the standard typing rules. This was motivated by our earlier proof of soundness, which relied on a saturated-set style term model, but we believe that this view is still valid with the new proof.

In this paper we study Martin-Löf's Logical Framework. Although there are important differences in philosophy between this system and the Edinburgh Logical Framework, technically the work needed to establish results about the two systems in their pure form is very similar.

We see this paper as part of a larger program to redevelop the computational foundations of type theory, replacing the Tait–Girard saturated sets or candidates of reducibility proof by the simpler proof using typed operational semantics.

The structure of the rest of the paper is as follows. In Section 2 we give a short presentation of the Logical Framework. In Section 3 we present the typed operational semantics for the Logical Framework and briefly discuss the motivations for the system. In Section 4 we develop the basic metatheory of the typed operational semantics. In Section 5 we prove the main lemma for the admissibility of substitution and application, and use this to show soundness of the Logical Framework for its typed operational semantics. Because the admissibility lemma is the most important technical contribution of this paper, we give the proof of the result in full detail. Finally, in Section 6 we summarize the contributions of the paper and mention possible further work.

2 The Logical Framework

In this section we give a brief introduction to Martin-Löf's Logical Framework. Our intention is only to give the basic definitions necessary for the technical development of this paper. For an introduction to the philosophy and intended use of the type theory, the interested reader should consult one of the more extensive references [11, 13]. To help the reader, we use standard λ-calculus notation rather than the usual notation for the Logical Framework.

The terms of the Logical Framework are defined by the following grammar:

$$\begin{aligned} A, B, C \in K &::= \mathrm{Type} \mid \mathrm{El}(M) \mid \Pi x{:}A.B \\ M, N, P \in T &::= x \mid \lambda x{:}A.M \mid M(N) \\ \Gamma, \Delta, \Phi \in C &::= () \mid \Gamma, x{:}A \end{aligned}$$

The elements $A, B, C \in K$ are called *kinds*. The kind Type represents the possible types of the type theory, and the operator $\mathrm{El}(M)$ represents the kind of elements of type M, if M is a type. This introduces an interdependency between terms and kinds.

We identify terms that are equivalent up to the renaming of bound variables and write $M \equiv N$ if M and N are equal in this way. We write $\mathrm{FV}(M)$ for the free variables in a term M, those variables not bound by abstractions. We write $[N/x]M$ for the usual capture-free substitution of N for the free variable x in M. Each of these operations is lifted to kinds and contexts in the natural way.

We say that a context $\Gamma \equiv x_1{:}A_1, \ldots, x_n{:}A_n$ such that the x_i are distinct and $\mathrm{FV}(A_i) \subseteq \{x_1, \ldots, x_{i-1}\}$ is *consistent*. We write $dom(\Gamma)$ for the set $\{x_1, \ldots, x_n\}$.

2.1 Basic Rules of Inference

The Logical Framework has five judgement forms:

- $\Gamma \vdash \mathrm{ok}$, meaning that Γ is a well-formed context of assumptions,
- $\Gamma \vdash A$ kind, meaning that A is a kind under assumptions Γ,
- $\Gamma \vdash A = B$, meaning that A and B are kinds and are equal under the assumptions Γ,
- $\Gamma \vdash M : A$, meaning that A is a kind and that M is in A, under assumptions Γ, and
- $\Gamma \vdash M = N : A$, meaning that A is a kind, that M and N are in A, and that they are equal in A, under assumptions Γ.

These judgements are defined inductively by the following rules of inference.

Valid Contexts

$$\textsc{Emp}\ \frac{}{() \vdash \mathrm{ok}} \qquad \textsc{Weak}\ \frac{\Gamma \vdash A\ \mathrm{kind} \quad x \notin dom(\Gamma)}{\Gamma, x{:}A \vdash \mathrm{ok}}$$

Types

$$\mathrm{Type}\ \frac{\Gamma \vdash \mathrm{ok}}{\Gamma \vdash \mathrm{Type\ kind}} \qquad \mathrm{El}\ \frac{\Gamma \vdash M : \mathrm{Type}}{\Gamma \vdash \mathrm{El}(M)\ \mathrm{kind}} \qquad \Pi\ \frac{\Gamma, x{:}A_1 \vdash A_2\ \mathrm{kind}}{\Gamma \vdash \Pi x{:}A_1.A_2\ \mathrm{kind}}$$

Type Equality

$$\text{KRefl}\ \frac{\Gamma \vdash A \text{ kind}}{\Gamma \vdash A = A} \qquad \text{KSym}\ \frac{\Gamma \vdash A = B}{\Gamma \vdash B = A}$$

$$\text{KTrans}\ \frac{\Gamma \vdash A = B \quad \Gamma \vdash B = C}{\Gamma \vdash A = C}$$

$$\text{El-Eq}\ \frac{\Gamma \vdash M = N : \text{Type}}{\Gamma \vdash \text{El}(M) = \text{El}(N)} \qquad \Pi\text{-Eq}\ \frac{\Gamma \vdash A_1 = B_1 \quad \Gamma, x{:}A_1 \vdash A_2 = B_2}{\Gamma \vdash \Pi x{:}A_1.A_2 = \Pi x{:}B_1.B_2}$$

Terms

$$\text{Var}\ \frac{\Gamma_0, x{:}A, \Gamma_1 \vdash \text{ok}}{\Gamma_0, x{:}A, \Gamma_1 \vdash x : A} \qquad \text{Eq}\ \frac{\Gamma \vdash M : A \quad \Gamma \vdash A = B}{\Gamma \vdash M : B}$$

$$\lambda\ \frac{\Gamma, x{:}A_1 \vdash M_0 : A_2}{\Gamma \vdash \lambda x{:}A_1.M_0 : \Pi x{:}A_1.A_2} \qquad \text{App}\ \frac{\Gamma \vdash M_1 : \Pi x{:}A_1.A_2 \quad \Gamma \vdash M_2 : A_1}{\Gamma \vdash M_1(M_2) : [M_2/x]A_2}$$

Term Equality

$$\text{Refl}\ \frac{\Gamma \vdash M : A}{\Gamma \vdash M = M : A} \qquad \text{Sym}\ \frac{\Gamma \vdash M = N : A}{\Gamma \vdash N = M : A}$$

$$\text{Trans}\ \frac{\Gamma \vdash M = N : A \quad \Gamma \vdash N = P : A}{\Gamma \vdash M = P : A}$$

$$=\text{R}\ \frac{\Gamma \vdash M = N : A \quad \Gamma \vdash A = B}{\Gamma \vdash M = N : B}$$

$$\lambda\text{-Eq}\ \frac{\Gamma \vdash A_1 = B_1 \quad \Gamma, x{:}A_1 \vdash M_0 = N_0 : A_2}{\Gamma \vdash \lambda x{:}A_1.M_0 = \lambda x{:}B_1.N_0 : \Pi x{:}A_1.A_2}$$

$$\text{App-Eq}\ \frac{\Gamma \vdash M_1 = N_1 : \Pi x{:}A_1.A_2 \quad \Gamma \vdash M_2 = N_2 : A_1}{\Gamma \vdash M_1(M_2) = N_1(N_2) : [M_2/x]A_2}$$

$$\beta\ \frac{\Gamma, x{:}A_1 \vdash M_0 : A_2 \quad \Gamma \vdash M_2 : A_1}{\Gamma \vdash (\lambda x{:}A_1.M_0)(M_2) = [M_2/x]M_0 : [M_2/x]A_2}$$

2.2 Structural Rules of Inference

The rules in this section are separated out because they are admissible. We write $\Gamma \vdash J$ for judgements derived in the full system including these rules, and $\Gamma \vdash^- J$ for judgements derived in the system without these rules. We shall prove the admissibility of the rules through the equivalence with the typed operational semantics.

Substitution Rules

$$\frac{\Gamma, x{:}A, \Gamma' \text{ ok} \quad \Gamma \vdash P : A}{\Gamma, [P/x]\Gamma' \text{ ok}}$$

$$\frac{\Gamma, x{:}A, \Gamma' \vdash B \text{ kind} \quad \Gamma \vdash P : A}{\Gamma, [P/x]\Gamma' \vdash [P/x]B \text{ kind}} \qquad \frac{\Gamma, x{:}A, \Gamma' \vdash B \text{ kind} \quad \Gamma \vdash N = P : A}{\Gamma, [M/x]\Gamma' \vdash [N/x]B = [P/x]B}$$

$$\frac{\Gamma, x{:}A, \Gamma' \vdash M : B \quad \Gamma \vdash P : A}{\Gamma, [P/x]\Gamma' \vdash [P/x]M : [P/x]B} \qquad \frac{\Gamma, x{:}A, \Gamma' \vdash M : B \quad \Gamma \vdash N = P : A}{\Gamma, [N/x]\Gamma' \vdash [N/x]M = [P/x]M : [N/x]B}$$

$$\frac{\Gamma, x{:}A, \Gamma' \vdash B = C \quad \Gamma \vdash P : A}{\Gamma, [P/x]\Gamma' \vdash [P/x]B = [P/x]C} \qquad \frac{\Gamma, x{:}A, \Gamma' \vdash M = N : B \quad \Gamma \vdash P : A}{\Gamma, [P/x]\Gamma' \vdash [P/x]M = [P/x]N : [P/x]B}$$

Thinning

$$\frac{\Gamma, \Gamma' \text{ ok} \quad \Gamma \vdash A \text{ kind} \quad x \notin FV(\Gamma, \Gamma')}{\Gamma, x{:}A, \Gamma' \text{ ok}}$$

$$\frac{\Gamma, \Gamma' \vdash B \text{ kind} \quad \Gamma \vdash A \text{ kind} \quad x \notin FV(\Gamma, \Gamma')}{\Gamma, x{:}A, \Gamma' \vdash B \text{ kind}}$$

$$\frac{\Gamma, \Gamma' \vdash M : B \quad \Gamma \vdash A \text{ kind} \quad x \notin FV(\Gamma, \Gamma')}{\Gamma, x{:}A, \Gamma' \vdash M : B}$$

$$\frac{\Gamma, \Gamma' \vdash B = C \quad \Gamma \vdash A \text{ kind} \quad x \notin FV(\Gamma, \Gamma')}{\Gamma, x{:}A, \Gamma' \vdash B = C}$$

$$\frac{\Gamma, \Gamma' \vdash M = N : B \quad \Gamma \vdash A \text{ kind} \quad x \notin FV(\Gamma, \Gamma')}{\Gamma, x{:}A, \Gamma' \vdash M = N : B}$$

Context Replacement

$$\frac{\Gamma, x{:}A, \Gamma' \text{ ok} \quad \Gamma \vdash A = B}{\Gamma, x{:}B, \Gamma' \text{ ok}}$$

$$\frac{\Gamma, x{:}A, \Gamma' \vdash C \text{ kind} \quad \Gamma \vdash A = B}{\Gamma, x{:}B, \Gamma' \vdash C \text{ kind}} \qquad \frac{\Gamma, x{:}A, \Gamma' \vdash M : C \quad \Gamma \vdash A = B}{\Gamma, x{:}B, \Gamma' \vdash M : C}$$

$$\frac{\Gamma, x{:}A, \Gamma' \vdash C = D \quad \Gamma \vdash A = B}{\Gamma, x{:}B, \Gamma' \vdash C = D} \qquad \frac{\Gamma, x{:}A, \Gamma' \vdash M = N : C \quad \Gamma \vdash A = B}{\Gamma, x{:}B, \Gamma' \vdash M = N : C}$$

Presuppositions

$$\frac{\Gamma \vdash J}{\Gamma \vdash \text{ok}} \quad \frac{\Gamma \vdash A = B}{\Gamma \vdash A \text{ kind}} \quad \frac{\Gamma \vdash M = N : A}{\Gamma \vdash M : A} \quad \frac{\Gamma \vdash M : A}{\Gamma \vdash A \text{ kind}}$$

3 A Typed Operational Semantics for the Logical Framework

The typed operational semantics for the Logical Framework has the following judgement forms with associated informal meaning:

- $\models \Gamma \to \Delta$, meaning that Γ has normal form Δ.
- $\Gamma \models A \to B$, meaning that A is well-formed under assumptions Γ and has normal form B.
- $\Gamma \models M \to N \to P\colon A$, meaning that M, N and P are well-formed of kind A under assumptions Γ, and M has weak-head normal form N and normal form P.

We also use the following abbreviations:

- $\Gamma \models \mathrm{ok}$ for $\models \Gamma \to \Delta$ when Δ is not relevant.
- $\Gamma \models M \to_{\mathrm{w}} N\colon A$ for $\Gamma \models M \to N \to P\colon A$ when P is not relevant.
- $\Gamma \models M \to_{\mathrm{n}} P\colon A$ for $\Gamma \models M \to N \to P\colon A$ when N is not relevant.
- $\Gamma \models M\colon A$ for $\Gamma \models M \to N \to P\colon A$ when N and P are not relevant.

We say that the meanings of the judgements are informal because the rules do not depend on weak-head or normal forms: the demonstration that Δ is the normal form of Γ in $\models \Gamma \to \Delta$, for example, is left to Lemma 11.

The typed operational semantics is defined inductively by the following rules of inference.

Contexts

$$\textsc{Emp}\ \frac{}{\models () \to ()} \qquad \textsc{Weak}\ \frac{\models \Gamma \to \Delta \qquad \Gamma \models A \to B \qquad x \notin dom(\Gamma)}{\models \Gamma, x{:}A \to \Delta, x{:}B}$$

Kinds

$$\mathrm{Type}\ \frac{\Gamma \models \mathrm{ok}}{\Gamma \models \mathrm{Type} \to \mathrm{Type}} \qquad \mathrm{El}\ \frac{\Gamma \models M \to N \to P\colon \mathrm{Type}}{\Gamma \models \mathrm{El}(M) \to \mathrm{El}(P)}$$

$$\Pi\ \frac{\Gamma \models A_1 \to B_1 \qquad \Gamma, x{:}A_1 \models A_2 \to B_2}{\Gamma \models \Pi x{:}A_1.A_2 \to \Pi x{:}B_1.B_2}$$

Terms

$$\textsc{Var}\quad \frac{\Gamma_0, x{:}A, \Gamma_1 \models A \to B}{\Gamma_0, x{:}A, \Gamma_1 \models x \to x \to x : B}$$

$$\eta\quad \frac{\begin{array}{c} \Gamma, x{:}A_1 \models M_0 \to_{\mathrm{n}} P(x) : B_2 \\ \Gamma \models A_1 \to B_1 \qquad \Gamma \models P \to P \to P : \Pi x{:}B_1.B_2 \end{array}}{\Gamma \models \lambda x{:}A_1.M_0 \to \lambda x{:}A_1.M_0 \to P : \Pi x{:}B_1.B_2}$$

$$\lambda\quad \frac{\begin{array}{l} \Gamma \models A_1 \to B_1 \\ \Gamma, x{:}A_1 \models M_0 \to_{\mathrm{n}} P_0 : B_2 \qquad \lambda x{:}B_1.P_0 \text{ not an } \eta\text{-redex} \end{array}}{\Gamma \models \lambda x{:}A_1.M_0 \to \lambda x{:}A_1.M_0 \to \lambda x{:}B_1.P_0 : \Pi x{:}B_1.B_2}$$

$$\textsc{Base}\quad \frac{\begin{array}{l} \Gamma \models M_1 \to N_1 \to P_1 : \Pi x{:}B_1.B_2 \\ \Gamma \models M_2 \to N_2 \to P_2 : B_1 \qquad \Gamma \models [M_2/x]B_2 \to C N_1 \text{ not an abstraction} \end{array}}{\Gamma \models M_1(M_2) \to N_1(M_2) \to P_1(P_2) : C}$$

$$\beta\quad \frac{\begin{array}{ll} \Gamma \models M_1 \to_{\mathrm{w}} \lambda x{:}A_1.N_0 : \Pi x{:}B_1.B_2 & \Gamma \models M_2 : B_1 \\ \Gamma \models [M_2/x]N_0 \to P \to Q : C & \Gamma \models [M_2/x]B_2 \to C \end{array}}{\Gamma \models M_1(M_2) \to P \to Q : C}$$

The typed operational semantics can be viewed as an alternative induction principle for the well-typed terms of the Logical Framework. We have chosen the rules of inference so that the induction principle be as powerful as possible, with the particular criterion that the completeness theorem, Theorem 1, follow as simply as possible.

The system is not simply a reduction relation with added type information. Each of the rules involving application requires the normal forms of the domain kind in the function to be identical to the kind of the argument, replacing the rules for kind equality. The relationship between judgements and derivations is therefore much closer than in the declarative presentation of the Logical Framework of Section 2, and we always know what the last rule of inference must be based on the structure of the judgement.

4 Metatheoretic Properties

In this section, we give an outline of the proofs of the metatheoretic properties of the typed operational semantics. We divide this into two subsections, one for results about typing and the other for results about reduction. As most of the results and proofs in this section have been published elsewhere [2,4] for similar systems, we avoid giving many details.

We shall use "inversion" on a derivation of a judgement to mean a case analysis on the possible last rules of inference for that judgement. Hence, inversion of a derivation of $\Gamma \models x \to N \to P : B$ gives us that $N \equiv x$, $P \equiv x$, $x{:}A \in \Gamma$ and $\Gamma \models A \to B$ (we also know that the derivation used Var as the last rule of inference, but this is usually not important). This is similar to Generation as used for PTS [1], and can be automatized as done in the proof assistants Lego and Coq.

4.1 Typing

Lemma 1 (Subcontext). *If $\Gamma_0, \Gamma_1 \models J$ then there is a not necessarily strict subderivation of $\Gamma_0 \models \mathrm{ok}$.*

Lemma 2 (Contexts). *If $\Gamma \models J$ then Γ is consistent. Furthermore:*

- *If $\Gamma \models A \to B$ then $\mathrm{FV}(A) \cup \mathrm{FV}(B) \subseteq dom(\Gamma)$.*
- *If $\Gamma \models M \to N \to P\colon A$ then $\mathrm{FV}(M) \cup \mathrm{FV}(N) \cup \mathrm{FV}(P) \cup \mathrm{FV}(A) \subseteq dom(\Gamma)$.*

Definition 1 (Renaming). *A map γ is a* substitution *from Δ to Γ if $\Delta \models \mathrm{ok}$, $dom(\gamma) = dom(\Gamma)$ and $x{:}A \in \Gamma$ implies $\Delta \models A[\gamma] \to B$ and $\Delta \models \gamma(x)\colon B$.*

A renaming *is a parallel substitution γ from Δ to Γ such that for each $x{:}A \in \Gamma$ we have $\gamma(x) \equiv y$ and $y{:}A[\gamma] \in \Delta$. We write* $\mathrm{weak}_\Gamma^\Delta$ *for the identity map over $dom(\Gamma)$ if Δ has all components of Γ.*

Lemma 3 (Renaming). *If γ is a renaming from Δ to Γ then:*

- *If $\Gamma \models A \to B$ then $\Delta \models A[\gamma] \to B[\gamma]$.*
- *If $\Gamma \models M \to N \to P\colon A$ then $\Delta \models M[\gamma] \to N[\gamma] \to P[\gamma]\colon A[\gamma]$.*

Proof. By induction on derivations, using Contexts (Lemma 2) for Π, λ and η.

Lemma 4. *If $\Gamma \models \mathrm{ok}$, $\Delta \models \mathrm{ok}$ and Δ has all components of Γ then* $\mathrm{weak}_\Gamma^\Delta$ *is a substitution from Δ to Γ.*

Corollary 1 (Thinning). *If $\Gamma \models J$ and Δ is a valid context with all components of Γ then $\Delta \models J$.*

The above proof of Thinning was inspired by McKinna and Pollack's [12] treatment of α-equivalence. The more complex treatment of Thinning is necessary because the new variables occurring in Δ may have been used for the bound variables of the subject in $\Gamma \models J$. This is also a problem for the traditional proof of Thinning for PTS [1]. See [12, 4] for more details.

In practice, we use the simpler result of Weakening, which simply says that if $\Gamma \models J$ and $\Gamma, \Delta \models \mathrm{ok}$ then $\Gamma, \Delta \models J$; this follows as a corollary to Thinning.

Lemma 5 (Determinacy).

- *If $\models \Gamma \to \Delta$ and $\models \Gamma \to \Phi$ then $\Delta \equiv \Phi$.*
- *If $\Gamma \models A \to B$ and $\Gamma \models A \to C$ then $B \equiv C$.*
- *If $\Gamma \models M \to N \to P\colon B$ and $\Gamma \models M \to Q \to R\colon C$ then $N \equiv Q$, $P \equiv R$, and $B \equiv C$.*

Proof. By simultaneous induction on derivations.

We consider case Π. By inversion of $\Gamma \models \Pi x{:}A_1.A_2 \to \Pi y{:}C_1.C_2$ we know that $\Gamma \models A_1 \to C_1$ and $\Gamma, y{:}A_1 \models [y/x]A_2 \to C_2$. By the induction hypothesis $B_1 \equiv C_1$. Furthermore, by Renaming $\Gamma, x{:}A_1 \models A_2 \to [x/y]C_2$, so by the induction hypothesis again $B_2 \equiv [x/y]C_2$, and so $\Pi x{:}B_1.B_2 \equiv \Pi y{:}C_1.C_2$.

Cases λ and η use Renaming similarly.

Strengthening is often the most difficult of the metatheoretic results for a type theory with an equality rule for types. For the typed operational semantics this result is straightforward, because kind equality is taken care of in the individual rules. The kind in each judgement about terms must always be normal, and although we do not explicitly use this fact it ensures that the strengthened variable never occurs in the kind. This is different from the situation in the Logical Framework, where variables that do not occur in the term can be introduced into the kind by kind equality.

Lemma 6 (Strengthening). *Suppose z is a variable such that $z \notin \mathrm{FV}(\Gamma_1)$. Then:*

- *If $\Gamma_0, z{:}C, \Gamma_1 \models \mathrm{ok}$ then $\Gamma_0, \Gamma_1 \models \mathrm{ok}$.*
- *If $\Gamma_0, z{:}C, \Gamma_1 \models A \to B$ and $z \notin \mathrm{FV}(A)$ then $\Gamma_0, \Gamma_1 \models A \to B$.*
- *If $\Gamma_0, z{:}C, \Gamma_1 \models M \to N \to P\colon A$ and $z \notin \mathrm{FV}(M)$ then $\Gamma_0, \Gamma_1 \models M \to N \to P\colon A$.*

Proof. By simultaneous induction on derivations, using Contexts for rule η.

We now show that the typed operational semantics is complete for the Logical Framework. We remind the reader that $\Gamma \vdash^- J$ represents judgements derived in the system without the structural rules in Section 2.2.

Theorem 1 (Completeness for LF^-).

- *If $\Gamma \models \mathrm{ok}$ then $\vdash^- \Gamma$.*
- *If $\Gamma \models A \to B$ then $\Gamma \vdash^- A$ kind and $\Gamma \vdash^- A = B$.*
- *If $\Gamma \models M \to N \to P\colon A$ then $\Gamma \vdash^- M : A$, $\Gamma \vdash^- M = N : A$, $\Gamma \vdash^- M = P : A$ and $\Gamma \vdash^- A = A$.*

Proof. By simultaneous induction on derivations.

We consider Var. By the induction hypothesis, $\Gamma_0, x{:}A, \Gamma_1 \vdash^- A$ kind and $\Gamma_0, x{:}A, \Gamma_1 \vdash^- A = B$. By Subcontext $\Gamma_0, x{:}A, \Gamma_1 \models \mathrm{ok}$, so by the induction hypothesis $\vdash^- \Gamma_0, x{:}A, \Gamma_1$. By Var $\Gamma_0, x{:}A, \Gamma_1 \vdash^- x : A$, and by Eq $\Gamma_0, x{:}A, \Gamma_1 \vdash^- x : B$. Furthermore, by Refl $\Gamma_0, x{:}A, \Gamma_1 \vdash^- x = x : B$. Finally, by KSym and KTrans $\Gamma_0, x{:}A, \Gamma_1 \vdash^- B = B$.

4.2 Untyped Reduction

Untyped reduction is an essential component of the presentation of some type theories, for example Pure Type Systems, where the equality relation is defined as the least equivalence relation containing untyped reduction. We have instead followed the Martin-Löf style presentation using judgemental equality, which ensures that all intermediate terms in a proof of equality are well-formed. Our formal presentation of the Logical Framework does not rely on untyped reduction in any way.

However, untyped reduction is still an essential component of our development of the metatheory of the Logical Framework, because the proof of soundness of the usual typing rules for the typed operational semantics relies on the result of subject reduction. In particular, we establish properties such as that if $\Gamma \models M \to_{\mathrm{n}} P : A$ then $\Gamma \models P \to P \to P : A$, and if $\Gamma \models [N/x]A \to C$ and $\Gamma \models A \to B$ then $\Gamma \models [N/x]B \to C$, using Adequacy, properties of untyped reduction such as substitution, and subject reduction.

We introduce the following one-step reduction relations:

$$(\lambda x{:}A_1.M_0)(M_2)\ \beta\ [M_2/x]M_0$$
$$\lambda x{:}A_1.M(x)\ \eta\ M \qquad\qquad x \notin \mathrm{FV}(M)$$

A term M is a *redex* if there is an N such that $M\beta N$ or $M\eta N$. Let *untyped reduction*, or just *reduction*, written $M \rhd N$, be the compatible closure of all of the above rules. We write $M \rhd^{+} N$ for the transitive closure of reduction and $M \rhd^{*} N$ for the reflexive, transitive closure of reduction.

Lemma 7 (Adequacy for Untyped Reduction).

- *If $\Gamma \models M \to N \to P : A$ then there is an N' such that $M \rhd^{*}_{\beta} N \rhd^{*}_{\beta} N' \rhd^{*}_{\eta} P$.*
- *If $\Gamma \models A \to C$ then there is a B such that $A \rhd^{*}_{\beta} B \rhd^{*}_{\eta} C$.*

Proof. By simultaneous induction on derivations, using Contexts for rule η.

Lemma 8. *If $\Gamma \models M \to_{\mathrm{w}} Q : \Pi x{:}A.B$, $\Gamma \models N \to_{\mathrm{w}} Q : \Pi x{:}A.B$ and $\Gamma \models M(P) \to R \to S : C$ then $\Gamma \models N(P) \to R \to S : C$.*

Proof. By inversion of the derivation of $\Gamma \models M(P) \to R \to S : C$, using Determinacy.

We now give some basic definitions and lemmas about weak-head normal and normal forms.

Definition 2 (Head Variable). *We say that x has head variable x, and that $M(N)$ has head variable x if M has head variable x.*

Definition 3 (Weak-Head Normal and Normal). *We say that x is weak-head normal, that $\lambda x{:}A.M$ is weak-head normal and that $M(N)$ is weak-head normal if M is weak-head normal and not an abstraction.*

We say that x is normal, that $\lambda x{:}A.M$ is normal if it is not an η-redex and A and M are normal, and that $M(N)$ is normal if M and N are normal and M is not an abstraction.

Normal forms lift to kinds in the natural way.

Lemma 9. *M is normal if and only if M has no reductions.*

Lemma 10. *There is an x such that M has head variable x if and only if M is weak-head normal and not an abstraction.*

Lemma 11 (Weak-head and Normal Forms).

- *If* $\models \Gamma \to \Delta$ *then* Δ *is normal.*
- *If* $\Gamma \models A \to B$ *then* B *is normal.*
- *If* $\Gamma \models M \to N \to P\colon A$ *then* N *is weak-head normal and* P *and* A *are normal.*

We define a simple ordering on kinds by the natural length function.

Definition 4. *We define the* length *of a kind* A, *written* $|A|$, *by structural recursion on* A*:*

$$|\mathrm{Type}| =_{\mathrm{df}} 0 \qquad |\mathrm{El}(M)| =_{\mathrm{df}} 0 \qquad |\Pi x{:}A_1.A_2| =_{\mathrm{df}} |A_1| + |A_2| + 1$$

Lemma 12. *If* $\Gamma \models M \to_{\mathrm{w}} N\colon A$, N *not an abstraction,* $x{:}B \in \Gamma$ *and the head variable of* N *is* x *then* $|A| \leq |B|$.

Proof. By induction on derivations of $\Gamma \models M \to_{\mathrm{w}} N\colon A$.

Lemma 13. *If* $\Gamma \models A \to B$ *then* $|A| = |B|$.

Lemma 14. *If* $\Gamma \models M \to_{\mathrm{w}} N\colon A$ *and* M *is weak-head normal then* $M \equiv N$.

Proof. By induction on derivations that $\Gamma \models M \to_{\mathrm{w}} N\colon A$.

For the β case, if $M_1(M_2)$ is weak-head normal then M_1 is weak-head normal, so by the induction hypothesis $M_1 \equiv \lambda x{:}A_1.M_0$, which is impossible by the definition of $M_1(M_2)$ being weak-head normal.

Lemma 15. *If* M *has head variable* x *and* $x \neq y$ *then* $[N/y]M$ *has head variable* x.

Lemma 16. *If* M *is weak-head normal and not an abstraction and* $M \rhd^{*}_{\beta\eta} N$ *then* N *is weak-head normal and not an abstraction.*

We write $\models \Gamma \downarrow \Delta$ if there is a Φ such that $\models \Gamma \to \Phi$ and $\models \Delta \to \Phi$.

Lemma 17 (Context Conversion). *If* $\Gamma \models J$ *and* $\models \Gamma \downarrow \Delta$ *then* $\Delta \models J$.

Proof. By simultaneous induction on derivations.

We consider VAR. If $\Gamma_0, x{:}A, \Gamma_1 \equiv \Delta$ then $\Delta \equiv \Delta_0, x{:}C, \Delta_1$, with $\Gamma_0 \models A \to B$ and $\Delta_0 \models C \to B$. By Weakening $\Delta_0, x{:}C, \Delta_1 \models C \to B$, so $\Delta_0, x{:}C, \Delta_1 \models x \to x \to x\colon B$ by VAR.

We can now show Subject Reduction for η-reduction.

Lemma 18 (Subject Reduction for η).
If $\Gamma \models \lambda x{:}A_1.M(x) \to N \to P\colon \Pi x{:}B_1.B_2$ *and* $x \notin \mathrm{FV}(M)$ *then there is a* N' *such that* $\Gamma \models M \to N' \to P\colon \Pi x{:}B_1.B_2$.

Proof. We first prove that if $\Gamma \models M \to N \to P\colon \Pi x{:}B_1.C$, $\Gamma \models A_1 \to B_1$ and $\Gamma, x{:}A_1 \models C \to B_2$ then $\Gamma \models \lambda x{:}A_1.M(x) \to Q \to P\colon \Pi x{:}B_1.B_2$ for some Q. This follows by case analysis of N, using inversion, Context Conversion and Weakening if N is an abstraction and Weakening if it is not.

Then, using inversion twice with Determinacy, we know $\Gamma, x{:}A_1 \models M \to Q \to R\colon \Pi x{:}B_1.C$ and $\Gamma, x{:}A_1 \models C \to B_2$ for some Q, R and C, so by Strengthening $\Gamma \models M \to Q \to R\colon \Pi x{:}B_1.C$, so $\Gamma \models \lambda x{:}A_1.M(x) \to S \to R\colon \Pi x{:}B_1.C$ for some S by the above. By Determinacy $R \equiv P$ and $C \equiv B_2$.

The idea underlying the proof of strong normalization for typed operational semantics is the same as that for all other such proofs for λ-calculus, which depends on bounding the length of weak-head reduction sequences. Then, because internal reduction is bounded by the induction hypothesis, and because internal reduction does not generate new weak-head reductions (the property of quasi-commuting from term rewriting), we know that strong normalization holds.

However, for technical reasons we use a slightly different presentation. Instead of defining internal reduction and showing that weak-head and internal reduction quasi-commute, we use the reflexive, transitive closure of weak-head reduction (as defined by the typed operational semantics) and parallel reduction. This allows us to prove subject reduction and Church–Rosser at the same time as establishing the necessary relationship between ordinary and weak-head reduction.

We now give a brief development of parallel reduction, necessary for the proof of subject reduction and strong normalization. Tait and Martin-Löf's proof of Church–Rosser using it highlighted the notion; an elegant presentation is given by Takahashi [16].

Definition 5 (Parallel Reduction). *We define* parallel reduction *as the least relation closed under the following rules of inference:*

$$\text{VAR}\ \frac{}{x \Rightarrow x} \qquad \lambda\ \frac{A \Rightarrow A' \quad M \Rightarrow M'}{\lambda x{:}A.M \Rightarrow \lambda x{:}A'.M'} \qquad \text{APP}\ \frac{M \Rightarrow M' \quad N \Rightarrow N'}{M(N) \Rightarrow M'(N')}$$

$$\beta\ \frac{M \Rightarrow M' \quad N \Rightarrow N'}{(\lambda x{:}A.M)(N) \Rightarrow [N'/x]M'} \qquad \eta\ \frac{M_0 \Rightarrow M'(x) \qquad x \notin \mathrm{FV}(M')}{\lambda x{:}A_1.M_0 \Rightarrow M'}$$

We extend the reduction in the obvious way to kinds and contexts.

Parallel reduction has some simple properties. First, we know $M \Rightarrow M$ for all M. Furthermore, if $M \rhd N$ then $M \Rightarrow N$, and if $M \Rightarrow N$ then $M \rhd^* N$. Finally, if $M \Rightarrow M'$ and $N \Rightarrow N'$ then $[N/x]M \Rightarrow [N'/x]M'$.

Lemma 19 (Parallel Subject Reduction).

- *If* $\models \Gamma \to \Delta$ *and* $\Gamma \Rightarrow \Gamma'$ *then* $\models \Gamma' \to \Delta$.
- *If* $\Gamma \models A \to B$, $\Gamma \Rightarrow \Gamma'$ *and* $A \Rightarrow A'$ *then* $\Gamma' \models A' \to B$.
- *If* $\Gamma \models M \to N \to P\colon A$, $\Gamma \Rightarrow \Gamma'$ *and* $M \Rightarrow M'$ *then there are* N' *and* N'' *such that* $\Gamma' \models M' \to N'' \to P\colon A$, $N \Rightarrow N'$ *and* $\Gamma' \models N' \to N'' \to P\colon A$.

Proof. By simultaneous induction on derivations.

Lemma 20 (Subject Reduction). *If* $\Gamma \models M \to N \to P\colon A$, $\Gamma \Rightarrow \Gamma'$ *and* $M \rhd M'$ *then there is an* N' *such that* $\Gamma \models M' \to N' \to P\colon A$ *and* $N \rhd^* N'$.

Notice that this lemma captures both subject reduction and Church–Rosser, because the full judgement and in particular the normal form is stable under one-step reduction.

Definition 6 (Strong Normalization). *Strong normalization for kinds, written* SN(A), *is the least predicate closed under the following rule of inference:*

$$\text{SN-I}\ \frac{\textit{for all } B.(A \rhd B) \Rightarrow \mathrm{SN}(B)}{\mathrm{SN}(A)}$$

and similarly for kinds.

Lemma 21. *If* $\lambda x{:}A_1.M_0 \Rightarrow N$ *and* $\Gamma \models N \to_{\mathrm{w}} \lambda x{:}B.N_0\colon C$ *then* $M_0 \rhd^* N_0$.

Lemma 22 (Strong Normalization). *If* $\Gamma \models M \to N \to P\colon A$ *then* M *is strongly normalizing. Similarly, if* $\Gamma \models A \to B$ *then* A *is strongly normalizing.*

Proof. By simultaneous induction on derivations.

We consider the case Base. By the induction hypothesis we know that M_1 and M_2 are strongly normalizing, and by assumption $\Gamma \models M_1 \to_{\mathrm{w}} N_1\colon \Pi x{:}B_1.B_2$ where N_1 is not an abstraction. By induction on the maximal length of reductions for M_1 and M_2, we show that if $\Gamma \models M_1 \to_{\mathrm{w}} N_1\colon \Pi x{:}B_1.B_2$ where N_1 is not an abstraction then $M_1(M_2)$ is strongly normalizing.

Then, by SN-I, we need that if $M_1(M_2) \rhd P$ then P is strongly normalizing. We consider the possible reductions:

- $M_1 \rhd M_1'$. Then by Parallel Subject Reduction there are N_1' and N_1'' such that $N_1 \Rightarrow N_1'$, $\Gamma \models N_1' \to_{\mathrm{w}} N_1''\colon \Pi x{:}B_1.B_2$, and $\Gamma \models M_1' \to_{\mathrm{w}} N_1''\colon \Pi x{:}B_1.B_2$. Hence by Lemma 16 N_1'' is not an abstraction, so by the induction hypothesis $M_1'(M_2)$ is strongly normalizing.
- $M_2 \rhd M_2'$. By the induction hypothesis.

Case β uses Lemma 21, Parallel Subject Reduction, the closure of Strong Normalization under reduction and the closure of reduction under substitution.

Finally, the following admissible rule is useful in the proof of Soundness, Theorem 2. It differs from the rule η by replacing the premise $\Gamma \models P\colon \Pi x{:}B_1.B_2$ by the requirement that $x \notin \mathrm{FV}(P)$.

Proposition 1 (Admissibility of η'). *If* $\Gamma \models A_1 \to B_1$ *and* $\Gamma, x{:}A_1 \models M_0 \to_{\mathrm{n}} P(x)\colon B_2$, *with* $x \notin \mathrm{FV}(P)$, *then there is an* N' *such that* $\Gamma \models \lambda x{:}A_1.M_0 \to N' \to P\colon \Pi x{:}B_1.B_2$.

4.3 Comparison with Other Work

We have introduced a new class of formal systems, typed operational semantics, which present type theory from a computational perspective. In doing so, we have arrived at a strategy for studying reduction in type theory opposite to the frequently adopted approach of removing type information from the proof of normalization. The specific typed operational semantics that we study, based on standard reduction, itself gives a precise, coherent description of the relationship between typing and reduction. We have demonstrated that this system gives a new treatment of fundamental results, including Church–Rosser, subject reduction and strong normalization.

The alternative presentation introduced by van Raamsdonk and Severi [18] uses sequences of applications to isolate the one-step weak-head reductions. This formulation hides the use of the diagram stating that weak-head and internal reduction commute in the proof of strong normalization, by giving an explicit description of each of the weak-head reduction steps. However, the basic argument for strong normalization is exactly the same: the number of weak-head reductions of a term is bounded by the derivation, and the internal reductions are bounded by the induction hypothesis, so the term is strongly normalizing.

In our approach, we incorporate the commuting diagram into the proof of Subject Reduction, and we prove Church–Rosser at the same time. Moreover, our use of a more abstract approach based on weak-head and parallel reduction means that it is more generally applicable. For example, when Joachimski and Matthes show strong normalization for Gödel's System T, they need to overload the syntax of application in order to maintain the validity of their rules of inference, instead of extending the reduction relations in the natural way as we would be able to do.

Parallel reduction is a tool we use in the proof of Subject Reduction, which would lead to a clear proof of Subject Reduction for the presentation with sequences of applications as well. The problem in both approaches is the same: the induction hypothesis needs to be strong enough to accommodate the closure of reduction under substitution. One-step reduction does not satisfy this property, and parallel reduction is the simplest reduction relation that does. This problem has not been faced in the alternative approach because that approach has not been used to study typing.

Finally, we have chosen a judgement form, $\Gamma \models M \to N \to P\colon A$, that includes weak-head normal forms. An alternative presentation used elsewhere [4, 5] uses two judgement forms $\Gamma \models M \to_{\mathrm{n}} P\colon A$ and $\Gamma \models M \to_{\mathrm{w}} N\colon A$, where the first indicates that M has normal form P and the second indicates that N is a *one-step* weak-head reduct of M. We prefer the former presentation because it involves fewer judgements and rules of inference, and because it extends naturally to systems of subtyping where the weak-head normal form is important [2], but the development discussed here can be adapted to the latter presentation without difficulty.

5 Soundness

We now show the admissibility of substitution and application in the typed operational semantics.

Intuitively, application and substitution are closely related. If we can substitute at a kind, then we can also apply at that kind, by analysis of the weak-head normal form of the applicator: if it is an abstraction then we substitute, and otherwise we use the rule BASE. Hence, we prove the admissibility of substitution and application simultaneously. Furthermore, values at the base kind can safely be substituted, because they have no applicative behavior. The admissibility of substitution and application can be lifted to higher kinds by induction, using the admissibility of application at smaller kinds.

By complete induction we mean the principle that:

$$\text{IND}\ \frac{\forall m.(\forall n.n < m \Rightarrow \phi(n)) \Rightarrow \phi(m)}{\forall m.\phi(m)}$$

Lemma 23. *Suppose $\Gamma \models N \to Q \to S \colon A'$ and $\Gamma \models A \to A'$. Then:*

1. *Substitution is admissible:*
   (a) *If $\models \Gamma, x{:}A, \Delta \to \Phi$ then there is a Ψ such that $\models \Gamma, [N/x]\Delta \to \Psi$.*
   (b) *If $\Gamma, x{:}A, \Delta \models B \to C$ then there is a D such that $\Gamma, [N/x]\Delta \models [N/x]B \to D$.*
   (c) *If $\Gamma, x{:}A, \Delta \models M \to P \to R \colon B$ then there are T, U and C such that $\Gamma, [N/x]\Delta \models [N/x]M \to T \to U \colon C$, $\Gamma, [N/x]\Delta \models [N/x]P \to T \to U \colon C$ and $\Gamma, [N/x]\Delta \models [N/x]B \to C$.*
2. *Application is admissible: if $\Gamma \models M \to P \to R \colon \Pi x{:}A'.B$ then there are T, U and C such that $\Gamma \models M(N) \to T \to U \colon C$ and $\Gamma \models [N/x]B \to C$.*

Proof. We prove the two cases simultaneously by complete induction on $|A'|$.

1. This follows by simultaneous induction on derivations that $\Gamma, x{:}A, \Delta \models J$.
   - EMP. Immediate.
   - WK. If $\Delta \equiv ()$ then $\models \Gamma \to \Psi$ by Subcontext. If $\Delta \equiv \Delta_0, z{:}B$ then $\Gamma, [N/x]\Delta_0 \models [N/x]B \to C$ by the induction hypothesis, and $\models \Gamma, [N/x]\Delta_0 \to \Psi$ by Subcontext, so $\models \Gamma, [N/x]\Delta_0, z{:}[N/x]B \to \Psi, z{:}C$.
   - Type. By the induction hypothesis $\Gamma, [N/x]\Delta \models$ ok, so $\Gamma, [N/x]\Delta \models \text{Type} \to \text{Type}$.
   - El. By the induction hypothesis $\Gamma, [N/x]\Delta \models [N/x]M \to T \to U \colon C$, where $\Gamma, [N/x]\Delta \models \text{Type} \to C$ implies $C \equiv \text{Type}$ by inversion. Hence $\Gamma, [N/x]\Delta \models [N/x]\text{El}(M) \to \text{El}(U)$.
   - Π. By the induction hypothesis $\Gamma, [N/x]\Delta \models [N/x]A_1 \to C_1$ and $\Gamma, [N/x]\Delta, z{:}[N/x]A_1 \models [N/x]A_2 \to C_2$, so $\Gamma, [N/x]\Delta \models [N/x](\Pi z{:}A_1.A_2) \to \Pi z{:}C_1.C_2$ *by* Π.
   - VAR. Then $M \equiv y$, $P \equiv y$ and $R \equiv y$. There are two cases:
     - $x = y$. We have the premise that $\Gamma, x{:}A, \Delta \models A \to B$, and $\Gamma \models N \to Q \to S \colon A'$ and $\Gamma \models A \to A'$ by assumption. By Subcontext we have a subderivation of $\Gamma, x{:}A, \Delta \models$ ok, by Weakening $\Gamma, x{:}A, \Delta \models A \to$

A', and so by Determinacy $A' \equiv B$. By the induction hypothesis $\Gamma, [N/x]\Delta \models \text{ok}$, and so by Weakening again $\Gamma, [N/x]\Delta \models N \to Q \to S\colon A'$. Furthermore, by Free Variables $x \notin \mathrm{FV}(A) \cup \mathrm{FV}(A')$, so $A' \equiv [N/x]A'$, and $\Gamma, [N/x]\Delta \models [N/x]A' \equiv A' \to A'$ by Adequacy, Subject Reduction and Weakening.

- $x \neq y$. We have the premise that $\Gamma, x{:}A, \Delta \models C \to B$, with $y{:}C \in \Gamma, x{:}A, \Delta$. By the induction hypothesis $\Gamma, [N/x]\Delta \models [N/x]C \to D$, so $\Gamma, [N/x]\Delta \models y \to y \to y\colon D$. Finally, by Adequacy and Subject Reduction $\Gamma, [N/x]\Delta \models [N/x]B \to D$.

– Base. We have premisses $\Gamma, x{:}A, \Delta \models M_1 \to P_1 \to R_1\colon \Pi z{:}B_1.B_2$ with P_1 not an abstraction, $\Gamma, x{:}A, \Delta \models M_2\colon B_1$, and $\Gamma, x{:}A, \Delta \models [M_2/x]B_2 \to B$. By Lemma 11 P_1 is weak-head normal, so by Lemma 10 there is a y that is the head variable of P_1.
Also, by the induction hypothesis there are T_1, U_1 and C such that $\Gamma, [N/x]\Delta \models [N/x]M_1 \to T_1 \to U_1\colon C$, $\Gamma, [N/x]\Delta \models [N/x]P_1 \to T_1 \to U_1\colon C$ and $\Gamma, [N/x]\Delta \models [N/x](\Pi z{:}B_1.B_2) \to C$, and also there are U_2 and C_1 such that $\Gamma, [N/x]\Delta \models [N/x]M_2 \to_{\mathrm{n}} U_2\colon C_1$ and $\Gamma, [N/x]\Delta \models [N/x]B_1 \to C_1$. Furthermore, $\Gamma, [N/x]\Delta \models [N/x][M_2/z]B_2 \to D$, again by the induction hypothesis.
We know $C \equiv \Pi z{:}C_1'.C_2$, $\Gamma, [N/x]\Delta \models [N/x]B_1 \to C_1'$ and $\Gamma, [N/x]\Delta, z{:}[N/x]B_1 \models [N/x]B_2 \to C_2$ by inversion. Hence by Determinacy $C_1 \equiv C_1'$. Furthermore, by Adequacy $[N/x]B_2 \rhd^* C_2$, and $[N/x][M_2/z]B_2 \equiv [[N/x]M_2/z][N/x]B_2$, so by Subject Reduction $\Gamma, [N/x]\Delta \models [[N/x]M_2/z]C_2 \to D$. Also, $\Gamma, x{:}A, \Delta \models [M_2/z]B_2 \to B$ implies $[M_2/z]B_2 \rhd^* B$ by Adequacy, and so $\Gamma, [N/x]\Delta \models [N/x]B \to D$ by Subject Reduction.
We have two cases:

- $x = y$. Then $|\Pi z{:}B_1.B_2| \leq |A'|$ by Lemma 12, so $|B_1| < |A'|$ and $|C_1| < |A'|$ by Lemma 13. Hence by the induction hypothesis there are T, U and D' such that $\Gamma, [N/x]\Delta \models ([N/x]M_1)([N/x]M_2) \to T \to U\colon D'$ and $\Gamma, [N/x]\Delta \models [[N/x]M_2/z]C_2 \to D'$. By Determinacy $D \equiv D'$, and $\Gamma, [N/x]\Delta \models ([N/x]P_1)([N/x]M_2) \to T \to U\colon D$ by Lemma 8.
- $x \neq y$. Then by Lemma 15 $[N/x]P_1$ has head variable y, so $[N/x]P_1 \equiv T$ by Lemma 14 because $\Gamma, [N/x]\Delta \models [N/x]P_1 \to T \to U\colon \Pi z{:}C_1.C_2$. We know $[N/x]P_1$ is not an abstraction by Lemma 10, so by Base

$$\Gamma, [N/x]\Delta \models ([N/x]M_1)([N/x]M_2) \to_{\mathrm{w}} ([N/x]P_1)([N/x]M_2)\colon D,$$
$$\text{and } \Gamma, [N/x]\Delta \models ([N/x]P_1)([N/x]M_2) \to_{\mathrm{w}} ([N/x]P_1)([N/x]M_2)\colon D$$

– λ, η. By the induction hypothesis there is a C_1 such that $\Gamma, [N/x]\Delta \models [N/x]A_1 \to C_1$, and there are U and C_2 such that $\Gamma, [N/x]\Delta, y{:}[N/x]A_1 \models [N/x]M_0 \to_{\mathrm{n}} U_0\colon C_2$ and $\Gamma, [N/x]\Delta, y{:}[N/x]A_1 \models [N/x]A_2 \to C_2$. If $U_0 \equiv U(x)$ with $x \notin \mathrm{FV}(U)$ then $\Gamma, [N/x]\Delta \models [N/x](\lambda y{:}A_1.M_0) \to_{\mathrm{n}} U\colon \Pi z{:}C_1.C_2$ by η', and otherwise $\Gamma, [N/x]\Delta \models [N/x](\lambda y{:}A_1.M_0) \to_{\mathrm{n}} \lambda y{:}C_1.U_0\colon \Pi z{:}C_1.C_2$ by λ. Furthermore, $\Gamma, [N/x]\Delta \models [N/x](\Pi z{:}A_1.A_2) \to \Pi z{:}C_1.C_2$ by Π.

- β. By the induction hypothesis there are T_1 and C such that $\Gamma, [N/x]\Delta \models [N/x]M_1 \rightarrow_{\mathrm{w}} T_1 : C$, $\Gamma, [N/x]\Delta \models [N/x](\lambda y{:}A_1.M_0) \rightarrow_{\mathrm{w}} T_1 : C$, and $\Gamma, [N/x]\Delta \models [N/x](\Pi z{:}B_1.B_2) \rightarrow C$. Also, there is a C_1 such that $\Gamma, [N/x]\Delta \models [N/x]M_2 : C_1$ and $\Gamma, [N/x]\Delta \models [N/x]B_1 \rightarrow C_1$. Furthermore, there are T, U and D such that

$$\Gamma, [N/x]\Delta \models [N/x][M_2/y]M_0 \equiv [[N/x]M_2/y][N/x]M_0 \rightarrow T \rightarrow U : D,$$
$$\Gamma, [N/x]\Delta \models [N/x]P \rightarrow T \rightarrow U : D, \text{ and } \Gamma, [N/x]\Delta \models [N/x]B \rightarrow D$$

  Also, there is a D' such that $\Gamma, [N/x]\Delta \models [N/x][M_2/z]B_2 \rightarrow D'$.
  We know $T_1 \equiv [N/x](\lambda y{:}A_1.M_0)$ by Lemma 14. By inversion $C \equiv \Pi z{:}C_1'.C_2$ with $\Gamma, [N/x]\Delta \models [N/x]B_1 \rightarrow C_1'$ and $\Gamma, [N/x]\Delta, z{:}[N/x]B_2 \models [N/x]B_2 \rightarrow C_2$, so by Determinacy $C_1 \equiv C_1'$. By Adequacy $[M_2/z]B_2 \rhd^\star B$, so by Subject Reduction $\Gamma, [N/x]\Delta \models [N/x]B \rightarrow D'$, and so by Determinacy $D \equiv D'$.
  Finally, $[N/x](\lambda z{:}A_1.M_0) \equiv \lambda z{:}[N/x]A_1.[N/x]M_0$ and $[N/x][M_2/z]M_0 \equiv [[N/x]M_2/z][N/x]M_0$, so by β we get $\Gamma, [N/x]\Delta \models [N/x](M_1(M_2)) \rightarrow T \rightarrow U : D$.

2. This follows by induction on derivations that $\Gamma \models M \rightarrow P \rightarrow R : \Pi x{:}A'.B$.
   - VAR. We have $y{:}D \in \Gamma$ and $\Gamma \models D \rightarrow \Pi x{:}A'.B$. By inversion $D \equiv \Pi x{:}D_1.D_2$ and $\Gamma, x{:}D_1 \models D_2 \rightarrow B$. By Case 1 we know $\Gamma \models [N/x]D_2 \rightarrow C$ for some C, and by Adequacy and Subject Reduction $\Gamma \models [N/x]B \rightarrow C$.
   - BASE. $P_1(M_2)$ is not an abstraction. Hence, by BASE $\Gamma \models M_1(M_2)(N) \rightarrow P_1(M_2)(N) \rightarrow R_1(R_2)(S) : C$, where $\Gamma \models [N/x]B \rightarrow C$ by Case 1, Adequacy and Subject Reduction.
   - λ, η. We have the premisses $\Gamma, x{:}A \models M_0 : B$ and $\Gamma \models A \rightarrow A'$. Hence by Case 1 there are T, U and C such that $\Gamma \models [N/x]M_0 \rightarrow T \rightarrow U : C$ and $\Gamma \models [N/x]B \rightarrow C$, so $\Gamma \models (\lambda x{:}A.M_0)(N) \rightarrow T \rightarrow U : C$ by β.
   - β. By the induction hypothesis there are T, U and C such that $\Gamma \models R(N) \rightarrow T \rightarrow U : C$ and $\Gamma \models [N/x]B \rightarrow C$, so $\Gamma \models M_1(M_2)(N) \rightarrow T \rightarrow U : C$ by Lemma 8.

The typed operational semantics can now be shown sound for the Logical Framework.

Theorem 2 (Soundness).

- *If* $\Gamma \vdash \mathrm{ok}$ *then there is a* Δ *such that* $\models \Gamma \rightarrow \Delta$.
- *If* $\Gamma \vdash A$ kind *then there is a* B *such that* $\Gamma \models A \rightarrow B$.
- *If* $\Gamma \vdash A = B$ *then there is a* C *such that* $\Gamma \models A \rightarrow C$ *and* $\Gamma \models B \rightarrow C$.
- *If* $\Gamma \vdash M : A$ *then there are* P *and* B *such that* $\Gamma \models A \rightarrow B$ *and* $\Gamma \models M \rightarrow_{\mathrm{n}} P : B$.
- *If* $\Gamma \vdash M = N : A$ *then there are* P *and* B *such that* $\Gamma \models A \rightarrow B$, $\Gamma \models M \rightarrow_{\mathrm{n}} P : B$ *and* $\Gamma \models N \rightarrow_{\mathrm{n}} P : B$.

Proof. By simultaneous induction on derivations. We consider several cases. Notice that we need to consider all structural rules at this point.

- El. By the induction hypothesis there are P and B such that $\Gamma \models M \to_{\mathrm{n}} P\colon B$ and $\Gamma \models \mathrm{Type} \to B$. By inversion $B \equiv \mathrm{Type}$, so $\Gamma \models \mathrm{El}(M) \to \mathrm{El}(P)$ by El.
- Var. By the induction hypothesis $\Gamma_0, x{:}A, \Gamma_1 \models \mathrm{ok}$. By Subcontext $\Gamma_0, x{:}A \models \mathrm{ok}$, so by inversion there is a B such that $\Gamma_0 \models A \to B$. Hence, by Weakening $\Gamma_0, x{:}A, \Gamma_1 \models A \to B$, and by Var $\Gamma_0, x{:}A, \Gamma_1 \models x \to x \to x\colon B$.
- λ. By the induction hypothesis there are P_0 and B_2 such that $\Gamma, x{:}A_1 \models M_0 \to_{\mathrm{n}} P_0\colon B_2$ and $\Gamma, x{:}A_1 \models A_2 \to B_2$. Furthermore, by Subcontext and inversion there is a B_1 such that $\Gamma \models A_1 \to B_1$. We know $\Gamma \models \Pi x{:}A_1.A_2 \to \Pi x{:}B_1.B_2$ by Π. We then have two cases:
  - $P_0 \equiv P(x)$ with $x \notin \mathrm{FV}(P)$. Then $\Gamma \models \lambda x{:}A_1.M_0 \to_{\mathrm{n}} P\colon \Pi x{:}B_1.B_2$ by η'.
  - $\lambda x{:}B_1.P_0$ is not an η-redex. Then $\Gamma \models \lambda x{:}A_1.M_0 \to_{\mathrm{n}} \lambda x{:}B_1.P_0\colon \Pi x{:}B_1.B_2$ by λ.
- App-Eq. By the induction hypothesis we know that $\Gamma \models M_1 \to_{\mathrm{n}} P_1\colon B$, $\Gamma \models N_1 \to_{\mathrm{n}} P_1\colon B$, and $\Gamma \models \Pi x{:}A_1.A_2 \to B$. Also, $\Gamma \models M_2 \to_{\mathrm{n}} P_2\colon B_1'$, $\Gamma \models M_2 \to_{\mathrm{n}} P_2\colon B_1'$, and $\Gamma \models A_1 \to B_1'$. By inversion of $\Gamma \models \Pi x{:}A_1.A_2 \to B$ we know $B \equiv \Pi x{:}B_1.B_2$, $\Gamma \models A_1 \to B_1$ and $\Gamma, x{:}A_1 \models A_2 \to B_2$. By Determinacy $B_1 \equiv B_1'$.
  By Lemma 23 Case 2 we know $\Gamma \models M_1(M_2) \to_{\mathrm{n}} U\colon C$ and $\Gamma \models [M_2/x]B_2 \to C$, and $\Gamma \models N_1(N_2) \to_{\mathrm{n}} U'\colon C'$ and $\Gamma \models [N_2/x]B_2 \to C'$. By Adequacy and Subject Reduction $\Gamma \models P_1(P_2) \to_{\mathrm{n}} U\colon C$ and $\Gamma \models [P_2/x]B_2 \to C$, and $\Gamma \models P_1(P_2) \to_{\mathrm{n}} U'\colon C'$ and $\Gamma \models [P_2/x]B_2 \to C'$. Hence, by Determinacy $U \equiv U'$ and $C \equiv C'$.
  Finally, we know $\Gamma, x{:}A_1 \models A_2 \to B_2$, so by Lemma 23 Case 1 $\Gamma \models [M_2/x]A_2 \to C''$. By Adequacy and Subject Reduction $\Gamma \models [M_2/x]B_2 \to C''$, so by Determinacy $C \equiv C''$.
- We consider the classical substitution rule, where other structural rules are similar. By the induction hypothesis $\Gamma_0, x{:}A, \Gamma_1 \models M \to_{\mathrm{n}} P\colon D$ with $\Gamma_0, x{:}A, \Gamma_1 \models B \to D$, and $\Gamma_0 \models N \to_{\mathrm{n}} Q\colon C$ with $\Gamma_0 \models A \to C$. By Lemma 23 Case 1 we know $\Gamma_0, [N/x]\Gamma_1 \models [N/x]M \to_{\mathrm{n}} R\colon E$ and $\Gamma_0, [N/x]\Gamma_1 \models [N/x]D \to E$ for some R and E, and $\Gamma_0, [N/x]\Gamma_1 \models [N/x]B \to F$ for some F. By Adequacy and Subject Reduction $\Gamma_0, [N/x]\Gamma_1 \models [N/x]D \to F$, so by Determinacy $E \equiv F$.

Corollary 2 (Equivalence).

- $\Gamma \vdash \mathrm{ok}$ *iff* $\Gamma \models \mathrm{ok}$.
- $\Gamma \vdash A$ kind *iff there is a* B *such that* $\Gamma \models A \to B$.
- $\Gamma \vdash M\ :\ A$ *iff there is a* B *such that* $A \models B \to$ *and* $\Gamma \models M\colon B$.
- $\Gamma \models A \downarrow B$ *iff there is a* C *such that* $\Gamma \models A \to C$ *and* $\Gamma \models B \to C$.
- $\Gamma \models M \downarrow N\colon A$ *iff there are* B *and* P *such that* $\Gamma \models A \to B$, $\Gamma \models M \to_{\mathrm{n}} P\colon B$ *and* $\Gamma \models N \to_{\mathrm{n}} P\colon B$.

By the equivalence of the Logical Framework and its typed operational semantics, we can straightforwardly transfer the results of Church–Rosser, subject reduction and strong normalization to the Logical Framework. Furthermore, as the typed operational semantics is sound for judgements $\Gamma \vdash J$ and complete for judgements $\Gamma \vdash^- J$, we have also demonstrated the admissibility of the structural rules in Section 2.2, using the trivial inclusion from $\Gamma \vdash^- J$ to $\Gamma \vdash J$.

6 Conclusions

We have showed that the simpler proof of strong normalization developed by Joachimski and Matthes for the simply-typed λ-calculus lifts naturally to the typed operational semantics for the Logical Framework. We have also given an elegant development of the full metatheory of the Logical Framework using typed operational semantics, and we have discussed the benefits of various design decisions that differentiate typed operational semantics from the operational definition of strong normalization used elsewhere in the literature.

We believe that extending this proof technique to the Calculus of Constructions is an important project for the type theory community. Given that we have now demonstrated the successful use of this technique for the dependent-type corner of Barendregt's cube, the most challenging outstanding problem seems to be studying the proof for System F.

We would also like to prove soundness without the use of untyped reduction. For the simply-typed λ-calculus this follows naturally, but for the Logical Framework there is a subtle interaction between application, substitution and binders that makes the straightforward proof technique fail. As untyped reduction plays no role in defining the Logical Framework, it seems natural to expect that it can also be removed from the metatheory. In addition to the philosophical interest of this question, it has practical consequences for the metatheory of systems where untyped reduction may not be well-behaved, for example in the Logical Framework with coercions [7].

Acknowledgments

I would like to thank Felix Joachimski and Ralph Matthes for discussions about their new proof of strong normalization for the simply-typed λ-calculus, and Adriana Compagnoni, Alex Jones, Pierre Leleu, Zhaohui Luo and James McKinna for useful comments about typed operational semantics that led to the presentation in this paper. The anonymous referees also had useful comments.

References

1. H. Barendregt. Lambda calculi with types. In S. Abramsky, D. M. Gabbai, and T. S. E. Maibaum, editors, *Handbook of Logic in Computer Science*, volume 2. Oxford University Press, 1991.
2. A. Compagnoni and H. Goguen. Typed operational semantics for higher order subtyping. Technical Report ECS-LFCS-97-361, Edinburgh Univ., July 1997. Submitted to *Information and Computation*.

3. H. Geuvers. *Logics and Type Systems.* PhD thesis, Katholieke Universiteit Nijmegen, Sept. 1993.
4. H. Goguen. *A Typed Operational Semantics for Type Theory.* PhD thesis, Edinburgh Univ., Aug. 1994.
5. H. Goguen. Typed operational semantics. In *Proceedings of TLCA*, volume 902 of *Lecture Notes in Computer Science*, pages 186–200. Springer–Verlag, Apr. 1995.
6. F. Joachimski and R. Matthes. Short proofs of strong normalization, 1998. Draft, submitted for publication.
7. A. Jones, Z. Luo, and S. Soloviev. Some algorithmic and proof-theoretical aspects of coercive subtyping. In *Proceedings of TYPES'96*, 1996. To appear.
8. P. Leleu. Metatheoretic results for a modal lambda calculus. Technical Report RR-3361, INRIA, 1998.
9. R. Loader. Normalisation by translation. Note distributed on TYPES list, Apr. 1995.
10. Z. Luo. *An Extended Calculus of Constructions.* PhD thesis, Edinburgh Univ., Nov. 1990.
11. Z. Luo. *Computation and Reasoning.* Oxford University Press, 1994.
12. J. McKinna and R. Pollack. Pure type systems formalized. In M. Bezem and J. F. Groote, editors, *Proceedings of TLCA*, pages 289–305. Springer–Verlag, LNCS 664, Mar. 1993.
13. B. Nordström, K. Petersson, and J. Smith. *Programming in Martin-Löf's Type Theory: An Introduction.* Oxford University Press, 1990.
14. G. Pottinger. Strong normalization for the terms of the theory of constructions. Technical Report TR 11-7, Odyssey Research Associates, Inc., 1987.
15. A. Salvesen. The Church-Rosser property for pure type systems with $\beta\eta$-reduction, Nov. 1991. Unpublished manuscript.
16. M. Takahashi. Parallel reductions in λ-calculus. *Information and Computation*, 118:120–127, 1995.
17. L. van Benthem Jutting, J. McKinna, and R. Pollack. Typechecking in pure type systems. In H. Barendregt and T. Nipkow, editors, *Types for Proofs and Programming.* Springer–Verlag, 1993.
18. F. van Raamsdonk and P. Severi. On normalisation. Technical Report CS-R9545, CWI Amsterdam, 1995.
19. B. Werner. *Une Théorie des Constructions Inductives.* PhD thesis, Université Paris 7, 1994.

Logical Predicates for Intuitionistic Linear Type Theories

Masahito Hasegawa

RIMS, Kyoto University, Kyoto 606-8502, Japan

Abstract. We develop a notion of Kripke-like parameterized logical predicates for two fragments of intuitionistic linear logic (MILL and DILL) in terms of their category-theoretic models. Such logical predicates are derived from the categorical glueing construction combined with the free symmetric monoidal cocompletion. As applications, we obtain full completeness results of translations between linear type theories.

1 Introduction

Suppose that a model of *Multiplicative Intuitionistic Linear Logic (MILL)* – the propositional fragment of linear logic [12] with I, $\otimes$ and $\multimap$ – is given. Also suppose that there is a property on elements of the model which is closed under tensor product and composition (cut) and other structural rules, and covers the interpretations of base types and constants. We show that such a property can be extended to the interpretation of all types so that it covers all MILL-definable elements. We also give a parallel result for *Dual Intuitionistic Linear Logic (DILL)* of Barber and Plotkin [5], which is an extension of MILL with the modality !. To achieve such results, we first give a suitable notion of such "predicates" on models of MILL and DILL, upon which we develop logical predicates and state the Basic Lemma. We then show that the construction above is an instance of our logical predicates.

To see why we need to introduce a property closed under tensor and so on, it would be instructive to observe that the standard logical predicates for models of simply typed lambda calculus do not work well with the linear calculi and their models. We may have a predicate $P_b \subseteq \mathcal{A}_b$ for each base type b, where $\mathcal{A}_\sigma$ is a set in which the closed terms of type σ are interpreted. As the standard logical predicates, we hope to define a predicate $P_\sigma \subseteq \mathcal{A}_\sigma$ for every type σ in an inductive way. However, we soon face a difficulty in constructing $P_{\sigma\otimes\tau}$ from P_σ and P_τ. The naive construction $P_{\sigma\otimes\tau} = \{a\otimes b \mid a \in P_\sigma, b \in P_\tau\}$ makes sense but can miss some interesting "undecomposable" elements of $\mathcal{A}_{\sigma\otimes\tau}$; in particular assume a constant of type $\sigma\otimes\tau$, then its interpretation may not belong to $P_{\sigma\otimes\tau}$ for any P_σ and P_τ. The same trouble appears when we construct $P_{!\sigma}$ from P_σ.

We solve this problem by parameterizing the predicates on the tensor-closed property (in the similar way to the Kripke logical relations [2]), so that the parameter indicates the linearly used resource (or the linear context). Such parameterized predicates form a model of MILL and serve as a basis for constructing logical predicates for MILL. The problem of tensor types disappears if each interesting element satisfies the tensor-closed property.

The construction is based on a few category-theoretic tools, specifically the presheaf construction (*free symmetric monoidal cocompletion* [15]) for symmetric monoidal categories and also a *glueing* (*sconing*, *Freyd covering*) construction [16, 21] on symmetric monoidal closed categories. It is known that a setting for standard logical predicates can be obtained by glueing a cartesian closed category to **Set** [21, 14]; ours is derived by glueing a symmetric monoidal closed category to the presheaf category of a small symmetric monoidal category (which specifies the tensor-closed property mentioned above). For DILL we further use a glueing construction of symmetric monoidal adjunction to accommodate the modality. However in this paper we leave these abstract idea rather implicit (except in Sect. 4) and describe all constructions concretely.

By applying our logical predicates method, we obtain the full completeness of syntactic translations between linear type theories. For instance, it is an immediate corollary of the Basic Lemma that MILL is a *full* fragment of DILL (Example 3), in the sense that, for any DILL-term $\emptyset\ ;\ \Delta \vdash M : \sigma$ with no ! in Δ nor σ, there always exists an MILL-term $\Delta \vdash N : \sigma$ such that $\emptyset\ ;\ \Delta \vdash M = N : \sigma$ holds. See Example 2 and 4 for other examples.

Though the existing syntax for linear type theories are rather diverging, their semantic models are now well-established and related each other, in terms of symmetric monoidal (closed) categories and adjunctions [6, 8, 5], and our approach based on such categorical models is likely to apply to many other linear type theories as well. In fact it is routine to modify our technique for non-commutative linear logic and monoidal (bi)closed categories (see [17]). Furthermore, by combining our approach with Hyland and Tan's double glueing construction [23] (see Example 5) we can deal with a classical linear type theory (MLL). These results, proofs and further category-theoretic analysis are reported in the full paper [13].

Also it might be fruitful to adapt our method to programming languages, see for example the complexity-parameterized logical relation used in [11]. Another interesting direction is to combine our approach to other techniques of specifying properties of semantic categories, for instance that of specification structures [1].

Acknowledgements I thank Gordon Plotkin for discussions at the initial stage of this work.

2 Multiplicative Intuitionistic Linear Logic

We recall a simple fragment of intuitionistic linear logic (Multiplicative Intuitionistic Linear Logic, MILL) together with the associated term calculus. The category-theoretic models are given as symmetric monoidal closed categories, for which soundness and completeness are known (e.g. [7]). See [10, 8] for the category-theoretic concepts used in this paper.

2.1 Syntax of MILL

We briefly recall the syntax of MILL. The detail is discussed e.g. in [7]; our presentation is chosen so that it will be compatible with DILL (Sect. 5). A set of base types (write b for one) and also a set of constants are fixed throughout this paper.

Types and Terms

$$\begin{aligned}\sigma &::= b \mid I \mid \sigma \otimes \sigma \mid \sigma \multimap \sigma \\ M &::= c(M) \mid x \mid * \mid \mathsf{let}\ *\ \mathsf{be}\ M\ \mathsf{in}\ M \mid M \otimes M \mid \mathsf{let}\ x \otimes x\ \mathsf{be}\ M\ \mathsf{in}\ M \mid \\ &\quad\ \lambda x.M \mid MM\end{aligned}$$

We assume that each constant c has a fixed arity $\sigma \to \tau$, where σ and τ are types which do not involve $\multimap$. (This restriction on arity is for ease of presentation and not essential.)

Typing

$$\frac{c : \sigma \to \tau \quad \Delta \vdash M : \sigma}{\Delta \vdash c(M) : \tau}\ (\textbf{Constant}) \qquad \frac{}{x : \sigma \vdash x : \sigma}\ (\textbf{Variable})$$

$$\frac{}{\vdash * : I}\ (I\textbf{I}) \qquad \frac{\Delta_1 \vdash M : I \quad \Delta_2 \vdash N : \sigma}{\Delta_1 \sharp \Delta_2 \vdash \mathsf{let}\ *\ \mathsf{be}\ M\ \mathsf{in}\ N : \sigma}\ (I\textbf{E})$$

$$\frac{\Delta_1 \vdash M : \sigma \quad \Delta_2 \vdash N : \tau}{\Delta_1 \sharp \Delta_2 \vdash M \otimes N : \sigma \otimes \tau}\ (\otimes\textbf{I}) \qquad \frac{\Delta_1 \vdash M : \sigma \otimes \tau \quad \Delta_2, x : \sigma, y : \tau \vdash N : \theta}{\Delta_1 \sharp \Delta_2 \vdash \mathsf{let}\ x \otimes y\ \mathsf{be}\ M\ \mathsf{in}\ N : \theta}\ (\otimes\textbf{E})$$

$$\frac{\Delta, x : \sigma \vdash M : \tau}{\Delta \vdash \lambda x.M : \sigma \multimap \tau}\ (\multimap\textbf{I}) \qquad \frac{\Delta_1 \vdash M : \sigma \multimap \tau \quad \Delta_2 \vdash N : \sigma}{\Delta_1 \sharp \Delta_2 \vdash MN : \tau}\ (\multimap\textbf{E})$$

where $\Delta_1 \sharp \Delta_2$ is a merge of Δ_1 and Δ_2 (this notation is taken from [5]). We note that any typing judgement has a unique derivation.

Axioms

$$\begin{aligned}\mathsf{let}\ *\ \mathsf{be}\ *\ \mathsf{in}\ M &= M & \mathsf{let}\ *\ \mathsf{be}\ M\ \mathsf{in}\ * &= M \\ \mathsf{let}\ x \otimes y\ \mathsf{be}\ M \otimes N\ \mathsf{in}\ L &= L[M/x, N/y] & \mathsf{let}\ x \otimes y\ \mathsf{be}\ M\ \mathsf{in}\ x \otimes y &= M \\ (\lambda x.M)N &= M[N/x] & \lambda x.Mx &= M\end{aligned}$$

$$\begin{aligned}C[\mathsf{let}\ *\ \mathsf{be}\ M\ \mathsf{in}\ N] &= \mathsf{let}\ *\ \mathsf{be}\ M\ \mathsf{in}\ C[N] \\ C[\mathsf{let}\ x \otimes y\ \mathsf{be}\ M\ \mathsf{in}\ N] &= \mathsf{let}\ x \otimes y\ \mathsf{be}\ M\ \mathsf{in}\ C[N]\end{aligned}$$

In the above $C[-]$ indicates a (well-typed) context – we assume suitable conditions on variables for avoiding undesirable captures. The equational theory of MILL is defined as the congruence relation on the terms with typing judgement generated from these axioms.

2.2 Semantics of MILL

Let $\mathbb{C}$ be a symmetric monoidal closed category with tensor product $\otimes$, unit object I and exponent $\multimap$. Assume that there is an object $[\![b]\!]$ for each base type b and an arrow $[\![c]\!] : [\![\sigma]\!] \to [\![\tau]\!]$ for each constant $c : \sigma \to \tau$, where $[\![\sigma]\!]$ is defined by $[\![I]\!] = I$, $[\![\sigma\otimes\tau]\!] = [\![\sigma]\!]\otimes[\![\tau]\!]$ and $[\![\sigma \multimap \tau]\!] = [\![\sigma]\!] \multimap [\![\tau]\!]$. For each typing judgement $\Delta \vdash M : \tau$, we define its interpretation $[\![\Delta \vdash M : \tau]\!] : [\![|\Delta|]\!] \to [\![\tau]\!]$ in $\mathbb{C}$ as follows, where $|\Delta| = (\ldots([\![\sigma_1]\!] \otimes [\![\sigma_2]\!])\ldots) \otimes [\![\sigma_n]\!]$ for $\Delta \equiv x_1 : \sigma_1, x_2 : \sigma_2, \ldots, x_n : \sigma_n$.

$$[\![\Delta \vdash c(M) : \tau]\!] = [\![|\Delta|]\!] \xrightarrow{[\![\Delta\vdash M:\sigma]\!]} [\![\sigma]\!] \xrightarrow{[\![c]\!]} [\![\tau]\!]$$

$$[\![x : \sigma \vdash x : \sigma]\!] = [\![\sigma]\!] \xrightarrow{id_{[\![\sigma]\!]}} [\![\sigma]\!]$$

$$[\![\vdash * : I]\!] = I \xrightarrow{id_I} I$$

$$[\![\Delta_1\sharp\Delta_2 \vdash \mathsf{let}\ *\ \mathsf{be}\ M\ \mathsf{in}\ N : \sigma]\!] =$$

$$[\![|\Delta_1\sharp\Delta_2|]\!] \stackrel{\simeq}{\to} [\![|\Delta_1|]\!] \otimes [\![|\Delta_2|]\!] \xrightarrow{[\![\Delta_1\vdash M:I]\!]\otimes[\![\Delta_2\vdash N:\sigma]\!]} I \otimes [\![\sigma]\!] \stackrel{\simeq}{\to} [\![\sigma]\!]$$

$$[\![\Delta_1\sharp\Delta_2 \vdash M\otimes N : \sigma \otimes \tau]\!] =$$

$$[\![|\Delta_1\sharp\Delta_2|]\!] \stackrel{\simeq}{\to} [\![|\Delta_1|]\!] \otimes [\![|\Delta_2|]\!] \xrightarrow{[\![\Delta_1\vdash M:\sigma]\!]\otimes[\![\Delta_2\vdash N:\tau]\!]} [\![\sigma]\!] \otimes [\![\tau]\!]$$

$$[\![\Delta_1\sharp\Delta_2 \vdash \mathsf{let}\ x \otimes y\ \mathsf{be}\ M\ \mathsf{in}\ N : \theta]\!] =$$

$$[\![|\Delta_1\sharp\Delta_2|]\!] \stackrel{\simeq}{\to} [\![|\Delta_1|]\!] \otimes [\![|\Delta_2|]\!] \xrightarrow{[\![\Delta_1\vdash M:\sigma\otimes\tau]\!]\otimes id_{[\![|\Delta_2|]\!]}}$$

$$([\![\sigma]\!] \otimes [\![\tau]\!]) \otimes [\![|\Delta_2|]\!] \stackrel{\simeq}{\to} ([\![|\Delta_2|]\!] \otimes [\![\sigma]\!]) \otimes [\![\tau]\!] \xrightarrow{[\![\Delta_2,x:\sigma,y:\tau\vdash N:\theta]\!]} [\![\theta]\!]$$

$$[\![\Delta \vdash \lambda x.M : \sigma \multimap \tau]\!] = [\![|\Delta|]\!] \xrightarrow{\Lambda([\![\Delta,x:\sigma\vdash M:\tau]\!])} [\![\sigma]\!] \multimap [\![\tau]\!]$$

$$[\![\Delta_1\sharp\Delta_2 \vdash MN : \tau]\!] = [\![|\Delta_1\sharp\Delta_2|]\!] \stackrel{\simeq}{\to}$$

$$[\![|\Delta_1|]\!] \otimes [\![|\Delta_2|]\!] \xrightarrow{[\![\Delta_1\vdash M:\sigma\multimap\tau]\!]\otimes[\![\Delta_2\vdash N:\sigma]\!]} ([\![\sigma]\!] \multimap [\![\tau]\!]) \otimes [\![\sigma]\!] \xrightarrow{\mathsf{ev}} [\![\tau]\!]$$

where "$\simeq$" denotes a (uniquely determined) canonical isomorphism. We write ev for the counit of the adjunction $- \otimes C \dashv C \multimap -$, and $\Lambda(f) : A \to C \multimap B$ for the adjoint mate of $f : A \otimes C \to B$.

Proposition 1. *This semantics is sound and complete.* □

3 Logical Predicates for MILL

We introduce parameterized predicates on objects of a symmetric monoidal closed category, and show that such predicates give rise to another symmetric monoidal closed category. We then define the logical predicates as type-indexed families of the predicates (inductively determined on the type structure), and state the Basic Lemma. We also give the canonically determined logical predicate which is used in showing full completeness of translations between linear type theories. We conclude this section by sketching the generalization to logical relations.

3.1 $\mathbb{C}_0$-Predicates

Let $\mathbb{C}_0$ be a small symmetric monoidal category, $\mathbb{C}_1$ a locally small symmetric monoidal closed category and $\mathbb{I}$ be a strict symmetric monoidal functor from $\mathbb{C}_0$ to $\mathbb{C}_1$.

Definition 1. *An* $\mathsf{Obj}(\mathbb{C}_0)$*-indexed set* $P = \{P(X)\}_{X \in \mathbb{C}_0}$ *is a* $\mathbb{C}_0$-predicate on $A \in \mathbb{C}_1$ *when*

- $P(X) \subseteq \mathbb{C}_1(\mathbb{I}X, A)$ *for* $X \in \mathbb{C}_0$*, and*
- *for* $f \in \mathbb{C}_0(X, Y)$*,* $g \in P(Y)$ *implies* $g \circ \mathbb{I}f \in P(X)$*.* □

We may intuitively think that $\mathbb{C}_1(\mathbb{I}X, A)$ represents the set of proofs of a sequent $X \vdash A$, and $\mathbb{C}_0$ (imported into $\mathbb{C}_1$ via $\mathbb{I}$) determines a property on proofs which is closed under tensor, composition and structural constructions. Unlike the traditional non-linear calculi and logical predicates over them, we explicitly state the "resource" X, which plays some significant role in our work. Then, for a $\mathbb{C}_0$-predicate P on A, $P(X)$ is a predicate on the proofs of $X \vdash A$. The second condition tells us that P is stable under the change of resource along a proof of $X \vdash Y$, provided that it satisfies the property $\mathbb{C}_0$.

Definition 2. *Define the category of* $\mathbb{C}_0$*-predicates* $\mathbb{C}_0\mathsf{PRED}$ *as follows:*

- *an object of* $\mathbb{C}_0\mathsf{PRED}$ *is a pair* (P, A) *where* P *is a* $\mathbb{C}_0$*-predicate on* $A \in \mathbb{C}_1$*;*
- *an arrow from* (P, A) *to* (Q, B) *is an arrow* $h \in \mathbb{C}_1(A, B)$ *such that* $g \in P(X)$ *implies* $h \circ g \in Q(X)$*.* □

Definition 3. *For* $\mathbb{C}_0$*-predicates* P *on* A *and* Q *on* B*, define* $\mathbb{C}_0$*-predicates* $P \otimes Q$ *on* $A \otimes B$ *and* $P \multimap Q$ *on* $A \multimap B$ *as follows.*

$$(P \otimes Q)(X) = \left\{ ((g \otimes h) \circ \mathbb{I}f) \;\middle|\; \begin{array}{l} \exists Y, Z \in \mathbb{C}_0 \;\; f \in \mathbb{C}_0(X, Y \otimes Z), \\ g \in P(Y), \;\; h \in Q(Z) \end{array} \right\}$$

$$(P \multimap Q)(X) = \left\{ f \in \mathbb{C}_1(\mathbb{I}X, A \multimap B) \;\middle|\; \begin{array}{l} \forall Y \in \mathbb{C}_0 \;\; a \in P(Y) \; \textit{implies} \\ \mathsf{ev} \circ (f \otimes a) \in Q(X \otimes Y) \end{array} \right\}$$

□

The definition of $P \otimes Q$ above is derived from a few category-theoretic tools, which will be explained in Sect. 4; for now, we shall give a proof-theoretic explanation. A sequent $X \vdash A \otimes B$ can be derived as

$$\dfrac{\begin{array}{c} \Pi_f \\ \vdots \\ X \vdash Y \otimes Z \end{array} \qquad \dfrac{\begin{array}{cc} \Pi_g & \Pi_h \\ \vdots & \vdots \\ Y \vdash A & Z \vdash B \end{array}}{Y, Z \vdash A \otimes B}\,(\otimes\mathbf{I})}{X \vdash A \otimes B}\,(\otimes\mathbf{E})$$

where $X \vdash Y \otimes Z$ splits a resource X to Y and Z which are used to prove A and B respectively. In general, such a splitting of resource is not unique, so we consider all possible cases such that (i) the proof Π_f of the splitting satisfies the "tensor-closed property" $\mathbb{C}_0$ and (ii) the proofs Π_g of $Y \vdash A$ and Π_h of $Z \vdash B$ satisfy the predicates $P(Y)$ and $Q(Z)$ respectively – in such cases we say that the derivation satisfies the property $(P \otimes Q)(X)$.

The definition of $P \multimap Q$ is in spirit the same as the usual definition of logical predicates; $M : A \Rightarrow B$ satisfies $P \Rightarrow Q$ if and only if $MN : B$ belongs to Q for any $N : A$ satisfying P. However, since our type theory is linear, we have to deal with the resources of terms linearly, and we explicitly state them in the definition: intuitively, $\Delta \vdash M : A \multimap B$ satisfies $P \multimap Q$ if and only if $\Delta, \Delta' \vdash MN : B$ satisfies Q for any $\Delta' \vdash N : A$ satisfying P.

Lemma 1. *For each $X, A \in \mathbb{C}_0$ define $\mathbb{P}_A(X) = \{\mathbb{I}f \mid f \in \mathbb{C}_0(X, A)\}$. Then*

- *$\mathbb{P}_A$ is a $\mathbb{C}_0$-predicate on $\mathbb{I}A$.*
- *$f : (\mathbb{P}_A, \mathbb{I}A) \to (\mathbb{P}_B, \mathbb{I}B)$ in $\mathbb{C}_0$PRED iff $f = \mathbb{I}g$ for some $g \in \mathbb{C}_0(A, B)$.*
- *$\mathbb{P}_A \otimes \mathbb{P}_B = \mathbb{P}_{A \otimes B}$.* □

Proposition 2. *$\mathbb{C}_0$PRED forms a symmetric monoidal closed category by the following data: the unit object is $(\mathbb{P}_I, I)$, tensor is given by $(P, A) \otimes (Q, B) = (P \otimes Q, A \otimes B)$, and exponent $(P, A) \multimap (Q, B) = (P \multimap Q, A \multimap B)$. Moreover $\mathbb{P}$ extends to a strict symmetric monoidal functor from $\mathbb{C}_0$ to $\mathbb{C}_0$PRED which is full.* □

Remark 1. If $\mathbb{C}_0$ is closed and $\mathbb{I}$ preserves exponents strictly, then so is $\mathbb{P}$ – in particular we have $\mathbb{P}_{A \multimap B} = \mathbb{P}_A \multimap \mathbb{P}_B$. □

Example 1 (Subsconing). If $\mathbb{C}_0$ is equivalent to the one object one arrow category, a $\mathbb{C}_0$-predicate on A is just a subset of $\mathbb{C}_1(I, A)$, thus is a predicate on the global elements of A. For predicates P on A and Q on B, we have

$$P \otimes Q = \{(g \otimes h) \circ \simeq \mid g \in P, h \in Q\}$$
$$P \multimap Q = \{f \in \mathbb{C}_1(I, A \multimap B) \mid \mathsf{ev} \circ (f \otimes g) \circ \simeq \in Q \text{ for any } g \in P\}$$

where $\simeq$ indicates the canonical isomorphism $I \xrightarrow{\simeq} I \otimes I$. Following [21] we call this category of predicates the *subsconing* of $\mathbb{C}_1$ and write $\widetilde{\mathbb{C}_1}$ for it. □

3.2 Logical $\mathbb{C}_0$-Predicates

Suppose that we have $\mathbb{C}_0$, $\mathbb{C}_1$ and $\mathbb{I} : \mathbb{C}_0 \to \mathbb{C}_1$ as before. Also we fix an interpretation $[\![-]\!]_1$ of MILL in $\mathbb{C}_1$.

Definition 4. *A type-indexed family $\{P_\sigma\}$ is a* logical $\mathbb{C}_0$-predicate *if*

- *P_σ is a $\mathbb{C}_0$-predicate on $[\![\sigma]\!]_1$,*
- *$P_I = \mathbb{P}_I$, $P_{\sigma \otimes \tau} = P_\sigma \otimes P_\tau$, $P_{\sigma \multimap \tau} = P_\sigma \multimap P_\tau$, and*
- *$[\![c]\!]_1 : (P_\sigma, [\![\sigma]\!]_1) \to (P_\tau, [\![\tau]\!]_1)$ for each constant $c : \sigma \to \tau$.* □

Note that a logical $\mathbb{C}_0$-predicate is determined by its instances at base types. Given a logical $\mathbb{C}_0$-predicate $\{P_\sigma\}$, we can interpret MILL in $\mathbb{C}_0\mathsf{PRED}$ by $[\![b]\!] = (P_b, [\![b]\!]_1)$ for each base type b and $[\![c]\!] = [\![c]\!]_1 : (P_\sigma, [\![\sigma]\!]_1) \to (P_\tau, [\![\tau]\!]_1)$ for each constant $c : \sigma \to \tau$. Thus we have

Lemma 2 (Basic Lemma for MILL). *Let $\{P_\sigma\}$ be a logical $\mathbb{C}_0$-predicate. Then, for any term $\Delta \vdash M : \tau$, $[\![\Delta \vdash M : \tau]\!]_1 : (P_{|\Delta|}, [\![|\Delta|]\!]_1) \to (P_\tau, [\![\tau]\!]_1)$ holds.* □

$\mathbb{C}_0$ itself determines a logical $\mathbb{C}_0$-predicate in a canonical way, provided that

- for each base type b there is an object $[\![b]\!]_0 \in \mathbb{C}_0$, and
- for each constant $c : \sigma \to \tau$ there is an arrow $[\![c]\!]_0 \in \mathbb{C}_0([\![\sigma]\!]_0, [\![\tau]\!]_0)$

where $[\![\sigma]\!]_0$ is defined inductively by $[\![I]\!]_0 = I$ and $[\![\sigma \otimes \tau]\!]_0 = [\![\sigma]\!]_0 \otimes [\![\tau]\!]_0$. Then we automatically have an interpretation $[\![-]\!]_1$ in $\mathbb{C}_1$ determined by $[\![b]\!]_1 = \mathbb{I}([\![b]\!]_0)$ and $[\![c]\!]_1 = \mathbb{I}([\![c]\!]_0)$. Now define the *canonical logical* $\mathbb{C}_0$*-predicate* $\{\mathbb{P}^*_\sigma\}$ by $\mathbb{P}^*_b = \mathbb{P}_{[\![b]\!]_0}$. Basic Lemma for the canonical logical $\mathbb{C}_0$-predicate implies that, at $\multimap$-free types (at any types if $\mathbb{C}_0$ and $\mathbb{I}$ are closed) a definable element is in the image of $\mathbb{I}$.

3.3 Binary Logical $\mathbb{C}_0$-Relations

It is straightforward to generalize (or specialize) our logical predicates to multiple arguments, i.e. *logical relations*, in the same way as demonstrated in [21]. Here we spell out the case of binary ones. Suppose that $\mathbb{C}_0$ is a small symmetric monoidal category, $\mathbb{C}_1$ and $\mathbb{C}_2$ are locally small symmetric monoidal closed categories and that $\mathbb{I}_1 : \mathbb{C}_0 \to \mathbb{C}_1$ and $\mathbb{I}_2 : \mathbb{C}_0 \to \mathbb{C}_2$ are strict symmetric monoidal functors. A binary $\mathbb{C}_0$-relation is just a $\mathbb{C}_0$-predicate obtained by replacing $\mathbb{C}_1$ by $\mathbb{C}_1 \times \mathbb{C}_2$ and $\mathbb{I}$ by $\langle \mathbb{I}_1, \mathbb{I}_2 \rangle : \mathbb{C}_0 \to \mathbb{C}_1 \times \mathbb{C}_2$. Explicitly:

Definition 5. *An $\mathsf{Obj}(\mathbb{C}_0)$-indexed set $R = \{R(X)\}_{X \in \mathbb{C}_0}$ is a* $\mathbb{C}_0$-relation *on $(A, B) \in \mathbb{C}_1 \times \mathbb{C}_2$ when $R(X) \subseteq \mathbb{C}_1(\mathbb{I}_1 X, A) \times \mathbb{C}_2(\mathbb{I}_2 X, B)$ for $X \in \mathbb{C}_0$, and, for $f \in \mathbb{C}_0(X, Y)$, $(g, h) \in P(Y)$ implies $(g \circ \mathbb{I}_1 f, h \circ \mathbb{I}_2 f) \in P(X)$.* □

Definition 6. *Define the category of $\mathbb{C}_0$-relations $\mathbb{C}_0\mathsf{REL}$ as follows: an object of $\mathbb{C}_0\mathsf{REL}$ is a triple (A, B, R) where R is a $\mathbb{C}_0$-relation on (A, B); and an arrow from (A, B, R) to (A', B', R') is a pair $(h \in \mathbb{C}_1(A, A'), k \in \mathbb{C}_2(B, B'))$ such that $(f, g) \in R(X)$ implies $(h \circ f, k \circ g) \in R'(X)$.* □

Proposition 2 tells us that $\mathbb{C}_0\mathsf{REL}$ is a symmetric monoidal closed category. More explicitly, for $\mathbb{C}_0$-relations R on (A, B) and R' on (A', B'), we have $\mathbb{C}_0$-relations $R \otimes R'$ on $(A \otimes A', B \otimes B')$ and $R \multimap R'$ on $(A \multimap A', B \multimap B')$ as follows.

$$(R \otimes R')(X) = \left\{ ((g \otimes g') \circ \mathbb{I}_1 f, (h \otimes h') \circ \mathbb{I}_2 f) \;\middle|\; \begin{array}{l} \exists Y, Z \in \mathbb{C}_0\ f \in \mathbb{C}_0(X, Y \otimes Z), \\ (g, h) \in R(Y), (g', h') \in R'(Z) \end{array} \right\}$$

$$(R \multimap R')(X) = \left\{ (f, g) \;\middle|\; \begin{array}{l} \forall Y \in \mathbb{C}_0\ \ (a, b) \in R(Y) \text{ implies} \\ (\mathsf{ev} \circ (f \otimes a), \mathsf{ev} \circ (g \otimes b)) \in R'(X \otimes Y) \end{array} \right\}$$

Now fix interpretations $[\![-]\!]_1$ and $[\![-]\!]_2$ of MILL in $\mathbb{C}_1$ and $\mathbb{C}_2$ respectively.

Definition 7. *A type-indexed family* $\{R_\sigma\}$ *is a* logical $\mathbb{C}_0$-relation *if*

- R_σ *is a* $\mathbb{C}_0$*-relation on* $([\![\sigma]\!]_1, [\![\sigma]\!]_2)$,
- $R_I(X) = \{(\mathbb{I}_1 f, \mathbb{I}_2 f) \mid f \in \mathbb{C}_0(X, I)\}$, $R_{\sigma\otimes\tau} = R_\sigma \otimes R_\tau$, $R_{\sigma\multimap\tau} = R_\sigma \multimap R_\tau$ *and*
- $([\![c]\!]_1, [\![c]\!]_2) : ([\![\sigma]\!]_1, [\![\sigma]\!]_2, R_\sigma) \to ([\![\tau]\!]_1, [\![\tau]\!]_2, R_\tau)$ *for each constant* $c : \sigma \to \tau$. □

Lemma 3 (Basic Lemma, binary version). *Let* $\{R_\sigma\}$ *be a logical* $\mathbb{C}_0$*-relation. Then, for any* $\Delta \vdash M : \tau$, $([\![\Delta \vdash M : \tau]\!]_1, [\![\Delta \vdash M : \tau]\!]_2) : ([\![|\Delta|]\!]_1, [\![|\Delta|]\!]_2, R_{|\Delta|}) \to ([\![\tau]\!]_1, [\![\tau]\!]_2, R_\tau)$ *holds.* □

4 Categorical Glueing

We sketch the categorical glueing constructions used in our development; the detailed category-theoretic analysis is found in [13].

We write $(\mathbb{D} \downarrow \Gamma)$ for the comma category [19] (or the "glued category") of a functor $\Gamma : \mathbb{C} \to \mathbb{D}$. An object of $(\mathbb{D} \downarrow \Gamma)$ is a triple $(D \in \mathbb{D}, C \in \mathbb{C}, f : D \to \Gamma C)$. An arrow from (D, C, f) to (D', C', f') is a pair $(d : D \to D', c : C \to C')$ satisfying $\Gamma c \circ f = f' \circ d$. We note that there is a projection functor $p : (\mathbb{D} \downarrow \Gamma) \to \mathbb{C}$ given by $p(D, C, f) = C$ and $p(d, c) = c$.

Lemma 4. *Suppose that* $\mathbb{C}$ *and* $\mathbb{D}$ *are symmetric monoidal closed categories and that* $\Gamma : \mathbb{C} \to \mathbb{D}$ *is a symmetric monoidal functor. Moreover suppose that* $\mathbb{D}$ *has pullbacks. Then the comma category* $\mathcal{G} \equiv (\mathbb{D} \downarrow \Gamma)$ *can be given a symmetric monoidal closed structure, so that the projection* $p : \mathcal{G} \to \mathbb{C}$ *is strict symmetric monoidal closed.*

Proof (sketch). We define the symmetric monoidal structure on $\mathcal{G}$ by

$$\begin{aligned} I_\mathcal{G} &\equiv (I_\mathbb{D}, I_\mathbb{C}, m_I) \\ (D, C, f) \otimes (D', C', f') &\equiv (D \otimes D', C \otimes C', m_{C,C'} \circ (f \otimes f')) \\ (d, c) \otimes (d', c') &\equiv (d \otimes d', c \otimes c') \end{aligned}$$

where $m_I : I_\mathbb{D} \to \Gamma I_\mathbb{C}$ and $m_{C,C'} : \Gamma C \otimes \Gamma C' \to \Gamma(C \otimes C')$ are the coherent morphisms of the symmetric monoidal functor Γ. Exponents are defined as

$$(D, C, f) \multimap (D', C', f') \;\equiv\; ((D \multimap D') \times_{D \multimap \Gamma C'} \Gamma(C \multimap C'),\; C \multimap C',\; \pi_2)$$

which is given by the following pullback in $\mathbb{D}$.

$$\begin{array}{ccc} (D \multimap D') \times_{D \multimap \Gamma C'} \Gamma(C \multimap C') & \xrightarrow{\pi_2} & \Gamma(C \multimap C') \\ & & \big\downarrow {\scriptstyle \Lambda(\Gamma \mathrm{ev}_{C,C'} \circ m_{C \multimap C', C})} \\ {\scriptstyle \pi_1} \big\downarrow & & \Gamma C \multimap \Gamma C' \\ & & \big\downarrow {\scriptstyle f \multimap \Gamma C'} \\ D \multimap D' & \xrightarrow[D \multimap f']{} & D \multimap \Gamma C' \end{array}$$

□

This result seems to be folklore. Notice that the glueing functor Γ does not have to be strong.

In the situation of the last section, by letting $\Gamma : \mathbb{C}_1 \to \mathbf{Set}^{\mathbb{C}_0^{op}}$ be the functor which sends X to $\mathbb{C}_1(\mathbb{I}-, X)$, we obtain the setting for the category of $\mathbb{C}_0$-predicates. The symmetric monoidal closed structure of $\mathbf{Set}^{\mathbb{C}_0^{op}}$ is given by $I(-) = \mathbb{C}_0(-, I)$, $(F \otimes G)(-) = \int^{X,Y} FX \times GY \times \mathbb{C}_0(-, X \otimes Y)$ and $(F \multimap G)(-) = \mathbf{Set}^{\mathbb{C}_0^{op}}(F(=), G(- \otimes =))$ (see [15]), for which Γ becomes symmetric monoidal. For describing the predicates, we are interested in the full subcategory of the glued category whose objects are subobjects in $\mathbf{Set}^{\mathbb{C}_0^{op}}$. This is precisely the category $\mathbb{C}_0\mathsf{PRED}$, which is again symmetric monoidal closed; the definition of unit and tensor are patched in the obvious way (this is possible because $\mathbf{Set}^{\mathbb{C}_0^{op}}$ admits epi-mono factorization), resulting the concrete descriptions in Sect. 3.

Lafont has shown that, using the glueing for cartesian closed categories, a small cartesian category fully and faithfully embeds to the cartesian closed category freely generated from the former [16]. We can use $\mathbb{C}_0\mathsf{PRED}$ for showing a parallel result:

Example 2. Let $\mathbb{C}_0$ be a small symmetric monoidal category and $\mathbb{C}_1$ be the symmetric monoidal closed category freely generated from $\mathbb{C}_0$. Then the embedding $\mathbb{I} : \mathbb{C}_0 \to \mathbb{C}_1$ is full faithful. Faithfulness is easily shown by constructing a symmetric monoidal closed category to which $\mathbb{C}_0$ faithfully embeds. Fullness follows from the commutative diagrams

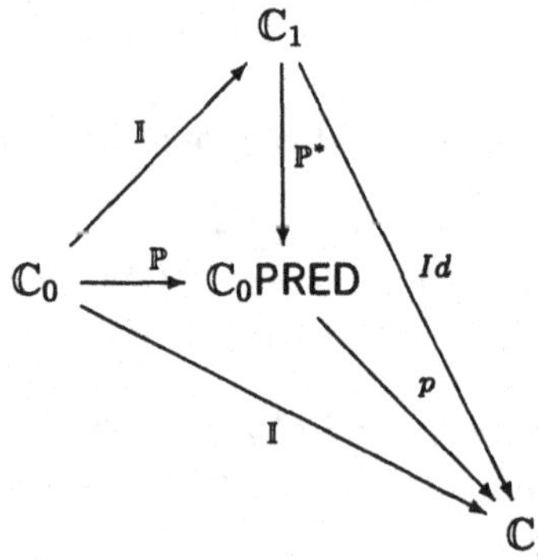

where $\mathbb{P}^*$ is the uniquely determined strict symmetric monoidal closed functor making the upper triangle commute, and the right triangle commutes because of the universal property of $\mathbb{I}$. Since both $\mathbb{P}$ and p are full, so is $\mathbb{I} = p \circ \mathbb{P}$. □

Syntactically, this implies that the I, $\otimes$-fragment of MILL is full in MILL; we can show it by applying the Basic Lemma to the canonical logical predicate (where $\mathbb{C}_0$ is the term model of the I, $\otimes$-fragment), which in fact is a concrete reworking of Example 2.

For interpreting the modality ! of DILL in the following section, we will need to determine a symmetric monoidal adjunction between the glued categories:

Lemma 5. *Suppose that* $\mathbb{C}_1 \underset{U}{\overset{F}{\underset{\leftarrow}{\overset{\rightarrow}{\perp}}}} \mathbb{C}_2$ *and* $\mathbb{D}_1 \underset{U'}{\overset{F'}{\underset{\leftarrow}{\overset{\rightarrow}{\perp}}}} \mathbb{D}_2$ *are (symmetric monoidal) adjunctions, with (symmetric monoidal) functors* $\Gamma_1 : \mathbb{C}_1 \to \mathbb{D}_1$ *and* $\Gamma_2 : \mathbb{C}_2 \to \mathbb{D}_2$ *together with a (monoidal) natural isomorphism* $\tau : U'\Gamma_2 \simeq \Gamma_1 U$. *For* $\mathcal{G}_1 \equiv (\mathbb{D}_1 \downarrow \Gamma_1)$ *and* $\mathcal{G}_2 \equiv (\mathbb{D}_2 \downarrow \Gamma_2)$, *there are functors* $\mathcal{F} : \mathcal{G}_1 \to \mathcal{G}_2$ *and* $\mathcal{U} : \mathcal{G}_2 \to \mathcal{G}_1$ *given by*

$$\begin{aligned} \mathcal{F}(D,C,f) &= (F'D,\ FC,\ \sigma_C \circ F'f), & \mathcal{F}(d,c) &= (F'd, Fc), \\ \mathcal{U}(Y,X,g) &= (U'Y,\ UX,\ \tau_X \circ U'g), & \mathcal{U}(y,x) &= (U'y, Ux) \end{aligned}$$

where $\sigma_C = \varepsilon'_{\Gamma_2 FC} \circ F'\tau^{-1}_{FC} \circ F'\Gamma_1\eta_C : F'\Gamma_1 C \to \Gamma_2 FC$ *(η is the unit of $F \dashv U$ and ε' is the counit of $F' \dashv U'$). $\mathcal{F}$ is (strong symmetric monoidal and) left adjoint to $\mathcal{U}$. Moreover the projections* $p_1 : \mathcal{G}_1 \to \mathbb{C}_1$ *and* $p_2 : \mathcal{G}_2 \to \mathbb{C}_2$ *give a map of adjunction* [19] *from* $\mathcal{G}_1 \underset{\mathcal{U}}{\overset{\mathcal{F}}{\underset{\leftarrow}{\overset{\rightarrow}{\perp}}}} \mathcal{G}_2$ *to* $\mathbb{C}_1 \underset{U}{\overset{F}{\underset{\leftarrow}{\overset{\rightarrow}{\perp}}}} \mathbb{C}_2$. □

5 Dual Intuitionistic Linear Logic

Now we enrich our logic and calculus with the modality !. There are many possible choices for this, see for instance [7]. Here we choose the formulation due to Barber and Plotkin, called Dual Intuitionistic Linear Logic (DILL) [5] for its simple syntax and equational theory, as well as for the well-established category-theoretic models of DILL in terms of symmetric monoidal adjunctions. Alternatively we could use Benton's Linear Non-Linear Logic (LNL Logic) [6] which has essentially the same class of category-theoretic models as DILL. In DILL a typing judgement takes the form $\Gamma\ ;\ \Delta \vdash M : \sigma$ in which Γ represents an intuitionistic (or additive) context whereas Δ is a linear (multiplicative) context.

5.1 Syntax of DILL

Types and Terms

$$\begin{aligned} \sigma &::= b \mid I \mid \sigma \otimes \sigma \mid \sigma \multimap \sigma \mid\ !\sigma \\ M &::= c(M) \mid x \mid * \mid \mathsf{let}\ *\ \mathsf{be}\ M\ \mathsf{in}\ M \mid M \otimes M \mid \mathsf{let}\ x \otimes x\ \mathsf{be}\ M\ \mathsf{in}\ M \mid \\ &\quad\ \ \lambda x.M \mid MM \mid\ !M \mid \mathsf{let}\ !x\ \mathsf{be}\ M\ \mathsf{in}\ M \end{aligned}$$

Typing

$$\frac{c : \sigma \to \tau \quad \Gamma\ ;\ \Delta \vdash M : \sigma}{\Gamma\ ;\ \Delta \vdash c(M) : \tau}\ (\textbf{Constant}) \qquad \frac{}{\Gamma\ ;\ x : \sigma \vdash x : \sigma}\ (\textbf{Variable}_{\textbf{lin}})$$

$$\frac{}{\Gamma\ ;\ \emptyset \vdash * : I}\ (I\textbf{I}) \qquad \frac{\Gamma\ ;\ \Delta_1 \vdash M : I \quad \Gamma\ ;\ \Delta_2 \vdash N : \sigma}{\Gamma\ ;\ \Delta_1 \sharp \Delta_2 \vdash \mathsf{let}\ *\ \mathsf{be}\ M\ \mathsf{in}\ N : \sigma}\ (I\textbf{E})$$

$$\frac{\Gamma\ ;\ \Delta_1 \vdash M : \sigma \quad \Gamma\ ;\ \Delta_2 \vdash N : \tau}{\Gamma\ ;\ \Delta_1 \sharp \Delta_2 \vdash M \otimes N : \sigma \otimes \tau}\ (\otimes\textbf{I}) \qquad \frac{\Gamma\ ;\ \Delta_1 \vdash M : \sigma \otimes \tau \quad \Gamma\ ;\ \Delta_2, x : \sigma, y : \tau \vdash N : \theta}{\Gamma\ ;\ \Delta_1 \sharp \Delta_2 \vdash \mathsf{let}\ x \otimes y\ \mathsf{be}\ M\ \mathsf{in}\ N : \theta}\ (\otimes\textbf{E})$$

$$\frac{\Gamma\ ;\ \Delta, x : \sigma \vdash M : \tau}{\Gamma\ ;\ \Delta \vdash \lambda x.M : \sigma \multimap \tau}\ (\multimap\textbf{I}) \qquad \frac{\Gamma\ ;\ \Delta_1 \vdash M : \sigma \multimap \tau \quad \Gamma\ ;\ \Delta_2 \vdash N : \sigma}{\Gamma\ ;\ \Delta_1 \sharp \Delta_2 \vdash MN : \tau}\ (\multimap\textbf{E})$$

$$\frac{}{\Gamma_1, x:\sigma, \Gamma_2 \,;\, \emptyset \vdash x:\sigma}\ (\mathbf{Variable_{int}})$$

$$\frac{\Gamma \,;\, \emptyset \vdash M:\sigma}{\Gamma \,;\, \emptyset \vdash !M : !\sigma}\ (!\mathrm{I}) \qquad \frac{\Gamma \,;\, \Delta_1 \vdash M : !\sigma \quad \Gamma, x:\sigma \,;\, \Delta_2 \vdash N:\tau}{\Gamma \,;\, \Delta_1 \sharp \Delta_2 \vdash \mathsf{let}\ !x\ \mathsf{be}\ M\ \mathsf{in}\ N:\tau}\ (!\mathrm{E})$$

Axioms

$$\begin{array}{rlrl} \mathsf{let}\ *\ \mathsf{be}\ *\ \mathsf{in}\ M &= M & \mathsf{let}\ *\ \mathsf{be}\ M\ \mathsf{in}\ * &= M \\ \mathsf{let}\ x\otimes y\ \mathsf{be}\ M\otimes N\ \mathsf{in}\ L &= L[M/x, N/y] & \mathsf{let}\ x\otimes y\ \mathsf{be}\ M\ \mathsf{in}\ x\otimes y &= M \\ (\lambda x.M)N &= M[N/x] & \lambda x.Mx &= M \\ \mathsf{let}\ !x\ \mathsf{be}\ !M\ \mathsf{in}\ N &= N[M/x] & \mathsf{let}\ !x\ \mathsf{be}\ M\ \mathsf{in}\ !x &= M \end{array}$$

$$\begin{array}{rl} C[\mathsf{let}\ *\ \mathsf{be}\ M\ \mathsf{in}\ N] &= \mathsf{let}\ *\ \mathsf{be}\ M\ \mathsf{in}\ C[N] \\ C[\mathsf{let}\ x\otimes y\ \mathsf{be}\ M\ \mathsf{in}\ N] &= \mathsf{let}\ x\otimes y\ \mathsf{be}\ M\ \mathsf{in}\ C[N] \\ C[\mathsf{let}\ !x\ \mathsf{be}\ M\ \mathsf{in}\ N] &= \mathsf{let}\ !x\ \mathsf{be}\ M\ \mathsf{in}\ C[N] \end{array}$$

where $C[-]$ is a linear context (no ! binds $[-]$).

5.2 Semantics of DILL

Let $\mathbb{C}$ be a cartesian category (category with finite products), $\mathbb{D}$ a symmetric monoidal closed category and $\mathbb{C} \underset{U}{\overset{F}{\rightleftarrows}} \mathbb{D}$ a symmetric monoidal adjunction; we understand that the symmetric monoidal structure on $\mathbb{C}$ is given by (a choice of) the terminal object and binary product. Assume that there is an object $[\![b]\!] \in \mathbb{D}$ for each base type b and an arrow $[\![c]\!] \in \mathbb{D}([\![\sigma]\!], [\![\tau]\!])$ for each constant $c : \sigma \to \tau$, where $[\![\sigma]\!] \in \mathbb{D}$ is inductively defined by $[\![I]\!] = I$, $[\![\sigma\otimes\tau]\!] = [\![\sigma]\!] \otimes [\![\tau]\!]$, $[\![\sigma \multimap \tau]\!] = [\![\sigma]\!] \multimap [\![\tau]\!]$ and $[\![!\sigma]\!] = FU[\![\sigma]\!]$. For each typing judgement $\Gamma \,;\, \Delta \vdash M : \sigma$, we define $[\![\Gamma \,;\, \Delta \vdash M : \sigma]\!] : [\![|\Gamma \,;\, \Delta|]\!] \to [\![\tau]\!]$ in $\mathbb{D}$ as follows, where $|\Gamma \,;\, \Delta| = |!\Gamma, \Delta|$ in which $!\Gamma = x_1 : !\sigma_1, \ldots, x_n : !\sigma_n$ for $\Gamma \equiv x_1 : \sigma_1, \ldots, x_n : \sigma_n$. First eight cases are dealt with as in MILL, with care for discarding or duplicating the intuitionistic context, using

$$\begin{array}{ll} \mathsf{discard}_{\Gamma,\Delta} &: [\![|\Gamma \,;\, \Delta|]\!] \to [\![|\Delta|]\!] \\ \mathsf{split}_{\Gamma,\Delta_1,\Delta_2} &: [\![|\Gamma \,;\, \Delta_1\sharp\Delta_2|]\!] \to [\![|\Gamma \,;\, \Delta_1|]\!] \otimes [\![|\Gamma \,;\, \Delta_2|]\!] \end{array}$$

which are defined in terms of projections and diagonal maps in $\mathbb{C}$ and imported into $\mathbb{D}$ via F. For last three cases we have

$$[\![\Gamma_1, x:\sigma, \Gamma_2 \,;\, \emptyset \vdash x : \sigma]\!] = \\ [\![|\Gamma_1, x:\sigma, \Gamma_2|]\!] \stackrel{\simeq}{\to} F(\ldots \times U[\![\sigma]\!] \times \ldots) \xrightarrow{F\mathrm{proj}} FU[\![\sigma]\!] \xrightarrow{\varepsilon} [\![\sigma]\!]$$

$$[\![\Gamma \,;\, \emptyset \vdash !M : !\sigma]\!] = [\![|\Gamma \,;\, \emptyset|]\!] \stackrel{\simeq}{\to} \textstyle\bigotimes_i FU[\![\sigma_i]\!] \xrightarrow{\bigotimes_i \delta} \bigotimes_i FUFU[\![\sigma_i]\!] \xrightarrow{m} \\ FU(\textstyle\bigotimes_i FU[\![\sigma_i]\!]) \stackrel{\simeq}{\to} FU[\![|\Gamma \,;\, \emptyset|]\!] \xrightarrow{FU[\![\Gamma \,;\, \emptyset \vdash M:\sigma]\!]} FU[\![\sigma]\!]$$

$$[\![\Gamma \,;\, \Delta_1\sharp\Delta_2 \vdash \mathsf{let}\ !x\ \mathsf{be}\ M\ \mathsf{in}\ N : \tau]\!] = \\ [\![|\Gamma \,;\, \Delta_1\sharp\Delta_2|]\!] \xrightarrow{\mathsf{split}} [\![|\Gamma \,;\, \Delta_1|]\!] \otimes [\![|\Gamma \,;\, \Delta_2|]\!] \xrightarrow{[\![\Gamma \,;\, \Delta_1 \vdash M : !\sigma]\!] \otimes id} \\ [\![!\sigma]\!] \otimes [\![|\Gamma \,;\, \Delta_2|]\!] \stackrel{\simeq}{\to} [\![|\Gamma, x:\sigma \,;\, \Delta_2|]\!] \xrightarrow{[\![\Gamma, x:\sigma \,;\, \Delta_2 \vdash N:\tau]\!]} [\![\tau]\!]$$

where proj is a suitable projection in $\mathbb{C}$, ε and δ are the counit and comultiplication of the comonad FU while m is an induced coherent morphism.

Proposition 3. *This semantics is sound and complete* [5]. □

5.3 Logical Predicates for DILL

Consider the following commutative diagram of functors

$$\begin{array}{ccc} \mathbb{C}_0 & \xrightarrow{F_0} & \mathbb{D}_0 \\ \Big\downarrow{\scriptstyle \mathbb{I}} & & \Big\downarrow{\scriptstyle \mathbb{J}} \\ \mathbb{C}_1 & \xrightarrow[F_1]{} & \mathbb{D}_1 \end{array}$$

in which $\mathbb{C}_0$ and $\mathbb{C}_1$ are cartesian categories, $\mathbb{D}_0$ symmetric monoidal and $\mathbb{D}_1$ symmetric monoidal closed; and F_0, F_1 are strong symmetric monoidal while $\mathbb{I}$, $\mathbb{J}$ are strict symmetric monoidal. Moreover assume that F_1 has a right adjoint $U_1 : \mathbb{D}_1 \to \mathbb{C}_1$.

As in Sect. 3, we define the categories of $\mathbb{C}_0$- and $\mathbb{D}_0$-predicates – let us call them $\mathbb{C}_0\mathsf{PRED}$ and $\mathbb{D}_0\mathsf{PRED}$ respectively. Note that $\mathbb{C}_0\mathsf{PRED}$ is a cartesian category with products given by $(P \times Q)(X) = \{\langle f, g\rangle \mid f \in P(X), g \in Q(X)\}$ for $\mathbb{C}_0$-predicates P and Q (which coincides with $P \otimes Q$ in Definition 3).

Now we give functors between $\mathbb{C}_0\mathsf{PRED}$ and $\mathbb{D}_0\mathsf{PRED}$. For a $\mathbb{C}_0$-predicate P on $A \in \mathbb{C}_1$, define a $\mathbb{D}_0$-predicate $L(P)$ on $F_1 A \in \mathbb{D}_1$ by

$$L(P)(Y) \;=\; \{F_1 g \circ \mathbb{J} f \mid \exists X \in \mathbb{C}_0\ f \in \mathbb{D}_0(Y, F_0 X), g \in P(X)\}$$

and, for a $\mathbb{D}_0$-predicate Q on $B \in \mathbb{D}_1$, a $\mathbb{C}_0$-predicate $\widehat{F_0}(Q)$ on $U_1 B \in \mathbb{C}_1$ by

$$\widehat{F_0}(Q)(X) = \{f^* \in \mathbb{C}_1(\mathbb{I}X, U_1 B) \mid f \in Q(F_0 X) \subseteq \mathbb{D}_1(\mathbb{J}F_0 X, B) = \mathbb{D}_1(F_1 \mathbb{I} X, B)\}$$

where $f^* : \mathbb{I}X \to U_1 B$ is the adjoint mate of $f : F_1 \mathbb{I} X \to B$.

Proposition 4. *L and $\widehat{F_0}$ extend to functors between $\mathbb{C}_0\mathsf{PRED}$ and $\mathbb{D}_0\mathsf{PRED}$. Moreover L is strong symmetric monoidal, and left adjoint to $\widehat{F_0}$.* □

Therefore we have a symmetric monoidal adjunction between a cartesian category $\mathbb{C}_0\mathsf{PRED}$ and a symmetric monoidal closed category $\mathbb{D}_0\mathsf{PRED}$. Let $!$ be the induced comonad on $\mathbb{D}_0\mathsf{PRED}$, that is, we define a $\mathbb{D}_0$-predicate $!P$ on $F_1 U_1 A$ by

$$(!P)(Y) \;=\; \{F_1 g^* \circ \mathbb{J} f \mid \exists X \in \mathbb{C}_0\ f \in \mathbb{D}_0(Y, F_0 X), g \in P(F_0 X)\}$$

for a $\mathbb{D}_0$-predicate P on A. These are derived from a category-theoretic construction (left Kan extension [19] gives a left adjoint of $(-) \circ F_0 : \mathbf{Set}^{\mathbb{D}_0^{op}} \to \mathbf{Set}^{\mathbb{C}_0^{op}}$) together with Lemma 5 (for glueing $\mathbb{C}_1 \underset{U_1}{\overset{F_1}{\rightleftarrows}} \mathbb{D}_1$ to $\mathbf{Set}^{\mathbb{C}_0^{op}} \underset{(-)\circ F_0}{\rightleftarrows} \mathbf{Set}^{\mathbb{D}_0^{op}}$), but here let us motivate $!P$ more intuitively. A sequent $\emptyset\ ;\ Y \vdash !A$ can be proved as

$$\dfrac{\begin{array}{c}\Pi_f\\ \vdots\\ \emptyset\ ;\ Y \vdash !X\end{array} \quad \dfrac{\begin{array}{c}\Pi_g\\ \vdots\\ X\ ;\ \emptyset \vdash A\end{array}}{X\ ;\ \emptyset \vdash !A}\,(!\mathrm{I})}{\emptyset\ ;\ Y \vdash !A}\,(!\mathrm{E})$$

where $\emptyset\ ;\ Y \vdash !X$ converts a linear resource Y to $!X$ which is used non-linearly in $X\ ;\ \emptyset \vdash !A$ to produce $!A$. Taking all such possible cases into account, we say that the proof satisfies $(!P)(Y)$ when Π_f belongs to $\mathbb{D}_0$ and Π_g satisfies $P(X)$.

Now let us fix an interpretation $[\![-]\!]_1$ of DILL in $\mathbb{C}_1 \underset{U_1}{\overset{F_1}{\rightleftarrows}} \mathbb{D}_1$.

Definition 8. *A type-indexed family* $\{P_\sigma\}$ *is a* logical $(\mathbb{C}_0 \overset{F_0}{\to} \mathbb{D}_0)$-predicate *if*

- P_σ *is a* $\mathbb{D}_0$*-predicate on* $[\![\sigma]\!]_1$,
- $P_I = \mathbb{P}_I$, $P_{\sigma\otimes\tau} = P_\sigma \otimes P_\tau$, $P_{\sigma\multimap\tau} = P_\sigma \multimap P_\tau$ *and* $P_{!\sigma} = !P_\sigma$ *hold, and*
- $[\![c]\!]_1 : (P_\sigma, [\![\sigma]\!]_1) \to (P_\tau, [\![\tau]\!]_1)$ *for each constant* $c : \sigma \to \tau$. □

Lemma 6 (Basic Lemma for DILL). *Let* $\{P_\sigma\}$ *be a logical* $(\mathbb{C}_0 \overset{F_0}{\to} \mathbb{D}_0)$*-predicate. Then, for* $\Gamma\ ;\ \Delta \vdash M : \tau$, $[\![\Gamma\ ;\ \Delta \vdash M : \tau]\!]_1 : (P_{|\Gamma\ ;\ \Delta|}, [\![|\Gamma\ ;\ \Delta|]\!]_1) \to (P_\tau, [\![\tau]\!]_1)$ *holds.* □

$(\mathbb{C}_0 \overset{F_0}{\to} \mathbb{D}_0)$ itself determines the *canonical logical* $(\mathbb{C}_0 \overset{F_0}{\to} \mathbb{D}_0)$*-predicate* when

- for each base type b there is an object $[\![b]\!]_0 \in \mathbb{D}_0$, and
- for each constant $c : \sigma \to \tau$ there is an arrow $[\![c]\!]_0 \in \mathbb{D}_0([\![\sigma]\!]_0, [\![\tau]\!]_0)$

where $[\![\sigma]\!]_0$ is defined inductively by $[\![I]\!]_0 = I$ and $[\![\sigma\otimes\tau]\!]_0 = [\![\sigma]\!]_0 \otimes [\![\tau]\!]_0$. In such cases we automatically have an interpretation $[\![-]\!]_1$ in $\mathbb{D}_1$ determined by $[\![b]\!]_1 = \mathbb{J}([\![b]\!]_0)$ and $[\![c]\!]_1 = \mathbb{J}([\![c]\!]_0)$, and the canonical logical $(\mathbb{C}_0 \overset{F_0}{\to} \mathbb{D}_0)$-predicate $\{\mathbb{P}^*_\sigma\}$ is determined by $\mathbb{P}^*_b = \mathbb{P}_{[\![b]\!]_0}$.

Example 3 (From MILL to DILL). Let $\mathbb{D}_0$ be the term model of MILL and $\mathbb{C}_0$ equivalent to the one object one arrow category, and $\mathbb{C}_1 \underset{U_1}{\overset{F_1}{\rightleftarrows}} \mathbb{D}_1$ be the term model of DILL with the same base types and constants. Applying the Basic Lemma to the canonical logical $(\mathbb{C}_0 \to \mathbb{D}_0)$-predicate it follows that MILL is a full fragment of DILL; note that $\mathbb{P}_{\sigma\multimap\tau} = \mathbb{P}_\sigma \multimap \mathbb{P}_\tau$ holds for !-free types σ and τ (see Remark 1). □

Example 4 (From action calculi to DILL). Suppose that $\mathbb{C}_0 \overset{F_0}{\to} \mathbb{D}_0$ is the term model of an *action calculus* [20, 22] and $\mathbb{C}_1 \underset{U_1}{\overset{F_1}{\rightleftarrows}} \mathbb{D}_1$ is that of the corresponding DILL (alternatively the LNL Logic of Benton [6]), with $\mathbb{I}$ and $\mathbb{J}$ induced by the translation from the action calculus to DILL. If we have only non-parameterized constants, Basic Lemma applied to the canonical logical predicate implies that the translation is full. In fact we can deal with parameterized constants (control operators) as well (see [13]), so together with the conservativity [4] we have the full completeness of DILL (LNL) over (static) action calculi. □

6 Related Work, Further Work

6.1 Categorical Logical Predicates

Our treatment of logical predicates in category-theoretic framework is inspired by Hermida's work on fibrations and logical predicates [14], and also influenced by Mitchell and others' work, in particular [21]. However, all these results are for typed lambda calculi. Blute and Scott [9] do consider a linear variant, and the intuition behind their work seems close to ours, though their work is on classical linear logic and better understood in connection with Tan's recent work (see below). We also note that Ambler [3] has studied some relevant idea. The fact that our construction yields (bi)fibrations has some significance in our glueing constructions; we leave this categorical analysis to the full paper [13].

6.2 Classical Linear Type Theories

So far we have only considered "intuitionistic" linear type theories. It is natural to expect that our construction works equally well in the settings with duality, i.e., classical linear theories. Here is a relevant construction explored by Tan:

Example 5 (Double Glueing). An attractive use of categorical glueing is developed in Tan's thesis [23]. Let $\mathbb{C}$ be a $*$-autonomous category (typically a compact closed category). Because of the duality, $\mathbb{C}^{\mathrm{op}}$ is also $*$-autonomous and we have subscones (Example 1) $\widetilde{\mathbb{C}}$ and $\widetilde{\mathbb{C}^{\mathrm{op}}}$ with projections $p_1 : \widetilde{\mathbb{C}} \to \mathbb{C}$ and $p_2 : \widetilde{\mathbb{C}^{\mathrm{op}}} \to \mathbb{C}^{\mathrm{op}}$. Hyland noticed that the category $\mathbf{G}\mathbb{C}$ obtained by the following pullback is a $*$-autonomous category.

$$
\begin{array}{ccc}
\mathbf{G}\mathbb{C} & \longrightarrow & \widetilde{\mathbb{C}^{\mathrm{op}}}^{\mathrm{op}} \\
\Big\downarrow & & \Big\downarrow {\scriptstyle p_2^{\mathrm{op}}} \\
\widetilde{\mathbb{C}} & \xrightarrow[p_1]{} & \mathbb{C}
\end{array}
$$

Explicitly, $\mathbf{G}\mathbb{C}$'s object is a triple $A = (|A| \in \mathbb{C}, A_s \subseteq \mathbb{C}(I, |A|), A_t \subseteq \mathbb{C}(|A|, I))$ and an arrow $f : A \to B$ in $\mathbf{G}\mathbb{C}$ is an arrow $f : |A| \to |B|$ in $\mathbb{C}$ satisfying $f \circ a \in B_s$ for $a \in A_s$ and also $b \circ f \in A_t$ for $b \in B_t$ (this generalizes Loader's "linear logical predicates" [18]). The duality between $\widetilde{\mathbb{C}}$ and $\widetilde{\mathbb{C}^{\mathrm{op}}}^{\mathrm{op}}$ induces a duality on $\mathbf{G}\mathbb{C}$ which determines a $*$-autonomous structure. Tan calls this construction a *double glueing*, from which she has obtained various full completeness results for multiplicative linear logic (MLL). □

In fact it makes sense to replace the subscones in double glueing by $\mathbb{C}_0\mathsf{PRED}$ for some suitably chosen symmetric monoidal category $\mathbb{C}_0$. Using this we can derive a notion of logical predicates for MLL and, for example, can show that MILL is a full fragment of MLL. See [13] for an exposition.

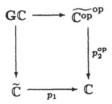

References

1. Abramsky, S., Gay, S.J., and Nagarajan, R. (1996), Specification structures and propositions-as-types for concurrency, *in* "Proceedings, 8th Banff Higher Order Workshop", Springer LNCS 1043, pp. 5–40.
2. Alimohamed, M. (1995), A characterization of lambda definability in categorical models of implicit polymorphism, *Theoret. Comp. Sci.* **146**, 5–23.
3. Ambler, S.J. (1992), "First Order Linear Logic in Symmetric Monoidal Closed Categories", Ph.D. thesis, ECS-LFCS-92-194, University of Edinburgh.
4. Barber, A., Gardner, P., Hasegawa, M., and Plotkin, G. (1998), From action calculi to linear logic, *in* "Computer Science Logic (CSL'97), Selected Papers", Springer LNCS 1414, pp. 78–97.
5. Barber, A., and Plotkin, G. (1997), Dual intuitionistic linear logic, submitted.
6. Benton, N. (1995), A mixed linear non-linear logic: proofs, terms and models, *in* "Computer Science Logic (CSL'94), Selected Papers", Springer LNCS 933, pp. 121–135.
7. Benton, N., Bierman, G.M., de Paiva, V., and Hyland, J.M.E. (1993), Linear lambda-calculus and categorical models revisited, *in* "Computer Science Logic (CSL'92), Selected Papers", Springer LNCS 702, pp. 61–84.
8. Bierman, G.M. (1995), What is a categorical model of intuitionistic linear logic? *in* "Proceedings, Typed Lambda Calculi and Applications (TLCA'95)", Springer LNCS 902, pp. 78–93.
9. Blute, R.F., Scott, P.J. (1996), Linear Läuchli semantics, *Ann. Pure Appl. Logic* **77**, 229–296.
10. Eilenberg, S., and Kelly, G.M. (1966), Closed categories, *in* "Proceedings, Categorical Algebra (La Jolla 1965)", pp. 421-562, Springer-Verlag.
11. Garrigue, J., and Minamide, Y. (1998), On the runtime complexity of type-directed unboxing, *in* "Proceedings, International Conference on Functional Programming (ICFP'98)", pp. 1–12, ACM Press.
12. Girard, J.-Y. (1987), Linear logic, *Theoret. Comp. Sci.* **50**, 1–102.
13. Hasegawa, M. (1999), Categorical glueing and logical predicates for models of linear logic, technical report, RIMS, Kyoto University.
14. Hermida, C. (1994), "Fibrations, Logical Predicates and Indeterminates", Ph.D. thesis, ECS-LFCS-93-277, University of Edinburgh.
15. Im, G.B., and Kelly, G.M. (1986), A universal property of the convolution monoidal structure, *J. Pure Appl. Algebra* **43**, 75–88.
16. Lafont, Y. (1988), "Logiques, Catégories et Machines", Thèse de Doctorat, Université Paris VII.
17. Lambek, J. (1995), Bilinear logic in algebra and linguistics, *in* "Advances in Linear Logic", pp. 43–59, Cambridge University Press.
18. Loader, R. (1994), "Models of Lambda Calculi and Linear Logic: Structural, Equational and Proof-Theoretic Characterisations", Ph.D. thesis, Oxford University.
19. Mac Lane, S. (1971), "Categories for the Working Mathematician", Graduate Texts in Mathematics 5, Springer-Verlag.
20. Milner, R. (1996), Calculi for interaction, *Acta Inform.* **33**(8), 707–737.
21. Mitchell, J.C., and Scedrov, A. (1992), Notes on sconing and relators, *in* "Computer Science Logic (CSL'92), Selected Papers", Springer LNCS 702, pp. 352–378.
22. Power, A.J. (1996), Elementary control structures, *in* "Proceedings, Concurrency Theory (CONCUR'96)", Springer LNCS 1119, pp. 115–130.
23. Tan, A.M. (1997), "Full Completeness for Models of Linear Logic", Ph.D. thesis, University of Cambridge.

Polarized Proof-Nets: Proof-Nets for LC
(Extended Abstract)

Olivier Laurent

Institut de Mathématiques de Luminy
CNRS-Marseille, France
olaurent@iml.univ-mrs.fr

Abstract. We define a notion of polarization in linear logic (**LL**) coming from the polarities of Jean-Yves Girard's classical sequent calculus **LC** [4]. This allows us to define a translation between the two systems. Then we study the application of this polar ization constraint to proof-nets for full linear logic described in [7]. This yields an important simplification of the correctness criterion for polarized proof-nets. In this way we obtain a system of proof-nets for **LC**.

The study of cut-elimination takes an important place in proof-theory. Much work is spent to deal with commutation of rules for cut-elimination in sequent calculi. The introduction of proof-nets (see [7] for instance) solves commutation problems and allows us to define a clear notion of reduction and complexity.

In [4], Jean-Yves Girard defines the sequent calculus **LC** using polarities. **LC** is a refinement of **LK** with a deterministic cut-elimination. J.-Y. Girard leaves open the following problem about the syntax:

"Find a better syntax (which would be to ***LC*** *what typed λ-calculus is to* ***LJ****) for normalization [...]. A kind of proof-nets could be the solution, and the fact that proof-nets are not available for full linear logic could be compensated by the fact that only certain linear configurations are used."*

In this paper we address this problem but the situation is now slightly different since proof-nets for full linear logic are given in [7]. In these proof-nets, the boxes for additives are replaced by weights on the nodes giving less sequentializatio n information. To use these proof-nets, we will first define a translation from **LC** to the fragment **LLP** of **LL** defined by restricting to polarized formulas. The "particular linear configurations" of **LC** correspond to the polarization of **LLP**.

We then turn to the study of proof-structures for **LLP** and show that the restriction to polarized formulas induces a natural orientation, the *orientation of polarization*, which is respected by the paths of **LL**'s correctness condition (Orientation Lemma). This yields a striking simplification of the correctness condition which allows us to get rid of the notion of switches. In particular it turns out to be cubic in the size of polarized proof-nets whereas the **LL** condition is immediately seen to be exponential.

1 Classical Logic: LC

Gentzen's classical sequent calculus **LK** has well known problems, such as the lack of a denotational semantics and the non determinism of cut-elimination. J.-Y. Girard proposed in [4] the calculus **LC** as a refinement of **LK** to solve these defects . The key point is the introduction of polarities for formulas. Let us just remind the syntax.

1.1 Formulas and Polarity

The formulas of **LC** are built from the atomic formulas and the constants V and F by using the connectives $\wedge$, $\vee$, $\neg$, $\exists$ and $\forall$. For each formula, we define its *polarity*: atomic formulas, V and F are positive; a s for the compound formulas we use the following table:

A B	$A \wedge B$	$A \vee B$	$\neg A$	$\exists x A$	$\forall x A$
$+$ $+$	$+$	$+$	$-$	$+$	$-$
$-$ $+$	$+$	$-$	$+$	$+$	$-$
$+$ $-$	$+$	$-$			
$-$ $-$	$-$	$-$			

In the sequel, P and Q will stand for positive formulas and N and M for negative ones.

1.2 Rules of the Sequent Calculus LC

To limit the number of rules, we will use one-sided sequents. The formulas will be defined modulo the De Morgan's laws. The sequents for **LC** are written $\vdash \Gamma; \Pi$ where Γ (the *body*) is a multi-set of formulas and Π (the *stoup*) is either empty or a unique positive formula.

Then the sequent calculus is defined by the following rules:

$$\frac{}{\vdash \neg P; P} \qquad \frac{\vdash \Gamma; P \quad \vdash \neg P, \Delta; \Pi}{\vdash \Gamma, \Delta; \Pi} \qquad \frac{\vdash \Gamma, N; \quad \vdash \neg N, \Delta; \Pi}{\vdash \Gamma, \Delta; \Pi}$$

$$\frac{\vdash \Gamma; P}{\vdash \Gamma, P;} \qquad \frac{\vdash \Gamma; \Pi}{\vdash \Gamma, A; \Pi} \qquad \frac{\vdash \Gamma, A, A; \Pi}{\vdash \Gamma, A; \Pi}$$

$$\frac{}{\vdash; V} \qquad \frac{}{\vdash \Gamma, \neg F; \Pi}$$

$$\frac{\vdash \Gamma; P \quad \vdash \Delta; Q}{\vdash \Gamma, \Delta; P \wedge Q} \qquad \frac{\vdash \Gamma; P \quad \vdash \Delta, N;}{\vdash \Gamma, \Delta; P \wedge N} \qquad \frac{\vdash \Gamma, M; \quad \vdash \Delta; Q}{\vdash \Gamma, \Delta; M \wedge Q}$$

$$\frac{\vdash \Gamma, M; \Pi \quad \vdash \Gamma, N; \Pi}{\vdash \Gamma, M \wedge N; \Pi}$$

$$\frac{\vdash \Gamma, A, B; \Pi}{\vdash \Gamma, A \vee B; \Pi} A \vee B \text{ negative} \qquad \frac{\vdash \Gamma; P}{\vdash \Gamma; P \vee Q} \qquad \frac{\vdash \Gamma; Q}{\vdash \Gamma; P \vee Q}$$

$$\frac{\vdash \Gamma, A; \Pi}{\vdash \Gamma, \forall x A; \Pi} x \notin \Gamma, \Pi \qquad \frac{\vdash \Gamma, N[^t/_x];}{\vdash \Gamma; \exists x N} \qquad \frac{\vdash \Gamma; P[^t/_x]}{\vdash \Gamma; \exists x P}$$

2 Linear Logic with Polarities

We can give a translation from **LC** to **LL** using the definition of the denotational semantics described in [4]. More precisely, we will define a polarized fragment of **LL** and we will show in which way it corresponds to **LC**. We start with the defin ition of two polarized fragments of **LL**.

The first notion of polarization for **LL** splits the connectives into reversible and non reversible ones.

Definition 1 (Polarized formula). *We define in the same time the* positive *(denoted by P, Q) and* negative *(denoted by M, N) formulas, starting from a set of atoms (denoted by A, B):*

$$P ::= \ !A \mid P \otimes P \mid P \oplus P \mid \exists x P \mid 1 \mid 0 \mid !N$$
$$N ::= ?A^{\perp} \mid N \parr N \mid N \,\&\, N \mid \forall x N \mid \perp \mid \top \mid ?P$$

A polarized formula *is either a positive one or a negative one.*

The second notion of polarization is more precise and corresponds to **LC**'s polarities. It will be used for studying translations between **LC** and **LL**.

Definition 2 (Strictly polarized formula). *We define in the same time the* strictly positive *(denoted by $\mathcal{P}$, $\mathcal{Q}$) and* strictly negative *(denoted by $\mathcal{M}$, $\mathcal{N}$) formulas, starting from a set of atoms (denoted by A, B):*

$$\mathcal{P} ::= \ !A \mid \mathcal{P} \otimes \mathcal{P} \mid \mathcal{P} \otimes !\mathcal{N} \mid !\mathcal{N} \otimes \mathcal{P} \mid \mathcal{P} \oplus \mathcal{P} \mid \exists x \mathcal{P} \mid \exists x !\mathcal{N} \mid 1 \mid 0$$
$$\mathcal{N} ::= ?A^{\perp} \mid \mathcal{N} \parr \mathcal{N} \mid \mathcal{N} \parr ?\mathcal{P} \mid ?\mathcal{P} \parr \mathcal{N} \mid \mathcal{N} \,\&\, \mathcal{N} \mid \forall x \mathcal{N} \mid \forall x ?\mathcal{P} \mid \perp \mid \top$$

A strictly polarized formula *is $\mathcal{P}$, $\mathcal{N}$, $?\mathcal{P}$ or $!\mathcal{N}$.*

Definition 3 (LLP and LLP$_C$). *The fragment **LLP** (resp. **LLP$_C$**) of **LL** is obtained by restricting to polarized (resp. strictly polarized) formulas and by adding the constraint that the $\top$-rule must introduce at most one positive formula.*

LLP$_C$ is a fragment of **LLP** (strictly polarized formulas are polarized) so all the results we will prove on **LLP** (about proof-nets,...) will be also true for **LLP$_C$**.

The constraint on the $\top$-rule is needed in particular for the next proposition.

Proposition 1. *If $\vdash \Gamma$ is provable in **LLP** then Γ has at most one positive formula.*

3 Translations between LC and LLP$_C$

We now prove the similarity of the two systems by defining two translations between **LC** and **LL**. More precisely these translations show that **LC** and **LLP$_C$** are almost isomorphic.

Definition 4. *LC^{rev} is the fragment of LC which refuses:*

- *structural rules on negative non atomic formulas;*
- *negative non atomic formulas in the context of the negative premise of:*
  - *negative cut-rule,*
  - *$\otimes$-rule between a negative and a positive formula,*
  - *$\exists$-rule on a negative formula.*

Every proof of **LC** can be transformed into a proof of $\mathbf{LC}^{\mathrm{rev}}$ by commuting some reversible rules with structural ones so we have no loss of provability in $\mathbf{LC}^{\mathrm{rev}}$. A study of these commutations of reversible rules has been done in a similar case by M. Quatrini and L. Tortora de Falco in [9] for translation of $\mathbf{LK}^{\eta,\rho}_{\mathrm{pol}}$ into **LL**.

3.1 $\mathbf{LC}^{\mathrm{rev}} \to \mathbf{LLP_C}$

Definition 5. *The translation $G \mapsto G^\bullet$ from LC^{rev} into LLP_C is defined on formulas by:*

$$\begin{array}{rclrcl} A^\bullet &=& !A & (\neg P)^\bullet &=& P^{\bullet\perp} \\ V^\bullet &=& 1 & F^\bullet &=& 0 \\ (P \wedge Q)^\bullet &=& P^\bullet \otimes Q^\bullet & (N \wedge M)^\bullet &=& N^\bullet \,\&\, M^\bullet \\ (P \wedge N)^\bullet &=& P^\bullet \otimes !N^\bullet & (N \wedge P)^\bullet &=& !N^\bullet \otimes P^\bullet \\ (\exists x P)^\bullet &=& \exists x P^\bullet & (\exists x N)^\bullet &=& \exists x !N^\bullet \end{array}$$

Given a sequent of LC, we can split the body into two parts: positive formulas and negative formulas, $\vdash \Gamma; \Pi = \vdash \Gamma^-, \Gamma^+; \Pi$. Then we can define the translation on sequents: $(\vdash \Gamma^-, \Gamma^+; \Pi)^\bullet = \vdash \Gamma^{-\bullet}, ?(\Gamma^+)^\bullet, \Pi^\bullet$.
The translation of proofs is defined rule by rule by introducing promotion rules on the negative premise before negative cut, before $\wedge$ between a positive and a negative formula and before $\exists$ for a negative formula. For example here is the case of the negative cut:

$$\frac{\vdash \Gamma^-, \Gamma^+, N; \quad \vdash \neg N, \Delta^-, \Delta^+; \Pi}{\vdash \Gamma^-, \Delta^-, \Gamma^+, \Delta^+; \Pi}$$

$$\downarrow$$

$$\frac{\dfrac{\vdash \Gamma^{-\bullet}, ?(\Gamma^+)^\bullet, N^\bullet}{\vdash \Gamma^{-\bullet}, ?(\Gamma^+)^\bullet, !N^\bullet} \quad \vdash ?(\neg N)^\bullet, \Delta^{-\bullet}, ?(\Delta^+)^\bullet, \Pi^\bullet}{\vdash \Gamma^{-\bullet}, \Delta^{-\bullet}, ?(\Gamma^+)^\bullet, ?(\Delta^+)^\bullet, \Pi^\bullet}$$

Remark 1. An empty stoup corresponds to a $?\Gamma$ context in **LL**, i.e. to a correct context for promotion.

LC accepts structural rules on non atomic negative formulas which are not translated by $?G$ formulas in **LL**. A solution is to add the constraints of $\mathbf{LC}^{\mathrm{rev}}$ to **LC** as we have done, but another one is to introduce cuts for the translation of these rules. This has been done with linear isomorphisms in Danos-Joinet-Schellinx [1].

3.2 $\mathbf{LLP_C} \to \mathbf{LC^{rev}}$

Definition 6. *The translation $G \mapsto G^*$ from $\boldsymbol{LLP_C}$ into $\boldsymbol{LC^{rev}}$ is defined on strictly polarized formulas by:*

$$
\begin{array}{rcl@{\qquad\qquad}rcl}
(!A)^* &=& A & (!\mathcal{N})^* &=& \mathcal{N}^* \\
1^* &=& V & 0^* &=& F \\
(\mathcal{P}\otimes\mathcal{Q})^* &=& \mathcal{P}^*\wedge\mathcal{Q}^* & (\mathcal{P}\oplus\mathcal{Q})^* &=& \mathcal{P}^*\vee\mathcal{Q}^* \\
(\mathcal{P}\otimes !\mathcal{N})^* &=& \mathcal{P}^*\wedge\mathcal{N}^* & (!\mathcal{N}\otimes\mathcal{P})^* &=& \mathcal{N}^*\wedge\mathcal{P}^* \\
(\exists x\mathcal{P})^* &=& \exists x\mathcal{P}^* & (\exists x!\mathcal{N})^* &=& \exists x\mathcal{N}^*
\end{array}
$$
$$(\mathcal{P}^\perp)^* = \neg\mathcal{P}^*$$

By Proposition 1, a sequent $\vdash \Gamma$ of $\boldsymbol{LLP_C}$ can be written $\vdash \Gamma', \Pi$ where Π is the unique strictly positive formula of Γ (if it exists). Then the translation is given on sequents by: $(\vdash \Gamma', \Pi)^* = \vdash \Gamma'^*; \Pi^*$. There is no problem for the translation of proofs, we just have to precise the translation of the promotion rule:

$$
\left(\begin{array}{c}\pi\\ \vdots\\ \dfrac{\vdash ?\Gamma, \mathcal{N}}{\vdash ?\Gamma, !\mathcal{N}}\end{array}\right)^* = \begin{array}{c}\pi^*\\ \vdots\\ \vdash \Gamma^*, \mathcal{N}^*;\end{array}
$$

Remark 2. This particular translation corresponds to the fact that a promotion is always followed by another rule: a *cut*-rule, a $\otimes$-rule or a $\exists$-rule. So promotion rules can be erased by the translation.

The translations $(.)^\bullet$ and $(.)^*$ are almost inverse of each other, more precisely:

- If G is a formula of $\mathbf{LC^{rev}}$, $G^{\bullet *} = G$.
- If $\mathcal{P}$ is a strictly positive formula of $\mathbf{LLP_C}$, $\mathcal{P}^{*\bullet} = \mathcal{P}$ and $(?\mathcal{P})^{*\bullet} = \mathcal{P}$.
- If $\mathcal{N}$ is a strictly negative formula of $\mathbf{LLP_C}$, $\mathcal{N}^{*\bullet} = \mathcal{N}$ and $(!\mathcal{N})^{*\bullet} = \mathcal{N}$.
- For the sequents: $(\vdash \Gamma; \Pi)^{\bullet *} = \vdash \Gamma; \Pi$ and $(\vdash \Gamma)^{*\bullet} = \vdash \Gamma$.
- If π is a proof in $\mathbf{LC^{rev}}$, $\pi^{\bullet *} = \pi$.

However the converse is wrong for proofs: $\pi^{*\bullet} \neq \pi$ because $\mathbf{LLP_C}$ is more flexible about the position of promotions. In the following example, the first **LL** proof puts weakening in between the promotion and its associated $\exists$-r ule whereas the third one, being translated from **LC**, has glued the promotion with the $\exists$-rule.

$$
\dfrac{\dfrac{\dfrac{\dfrac{\dfrac{}{\vdash !A, ?A^\perp}}{\vdash ?!A, ?A^\perp}}{\vdash ?!A, !?A^\perp}}{\vdash ?B^\perp, ?!A, !?A^\perp}}{\vdash ?B^\perp, ?!A, \exists x!?A^\perp}
\quad \overset{\bullet}{\to} \quad
\dfrac{\dfrac{\dfrac{\dfrac{}{\vdash \neg A; A}}{\vdash \neg A, A;}}{\vdash \neg B, \neg A, A;}}{\vdash \neg B, A; \exists x \neg A}
\quad \overset{\bullet}{\to} \quad
\dfrac{\dfrac{\dfrac{\dfrac{\dfrac{}{\vdash !A, ?A^\perp}}{\vdash ?!A, ?A^\perp}}{\vdash ?B^\perp, ?!A, ?A^\perp}}{\vdash ?B^\perp, ?!A, !?A^\perp}}{\vdash ?B^\perp, ?!A, \exists x!?A^\perp}
$$

4 Proof-Nets

Proof-nets have been introduced in [3] for the multiplicative case and then extended in [5] and [7] to full linear logic.

4.1 Proof-Structure

The following definitions come from [7] with just some modifications.

Definition 7 (Weight). *Given a set of* elementary weights, *i.e. boolean variables, (denoted by* p, q,... *), a* weight *is a product (conjunction) of elementary weights* p *and of negations of elementary weights* $\bar{p}$.

As a convention, we use 1 for the empty product and 0 for a product where p *and* $\bar{p}$ *appear. We also replace* $p.p$ *by* p. *With this convention we say that the weight* w depends on p *when* p *or* $\bar{p}$ *appears in* w.

A *proof-structure* is an oriented graph with pending edges, for which each edge is associated with an **LL** formula, constructed on the following set of nodes respecting the following typing constraints. The orientation is from top to bottom.

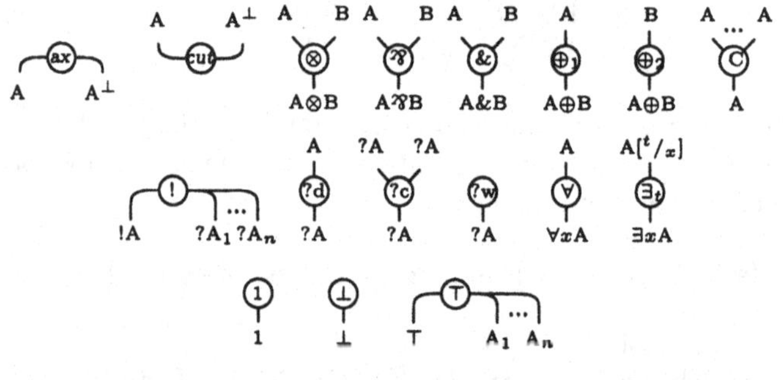

To avoid confusion with the other orientation that we will introduce later, this orientation will be called the *geographic orientation* and we will refer to it by the terms: *top, bottom, above, bellow, to go up, to go down, premise* of a node (edge just above the node), *conclusion* of a node (edge just bellow the node),...

A *unary node* is a node with only one premise and a *binary node* is a node with two premises. The C-nodes must have at least two premises.

In such a graph:

- we associate an elementary weight to each &-node called its *eigen weight*;
- the variable used in the quantification of a ∀-node is called its *eigen variable*;
- we associate a non empty set of nodes (different from *cut*) to each ⊥-node and ?w-node. These are called the *jumps* of the node.

Eigen weights and eigen variables are supposed to be different.

We associate a weight to each node with the constraint that if two nodes have a common edge, they must have the same weight except if the edge is a premise of a &-node or of a C-node (*additive contraction*). In these particular cases the weight changes:

- if w is the weight of a &-node and p is its eigen weight then w does not depend on p and its premise nodes must have weights $w.p$ and $w.\bar{p}$;
- if w is the weight of a C-node and $w_1, \ldots, w_n$ are the weights of its premise nodes then we must have $w = w_1 + \ldots + w_n$ and $w_i w_j = 0, \forall i \neq j$.

Then we can define the following notions:

- A node L with weight w is said *to depend on* p if w depends on p or if L is a C-node and one of the weights just above it depends on p.
- A node L is said *to depend on* an eigen variable x if x is free in the formula associated to the conclusion of L or if L is a $\exists_t$-node and x is free in t.

A proof-structure must also satisfy the following properties:

- a *conclusion node* (i.e. a node with pending edge) has weight 1;
- eigen variables are not free in the formulas associated to pending edges;
- if w is the weight of a &-node with eigen weight p and w' is a weight depending on p and appearing in the proof-structure then $w' \leq w$;
- if w is the weight of a $\forall$-node with eigen variable x and w' is the weight of a node depending on x then $w' \leq w$;
- if w is the weight of a $\perp$-node or of a ?w-node and w' is the weight of one of its jumps then $w \leq w'$.

With this definition we have a notion of proof-structures for full linear logic. Now to make it clear, let us look at the example of a proof-structure for $A \oplus B \multimap A \oplus B$:

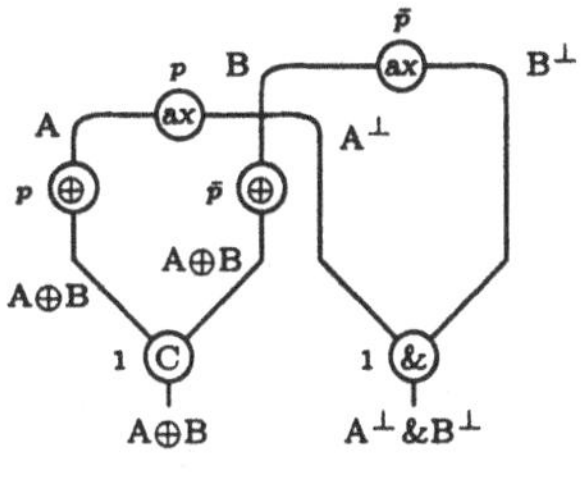

4.2 Sequentialization and Correctness

An important point in the study of proof-nets is the problem of correctness criterions that is the problem to know whether a proof-structure is a proof. More technically, can you inductively deconstruct a proof-structure?

There exist different correctness criterions for multiplicative proof-structures like [3] or [2] which lead to the criterion of [7] for the full case. We present here this general criterion.

Definition 8 (Sequentialization of a proof-structure). *The relation* "L sequentializes $\mathcal{R}$ into $\mathcal{E}$" *is defined for each possible L. $\mathcal{R}$ is a proof-structure, $\mathcal{E}$ is a set of proof-structures and L is a conclusion node of $\mathcal{R}$ or a cut.*

- ax, !, 1, $\top$: *if L is the only node of $\mathcal{R}$ then L sequentializes $\mathcal{R}$ into $\emptyset$;*

- *cut, $\otimes$: if it is possible to split the graph obtained by erasing L into two proof-structures $\mathcal{R}_1$ and $\mathcal{R}_2$ then L sequentializes $\mathcal{R}$ into $\{\mathcal{R}_1, \mathcal{R}_2\}$;*
- *$⅋$, $\oplus_1$, $\oplus_2$, $?d$, $?c$, $?w$, $\forall$, $\exists$, $\bot$: if when we erase L in $\mathcal{R}$, we obtain a proof-structure $\mathcal{R}_0$ then L sequentializes $\mathcal{R}$ into $\{\mathcal{R}_0\}$;*
- *&: let p be the eigen weight of L. The graph $\mathcal{R}_0$ (resp. $\mathcal{R}_1$) is obtained by giving to p the value 0 (resp. 1) and just keeping nodes with non zero weights and identifying the unary* C*-nodes to the node just above. If $\mathcal{R}_0$ (resp. R_1) is a proof-structure then L sequentializes $\mathcal{R}$ into $\{\mathcal{R}_0, \mathcal{R}_1\}$;*
- C*: a* C*-node never sequentializes a proof-structure.*

Definition 9 (Sequentializable proof-structure). *A proof-structure $\mathcal{R}$ is said to be* sequentializable *if one of its nodes sequentializes $\mathcal{R}$ into a set of sequentializable proof-structures or into the empty set.*

Definition 10 (Valuation). *A* valuation *φ for a proof-structure $\mathcal{R}$ is a function from the set of the eigen weights of $\mathcal{R}$ into $\{0,1\}$. Such a valuation can easily be seen as a function defined on the set of all the weights of $\mathcal{R}$.*

Definition 11 (Slice). *Given a valuation φ of a proof-structure $\mathcal{R}$, the* slice *$\varphi(\mathcal{R})$ is the proof-structure obtained from $\mathcal{R}$ by keeping only the nodes with weights w such that $\varphi(w) = 1$ and the edges bellow a kept node and by identifying the unary* C*-nodes with the upper node. A slice is not really a proof-structure according to definition of the Sect. 4.1 because unary &-nodes appear.*

Definition 12 (Switch). *Given a valuation φ of a proof-structure $\mathcal{R}$, a* switch *$\mathcal{S}$ of $\mathcal{R}$ is defined as a non oriented graph constructed with the nodes and the edges of $\varphi(\mathcal{R})$ with the modifications:*

- *for each $⅋$- or $?c$-node, we keep only one premise;*
- *for each &-node L, we erase the premise appearing in $\varphi(\mathcal{R})$ and we add an edge, called* dependency edge, *from a node depending on L to L (this may change nothing);*
- *for each $\forall$-node L, we erase the premise and we add an edge, called* dependency edge, *from a node depending on its eigen variable to L (this may change nothing);*
- *for each $?w$- or $\bot$-node L, we add an edge, called* jump edge, *from a jump of L to L.*

Definition 13 (Proof-net). *A proof-structure is a* proof-net *if all its switches are acyclic and connected.*

Theorem 1 (Sequentialization – *J.-Y. Girard* in [7]**).** *A proof-structure is sequentializable iff it is a proof-net.*

5 Polarized Proof-Nets

Now we restrict proof-nets to the polarized case. This strong constraint will allow us to define a new and simpler correctness criterion.

Definition 14 (Polarized proof-structure). *A* polarized proof-structure *is a proof-structure made only of polarized formulas and with the constraint that at most one of the formulas associated to the conclusions of a* $\top$*-node can be positive.*

In other words, a polarized proof-structure is a proof-structure typed by **LLP**. As $\mathbf{LLP_C}$ is a fragment of **LLP**, all the following results will give a notion of proof-nets for **LC** through the translations in Sect. 3.

Definition 15 (Edges). *We give here some new terminology on edges in a polarized proof-structure:*

- *a* positive *(resp.* negative*) edge is an edge with a positive (resp. negative) formula;*
- *a* principal edge *in a switch is an edge already appearing in the proof-structure; a* switching edge *is either a dependency edge or a jump edge. For switching edges, we extend the polarization and the geographic orientation by considerin g them negative and oriented towards the corresponding* &*-,* $\forall$*-,* $?w$*- or* $\bot$*-node.*

In the sequel, we will distinguish between two C-nodes: the C^+*-node* with positive premises and conclusion and the C^-*-node* with negative ones.

Definition 16 (Positive and negative nodes). *A* positive node *is a node with positive edges, that is* $\otimes$, $\oplus$, C^+, $\exists$ *and* 1, *and a* negative node *is a node with negative edges, that is* $⅋$, &, C^-, $?c$, $?w$, $\forall$ *and* $\bot$.

5.1 Towards Specific Criterions

The key point for the simplification of the correctness criterion in the case of polarized proof-nets is the existence of a specific orientation in these proof-nets as shown in Lemma 2. The use of this orientation allows us to forget the notion of switch and then also the notion of slice.

The idea of orientation linked to polarization in proof-nets has already been used. For example François Lamarche proposed in [8] a criterion for proof-nets for intuitionistic linear logic with Danos-Regnier polarities.

We define a new orientation on proof-structures, the *orientation of polarization* (or *p-orientation*): positive edges are oriented upwardly and negative edges downwardly. We will talk about this orientation using the terms: *to arrive to* , *to come from*, *incident edge*, *emergent edge*,...

Lemma 1. *In a switch of a polarized proof-structure, a node has at most one incident edge. Positive and negative nodes have exactly one incident edge.*

Proof. We study each node:

- the only nodes with incident switching edges are &, $\forall$, ?w and $\bot$ and by the definition of a switch these nodes have exactly one incident edge in a switch (either a premise or a switching edge);

- $\parr$- and ?c-nodes have just one premise in a switch so just one incident edge;
- positive nodes, ax, *cut* and ! have only principal edges in a switch and the only incident one is their positive conclusion (negative premise for *cut*);
- ?d-nodes have only emergent edges;
- $\top$-nodes with a positive conclusion are like ! and those with only negative conclusions have no incident edges;
- there are no C-nodes in a switch. □

Lemma 2 (Orientation lemma). *A non bouncing path in a switch of a polarized proof-structure starting accordingly to the p-orientation always respects this orientation.*

Proof. We prove the result by induction on the length of the path, the case of length 0 being given by the starting hypothesis. Now when the path arrives to a new node, this is only possible through the incident edge so when the path continues it must be by anot her edge, thus an emergent one (by Lemma 1) since it does not bounce. □

Lemma 3. *A non oriented cycle in a switch of a polarized proof-structure is p-oriented.*

Definition 17 (Correction graph). *The* correction graph *of a proof-structure $\mathcal{R}$ is the oriented graph obtained by putting on $\mathcal{R}$ the p-orientation and by adding some new edges:*

- *from each node depending on an eigen weight to the corresponding &-node;*
- *from each node depending on an eigen variable to the corresponding $\forall$-node;*
- *from the jumps to the nodes they are associated to.*

Lemma 4. *If there is a (non oriented) cycle in a switch of a proof-structure then there is a p-oriented cycle in its correction graph.*

Definition 18 (Initial and final nodes). *In a correction graph, a node is* initial *(resp.* final*) if all the edges starting from (resp. arriving to) it are pending edges.*

Remark 3. A final node is a conclusion node so its weight is always 1. A ?d-node is always initial.

5.2 Weak Criterion

We give here our first criterion for polarized proof-nets, which is simpler than the general one but equivalent. To obtain this result we still need to use the notion of slices.

Definition 19 (Slice of a correction graph). *A* slice *of a correction graph $\mathcal{G}$ is the sub-graph of $\mathcal{G}$ made only of the nodes and the edges of a slice of the proof-structure (in other terms it is the correction graph of the slice).*

Theorem 2 (Correctness criterion). *A polarized proof-structure has all its switches acyclic and connected iff all the slices of its correction graph are acyclic (with orientation), contain exactly one initial node and all the nodes of the slice are p-accessible from the initial one (in thi s case we say that the correction graph is* weakly correct*).*

Proof. By Theorem 1, a proof-structure with all its switches acyclic and connected is sequentializable and by an easy induction, a sequentializable polarized proof-structure has a weakly correct correction graph. Conversely if the correction graph is weakly correct, switches cannot contain any cycle by Lemma 4.

To finish, we can prove by induction on the sum Σ of the lengths of all the paths from the initial node i of the slice to a fixed node s that in all the switches of this slice there is a path between i and s.

- If $\Sigma = 0$ then $s = i$.
- If $\Sigma = n + 1$, s is not an initial node in the slice. We choose a switch $\mathcal{S}$, there exist a node s' and an edge a from s' to s such that a appears in $\mathcal{S}$ (by definition of a switch we always keep such an edge). Then by induction hyp othesis on s', there is a path in $\mathcal{S}$ between i and s' which can be extended with a into a path between i and s. □

We can apply to our polarized proof-structures all the results of the general case given in [7] about sequentialization, cut-elimination,...

5.3 Strong Criterion

Following the same direction we obtain a second and most important criterion which allows us to forget also slices.

Definition 20 (Strong correctness criterion). *The correction graph of a polarized proof-structure is* strongly correct *if it is acyclic and if for all pair of distinct initial nodes with weights w_i and w_j: $w_i.w_j = 0$.*

Theorem 3 (Strong criterion and weak criterion). *A strongly correct correction graph is weakly correct.*

Proof. No problem for acyclicity because a slice of a correction graph has less edges than the correction graph itself. Then by acyclicity of the slices we have at least one initial node in each slice. But also at most one because taking a slice does not create any initial node (a negative node is never initial and the other ones cannot lose the node under their conclusion) so the condition on initial nodes of the correction graph is sufficient.

For accessibility of nodes, we prove by induction on the sum Σ of the lengths of all the paths from an initial node to a fixed node s that s is p-accessible by the initial node in each slice where it appears:

- if $\Sigma = 0$ then s is initial;
- if $\Sigma = n+1$ then in a slice where s appears either it is initial and there is no problem or there is another node s' with an edge from s' to s. By induction hypothesis s' is accessible from the initial node in every slice where it ap pears. Thus in the slice we are looking at, s' is accessible and also s by adding the edge to a path arriving to s'. □

The converse is wrong, some proof-structures are weakly correct but rejected by the strong criterion because some cycles may come from the interactions between different slices. However we keep enough proof-structures to have proof-nets for all proofs of sequent calculus and the strong criterion is preserved by cut-elimination. We will see this in the Sects. 5.5 and 5.6.

5.4 Sequentialization

We will now give a proof of sequentializability of strongly correct proof-nets different from the one consisting in using the proof for the general criterion by Theorems 3, 2 and then 1.

Definition 21 (Positive tree). *A* positive tree *of a correction graph is a non empty connected set of positive nodes and positive edges maximal for inclusion.*

A positive tree $\mathcal{A}$ is terminal *when for each positive edge a of the correction graph if there is a path from $\mathcal{A}$ to a then a is in $\mathcal{A}$.*

Theorem 4 (Sequentialization). *A polarized proof-net is sequentializable.*

Proof. The first point is to sequentialize by all negative final nodes. We prove that if a ⅋-, &-, ?c-, ?w-, $\bot$- or $\forall$-node is final then it sequentializes the proof-net. We remark that $\otimes$-, $\oplus_i$, $\exists$-, C^+-, C^--, ?d- a nd *cut*-nodes are never final. So we have to sequentialize a proof-net with only ax, !, $\top$ and 1-nodes as final ones.

Lemma 5. *If the only final nodes of a polarized proof-net are* ax, !, $\top$ *and* 1 *then from each non final node there exists a path to a terminal positive tree.*

Definition 22 (Cut positive tree). *A positive tree is said to be* cut *if it has a cut-node hereditary above it.*

Proof (Theorem 4 – continued). Given a proof-net with only ax, !, $\top$ and 1-nodes as final ones, by Lemma 5 it contains a terminal positive tree. If there is no nodes under this tree, it can be sequentialized. Otherwise this is a cut positive tree and we show by termin ality of the tree that the *cut*-node under it sequentializes the proof-net. □

Proposition 2. *The criterion given by Theorem 4 has a cubic complexity in the size of the proof-net (i.e. the number of its nodes).*

5.5 Translation from Sequent Calculus

To show that the strong criterion keep enough proof-structures we have to define a translation from **LLP** to polarized proof-structures and to prove the correctness of the proof-structures built in this way.

When we talked about sequentialization we used proof-structures with !-nodes just seen as generalized axioms but to talk about the translation of proofs and about cut-elimination, we need to refine our definition of proof-structure.

Definition 23 (Proof-structure and proof-net with boxes). *We define a* proof-structure with boxes *by induction, it is:*

- *either a proof-structure with no* !*-nodes,*
- *or a proof-structure together with a proof-structure with boxes of conclusions* $A, ?B_1, \ldots, ?B_n$ *associated to each* !*-node of conclusions* $!A, ?B_1, \ldots, ?B_n$.

We can define in the same way proof-nets with boxes *from proof-nets.*

In the sequel we will use the term proof-structure (resp. proof-net) instead of proof-structure (resp. proof-net) with boxes.

Definition 24 (Translation of proofs). *We define the translation from* ***LLP*** *to polarized proof-structures by induction on the size of the proof:*

- *&: by induction we obtain two polarized proof-structures* $\mathcal{R}_1$ *and* $\mathcal{R}_2$ *from the two proofs of the premises of the &-rule. We choose a new elementary weight* p *and multiply all the weights of* $\mathcal{R}_1$ *by* p *and all the weights of* $\mathcal{R}_2$ *by* $\bar{p}$*. Then we add a &-node (with eigen weight* p*) between the two pending edges corresponding to the formulas used by the & and a* C*-node for each pair of formulas of the context coming from* $\mathcal{R}_1$ *and* $\mathcal{R}_2$*;*
- !*: the new proof-structure is just a single* !*-node introducing the conclusions* $!A, ?B_1, \ldots, ?B_n$ *of the rule and the proof-structure associated to it is the one obtained at the previous step with conclusions* $A, ?B_1, \ldots, ?B_n$*;*
- $?w$*: we just add a* $?w$*-node to the proof-structure* $\mathcal{R}$ *of the previous step with a set of jumps constituted of all the conclusion nodes of* $\mathcal{R}$*;*
- $\bot$*: same as* $?w$*;*

no problem for the other rules.

Theorem 5. *The previous translation is in fact from* ***LLP*** *to polarized proof-nets.*

5.6 Cut Elimination

Definition 25 (Reduction step). *The different cut-elimination steps are the following ones:*

- Axiom cut: *we erase the* ax- *and* cut-*nodes and replace them by an edge, the jumps coming from the* ax-*node are moved to the other node above the cut.*

- Multiplicative cut: *we erase the* ⅋ *and the* ⊗, *the cut is duplicated between the two pairs of premises. All the jumps are duplicated and moved up.*

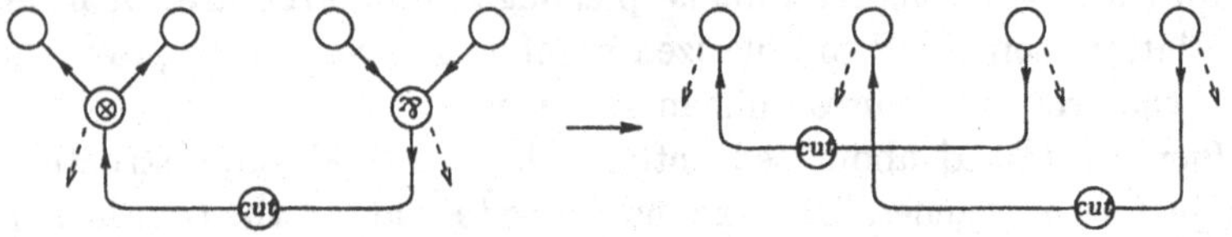

- Additive cut: *if the* $\oplus$*-node is a* $\oplus_1$*-node (resp.* $\oplus_2$*-node) we erase in the proof-structure all the nodes with null weights when* $p = 1$ *(resp.* $p = 0$*) and the cut moves up as the jumps.*

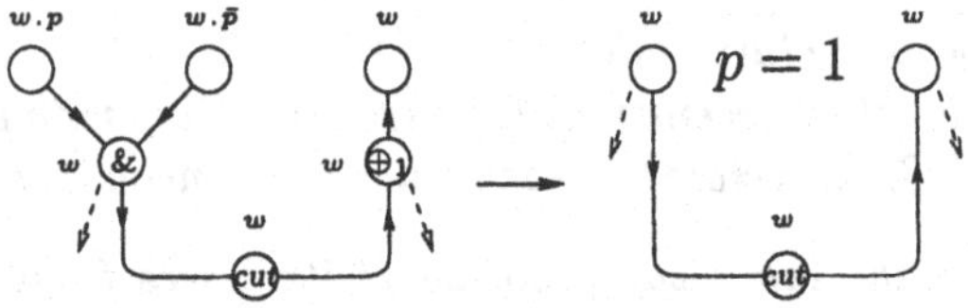

- Dereliction cut: *the box is opened and the cut moves up as the jumps.*
- Contraction cut: *the !-node is duplicated and also the cut to be put between each premise of the ?c and a box. New ?c-nodes are put between the pairs of conclusions of the !. Jumps from the ! and from the ?c are duplicated.*
- Weakening cut: *we just erase the box and put new ?w-nodes above its conclusions. The jumps of these new nodes are the jumps of the cut one.*
- Commutative exponential cut: *the box with the cut !-node comes into the other one and the other !-node is extended with the conclusions of the first one. All the jumps coming from the two !-nodes are put on the second one.*
- Quantifier cut: *we erase the two nodes* $\forall$ *and* $\exists_t$, *the cut goes up as the jumps. In all the proof-structure we make the substitution of* x *by* t.

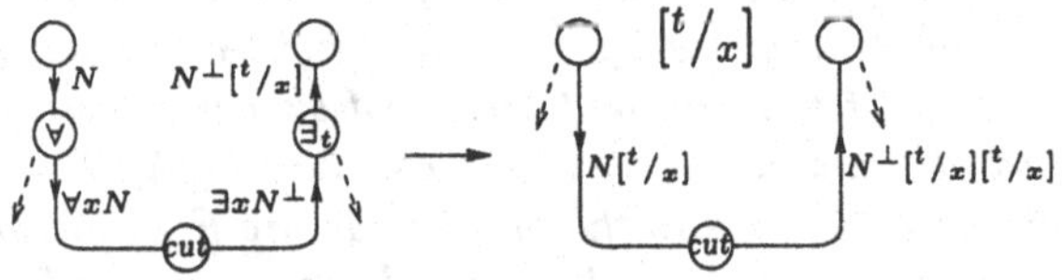

- Multiplicative constant cut: *we erase the three nodes:* 1, $\perp$ *and cut. The jumps starting from them are duplicated and moved to the jumps of* $\perp$.

The cases of a cut with a $\top$- or a C-node are still to be studied. A solution for the additive contraction is proposed in [7] but is not uniform with the other reduction steps. However with the restriction on the steps defined above, we have th e same result as in [7]:

Theorem 6. *A proof-net without* $\top$*-node and without &-connectives in the formulas associated to its pending edges, which cannot be reduce by any step described above, is in normal form (i.e. without cut-node).*

This has been already proved by J.-Y. Girard for the multiplicative-additive case but we give here a really different proof using the p-orientation.

Proof. If the proof-net contains no &-nodes, all the weights are 1 and there are no problems. Otherwise let L be a terminal &-node, that is with no paths to another &-node. By the hypothesis, there must be a *cut*-node (hereditary) under L. Then this *cut*-node can be reduced by terminality of L. □

Theorem 7 (Cut-elimination). *Strong correctness is preserved by the cut-elimination procedure.*

Proof. The steps are well defined in a proof-net (x is not free in $N^{\perp}[{}^{t}/_{x}]$ for the quantifier step by acyclicity). Then each step preserves the strong criterion. □

Conclusion

The polarization constraint, coming from **LC**, gives a system of proof-nets with a correctness criterion which is really simpler than the one in the general case [7]. Through the translation between **LC** and **LLP**, this gives proof-nets for the sequen t calculus **LC**, solving our starting problem.

The last section of this paper is devoted to cut-elimination where the problem of commutative additive contraction appears. A full solution has still to be found.

Much work is now possible such as an extension of our approach to second order quantifiers, the study of a geometry of interaction or of a game semantics for such proof-nets, the continuation of this work towards the intuitionistic polarities as defined i n [6],...

References

[1] Vincent Danos, Jean-Baptiste Joinet, and Harold Schellinx. Computational isomorphisms in classical logic (extended abstract). In Jean-Yves Girard, Mitsu Okada, and Andr Scedrov, editors, *Proceedings Linear Logic '96 Tokyo Meeting*, volume 3 of *Electronic Notes in Theoretical Computer Science*. Elsevier, Amsterdam, 1996.

[2] Vincent Danos and Laurent Regnier. The structure of multiplicatives. *Archive for Mathematical Logic*, 28:181–203, 1989.

[3] Jean-Yves Girard. Linear logic. *Theoretical Computer Science*, 50:1–102, 1987.

[4] Jean-Yves Girard. A new constructive logic : classical logic. *Mathematical Structures in Computer Science*, 1(3):255–296, 1991.

[5] Jean-Yves Girard. Quantifiers in linear logic II. In Corsi and Sambin, editors, *Nuovi problemi della logica e della filosofia della scienza*, pages 79–90, Bologna, 1991. CLUEB.

[6] Jean-Yves Girard. On the unity of logic. *Annals of Pure and Applied Logic*, 59:201–217, 1993.

[7] Jean-Yves Girard. Proof-nets : the parallel syntax for proof-theory. In Ursini and Agliano, editors, *Logic and Algebra*, New York, 1996. Marcel Dekker.

[8] Franois Lamarche. From proof nets to games (extended abstract). In Jean-Yves Girard, Mitsu Okada, and Andr Scedrov, editors, *Proceedings Linear Logic '96 Tokyo Meeting*, volume 3 of *Electronic Notes in Theoretical Computer Science*. Elsevier, Amsterdam, 1996.

[9] Myriam Quatrini and Lorenzo Tortora de Falco. Polarisation des preuves classiques et renversement. *Compte Rendu de l'Acadmie des Sciences de Paris*, 323:113–116, 1996.

Call-by-Push-Value: A Subsuming Paradigm
(Extended Abstract)

Paul Blain Levy*

Department of Computer Science, Queen Mary and Westfield College
LONDON E1 4NS
pbl@dcs.qmw.ac.uk

Abstract. Call-by-push-value is a new paradigm that subsumes the call-by-name and call-by-value paradigms, in the following sense: both operational and denotational semantics for those paradigms can be seen as arising, via translations that we will provide, from similar semantics for call-by-observable.
To explain call-by-observable, we first discuss general operational ideas, especially the distinction between values and computations, using the principle that "a value is, a computation does". Using an example program, we see that the lambda-calculus primitives can be understood as push/pop commands for an operand-stack.
We provide operational and denotational semantics for a range of computational effects and show their agreement. We hence obtain semantics for call-by-name and call-by-value, of which some are familiar, some are new and some were known but previously appeared mysterious.

1 Introduction

1.1 A Single Paradigm

In a recent invited lecture [Rey98], Reynolds, surveying over 30 years of programming language development, called for a common framework for typed call-by-name (CBN) and typed call-by-value (CBV). We consider this an important problem, as the existence of two separate paradigms is troubling:

- it makes each language appear arbitrary (whereas a unified language might be more canonical);
- on a more practical level, each time we create a new style of semantics, e.g. Scott semantics, operational semantics, game semantics, continuation semantics etc., we always need to do it twice—once for each paradigm.

We propose call-by-push-value (CBPV), a new typed paradigm based on Filinski's variant of Moggi's computational λ-calculus [Fil96,Mog91], as a solution to this problem. We will introduce a CBPV language, and give translations from CBN and CBV languages into it. We claim that, via these translations, CBPV "subsumes" CBN and CBV.

* supported by EPSRC research studentship no. 96308344

But what does it mean for one language to subsume another? After all, there are sound, adequate translations from CBN and CBV languages into each other [Plo76,HD97] and into other languages such as linear λ-calculus, Moggi's calculus [BW96,Mog91] and others [Mar98,MC88,JLST98,SJ98]. So we must explain in what sense our translations into CBPV go beyond these "classic" transforms, and why, consequently, CBPV is a solution to Reynolds' problem.

We therefore introduce the following informal criterion. A translation α from language L' into language L is *subsumptive* if every "naturally arising" denotational semantics, operational semantics or equation for L' arises, via α, from a "similar" denotational semantics, operational semantics or equation for L.

The importance of such a translation is that the semanticist need no longer attend to L', because its primitives can be seen as no more than syntactic sugar for complex constructs of L. We shall see in Sect. 1.2 that the classic translations mentioned above are not subsumptive.

The essence of Reynolds' problem can now be expressed as follows:

> Give subsumptive translations from CBN and CBV languages into a single language.

The key features of CBPV that enable it to solve this problem are that

1. it divides Moggi's type constructor T into two type constructors U and F, that give types of *thunks* and of *producers* respectively;
2. it distinguishes between *values* and *computations*;
3. writing $V`M$ for "M applied to V", the λ-calculus primitives can be understood as commands for an *operand-stack*:
   - $V`$ can be read as "push V";
   - $\lambda\mathtt{x}$ can be read as "pop $\mathtt{x}$".

(1) is reminiscent of the division of a monad into an adjunction. However, while an adjunction (with extra structure) gives rise to a model for CBPV, different (non-equivalent) adjunctions can give rise to the same model, because not all of the adjunctional structure is used. This is explained in [Lev98].

Feature (2) is shared with CBV, and feature (3) with CBN. (Indeed the push/pop reading is widely used in implementation of lazy languages [Jon92].)

That our translations into CBPV are subsumptive is too informal a claim to prove, but we have a diverse collection of examples to corroborate it:

- We can give operational semantics for CBPV in big-step, small-step or machine form, and recover standard operational semantics for CBV and CBN. These can be formulated to include various computational effects.
- We can give Scott semantics for CBPV, and recover those for CBN [Plo77] and for CBV [Plo85].
- We can give state-passing semantics for CBPV, and recover the mysterious CBN semantics of O'Hearn [O'H93], and a straightforward CBV semantics.
- We can give continuation semantics for CBPV, and recover the CBV semantics of [Plo76] and the CBN semantics of [SR96] (NB *not* that of [Plo76] which is not quite CBN, as it does not validate the η-law).

- We can give game semantics for CBPV, and recover the CBN game semantics of [HO94] and the CBV game semantics of [AM98].
- We can give an equational theory for CBPV. The equations that this gives us for CBN include the β- and η-laws for functions, which generally fail in CBV. The equations that we obtain for CBV include for example

$$\Gamma, \mathtt{x} : \mathtt{bool} \vdash M = \mathtt{if}\ \mathtt{x}\ \mathtt{then}\ M[\mathtt{true}/\mathtt{x}]\ \mathtt{else}\ M[\mathtt{false}/\mathtt{x}] \qquad (1)$$

  which generally fails in CBN. (1) is in fact a special case of the η-law for sum types.
- We can give a (rather messy) categorical semantics for CBPV. From a CBPV-structure we can construct for CBN a cartesian closed category, and for CBV a premonoidal category in the sense of [PR97].
- If we add sum types to both CBN and CBV languages, our translations into CBPV can be extended to include them. While both operational and denotational semantics for sum types differ between CBN and CBV, all the differences are recovered from their translation into CBPV.

After discussing related work, we give an operational account of the principles of CBPV. We add divergence and recursion to the basic language, and provide Scott semantics, which helps to motivate our translations from CBN and CBV into CBPV. Finally, we provide operational and denotational semantics for a range of computational effects.

Acknowledgements I am grateful to M. Fiore, M. Marz, E. Moggi, P. O'Hearn, S. Peyton Jones, J. Power, U. Reddy, J. Reynolds, E. Robinson, H. Thielecke, referees and others for their helpful comments on this and related material.

1.2 Related Work

We briefly give some ways in which other proposed translations from CBN and CBV, even those on which ours are based, are not subsumptive. Of course, the objectives that they were designed to achieve are different.

We first look at cases in which semantics for the source language does not, so far as we can see, extend along the translation.

- It is not evident how to provide operational semantics for the monadic target languages of [BW96,Fil96,Mog91] so as to recover standard operational semantics for the source languages.
- The monad language of [BW96,Mog91] does not provide semantics for CBN, because the translation from CBN into it—like the *thunking transform* from CBN to CBV [HD97,SJ98]—does not preserve the η-law for functions.
- As remarked in [BW96], the linear language used there assumes "commutativity" of effects, so that continuation models, for example, do not arise from it; likewise for the language of [Mar98,MC88].
- The CPS transforms of [SR96,Plo76] do not, of course, preserve Scott semantics.

More subtly, there are cases where a semantics for the source language does extend to one for the target, but not (as subsumptiveness requires) to a "similar" one—the semantics of type becomes more complicated.

- To decompose the CBV predomain[1] model of [Plo85] using $A \to_{\mathsf{CBV}} B = A \to TB$ [BW96,Mog91], we must drop the countable-base condition on predomains, because the *total function space* operation does not preserve it. For example, $\mathbb{N} \to \mathbb{N}$ is a flat, uncountable "predomain".
- The CBN game model of [HO94] can exhibit a linear decomposition $A \to_{\mathsf{CBN}} B = !A \multimap B$ [BW96,Gir87], but types must then denote *games* rather than *arenas*. (Some further problems with this linear approach are discussed in [McC96], and it is abandoned for technical reasons in [AHM98].)

2 Call-by-push-value

We introduce CBPV in this section using an operational account, because (as for CBN and CBV) the operational ideas remain essentially constant across different effects, whereas the range of models is wide.

2.1 Operational Principles and Types

In CBPV we distinguish between *computations* and *values*. Intuitively speaking, a computation *does*, while a value *is*. CBPV has two disjoint classes of types: a computation has a *computation type*, while a value has a *value type*. For clarity we underline computation types.

The two classes of types are given by

$$\begin{array}{rl} A ::= & U\underline{B} \mid \sum_{i\in I} A_i \mid 1 \mid A \times A \\ \underline{B} ::= & FA \mid \prod_{i\in I} \underline{B}_i \mid A \to \underline{B} \end{array} \tag{2}$$

where each set I of *tags* is countable (so the language is infinitary).

We explain the types as follows; notice how this explanation maintains the does/is principle. Throughout execution, there is an *operand-stack* of values and tags that is pushed onto and popped from.

- A value of type $U\underline{B}$ is a *thunk* of a computation of type $\underline{B}$.
- A value of type $\sum_{i\in I} A_i$ is a pair (i, V), where $i \in I$ and V is a value of type A_i.
- A value of type $A \times A'$ is a pair (V, V'), where V is a value of type A and V' is a value of type A'.
- A value of type 1 is the 0-tuple $()$. We largely omit further mention of this type, as it is entirely analogous to $\times$.

[1] A *predomain* $(X, \leqslant)$ is a countably based, algebraic directed-complete poset, with joins of all nonempty bounded subsets, in which the down-set $\{y \in X : y \leqslant x\}$ of each $x \in X$ has a least element. (The last condition is adapted from [AM98]). A *domain* $(X, \leqslant, \bot)$ is a predomain with a least element $\bot$.

- A computation of type FA *produces* a value of type A.
- A computation of type $\prod_{i \in I} \underline{B}_i$ *pops* a tag $i \in I$ from the operand-stack, and then behaves as a computation of type $\underline{B}_i$.
- A computation of type $A \to \underline{B}$ *pops* a value of type A from the operand-stack, and then behaves as a computation of type $\underline{B}$.

A computation can perform other effects besides popping and producing. For example, a computation M of type $A \to FA'$ might output, then pop a value of type A, then push a value of type C, then input, then pop a value of type C and finally produce a value of type A'. Or it might crash, diverge, make some choices, jump out etc. But it cannot perform any further effects after producing, for then another computation begins, using the value that M produced.

Values alone can be stored, input, output, pushed, popped or chosen. Identifiers can be bound to (or replaced by) values alone, and therefore they always have value type. A computation is too "active" for this, although a *thunk* of a computation M is a value, so it can be stored etc. Later the thunk can be *forced*, and M then happens. Of course, a single thunk can be forced several times.

We call a value type of the form $\sum_{i \in I} 1$ a *groundtype* and write $\underline{n}$ or even just n for $(n, ())$. In particular, we write `bool` and `nat` for the groundtypes $\sum_{i \in \{\mathsf{true},\mathsf{false}\}} 1$ and $\sum_{i \in \mathbb{N}} 1$ respectively. A computation of type $F\sum_{i \in I} 1$ is called a *ground producer* because it produces a ground value.

Moggi's type TA [Mog91] becomes in our type system UFA, because a value of type TA is a thunk of a computation that produces a value of type A.

2.2 The basic language

Definition 1. *A* context Γ *is a finite sequence of identifiers with value types* $\mathtt{x}_0 : A_0, \ldots, \mathtt{x}_{m-1} : A_{m-1}$. *Sometimes we omit the identifiers and write* Γ *as a list of value types.*

The calculus has two kinds of judgement

$$\Gamma \vdash^{\mathsf{c}} M : \underline{B} \qquad\qquad \Gamma \vdash^{\mathsf{v}} V : A$$

for computations and values respectively. The terms are defined by Fig. 1. We include `let`, although it could be regarded as sugar. Note that $\prod$ is a projection product, whereas $\times$ is a pattern-match product. The key computations are

- `produce` V, the trivial producer of V;
- M `to x in` M', the sequenced computation (called "generalized `let`" by Filinski [Fil96]) where firstly the producer M happens, and if it produces a value V then M' happens with `x` bound to V.

Imperatively, V‘ means "push V" and λ`x` means "pop `x`"; and there are similar interpretations for π_{i_0} and $\langle \cdots \rangle$. This reading is illustrated in Sect. 2.3.

$$\frac{}{\Gamma, \mathbf{x} : A, \Gamma' \vdash^{\mathsf{v}} \mathbf{x} : A} \qquad \frac{\Gamma \vdash^{\mathsf{v}} V : A \quad \Gamma, \mathbf{x} : A \vdash^{\mathsf{c}} M : \underline{B}}{\Gamma \vdash^{\mathsf{c}} \mathtt{let}\ \mathbf{x}\ \mathtt{be}\ V\ \mathtt{in}\ M : \underline{B}}$$

$$\frac{\Gamma \vdash^{\mathsf{v}} V : A}{\Gamma \vdash^{\mathsf{c}} \mathtt{produce}\ V : FA} \qquad \frac{\Gamma \vdash^{\mathsf{c}} M : FA \quad \Gamma, \mathbf{x} : A \vdash^{\mathsf{c}} N : \underline{B}}{\Gamma \vdash^{\mathsf{c}} M\ \mathtt{to}\ \mathbf{x}\ \mathtt{in}\ N : \underline{B}}$$

$$\frac{\Gamma \vdash^{\mathsf{c}} M : \underline{B}}{\Gamma \vdash^{\mathsf{v}} \mathtt{thunk}\ M : U\underline{B}} \qquad \frac{\Gamma \vdash^{\mathsf{v}} V : U\underline{B}}{\Gamma \vdash^{\mathsf{c}} \mathtt{force}\ V : \underline{B}}$$

$$\frac{\Gamma \vdash^{\mathsf{v}} V : A_{i_0}}{\Gamma \vdash^{\mathsf{v}} (i_0, V) : \sum_{i \in I} A_i} \qquad \frac{\Gamma \vdash^{\mathsf{v}} V : \sum_{i \in I} A_i \quad \cdots \quad \Gamma, \mathbf{x} : A_i \vdash^{\mathsf{c}} M_i : \underline{B} \quad \cdots}{\Gamma \vdash^{\mathsf{c}} \mathtt{pm}\ V\ \mathtt{as}\ \ldots, (i, \mathbf{x})\ \mathtt{in}\ M_i, \ldots : \underline{B}}$$

$$\frac{\Gamma \vdash^{\mathsf{v}} V : A \quad \Gamma \vdash^{\mathsf{v}} V' : A'}{\Gamma \vdash^{\mathsf{v}} (V, V') : A \times A'} \qquad \frac{\Gamma \vdash^{\mathsf{v}} V : A \times A' \quad \Gamma, \mathbf{x} : A, \mathbf{y} : A' \vdash^{\mathsf{c}} M : \underline{B}}{\Gamma \vdash^{\mathsf{c}} \mathtt{pm}\ V\ \mathtt{as}\ (\mathbf{x}, \mathbf{y})\ \mathtt{in}\ M : \underline{B}}$$

$$\frac{\cdots \quad \Gamma \vdash^{\mathsf{c}} M_i : \underline{B}_i \quad \cdots}{\Gamma \vdash^{\mathsf{c}} \langle \ldots, M_i, \ldots \rangle : \prod_{i \in I} \underline{B}_i} \qquad \frac{\Gamma \vdash^{\mathsf{c}} M : \prod_{i \in I} \underline{B}_i}{\Gamma \vdash^{\mathsf{c}} \pi_{i_0} M : \underline{B}_{i_0}}$$

$$\frac{\Gamma, \mathbf{x} : A \vdash^{\mathsf{c}} M : \underline{B}}{\Gamma \vdash^{\mathsf{c}} \lambda \mathbf{x} M : A \to \underline{B}} \qquad \frac{\Gamma \vdash^{\mathsf{v}} V : A \quad \Gamma \vdash^{\mathsf{c}} M : A \to \underline{B}}{\Gamma \vdash^{\mathsf{c}} V`M : \underline{B}}$$

`pm` is an abbreviation for `pattern − match`.

$$\frac{M[V/\mathbf{x}] \Downarrow T}{\mathtt{let}\ \mathbf{x}\ \mathtt{be}\ V\ \mathtt{in}\ M \Downarrow T}$$

$$\frac{}{\mathtt{produce}\ V \Downarrow \mathtt{produce}\ V} \qquad \frac{M \Downarrow \mathtt{produce}\ V \quad N[V/\mathbf{x}] \Downarrow T}{M\ \mathtt{to}\ \mathbf{x}\ \mathtt{in}\ N \Downarrow T}$$

$$\frac{M \Downarrow T}{\mathtt{force}\ \mathtt{thunk}\ M \Downarrow T}$$

$$\frac{M_{i_0}[V/\mathbf{x}] \Downarrow T}{\mathtt{pm}\ (i_0, V)\ \mathtt{as}\ \ldots, (i, \mathbf{x})\ \mathtt{in}\ M_i, \ldots \Downarrow T}$$

$$\frac{M[V/\mathbf{x}, V'/\mathbf{y}] \Downarrow T}{\mathtt{pm}\ (V, V')\ \mathtt{as}\ (\mathbf{x}, \mathbf{y})\ \mathtt{in}\ M \Downarrow T}$$

$$\frac{}{\langle \ldots, M_i, \ldots \rangle \Downarrow \langle \ldots, M_i, \ldots \rangle} \qquad \frac{M \Downarrow \langle \ldots, N_i, \ldots \rangle \quad N_{i_0} \Downarrow T}{\pi_{i_0} M \Downarrow T}$$

$$\frac{}{\lambda \mathbf{x} M \Downarrow \lambda \mathbf{x} M} \qquad \frac{M \Downarrow \lambda \mathbf{x} N \quad N[V/\mathbf{x}] \Downarrow T}{V`M \Downarrow T}$$

Fig. 1. Terms of Basic Language, and Big-Step Semantics

Remark 1. The reader may wonder why we have not included *complex values* such as $\mathtt{x} : A \times A' \vdash^{\mathsf{v}} \mathbf{pm}\ \mathtt{x}\ \mathbf{as}\ (\mathtt{y}, \mathtt{z})\ \mathbf{in}\ \mathtt{y} : A$ or arithmetic expressions. The reason is that they somewhat complicate the operational semantics, our presentation of which exploits the fact that values do not need to be evaluated. Consequently, and since they lie outside the range of our translations from CBN and CBV, we omit them, except in the example program of Sect. 2.3. Nonetheless, all our denotational and categorical models can interpret them straightforwardly.

2.3 Example Computation

The following example M illustrates the naive imperative reading of CBPV. To this end, we add to the language arithmetic expressions as values (Remark 1) and the facility to prefix a `print` command to any computation.

```
print "hello0";
let x be 3 in
let y be thunk (
            print "hello1";
            λz
            print "we just popped "z;
            produce x + z
          ) in
print "hello2";
( print "hello3";
  7'
  print "we just pushed 7";
  force y
) to w in
print "w is bound to "w;
produce w + 5
```

Note that if the word **thunk** were omitted, M would be ill-typed, because y can identify only a value, not a computation. The type of y is $U(\mathtt{nat} \to F\,\mathtt{nat})$, because y identifies a thunk of a computation that pops a natural number and then produces a natural number.

M outputs as follows

```
hello0
hello2
hello3
we just pushed 7
hello1
we just popped 7
w is bound to 10
```

and finally produces the value 15.

It is clear that if the lines `print "hello1"` and $\lambda\mathtt{z}$ were exchanged, or if the lines `print "hello3"` and 7' were exchanged, the behaviour of M would be unchanged. We say that "effects commute with λ and with ' ". A more familiar example of this phenomenon is the equivalence of $\lambda\mathtt{x}$ **diverge** and **diverge**. (We are assuming here that, as in our example, the global computation is a producer, so there is no danger that we will try to pop from an empty stack.)

2.4 Big-Step Operational Semantics

Terminal computations (a subset of closed computations) are given by

$$T ::= \ \mathtt{produce}\ V \mid \langle \ldots, M_i, \ldots \rangle \mid \lambda\mathtt{x}M \qquad (3)$$

Intuitively these are computations that cannot proceed if the operand-stack is empty. We write $\mathbb{C}_{\underline{B}}$ for the set of closed computations of type $\underline{B}$, $\mathbb{T}_{\underline{B}}$ for the set of terminal elements of $\mathbb{C}_{\underline{B}}$, and $\mathbb{V}_A$ for the set of closed values of type A.

For the basic language, we define in Fig. 1 a relation $\Downarrow$ from $\mathbb{C}_{\underline{B}}$ to $\mathbb{T}_{\underline{B}}$. It can be proved to be a total function. Note that only computations happen; values do not need to be evaluated.

2.5 Equations and Observational Equivalence

We form an equational theory whose axioms are all substitution instances of the equations in Fig. 2. Compare this theory to those of CBN and CBV.

- In CBV, equations such as η for + types hold because *an identifier can be bound only to a value.*
- In CBN, equations such as η for $\rightarrow$ types hold because *a term of* $\rightarrow$ *type can be evaluated only by applying it.*

Since CBPV has both of these features, it has both kinds of equation, which is essentially why it can subsume both paradigms.

Definition 2. *A* ground context $C[]$ *is a closed ground producer with zero or more occurrences of a hole which can be either a computation or a value.*

Definition 3. *We say that* $M \simeq M'$ *when for all ground contexts* $C[]$, $C[M] \Downarrow \mathtt{produce}\ \underline{n}$ *iff* $C[M'] \Downarrow \mathtt{produce}\ \underline{n}$.

In all of our CBPV languages (e.g. in Sect. 5.1) the equations of the theory hold as observational equivalences (for the appropriate variation on Def. 3). As usual, this will follow from the soundness and adequacy of our models.

It is worth noticing that, with our imperative understanding of V' and $\lambda\mathtt{x}$, the β-law for $\rightarrow$ equates "push V, then pop $\mathtt{x}$, then M" with $M[V/\mathtt{x}]$. Similarly, the η-law for $\rightarrow$ equates M (in which $\mathtt{x}$ is not free) with "pop $\mathtt{x}$, then push $\mathtt{x}$, then M". These are both intuitively compelling.

$$
\begin{array}{llll}
\Gamma \vdash^{\mathsf{c}} \mathtt{let}\ \mathtt{x}\ \mathtt{be}\ V\ \mathtt{in}\ M & = M[V/\mathtt{x}] & : \underline{B} & \\
\Gamma \vdash^{\mathsf{c}} (\mathtt{produce}\ V)\ \mathtt{to}\ \mathtt{x}\ \mathtt{in}\ M & = M[V/\mathtt{x}] & : \underline{B} & \\
\Gamma \vdash^{\mathsf{c}} \mathtt{force}\ \mathtt{thunk}\ M & = M & : \underline{B} & \\
\Gamma \vdash^{\mathsf{c}} \mathtt{pm}\ (\hat{\imath}_0, V)\ \mathtt{as}\ \ldots, (\hat{\imath}, \mathtt{x})\ \mathtt{in}\ M_{\hat{\imath}}, \ldots & = M_{\hat{\imath}_0}[V/\mathtt{x}] & : \underline{B} & \\
\Gamma \vdash^{\mathsf{c}} \mathtt{pm}\ (V, V')\ \mathtt{as}\ (\mathtt{x}, \mathtt{y})\ \mathtt{in}\ M & = M[V/\mathtt{x}, V'/\mathtt{y}] & : \underline{B} & \\
\Gamma \vdash^{\mathsf{c}} \pi_{\hat{\imath}_0} \langle \ldots, M_{\hat{\imath}}, \ldots \rangle & = M_{\hat{\imath}_0} & : \underline{B}_{\hat{\imath}_0} & \\
\Gamma \vdash^{\mathsf{c}} V`\lambda\mathtt{x}M & = M[V/\mathtt{x}] & : \underline{B} & \\
\Gamma \vdash^{\mathsf{c}} M = M\ \mathtt{to}\ \mathtt{x}\ \mathtt{in}\ \mathtt{produce}\ \mathtt{x} & & : FA & (\mathtt{x} \notin \Gamma) \\
\Gamma \vdash^{\mathsf{v}} V = \mathtt{thunk}\ \mathtt{force}\ V & & : U\underline{B} & \\
\Gamma, \mathtt{z} : \sum_{\hat{\imath} \in I} A_{\hat{\imath}} \vdash^{\mathsf{c}} M = \mathtt{pm}\ \mathtt{z}\ \mathtt{as}\ \ldots, (\hat{\imath}, \mathtt{x})\ \mathtt{in}\ M[(\hat{\imath}, \mathtt{x})/\mathtt{z}], \ldots & & : \underline{B} & (\mathtt{x} \notin \Gamma) \\
\Gamma, \mathtt{z} : A \times A' \vdash^{\mathsf{c}} M = \mathtt{pm}\ \mathtt{z}\ \mathtt{as}\ (\mathtt{x}, \mathtt{y})\ \mathtt{in}\ M[(\mathtt{x}, \mathtt{y})/\mathtt{z}] & & : \underline{B} & (\mathtt{x}, \mathtt{y} \notin \Gamma) \\
\Gamma \vdash^{\mathsf{c}} M = \langle \ldots, \pi_{\hat{\imath}} M, \ldots \rangle & & : \prod_{\hat{\imath} \in I} \underline{B}_{\hat{\imath}} & \\
\Gamma \vdash^{\mathsf{c}} M = \lambda\mathtt{x}\ \mathtt{x}`M & & : A \to \underline{B} & (\mathtt{x} \notin \Gamma) \\
\Gamma \vdash^{\mathsf{c}} (M\ \mathtt{to}\ \mathtt{x}\ \mathtt{in}\ M')\ \mathtt{to}\ \mathtt{y}\ \mathtt{in}\ M'' = M\ \mathtt{to}\ \mathtt{x}\ \mathtt{in}\ (M'\ \mathtt{to}\ \mathtt{y}\ \mathtt{in}\ M'') & & : \underline{B} & (\mathtt{x}, \mathtt{y} \notin \Gamma) \\
\Gamma \vdash^{\mathsf{c}} \pi_{\hat{\imath}_0}(M\ \mathtt{to}\ \mathtt{x}\ \mathtt{in}\ M') = M\ \mathtt{to}\ \mathtt{x}\ \mathtt{in}\ \pi_{\hat{\imath}_0} M' & & : \underline{B}_{\hat{\imath}_0} & (\mathtt{x} \notin \Gamma) \\
\Gamma \vdash^{\mathsf{c}} V`(M\ \mathtt{to}\ \mathtt{x}\ \mathtt{in}\ M') = M\ \mathtt{to}\ \mathtt{x}\ \mathtt{in}\ V`M' & & : \underline{B} & (\mathtt{x} \notin \Gamma)
\end{array}
$$

Fig. 2. β-laws, η-laws and other laws

3 Divergence, Recursion and Scott Semantics

As divergence is the computational effect most familiar to semanticists, we study it first. We add to the basic language the computations

$$
\frac{}{\Gamma \vdash^{\mathsf{c}} \mathtt{diverge} : \underline{B}} \qquad \frac{\Gamma, \mathtt{x} : U\underline{B} \vdash^{\mathsf{c}} M : \underline{B}}{\Gamma \vdash^{\mathsf{c}} \mu\mathtt{x}M : \underline{B}}
$$

and the big-step rules

$$
\frac{\mathtt{diverge} \Downarrow T}{\mathtt{diverge} \Downarrow T} \qquad \frac{M[\mathtt{thunk}\ \mu\mathtt{x}M/\mathtt{x}] \Downarrow T}{\mu\mathtt{x}M \Downarrow T}
$$

so that $\Downarrow$ is now a partial function from $\mathbb{C}_{\underline{B}}$ to $\mathbb{T}_{\underline{B}}$. The recursion binder $\mu\mathtt{x}$ can be read imperatively as "bind-to-a-thunk-of-the-present-computation x", and therefore $\mu\mathtt{x}M$ is a computation.

The Scott semantics for CBPV interprets value types (and hence contexts) as predomains and computation types as domains. For example,

- $[\![FA]\!]$ is the lift of $[\![A]\!]$;
- if $[\![\underline{B}]\!]$ is the domain $(X, \leqslant, \bot)$ then $[\![U\underline{B}]\!]$ is its underlying predomain $(X, \leqslant)$;
- $[\![A \to \underline{B}]\!]$ is the domain of continuous functions from $[\![A]\!]$ to $[\![\underline{B}]\!]$

Then to each computation $\Gamma \vdash^{\mathsf{c}} M : \underline{B}$ we associate a continuous function $[\![M]\!] : [\![\Gamma]\!] \to [\![\underline{B}]\!]$, and to each value $\Gamma \vdash^{\mathsf{v}} V : A$ we associate a continuous function $[\![V]\!] : [\![\Gamma]\!] \to [\![A]\!]$. For example, where $\rho \in [\![\Gamma]\!]$,

$$
[\![\mathtt{produce}\ V]\!]\rho = \mathsf{lift}\ ([\![V]\!]\rho)
$$

$$[\![M \texttt{ to x in } N]\!]\rho = \begin{cases} \bot & \text{if } [\![M]\!]\rho = \bot \\ [\![N]\!](\rho, \texttt{x} \mapsto x) & \text{if } [\![M]\!]\rho = \mathsf{lift}\ x \end{cases}$$

$$[\![\texttt{thunk } M]\!]\rho = [\![M]\!]\rho$$

$$[\![\texttt{force } V]\!]\rho = [\![V]\!]\rho$$

In particular, $[\![\texttt{thunk diverge}]\!]\rho$ is the least element of the predomain $[\![U\underline{B}]\!]$.

Proposition 1 (Soundness/Adequacy). *For any closed computation M,*

1. *if $M \Downarrow T$, then $[\![M]\!] = [\![T]\!]$;*
2. *if $[\![M]\!] > \bot$, then $M \Downarrow T$ for some T.*

4 Translating CBN and CBV into CBPV

As we would expect from the Scott semantics of Sect. 3, CBN types translate into computation types, while CBV types translate into value types. The most important type decomposition into CBPV is

$$\underline{B} \to_{\mathsf{CBN}} \underline{B}' = (U\underline{B}) \to \underline{B}' \tag{4}$$

This corresponds to the fact that in CBN a function is effectively applied to a thunk. Perhaps it is because the interpretation of U and of **thunk** is almost invisible in CBPV Scott semantics that this decomposition has remained hidden for so long.

Another important type decomposition into CBPV is

$$A \to_{\mathsf{CBV}} A' = U(A \to FA') \tag{5}$$

This is similar, and in a sense equivalent, to Moggi's decomposition [Mog91] as $A \to TB$, but notice that (5) avoids the countability problem mentioned in Sect. 1.1. It says that a CBV function from A to A' is a thunk of a computation that pops a value of type A and then produces a value of type A'.

The translations into CBPV are given in Fig. 3 and Fig. 4. The source languages of these translations are prototypical CBN and CBV languages like PCF and PCF_{v}, with sum types. They are equipped with Scott semantics $[\![-]\!]_{\mathsf{CBN}}$ and $[\![-]\!]_{\mathsf{CBV}}$ (together with a semantics $[\![-]\!]^{\mathsf{val}}_{\mathsf{CBV}}$ for CBV values) and big-step semantics $\Downarrow_{\mathsf{CBN}}$ and $\Downarrow_{\mathsf{CBV}}$. We omit presenting them in detail. For simplicity, we have supplied a projection product for CBN but a pattern-match product for CBV; although in principle one could have both kinds of product in each paradigm.

Some of the technical results for the CBN translation concern not the *function* $-^{\mathsf{n}}$ (which does not commute with substitution) but a *relation* $\mapsto^{\mathsf{n}}$ from CBN to CBPV terms. Informally, $M \mapsto^{\mathsf{n}} M'$ means that M' is M^{n} with possibly some extra `force thunk` prefixes. The direct inductive definition of $\mapsto^{\mathsf{n}}$ is comprised of one rule for each CBN term-constructor, e.g.

$$\frac{}{\texttt{x} \mapsto^{\mathsf{n}} \texttt{force x}} \qquad \frac{N \mapsto^{\mathsf{n}} N' \quad M \mapsto^{\mathsf{n}} M'}{N`M \mapsto^{\mathsf{n}} (\texttt{thunk } N')`M'}$$

C	C^{n} (a computation type)
bool	$F\sum_{b\in\{\mathsf{true},\mathsf{false}\}}1$
$A \to B$	$UA^{\mathsf{n}} \to B^{\mathsf{n}}$
$A \times B$	$A^{\mathsf{n}}\ \Pi\ B^{\mathsf{n}}$
$A + B$	$F(UA^{\mathsf{n}} + UB^{\mathsf{n}})$

$A_0, \ldots, A_{m-1} \vdash M : C$	$UA_0{}^{\mathsf{n}}, \ldots, UA_{m-1}{}^{\mathsf{n}} \vdash^{\mathsf{c}} M^{\mathsf{n}} : C^{\mathsf{n}}$
$\mathtt{x}$	$\mathtt{force\ x}$
$\mathtt{false}$	$\mathtt{produce}\ \underline{\mathsf{false}}$
$\mathtt{if}\ M\ \mathtt{then}\ N\ \mathtt{else}\ N'$	$M^{\mathsf{n}}\ \mathtt{to\ z\ in\ pm\ z\ as}\ \underline{\mathsf{true}}\ \mathtt{in}\ N^{\mathsf{n}}, \underline{\mathsf{false}}\ \mathtt{in}\ N'^{\mathsf{n}}$
$\lambda \mathtt{x} M$	$\lambda \mathtt{x} M^{\mathsf{n}}$
$N`M$	$(\mathtt{thunk}\ N^{\mathsf{n}})`M^{\mathsf{n}}$
$\langle M, M' \rangle$	$\langle M^{\mathsf{n}}, M'^{\mathsf{n}} \rangle$
πM	πM^{n}
$\mathtt{inl}\ M$	$\mathtt{produce\ inl\ thunk}\ M^{\mathsf{n}}$
$\mathtt{pm}\ M\ \mathtt{as\ inl\ x\ in}\ N, \mathtt{inr\ x\ in}\ N'$	$M\ \mathtt{to\ z\ in\ pm\ z\ as\ inl\ x\ in}\ N^{\mathsf{n}}, \mathtt{inr\ x\ in}\ N'^{\mathsf{n}}$
$\mu \mathtt{x} M$	$\mu \mathtt{x} M^{\mathsf{n}}$

Fig. 3. Translation of CBN types and terms

C	C^{v} (a value type)
bool	$\sum_{b\in\{\mathsf{true},\mathsf{false}\}}1$
$A \to B$	$U(A^{\mathsf{v}} \to FB^{\mathsf{v}})$
$A \times B$	$A^{\mathsf{v}} \times B^{\mathsf{v}}$
$A + B$	$A^{\mathsf{v}} + B^{\mathsf{v}}$

$A_0, \ldots, A_{m-1} \vdash V : C$	$A_0{}^{\mathsf{v}}, \ldots, A_{m-1}{}^{\mathsf{v}} \vdash^{\mathsf{v}} V^{\mathsf{val}} : C^{\mathsf{v}}$
$\mathtt{x}$	$\mathtt{x}$
$\mathtt{false}$	$\underline{\mathsf{false}}$
$\lambda \mathtt{x} M$	$\mathtt{thunk}\ \lambda \mathtt{x} M^{\mathsf{v}}$
$\mu \mathtt{y} \lambda \mathtt{x} M$	$\mathtt{thunk}\ \mu \mathtt{y} \lambda \mathtt{x} M^{\mathsf{v}}$
(V, V')	$(V^{\mathsf{val}}, V'^{\mathsf{val}})$
$\mathtt{inl}\ V$	$\mathtt{inl}\ V^{\mathsf{val}}$

$A_0, \ldots, A_{m-1} \vdash M : C$	$A_0{}^{\mathsf{v}}, \ldots, A_{m-1}{}^{\mathsf{v}} \vdash^{\mathsf{c}} M^{\mathsf{v}} : FC^{\mathsf{v}}$
V (a value)	$\mathtt{produce}\ V^{\mathsf{val}}$
$\mathtt{if}\ M\ \mathtt{then}\ N\ \mathtt{else}\ N'$	$M^{\mathsf{v}}\ \mathtt{to\ z\ in\ pm\ z\ as}\ \underline{\mathsf{true}}\ \mathtt{in}\ N^{\mathsf{n}}, \underline{\mathsf{false}}\ \mathtt{in}\ N'^{\mathsf{n}}$
MN (M first)	$M^{\mathsf{v}}\ \mathtt{to\ f\ in}\ N^{\mathsf{v}}\ \mathtt{to\ x\ in\ x`(force\ f)}$
$\mathtt{pm}\ M\ \mathtt{as\ (x,y)\ in}\ N$	$M^{\mathsf{v}}\ \mathtt{to\ z\ in\ pm\ z\ as\ (x,y)\ in}\ N'^{\mathsf{v}}$
$\mathtt{pm}\ M\ \mathtt{as\ inl\ x\ in}\ N, \mathtt{inr\ x\ in}\ N'$	$M^{\mathsf{v}}\ \mathtt{to\ z\ in\ pm\ z\ as\ inl\ x\ in}\ N^{\mathsf{v}}, \mathtt{inr\ y\ in}\ N'^{\mathsf{v}}$

Fig. 4. Translation of CBV types, values and terms

and the additional rule $$\frac{M \mapsto^{\mathsf{n}} M'}{M \mapsto^{\mathsf{n}} \texttt{force thunk } M'}$$

Proposition 2. *1.* $(M[V/\mathtt{x}])^{\mathsf{v}} = M^{\mathsf{v}}[V^{\mathsf{val}}/\mathtt{x}]$
2. If $M \mapsto^{\mathsf{n}} M'$ *and* $N \mapsto^{\mathsf{n}} N'$ *then* $M[N/\mathtt{x}] \mapsto^{\mathsf{n}} M'[\texttt{thunk } N'/\mathtt{x}]$

We are now in a position to describe the fundamental subsumption properties: that the Scott and big-step semantics of CBN and CBV can be recovered from those of CBPV.

The preservation of the Scott semantics is straightforward:

Proposition 3. *1. If* A *is a CBN type then* $[\![A]\!]_{\mathsf{CBN}} = [\![A^{\mathsf{n}}]\!]$
2. If $\Gamma \vdash M : A$ *is a CBN term and* $M \mapsto^{\mathsf{n}} M'$ *then* $[\![M]\!]_{\mathsf{CBN}} = [\![M']\!]$
3. If A *is a CBV type then* $[\![A]\!]_{\mathsf{CBV}} = [\![A^{\mathsf{v}}]\!]$
4. If $\Gamma \vdash V : A$ *is a CBV value then* $[\![V]\!]^{\mathsf{val}}_{\mathsf{CBV}} = [\![V^{\mathsf{val}}]\!]$
5. If $\Gamma \vdash M : A$ *is a CBV term then* $[\![M]\!]_{\mathsf{CBV}} = [\![M^{\mathsf{v}}]\!]$

That the equations of CBN/CBV are preserved follows from Prop. 2.

Proposition 4. *Suppose* M *is a closed CBN term, and* $M \mapsto^{\mathsf{n}} M'$.

1. *If* M' *is terminal then* M *is, and* M *is terminal iff* M^{n} *is.*
2. *If* $M \Downarrow_{\mathsf{CBN}} T$ *then, for some* T', $T \mapsto^{\mathsf{n}} T'$ *and* $M' \Downarrow T'$.
3. *If* $M^{\mathsf{n}} \Downarrow T'$, *then, for some* T, $T \mapsto^{\mathsf{n}} T'$ *and* $M \Downarrow_{\mathsf{CBN}} T$.

Proposition 5. *Suppose* M *is a closed CBV term.*

1. M *is terminal iff* M^{v} *is terminal.*
2. *If* $M \Downarrow_{\mathsf{CBV}} T$ *then* $M^{\mathsf{v}} \Downarrow T^{\mathsf{v}}$.
3. *If* $M^{\mathsf{v}} \Downarrow T'$, *then, for some* T, $T^{\mathsf{v}} = T'$, *and* $M \Downarrow_{\mathsf{CBV}} T$.

Parts (2) and (3) of these are proved by induction, primarily on the big-step derivation, and secondarily on $\mapsto^{\mathsf{n}}$ (for Prop. 4) or M (for Prop. 5).

5 Operational Semantics for Computational Effects

It is straightforward to adapt the big-step semantics of Sect. 2.4 to various computational effects (except for control effects, which require *machine semantics*, where the search for a redex is made explicit). We give two examples: global store and nondeterminism.

5.1 Global Groundtype Store

We will consider a single global storage cell $\mathtt{X}$ that stores a value of groundtype $\sum_{s \in S} 1$. We add to the basic langugage the computations

$$\frac{}{\Gamma \vdash^{\mathsf{c}} \texttt{deref X} : F\sum_{s\in S} 1} \qquad \frac{\Gamma \vdash^{\mathsf{v}} V : \sum_{s\in S} 1 \quad \Gamma \vdash^{\mathsf{c}} M : \underline{B}}{\Gamma \vdash^{\mathsf{c}} X := V; M : \underline{B}}$$

While it is possible to give type $F1$ to commands such as assignment and output, here we regard them as prefixes.

We define a relation $\Downarrow$ from $S \times \mathbb{C}_{\underline{B}}$ to $S \times \mathbb{T}_{\underline{B}}$, adapting the rules of Sect. 2.4 and adding rules for the new constructs. For example:

$$\frac{}{s, T \Downarrow s, T} \qquad \frac{s, M \Downarrow s', \lambda \mathtt{x} N \quad s', N[V/\mathtt{x}] \Downarrow s'', T}{s, V`M \Downarrow s'', T}$$

$$\frac{}{s, \mathtt{deref}\ \mathtt{X} \Downarrow s, \mathtt{produce}\ \underline{s}} \qquad \frac{s', M \Downarrow s'', T}{s, \mathtt{X} := \underline{s'}; M \Downarrow s'', T}$$

$\Downarrow$ can be proved to be a total function.

Finally, we say that $M \simeq M'$ when for all ground contexts $\mathcal{C}[]$ and $s, s' \in S$, $s, \mathcal{C}[M] \Downarrow s', \mathtt{produce}\ \underline{n}$ iff $s, \mathcal{C}[M'] \Downarrow s', \mathtt{produce}\ \underline{n}$.

5.2 Nondeterminism

We add to the basic language the divergence and recursion facilities of Sect. 3 together with the following term and big-step rule:

$$\frac{\Gamma, \mathtt{x} : A \vdash^{\mathsf{c}} M : \underline{B}}{\Gamma \vdash^{\mathsf{c}} \mathtt{choose}\ \mathtt{x}\ M : \underline{B}} \qquad \frac{M[V/\mathtt{x}] \Downarrow T}{\mathtt{choose}\ \mathtt{x}\ M \Downarrow T}$$

6 Denotational Semantics for Computational Effects

We describe denotational semantics for the effects of Sect. 5. Part is easy: a value type (or a context) should denote a set, with $\times$ and $\sum$ interpreted in the usual way, and a value $\Gamma \vdash^{\mathsf{v}} V : A$ should denote a function $[\![V]\!] : [\![\Gamma]\!] \longrightarrow [\![A]\!]$.

The remainder of the semantics differs between the effects. While logically we should present the various semantics first, and then state the soundness results, this makes the interpretation of type constructors appear ad hoc. So we will proceed in reverse order. For global store and nondeterminism, we will state first the soundness and adequacy theorems that we are aiming to achieve, even though they are not yet meaningful, and use this to motivate the semantics. We will also give continuation semantics for the basic language. (Using machine semantics, this can be similarly motivated.)

Proposition 6 (Soundness/Adequacy). *Let M be a closed computation.*

1. *For global store, if $s, M \Downarrow s', T$ then $[\![M]\!]s = [\![T]\!]s'$.*
2. *For nondeterminism, $[\![M]\!] = \bigcup_{M \Downarrow T} [\![T]\!]$.*

By looking at Prop. 6, we can guess the interpretation of a computation $\Gamma \vdash^{\mathsf{c}} M : \underline{B}$. (Recall that if $\underline{B} = FA$ then this judgement corresponds to a CBV term of type A, so its interpretation is familiar.)

- For global store, $[\![M]\!]$ will be a function from $S \times [\![\Gamma]\!]$ to $[\![\underline{B}]\!]$, where $[\![\underline{B}]\!]$ is a set. If $\underline{B} = FA$ then $[\![\underline{B}]\!] = S \times [\![A]\!]$, so that $[\![M]\!]$ is a function from $S \times [\![\Gamma]\!]$ to $S \times [\![A]\!]$.
- For nondeterminism, $[\![M]\!]$ will be a relation from $[\![\Gamma]\!]$ to $[\![\underline{B}]\!]$, where $[\![\underline{B}]\!]$ is a set. If $\underline{B} = FA$, then $[\![\underline{B}]\!] = [\![A]\!]$, so that $[\![M]\!]$ is a relation from $[\![\Gamma]\!]$ to $[\![A]\!]$.
- For continuation semantics, $[\![M]\!]$ will be a function from $[\![\Gamma]\!] \times [\![\underline{B}]\!]$ to Ans (a fixed set that we regard as the set of "answers"), where $[\![\underline{B}]\!]$ is a set. If $\underline{B} = FA$, then $[\![\underline{B}]\!] = [\![A]\!] \to \mathsf{Ans}$, so that $[\![M]\!]$ is a function from $[\![\Gamma]\!] \times ([\![A]\!] \to \mathsf{Ans})$ to Ans.

We next turn our attention to the interpretation of U, $\prod_{i \in I}$ and $\to$. For U, we know that values $\Gamma \vdash^{\mathsf{v}} U\underline{B}$ correspond to computations $\Gamma \vdash^{\mathsf{c}} \underline{B}$. Thus, in the case of global store, functions from $[\![\Gamma]\!]$ to $[\![U\underline{B}]\!]$ must correspond to functions from $S \times [\![\Gamma]\!]$ to $[\![\underline{B}]\!]$. Therefore we set $[\![U\underline{B}]\!] = S \to [\![\underline{B}]\!]$. Similarly we can determine the interpretation of U for each effect. As expected, it follows in each case that UFA denotes the same set as Moggi's type TA [Mog91]:

effect	U	F	$T = UF$
global store	$S \to -$	$S \times -$	$S \to (S \times -)$
nondeterminism	$\mathcal{P}$	$-$	$\mathcal{P}$
control	$- \to \mathsf{Ans}$	$- \to \mathsf{Ans}$	$(- \to \mathsf{Ans}) \to \mathsf{Ans}$

For $A \to \underline{B}$, we know that computations $\Gamma \vdash^{\mathsf{c}} A \to \underline{B}$ correspond to computations $\Gamma, A \vdash^{\mathsf{c}} \underline{B}$. Thus, in the case of nondeterminism, relations from $[\![\Gamma]\!]$ to $[\![A \to \underline{B}]\!]$ must correspond to relations from $[\![\Gamma]\!] \times [\![A]\!]$ to $[\![\underline{B}]\!]$. Therefore we set $[\![A \to \underline{B}]\!]$ to be $[\![A]\!] \times [\![\underline{B}]\!]$. Similar reasoning suggests interpretations for both $\to$ and $\prod_{i \in I}$ for each of our effects:

effect	$\prod_{i \in I}$	$\to$
global store	$\prod_{i \in I}$	$\to$
nondeterminism	$\sum_{i \in I}$	$\times$
control	$\sum_{i \in I}$	$\times$

We omit the straightforward semantics of terms.

Proposition 7. *These five denotational semantics for CBPV all validate the equations of Sect. 2.5. More precisely, if $M = M'$ is provable in the equational theory then $[\![M]\!] = [\![M']\!]$.*

Prop. 6 is now meaningful and can be proved. In particular, (1) is trivial.

All these models induce models for CBN and CBV. For CBV we recover the familiar continuation semantics of $A \to_{\mathsf{CBV}} A'$ as $(A \times (A' \to \mathsf{Ans})) \to \mathsf{Ans}$. For CBN we recover the continuation semantics of [SR96], and also, from our CBPV global store semantics, the state-passing semantics of [O'H93].

References

[AHM98] S. Abramsky, K. Honda, and G. McCusker. A fully abstract game semantics for general references. Proceedings, Thirteenth Annual IEEE Symposium on Logic in Computer Science, IEEE Computer Society Press, 1998.

[AM98] S. Abramsky and G. McCusker. Call-by-value games. In M. Nielsen and W. Thomas, editors, *Computer Science Logic: 11th International Workshop Proceedings*, Lecture Notes in Computer Science. Springer-Verlag, 1998.

[BW96] N. Benton and P. Wadler. Linear logic, monads and the lambda calculus. In *Proceedings, 11th Annual IEEE Symposium on Logic in Computer Science*, pages 420–431, New Brunswick, 1996. IEEE Computer Society Press.

[Fil96] A. Filinski. *Controlling Effects*. PhD thesis, School of Computer Science, Carnegie Mellon University, Pittsburgh, Pennsylvania, 1996.

[Gir87] J.-Y. Girard. Linear logic. *Theoretical Computer Science*, 50:1–102, 1987.

[HD97] J. Hatcliff and O. Danvy. Thunks and the λ-calculus. *Journal of Functional Programming*, 7(3):303–319, May 1997.

[HO94] M. Hyland and L. Ong. On full abstraction for PCF. submitted, 1994.

[JLST98] S. Peyton Jones, J. Launchbury, M. Shields, and A. Tolmach. Bridging the gulf: A common intermediate language for ML and Haskell. In *Proc. 25th ACM Symposium on Principles of Programming Languages*, San Diego, 1998.

[Jon92] S. L. Peyton Jones. Implementing lazy functional languages on stock hardware: the spineless tagless G-machine. *Journal of Functional Programming*, 2(2):127–202, July 1992.

[Lev98] P. B. Levy. Categorical aspects of call-by-push-value. draft, available at `http://www.dcs.qmw.ac.uk/ pbl/papers.html`, 1998.

[Mar98] M. Marz. A fully abstract model for sequential computation. draft, 1998.

[MC88] A. Meyer and S. Cosmodakis. Semantical Paradigms. In *Proc. Third Annual Symposium on Logic in Computer Science*. Computer Society Press, 1988.

[McC96] G. McCusker. *Games and Full Abstraction for a Functional Metalanguage with Recursive Types*. PhD thesis, University of London, 1996.

[Mog91] E. Moggi. Notions of computation and monads. *Information and Computation*, 93:55–92, 1991.

[O'H93] P. W. O'Hearn. Opaque types in algol-like languages. manuscript, 1993.

[Plo76] G. D. Plotkin. Call-by-name, call-by-value and the λ-calculus. *Theoretical Computer Science*, 1(1):125–159, 1976.

[Plo77] G. D. Plotkin. LCF as a programming language. *Theoretical Computer Science*, 5, 1977.

[Plo85] G. D. Plotkin. Lectures on predomains and partial functions. Course notes, Center for the Study of Language and Information, Stanford, 1985.

[PR97] A. J. Power and E. P. Robinson. Premonoidal categories and notions of computation. *Math. Struct. in Comp. Sci.*, 7(5):453–468, October 1997.

[Rey98] J. Reynolds. Where theory and practice meet: POPL past and future. Invited Lecture, 25th ACM SIGPLAN-SIGACT Symposium on Principles of Programming Languages, San Diego, California, January 19–21, 1998.

[SJ98] M. Shields and S. Peyton Jones. Bridging the gulf better. Draft, 1998.

[SR96] Th. Streicher and B. Reus. Continuation semantics, abstract machines and control operators. submitted to Journal of Functional Programming, 1996.

A Study of Abramsky's Linear Chemical Abstract Machine

Seikoh Mikami[1] and Yohji Akama[2]

[1] Department of Information Science, Tokyo University, Hongoh, 7-3-1, Tokyo, Japan, 113-0033
[2] Mathematical Institute, Tohoku University, Sendai Miyagi, JAPAN, 980-8578
akama@math.tohoku.ac.jp

Abstract. Abramsky's Linear Chemical Abstract Machine (LCHAM) is a term calculus which corresponds to Linear Logic, via the *Curry-Howard isomorphism*. We introduce a translation from a linear λ-calculus into LCHAM. The translation result can be well regarded as a black box with the i/o ports being atomic. We show that one step computation of LCHAM is equivalent to that of the linear λ-calculus. Then, we prove the *principal typing theorem* of LCHAM, which implies the decidability of type checking.

1 Introduction

There are attempts to regard concurrent computations as chemical reactions. Chemical Abstract Machine (CHAM) [5] is a model of concurrent computation in this line. CHAM influenced on various concurrent calculi such as π-calculus, ambient calculus [6] and join calculus [8].

The points of CHAM is the following:

- once a multiset of objects is applied by a rewriting rule, then the multiset will be consumed and will be transformed to a multiset of objects (in chemistry, a solution of molecules will changes according to chemical reaction laws). In fact, CHAM is resource-sensitive, like Linear Logic (LL).
- a multiset of objects is again an object (in chemistry, a solution encapsulated by a membrane often acts like a molecule). Inside the multiset, computations go through independently. This mechanism may enable us to describe computations inside a sub-network and/or dynamic structuring of networks. The 'membrane' plays an important role in mobile calculi such as ambient calculus and join calculus. CHAM's encapsulation mechanism of computation reminds us of the boxing operation of *proof net* (Girard [9]).

So, we are concerned with Linear Chemical Abstract Machine (Abramsky [1]), which corresponds to LL through *Curry-Howard isomorphism*. Linear Chemical Abstract Machine (LCHAM) consists of not only rewriting rules but also typing rules.

To investigate computational properties of LCHAM, we introduce a translation from a linear λ-calculus into LCHAM. A linear λ-calculus is a resource-sensitive refinement of λ-calculus. It is employed for analyzing functional programming languages with respect to evaluation strategy [12],[4] and/or resource allocation [7]. We are concerned with a linear λ-calculus which is introduced by Bierman [3], and we translate the terms into proof nets. Then we prove that one step reduction in the linear λ-calculus corresponds to one step reduction in LCHAM modulo a bisimulation.

To investigate type-theoretic properties of LL, we prove the principal typing theorem of LCHAM. The principal typing theorem is an indispensable theorem for implementing a functional language that has a polymorphic type-inference system, such as a programming language ML.

Related Work There are various versions of linear λ-calculi. Abramsky introduced a call-by-value linear λ-calculus [1], Chirimar-Gunter-Riecke introduced a linear λ-calculi with a fix point operator for non-linear function [7].

The linear λ-calculus of this paper was introduced in Bierman et al [3]. Their calculus does not suffer from the coherence problem. Furthermore it has a stable notion of commuting conversions. The commuting-convertible linear λ-terms is translated by our translation into the same proof net. Under the presence of the fix point operator, we don't know how to define the commuting conversion, and how the commuting conversion is related to the structure of proof net.

We introduce a translation from the linear λ-calculus into proof nets, and the translation satisfies the following property: The resulting proof nets can be well regarded as a black box with the i/o ports being 'atomic'. So, such black boxes can be easily connected through their ports. It is not the case in most translation of their multiplicative λ-calculus into proof nets (Bellin-Scott [2], Mackie [10], etc.)

Mackie [11] proved the principal typing theorem of Abramsky's linear λ-calculus. We prove the same theorem for LCHAM in this paper. In proving the principal typing theorems, the reconstruction algorithm of a derivation of a given typing assertion is essential. In the case of linear λ-calculus, the reconstruction algorithm will be deterministic. The type assertions are two-sided sequents $\Gamma \vdash t : A$, and we can only decompose t on their antecedents in reconstructing the derivations.

However, in the case of LCHAM, the reconstruction algorithm will be non-deterministic. Because the type assertions are one-sided sequents like $\vdash t_1 : A_1, \ldots, t_n : A_n$, the reconstruction algorithm choose non-deterministically t_i to decompose. Furthermore, some type-inference rules of LCHAM is another source of non-determinism. So, the existence proof of principal type is not trivial.

Organization In the next section, we review LCHAM [1], a rewriting system for a *proof expression*, which is a representation of a proof of LL. In Section 3, we review the *linear λ-calculus* (Benton et al [3]). We introduce a translation from the linear λ-calculus to LCHAM, and show that one step β-reduction in the linear λ-calculus 'roughly' corresponds to one step reaction rules in LCHAM. In Section 4, we prove the principal typing theorem of LCHAM. To prove this theorem, we introduce *locally correct assertions*, which correspond to proof structures in

LL [9], (ordinary type assertions correspond to proof nets of LL). The complete proofs in this paper can be found in [13].

2 Linear Chemical Abstract Machine

We begin by reviewing Linear Chemical Abstract Machine (LCHAM) [1], a rewriting system representing the cut-elimination procedure of a proof in LL.

*A proof expression (*PEXP*)* is an object to rewrite in LCHAM. *Proof expressions* are defined together with *terms* and *coequations* as follows. Letters $P, Q, \ldots$ stand for PEXPs, $t, u, \ldots$ for terms, $x, y, z, \ldots$ for *names*, and $\bar{x}, \bar{y}, \bar{z}, \ldots$ for lists of names. *Terms* are defined as

$$t ::= x \mid * \mid \odot \mid t_1 \otimes t_2 \mid t_1 \mathbin{⅋} t_2 \mid \mathsf{inl}(t) \mid \mathsf{inr}(t) \mid \bar{x}(P \parallel Q) \mid ?t \mid \_ \mid t_1 \mathbin{@} t_2 \mid \bar{x}(P).$$

We call a term of the form $\bar{x}(P)$ or $\bar{x}(P \parallel Q)$ a *closure* and $\bar{x}$ of $\bar{x}(\cdots)$ *binding names*. *Coequations* have the form $t \perp u$, where t and u are terms. Proof expressions have the form $\Theta; \bar{t}$, where Θ is a finite sequence of coequations and $\bar{t}$ is a finite sequence of terms.

2.1 Type inference

Types, ranged over by $A, B, C, \ldots$, are exactly the formulas of LL. For every formula A, its *linear negation* is denoted by $A^{\perp}$. We sometimes write $\bar{t} : \Gamma$ for $t_1 : A_1, \ldots, t_n : A_n$ where $\bar{t} = t_1, \ldots, t_n$ and $\Gamma = A_1, \ldots, A_n$. *Different names are introduced for each instance of the Axiom, With and OfCourse rules.*

$$\frac{\vdash \Theta; \Gamma, u : A, t : B, \Delta}{\vdash \Theta; \Gamma, t : A, u : B, \Delta}\ \text{Exchange} \qquad \frac{}{\vdash ;\ x : A^{\perp}, x : A}\ \text{Axiom}$$

$$\frac{\vdash \Theta;\ \Gamma, t : A \quad \vdash \Xi;\ \Delta, u : A^{\perp}}{\vdash \Theta,\ \Xi,\ t \perp u;\ \Gamma,\ \Delta}\ \text{Cut}$$

$$\frac{}{\vdash ;\ * : \mathbf{1}}\ \text{One} \qquad \frac{\vdash \Theta;\ \Gamma}{\vdash \Theta;\ \Gamma, \odot : \perp}\ \text{Bot}$$

$$\frac{\vdash \Theta;\ \Gamma, t : A \quad \vdash \Xi;\ \Delta, u : B}{\vdash \Theta, \Xi;\ \Gamma,\ \Delta,\ t \otimes u : A \otimes B}\ \text{Times} \qquad \frac{\vdash \Theta;\ \Gamma, t : A, u : B}{\vdash \Theta;\ \Gamma, t \mathbin{⅋} u : A \mathbin{⅋} B}\ \text{Par}$$

$$\frac{\vdash \Theta;\ \bar{t} : \Gamma, t : A \quad \vdash \Xi;\ \bar{u} : \Gamma, u : B}{\vdash ;\ \bar{x} : \Gamma,\ \bar{x}(\Theta; \bar{t}, t \parallel \Xi; \bar{u}, u) : A \,\&\, B}\ \text{With}$$

$$\frac{\vdash \Theta;\ \Gamma, t : A}{\vdash \Theta;\ \Gamma,\ \mathsf{inl}(t) : A \oplus B}\ \text{Plus-1} \qquad \frac{\vdash \Theta;\ \Gamma, t : B}{\vdash \Theta;\ \Gamma,\ \mathsf{inr}(t) : A \oplus B}\ \text{Plus-2}$$

$$\frac{\vdash \Theta;\ \Gamma}{\vdash \Theta;\ \Gamma,\ \_ : ?A}\ \text{Weakening} \qquad \frac{\vdash \Theta;\ \bar{t} : ?\Gamma, t : A}{\vdash ;\ \bar{x} : ?\Gamma,\ \bar{x}(\Theta; \bar{t}, t) : !A}\ \text{OfCourse}$$

$$\frac{\vdash \Theta;\ \Gamma, t : ?A, u : ?A}{\vdash \Theta;\ \Gamma,\ t \mathbin{@} u : ?A}\ \text{Contraction} \qquad \frac{\vdash \Theta;\ \Gamma, t : A}{\vdash \Theta;\ \Gamma,\ ?t : ?A}\ \text{Dereliction}$$

Remark 1. Note that we obtain the rules of LL from the type inference rules by ignoring PEXPs. We can say that $\vdash \Theta;\ \bar{t} : \Gamma$ corresponds to a proof net [9] such that

- the lowest nodes are Γ,
- the Cut-links are represented by Θ, and
- the closures are represented by the boxes.

For example, $\vdash\ \bar{x}(P \parallel Q) \perp \mathsf{inl}(y);\ \bar{x} : \Gamma, y : A$ represents the following proof net.

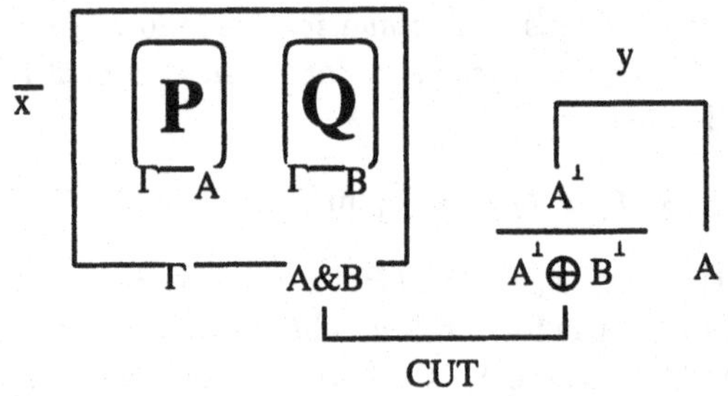

2.2 Reductions

Our discussion is limited to linear PEXPs, which we define slightly different from the ones in Abramsky [1]. In our definition, we consider occurrences of names in PEXP only outside closures and not ones in PEXPs inside closures. We consider binding names to be outside the closure.

Definition 1. *A* PEXP $\Theta;\bar{t}$ *is* linear *if and only if*

- *Each name occurring in* $\Theta;\bar{t}$, *does so exactly* twice;
- *If a closure* $\bar{x}(\cdots)$ *occurs in* $\Theta;\bar{t}$, *then none of the other occurrences of* $\bar{x}$ *are binding names; and*
- *Each* PEXP *inside a closure is linear.*

We say a PEXP $\Theta;\bar{t}$ is *typable* if and only if $\vdash \Theta;\bar{t} : \Gamma$ is derivable for some Γ. We note that every typable PEXP is linear. Intuitively, the linearity condition of a PEXP means that the PEXP can represent a skeleton of some proof structure [9].

Rewriting rules in LCHAM are classified into *reaction rules* and a *cleanup rule*. The reduction relation determined by the reaction rules is written as $\rightarrow_r$. The reduction relation determined by the cleanup rule is written as $\rightarrow_c$. The reaction rule rewrites only the 'coequations part' of a PEXP.

We regard Θ of a PEXP $\Theta;\bar{t}$ as a *multiset* of coequations, and identify coequations $t \perp u$ and $u \perp t$. We write $\Theta \equiv \Theta'$ if Θ and Θ' are equal in the sense described above. (This corresponds to the "structural rules" in Abramsky [1].) Hereafter, we simply identify Θ and Θ' if $\Theta \equiv \Theta'$.

The cleanup rule represents a contraction of a Cut-link involving an Axiom-link.

Cleanup rule.

$$\Theta,\ x \perp t;\ \bar{u} \rightarrow_c \Theta;\ \bar{u}[t/x]$$

where x is outside of closures and not a binding variable.

Reaction rule.

Communication $\quad t \perp x,\ x \perp u \rightarrow_r t \perp u$

Unit $\quad * \perp \odot \rightarrow_r$

Pair $\quad t \otimes u \perp t' \parr u' \rightarrow_r t \perp t',\ u \perp u'$

Case Left† $\quad \bar{x}(\Theta; \bar{t}, t \parallel \Xi; \bar{u}, u) \perp \mathsf{inl}(v) \rightarrow_r \Theta,\ \bar{x} \perp \bar{t},\ t \perp v$

Case Right $\quad \bar{x}(\Theta; \bar{t}, t \parallel \Xi; \bar{u}, u) \perp \mathsf{inr}(v) \rightarrow_r \Xi,\ \bar{x} \perp \bar{u},\ u \perp v$

Read $\quad \bar{x}(\Theta; \bar{t}, t) \perp ?u \rightarrow_r \Theta,\ \bar{x} \perp \bar{t},\ t \perp u$

Discard $\quad \bar{x}(P) \perp \_ \rightarrow_r x_1 \perp \_,\ \ldots,\ x_n \perp \_$

Copy‡ $\quad \bar{x}(P) \perp u \mathbin{@} v \rightarrow_r \bar{x} \perp (\bar{x}^l \mathbin{@} \bar{x}^r),\ \bar{x}(P)^l \perp u,\ \bar{x}(P)^r \perp v$

(†) $\bar{x} \perp \bar{t}$ denotes $x_1 \perp t_1,\ \ldots,\ x_n \perp t_n$ if $\bar{x} = x_1, \ldots, x_n$ and $\bar{t} = t_1, \ldots, t_n$.
(‡) $\bar{x}^l$ denotes a list of new names $x_1^l, \ldots, x_n^l$ if $\bar{x} = x_1, \ldots, x_n$, and $\bar{x}(P)^l$ denotes a term where small l's are attached to all names in $\bar{x}(P)$. $\bar{x}^r$ and $\bar{x}(P)^r$ are defined in the same way.

3 Translation From Linear λ-calculus to LCHAM

This section begins with a review of a linear λ-calculus which was introduced by Benton et al[3].

3.1 The Linear λ-calculus

We only consider the $(\multimap, \otimes, !)$-fragment of *intuitionistic linear logic* (ILL).

Types are either a type variable, $A_1 \otimes A_2$, $!A$, or a *linear implication* $A_1 \multimap A_2$.

Pre-linear λ-terms, ranged over by $t, u, \ldots$, are defined as:

$$\begin{aligned} t ::= {} & x \mid t_1 t_2 \mid \lambda x.t \mid t_1 \otimes t_2 \mid \mathsf{let}\ t_1\ \mathsf{be}\ x \otimes y\ \mathsf{in}\ t_2 \\ & \mid \mathsf{promote}\ t_1, \ldots, t_n\ \mathsf{for}\ x_1, \ldots, x_n\ \mathsf{in}\ u \mid \mathsf{derelict}(t) \\ & \mid \mathsf{discard}\ t_1\ \mathsf{in}\ t_2 \mid \mathsf{copy}\ t_1\ \mathsf{as}\ x, y\ \mathsf{in}\ t_2. \end{aligned}$$

Here, *bound* occurrence of variables are either (1) occurrences of x in $(\lambda x. \ldots)$, (2) occurrences of x or y in (let t_1 be $x \otimes y$ in ...) or (copy t_1 as x, y in ...), or (3) occurrences of $x_1, \ldots, x_n$ in (promote $t_1, \ldots, t_n$ for $x_1, \ldots, x_n$ in ...). An occurrence of a variable is called *free* if it is not bound. *A linear λ-term* is a pre-linear λ-term t such that each variable occurring free in t does so exactly *once*.

Type inference rules.

$$\frac{}{x : A \vdash x : A}\ \mathrm{Id}$$

$$\frac{\Gamma, x : A \vdash t : B}{\Gamma \vdash \lambda x.t : A \multimap B}\ {\multimap}\mathrm{I} \qquad \frac{\Gamma \vdash t : A \multimap B \quad \Delta \vdash u : A}{\Gamma, \Delta \vdash tu : B}\ {\multimap}\mathrm{E}$$

$$\frac{\Gamma \vdash t : A \quad \Delta \vdash u : B}{\Gamma, \Delta \vdash t \otimes u : A \otimes B}\ \otimes\mathrm{I} \qquad \frac{\Gamma \vdash t : A \otimes B \quad \Delta, x : A, y : B \vdash u : C}{\Gamma, \Delta \vdash \mathsf{let}\ t\ \mathsf{be}\ x \otimes y\ \mathsf{in}\ u : C}\ \otimes\mathrm{E}$$

$$\frac{\Delta_1 \vdash t_1 : !A_1 \quad \cdots \quad \Delta_n \vdash t_n : !A_n \quad x_1 : A_1, \ldots, x_n : A_n \vdash u : B}{\Delta_1, \ldots, \Delta_n \vdash \mathsf{promote}\ t_1, \ldots, t_n\ \mathsf{for}\ x_1, \ldots, x_n\ \mathsf{in}\ u : !B}\ \mathrm{Promotion}$$

$$\frac{\Gamma \vdash t : !A}{\Gamma \vdash \mathsf{derelict}(t) : A}\ \mathrm{Dereliction} \qquad \frac{\Gamma \vdash t : !A \quad \Delta \vdash u : B}{\Gamma, \Delta \vdash \mathsf{discard}\ t\ \mathsf{in}\ u : B}\ \mathrm{Weakening}$$

$$\frac{\Gamma \vdash t : !A \quad \Delta, x : !A, y : !A \vdash u : B}{\Gamma, \Delta \vdash \mathsf{copy}\ t\ \mathsf{as}\ x, y\ \mathsf{in}\ u : B}\ \mathrm{Contraction}$$

β-reduction of the linear λ-calculus is defined by the following five rewriting rules: $(\lambda x.t)u \to_\beta t[u/x]$,

$$\begin{gathered}
\mathsf{let}\ t \otimes u\ \mathsf{be}\ x \otimes y\ \mathsf{in}\ v \to_\beta v[t/x, u/y] \\
\mathsf{derelict}\ (\mathsf{promote}\ \bar{t}\ \mathsf{for}\ \bar{x}\ \mathsf{in}\ u) \to_\beta u[\bar{t}/\bar{x}], \\
\mathsf{discard}\ (\mathsf{promote}\ \bar{t}\ \mathsf{for}\ \bar{x}\ \mathsf{in}\ u)\ \mathsf{in}\ v \to_\beta \mathsf{discard}\ \bar{t}\ \mathsf{in}\ v, \text{ and} \\
\mathsf{copy}\ (\mathsf{promote}\ \bar{t}\ \mathsf{for}\ \bar{x}\ \mathsf{in}\ u)\ \mathsf{as}\ y^l, y^r\ \mathsf{in}\ s \\
\to_\beta\ \mathsf{copy}\ \bar{t}\ \mathsf{as}\ \bar{z}^l, \bar{z}^r \mathsf{in}\ s\,[\mathsf{promote}\ \bar{z}^l\ \mathsf{for}\ \bar{x}^l\ \mathsf{in}\ u^l\ /\ y^l,\ \mathsf{promote}\ \bar{z}^r\ \mathsf{for}\ \bar{x}^r\ \mathsf{in}\ u^r\ /\ y^r]
\end{gathered}$$

Here $\bar{t}, \bar{u}, \ldots$ stand for lists of linear λ-terms, $\bar{x}, \bar{y}, \ldots$ for lists of variables. And if $\bar{x} = x_1, \ldots, x_n$ and $\bar{t} = t_1, \ldots, t_n$, then

$$\begin{aligned}
\mathsf{promote}\ \bar{t}\ \mathsf{for}\ \bar{x}\ \mathsf{in}\ u &= \mathsf{promote}\ t_1, \ldots, t_n\ \mathsf{for}\ x_1, \ldots, x_n\ \mathsf{in}\ u \\
\mathsf{discard}\ \bar{x}\ \mathsf{in}\ t &= \mathsf{discard}\ x_1\ \mathsf{in}\ \cdots \mathsf{discard}\ x_n\ \mathsf{in}\ t \\
\mathsf{copy}\ \bar{t}\ \mathsf{as}\ \bar{x}, \bar{y}\ \mathsf{in}\ u &= \mathsf{copy}\ t_1\ \mathsf{as}\ x_1, y_1\ \mathsf{in}\ \cdots \mathsf{copy}\ t_n\ \mathsf{as}\ x_n, y_n\ \mathsf{in}\ u
\end{aligned}$$

3.2 Special Proof Expressions

We translate linear λ-terms into special PEXPs:

Definition 2 (Special Proof Expressions). *We call a* PEXP $\Theta; \bar{x}$ *a* special proof expression, *or a* special PEXP, *if* $\bar{x}$ *is a list of distinct names.*

The coequation part Θ is sufficient to determine the computational content of the special PEXP $\Theta; \bar{x}$. On the other hand, it is not the case for a usual PEXP; See the following quotation from Abramsky [1]:

> The "molecules" of the linear CHAM are the coequations. We refer to Θ in $\Theta; \bar{t}$ as the "solution", and to $\bar{t}$ as the "main body". The idea is that the computation is done in the solution, with the result recorded in the main body. One can think of each coequation either as a single sequential process, or as a tightly coupled synchronous parallel composition of two processes, proceeding in lockstep. (So coequations could be modelled by "membranes" in Berry and Boudol's terminology; but we shall not pursue this idea.)

We regard the main part $t_1, \ldots, t_n$ of PEXP $\Theta; t_1, \ldots, t_n$ as the ports, and we let the computation results be recorded not on $t_1, \ldots, t_n$ but be recorded in the coequation parts. Moreover, we allow PEXPs to connect each other through their ports. So, we restrict $t_1, \ldots, t_n$ being variables $\bar{x}$. Thus, $\bar{x}$ can be easily interpreted as a list of *port names* in concurrent calculi such as CCS [14]. Therefore, a clear translation of special PEXPs into, for example, agents of CCS may be made easily.

Thus, in the translation of λ-terms, $\Theta; \bar{x}, x'$ is interpreted as having $\bar{x}$ as *input ports* and x' as an *output port* as in the following figure.

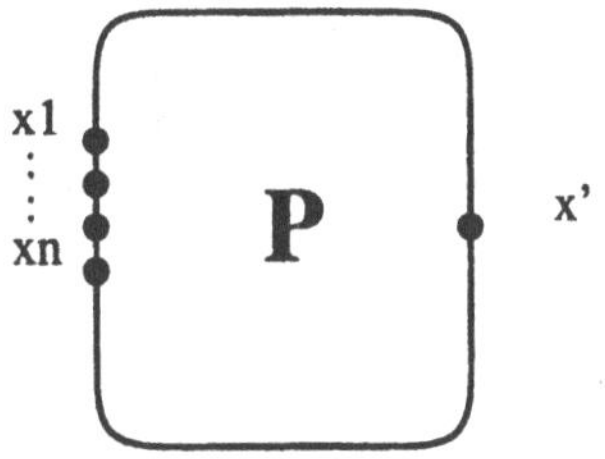

3.3 The Translation

Basic Idea. Linear λ-terms represent natural deduction style proofs in ILL, while PEXPs represent sequent calculus style proofs in LL (more precisely, an equivalence class of proofs where the equivalence is defined to be "the equality as proof nets"). We adapt Gentzen's translation of natural deduction style proofs into sequent calculus style proofs.

But, we employ a trick to make the translation result a special PEXP. For example, the ($\multimap$I) rule is translated into $\dfrac{\vdash \Gamma^\perp, A^\perp, B}{\vdash \Gamma^\perp, A^\perp \parr B}$ Par (In LL, $A_1 \multimap A_2$ is $A_1^\perp \parr A_2$.). If we assign terms to them, then $\dfrac{\bar{x} : \Gamma, x : A \vdash t : B}{\bar{x} : \Gamma \vdash \lambda x.t : A \multimap B} \multimap\mathrm{I}$ is translated into $\dfrac{\vdash \Theta;\ \bar{x} : \Gamma^\perp, x : A^\perp, x' : B}{\vdash \Theta;\ \bar{x} : \Gamma^\perp, x \parr x' : A^\perp \parr B}$ Par. However, the lower PEXP $\Theta; \bar{x}, x \parr x'$ in the last figure is not a special PEXP, so we let the translation result of $\lambda x.t$ be $\Theta,\ y' \perp x \parr x';\ \bar{x},\ y'$. with y' being a fresh variable.

This coincides with introducing Cut-rule:

$$\dfrac{\dfrac{\vdash \Theta;\ \bar{x}:\Gamma^{\perp},\ x:A^{\perp},\ x':B}{\vdash \Theta;\ \bar{x}:\Gamma^{\perp},\ x \parr x' : A^{\perp} \parr B}\ \text{Par} \qquad \vdash;\ y':(A^{\perp} \parr B)^{\perp},\ y':A^{\perp} \parr B}{\vdash \Theta,\ y' \perp x \parr x';\ \bar{x}:\Gamma^{\perp},\ y':A^{\perp} \parr B}\ \text{Cut}\ .$$

The Translation Rules. For a linear λ-term t, we define its translation result t° by induction on the construction of t. In a PEXP $\Theta; \bar{x}, x'$, we consider Θ to be a multiset of coequations, and $\bar{x}, x'$ as an ordered pair of a *multiset* of names $\bar{x}$ and a name x'.

$$\dfrac{}{x^{\circ} = x \perp x';\ x, x'}\ \text{Id}$$

$$\dfrac{t^{\circ} = \Theta;\ \bar{x}, x, x'}{(\lambda x.t)^{\circ} = \Theta,\ z' \perp x \parr x';\ \bar{x}, z'}\ {\multimap}I \qquad \dfrac{t^{\circ} = \Theta;\ \bar{x}, x' \quad u^{\circ} = \Xi;\ \bar{y}, y'}{(tu)^{\circ} = \Theta,\ \Xi,\ x' \perp y' \otimes z';\ \bar{x}, \bar{y}, z'}\ {\multimap}E$$

$$\dfrac{t^{\circ} = \Theta;\ \bar{x}, x' \quad u^{\circ} = \Xi;\ \bar{y}, y'}{(t \otimes u)^{\circ} = \Theta,\ \Xi,\ z' \perp x' \otimes y';\ \bar{x}, \bar{y}, z'}\ \otimes I$$

$$\dfrac{t^{\circ} = \Theta;\ \bar{x}, x' \quad u^{\circ} = \Xi;\ \bar{y}, y_1, y_2, y'}{(\mathsf{let}\ t\ \mathsf{be}\ y_1 \otimes y_2\ \mathsf{in}\ u)^{\circ} = \Theta,\ \Xi,\ x' \perp y_1 \parr y_2;\ \bar{x}, \bar{y}, y'}\ \otimes E$$

$$\dfrac{t_i^{\circ} = \Theta_i;\ \bar{y}_i, y_i'\ (i = 1, \ldots, n) \qquad u^{\circ} = \Xi;\ \bar{x}, x_1, \ldots, x_n, x'}{\begin{array}{l}(\mathsf{promote}\ t_1, \ldots, t_n\ \mathsf{for}\ x_1, \ldots, x_n\ \mathsf{in}\ u)^{\circ} = \\ \quad \Theta_1, \ldots, \Theta_n,\ x_1 \perp y_1', \ldots, x_n \perp y_n', \\ \quad z' \perp \bar{x}x_1 \cdots x_n(\Xi[\bar{z}/\bar{x}, z_1/x_1, \ldots, z_n/x_n]; \bar{z}, z_1, \ldots, z_n, x'); \\ \quad \bar{x}, \bar{y}_1, \ldots, \bar{y}_n, z'\end{array}}\ \text{Promotion}$$

$$\dfrac{t^{\circ} = \Theta;\ \bar{x}, x'}{(\mathsf{derelict}(t))^{\circ} = \Theta,\ x' \perp ?y';\ \bar{x}, y'}\ \text{Dereliction}$$

$$\dfrac{t^{\circ} = \Theta;\ \bar{x}, x' \quad u^{\circ} = \Xi;\ \bar{y}, y'}{(\mathsf{discard}\ t\ \mathsf{in}\ u)^{\circ} = \Theta,\ \Xi,\ x' \perp \_;\ \bar{x}, \bar{y}, y'}\ \text{Discard}$$

$$\dfrac{t^{\circ} = \Theta;\ \bar{x}, x' \quad u^{\circ} = \Xi;\ \bar{y}, y_1, y_2, y'}{(\mathsf{copy}\ t\ \mathsf{as}\ y_1, y_2\ \mathsf{in}\ u)^{\circ} = \Theta,\ \Xi,\ x' \perp y_1 @ y_2;\ \bar{x}, \bar{y}, y'}\ \text{Copy}$$

3.4 The Computational Properties

The set of all the special PEXPs is not closed under the cleanup rule, Fortunately, the cleanup rule is not so important when considering its computational meaning. Instead of the cleanup rule, we define several concepts about special PEXPs. *In the rest of this section, we consider only* linear *special* PEXP*s.*

Definition 3. *On the set of all the linear special* PEXP*s, we define* $\cong$ *to be the smallest equivalence relation satisfying:*

(1) $P[z/x] \cong P$, *for a fresh name* z. (2) $\Theta, y \perp z; \bar{x}, x' \cong \Theta[y/z]; \bar{x}, x'$.
(3) $\Theta, x \perp \_; \bar{x}, x, x' \cong \Theta; \bar{x}, x'$. (4) $\Theta, x \perp \odot; \bar{x}, x, x' \cong \Theta; \bar{x}, x'$.

Clause (2) is sufficient to handle a cleanup rule;

$\Theta, y' \perp x'; \bar{x}, y' \overset{by(2)}{\cong} \Theta[y'/x']; \bar{x}, y' \overset{by(1)}{\cong} \Theta; \bar{x}, x'$. Clauses (3) and (4) are required because free variables often disappear via β-reduction in λ-calculus. (In fact, clause (4) is not needed here, but if we accept clause (3), it is unnatural not to accept clause (4).)

Definition 4. *Define* $P \Rightarrow_r Q \overset{\text{def}}{\Longleftrightarrow} P \rightarrow_{r0}^{*} \rightarrow_{r1} \rightarrow_{r0}^{*} Q$. *Here* $\rightarrow_{r0}^{*}$ *is a reflexive and transitive closure of* $\rightarrow_{r0}$. *The* $\rightarrow_{r0}$ *is determined by the communication rule, and* $\rightarrow_{r1}$ *by the other reaction rules.*

Proposition 1. *The translation result of any linear* λ*-term is normal with respect to* $\rightarrow_{r1}$.

Proposition 2. $\cong$ *is a* bisimulation *with respect to* $\Rightarrow_r$, *that is, if* $P \cong Q$ *and* $P \Rightarrow_r P'$, *then some* Q' *satisfies* $P' \cong Q'$ *and* $Q \Rightarrow_r Q'$.

Proof. In view of the linearity of P and Q, it is easily shown that $\cong$ is a bisimulation with respect to $\rightarrow_{r0}^{*}$ and $\rightarrow_{r1}$, from which the proposition follows directly.

Corollary 1. *If* $P \cong\Rightarrow_r Q$, *then* $P \Rightarrow_r\cong Q$.

The relation $\cong$ is 'compatible' with the translation. For example, if $\Theta; \bar{x}, x, x' \cong \Xi; \bar{y}, x, y'$, then $\Theta,\ z' \perp x \parr x'; \bar{y}, z' \cong \Xi,\ z' \perp x \parr y'; \bar{y}, z'$ holds. In particular, if $t^\circ \cong u^\circ$, then $(\lambda x.t)^\circ \cong (\lambda x.u)^\circ$ and so forth.

Next, we prove the following theorem:

Theorem 1. *Let* t *and* u *be linear* λ*-terms. If* $t \rightarrow_\beta u$, *then* $t^\circ \Rightarrow_r\cong u^\circ$, *i.e,* t° *goes to a term which is* $\Rightarrow_r\cong$ *to* u°.

To verify the theorem, we define a concept which corresponds to substitution.

Definition 5. *For* $P = \Theta;\ \bar{x}, x, x'$ *and* $Q = \Xi;\ \bar{y}, y'$, *we define* $P[x \leftarrow Q] \overset{\text{def}}{=} \Theta,\ \Xi,\ x \perp y';\ \bar{x}, \bar{y}, x'$.

Intuitively, $P[x \leftarrow Q]$ is a process where an "input port" x of P is connected to the "output port" of Q. The following figure illustrates this.

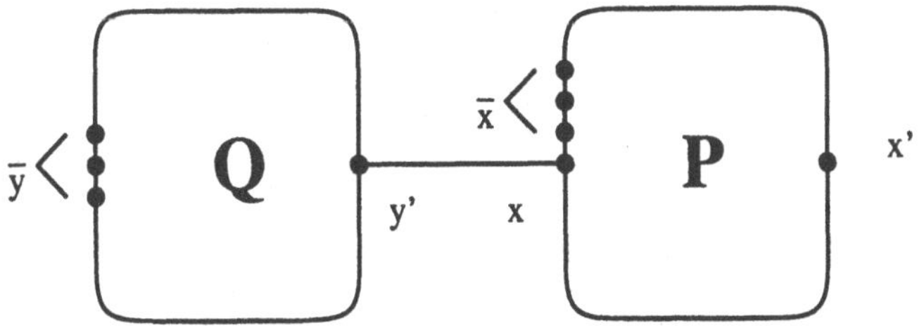

Proposition 3. *For all linear* λ*-terms* t *and* u, *and for all free variable* x *in* t, $(t[u/x])^\circ \cong t^\circ[x \leftarrow u^\circ]$.

Proof. Let $u^\circ = \Xi; \bar{y}, y'$. The proof is done by induction on the construction of t. Note that x occurs in t exactly once. In this proof, the 'compatibility' of $\cong$ with the translation described above is used.

Proposition 4. *For each rewriting rule $l \to_\beta r$ of β-reduction, we have $l^\circ \Rightarrow_r \cong r^\circ$. That is,*

$$
\begin{aligned}
((\lambda x.t)u)^\circ &\Rightarrow_r\cong (t[u/x])^\circ.\\
(\mathsf{let}\ t_1 \otimes t_2\ \mathsf{be}\ y_1 \otimes y_2\ \mathsf{in}\ u)^\circ &\Rightarrow_r\cong (u[t_1/y_1, t_2/y_2])^\circ.\\
(\mathsf{derelict}\ (\mathsf{promote}\ \bar{t}\ \mathsf{for}\ \bar{x}\ \mathsf{in}\ u))^\circ &\Rightarrow_r\cong (u[\bar{t}/\bar{x}])^\circ.\\
(\mathsf{discard}\ (\mathsf{promote}\ \bar{t}\ \mathsf{for}\ \bar{x}\ \mathsf{in}\ u)\ \mathsf{in}\ s)^\circ &\Rightarrow_r\cong (\mathsf{discard}\ \bar{t}\ \mathsf{in}\ s)^\circ.\\
(\mathsf{copy}\ (\mathsf{promote}\ \bar{t}\ \mathsf{for}\ \bar{x}\ \mathsf{in}\ u)\ \mathsf{as}\ y^l, y^r\ \mathsf{in}\ s)^\circ &\Rightarrow_r\cong (\mathsf{copy}\ \bar{t}\ \mathsf{as}\ \bar{z}^l, \bar{z}^r\ \mathsf{in}\\
&\qquad s[\mathsf{promote}\ \bar{z}^l\ \mathsf{for}\ \bar{x}^l\ \mathsf{in}\ u^l\ /\ y^l,\\
&\qquad \mathsf{promote}\ \bar{z}^r\ \mathsf{for}\ \bar{x}^r\ \mathsf{in}\ u^r\ /\ y^r])^\circ.
\end{aligned}
$$

Proof. For the proof of the first claim, let $t^\circ = \Theta;\ \bar{x}, x, x'$ and $u^\circ = \Xi;\ \bar{y}, y'$. Then, $(\lambda x.t)^\circ = \Theta;\ z' \perp x \parr x';\ \bar{x}, z'$ and so

$$
\begin{aligned}
((\lambda x.t)u)^\circ &= \Theta, \Xi, z' \perp x \parr x', z' \perp y' \otimes w';\ \bar{x}, \bar{y}, w'\\
&\to_{r0} \Theta, \Xi, x \parr x' \perp y' \otimes w';\ \bar{x}, \bar{y}, w'\\
&\to_{r1} \Theta, \Xi, x \perp y', x' \perp w';\ \bar{x}, \bar{y}, w'\\
&\cong \Theta, \Xi, x \perp y';\ \bar{x}, \bar{y}, x' = t^\circ[x \leftarrow u^\circ] \cong (t[u/x])^\circ
\end{aligned}
$$

The last is by Proposition 3. The other four claims can be proved similarly.

The proof of Theorem 1 is by induction on the derivation of $t \to_\beta u$.

From Theorem 1, we can conclude that the β-reductions in linear λ-calculus roughly correspond to the reaction rules except the communication rule in LCHAM.

The commuting conversion $\to_c$ is defined as follows. Let $f(t)$ stand for either $(\mathsf{let}\ s\ \mathsf{be}\ x \otimes y\ \mathsf{in}\ t)$, $(\mathsf{discard}\ s\ \mathsf{in}\ t)$, or $(\mathsf{copy}\ s\ \mathsf{as}\ x, y\ \mathsf{in}\ t)$. And let $g(t)$ stand for either (tu), $(\mathsf{let}\ t\ \mathsf{be}\ z_1 \otimes z_2\ \mathsf{in}\ u)$, $(\mathsf{discard}\ t\ \mathsf{in}\ u)$, $(\mathsf{copy}\ t\ \mathsf{as}\ z_1, z_2\ \mathsf{in}\ u)$, or $(\mathsf{derelict}(t))$. Then, the commuting conversion is by definition $g(f(t)) \to_c f(g(t))$. For example, $(\mathsf{let}\ s\ \mathsf{be}\ x \otimes y\ \mathsf{in}\ t)u \to_c \mathsf{let}\ s\ \mathsf{be}\ x \otimes y\ \mathsf{in}\ tu$. The commuting conversions expose 'hidden' redexes in terms.

We can prove that commuting-convertible linear λ-terms are identified when translated into PEXP. More precisely,

Proposition 5. *If t, u are linear λ-terms and $t \to_c u$, then $t^\circ \cong u^\circ$.*

Proof. We have only to check all the entries of commuting conversions. For example, if $s^\circ = \Theta; \bar{x}, x'$, $t^\circ = \Xi; \bar{y}, y'$ and $u^\circ = \Pi; \bar{z}, z'$, then both $((\mathsf{discard}\ s\ \mathsf{in}\ t)u)^\circ$ and $(\mathsf{discard}\ s\ \mathsf{in}\ tu)^\circ$ turn out to be $\Theta, \Xi, \Pi, x' \perp \_, y' \perp z' \otimes w';\ \bar{x}, \bar{y}, \bar{z}, w'$. Thus, the translation is preserved via a commuting conversion $(\mathsf{discard}\ s\ \mathsf{in}\ t)u \to_c \mathsf{discard}\ s\ \mathsf{in}\ tu$. The other cases are all done in the same way.

4 Principal Typing Theorem of LCHAM

Next, we prove the *principal typing theorem* of LCHAM:

Theorem 2 (Principal Typing). *There is an algorithm such that given a* PEXP P,

1. *if P is typable, then it computes a principal type,*
2. *or else it terminates by outputting "not typable".*

Here,

Definition 6 (Principal Typing). *We write* $\Vdash \Theta;\ \bar{t} : \Gamma$, *when for all* Δ, *these are equivalent: (1)* $\vdash \Theta;\ \bar{t} : \Delta$, *and (2)* $\Delta = \Gamma\sigma$ *for some substitution* σ.

Γ *is called a* principal type *of* $\Theta;\ \bar{t}$. *It is easy to see that* Γ *is unique up to renaming of type variables. Hereafter, we write* $\mathrm{pt}(\Theta;\ \bar{t})$ *to represent* Γ.

In LCHAM, a type-assertion may have many derivations, unlike a type system of λ-calculus. In particular, a type-assertion $\vdash \Theta, \Xi;\ \Gamma,\ \Delta,\ t \otimes u : A \otimes B$ can be inferred from $\vdash \Theta;\ \Gamma, t : A$ and $\vdash \Xi;\ \Delta, u : B$ by an inference rule R=Times, but it may also be inferred by R from another $\vdash \cdots, t : A$ and $\vdash \cdots, u : B$. The same annoyance arises when R is a Cut-rule.This is why the algorithm we will construct in the proof is non-deterministic, while the algorithm for principal types of λ-terms is deterministic.

In Subsection 4.3, we will present the algorithm, and will prove the termination property and the correctness. The correctness proof consists of the verification of Theorem 2 (1) and (2). Theorem 2 (1) will be proved by using the Principal Inference Lemma (i.e., Proposition 7 and 11) in Subsection 4.1, and (2) will be proved by using the Generation Lemma (i.e., Proposition 10 and Proposition 6) in Subsection 4.2.

Hereafter, for sequences Γ and Δ of formulas, we denote by $\mathrm{mgu}(\Gamma; \Delta)$ a most general unifier θ such that $\Gamma\theta = \Delta\theta$. Note that it is computable.

4.1 Easy Part of the Proof

Proposition 6 (Generation Lemma, part 1).

1. *If* $\vdash \Theta;\ \Gamma,\ t_1 \parr t_2 : C$, *then* C *is of the form* $A \parr B$ *and* $\vdash \Theta;\ \Gamma,\ t_1 : A,\ t_2 : B$.
2. *If* $\vdash \Theta;\ \Gamma,\ \mathsf{inl}(t) : C$, *then* C *is of the form* $A \oplus B$ *and* $\vdash \Theta;\ \Gamma, t : A$.
3. *If* $\vdash \Theta;\ \Gamma,\ \mathsf{inr}(t) : C$, *then* C *is of the form* $A \oplus B$ *and* $\vdash \Theta;\ \Gamma, t : B$.
4. *If* $\vdash \Theta;\ \Gamma,\ \odot : C$, *then* $C = \bot$ *and* $\vdash \Theta;\ \Gamma$.
5. *If* $\vdash \Theta;\ \Gamma,\ \_ : C$, *then* C *is of the form* $?A$ *and* $\vdash \Theta;\ \Gamma$.
6. *If* $\vdash \Theta;\ \Gamma,\ ?t : C$, *then* C *is of the form* $?A$ *and* $\vdash \Theta;\ \Gamma, t : A$.
7. *If* $\vdash \Theta;\ \Gamma,\ t_1 @ t_2 : C$, *then* C *is of the form* $?A$ *and* $\vdash \Theta;\ \Gamma,\ t_1 : C,\ t_2 : C$.
8. *If* $\vdash\ ;\ \bar{x} : \Gamma,\ \bar{x}(\Theta; \bar{t}, t) : A$, *then for some* Γ', A', *we have* $\Gamma = ?\Gamma'$ *and* $A = !A'$ *and* $\vdash \Theta;\ \bar{t} : \Gamma,\ t : A'$.
9. *If* $\vdash\ ;\ \bar{x} : \Gamma,\ \bar{x}(\Theta_1; \bar{t_1}, u_1 \parallel \Theta_2; \bar{t_2}, u_2) : C$, *then* C *is of the form* $A_1 \& A_2$, *and* $\vdash \Theta_i;\ \bar{t_i} : \Gamma,\ u_i : A_i$ *for* $i = 1, 2$.

Proposition 7 (Principal Type inference, part 1). *The following are admissible inference rules.*

$$\frac{\Vdash \Theta;\ \Gamma, t : A, u : B}{\Vdash \Theta;\ \Gamma, t \parr u : A \parr B} \quad \frac{\Vdash \Theta;\ \Gamma, t : A}{\Vdash \Theta;\ \Gamma, \mathsf{inl}(t) : A \oplus \alpha}\dagger \quad \frac{\Vdash \Theta;\ \Gamma, t : A}{\Vdash \Theta;\ \Gamma, \mathsf{inr}(t) : \alpha \oplus A}\dagger .$$

$$\frac{\Vdash \Theta;\ \Gamma}{\Vdash \Theta;\ \Gamma, \odot : \bot} \quad \frac{\Vdash \Theta;\ \Gamma}{\Vdash \Theta;\ \Gamma, \_ : ?\alpha}\dagger \quad \frac{\Vdash \Theta;\ \Gamma, t : A}{\Vdash \Theta;\ \Gamma,\ ?t : ?A.}$$

$$\frac{\Vdash \Theta;\ \Gamma, t : A, u : B \qquad \mu = \mathrm{mgu}(A; B),\ \nu = \mathrm{mgu}(?\alpha; A\mu)\ \textit{exist}}{\Vdash \Theta;\ \Gamma\mu\nu, t \,@\, u : A\mu\nu}\dagger .$$

$$\frac{\Vdash \Theta;\ \bar{t} : \Gamma, t : A \quad \mu = \mathrm{mgu}(\Gamma;\ ?\bar{\alpha})\ \textit{exists}}{\Vdash\ ;\ \bar{x} : \Gamma\mu,\ \bar{x}(\Theta; \bar{t}, t) : !A\mu}\dagger .$$

$$\frac{\Vdash \Theta;\ \bar{t} : \Gamma, t : A \quad \Vdash \Xi;\ \bar{u} : \Gamma', u : B \quad \mu = \mathrm{mgu}(\Gamma; \Gamma')\ \textit{exists}}{\Vdash\ ;\ \bar{x} : \Gamma\mu,\ \bar{x}(\Theta; \bar{t}, t \parallel \Xi; \bar{u}, u) : (A \,\&\, B)\mu}\P$$

(†) α is a fresh type variable. (¶) The premises of the form $\Vdash \cdots$ share no variables.

4.2 Difficult Part of the Proof

As we explained in Section 2, a linear PEXP represents a skeleton of a proof structure. It is well-known that correctness of a proof structure depends mainly on the *skeleton*. We introduce *locally correct assertions*, which correspond to 'proof structures.'

Definition 7 (Locally Correct Assertion). *An assertion $\vdash_l \Theta;\ \bar{t} : \Gamma$, which we call a* locally correct assertion, *holds if and only if it is derivable in the inference system* LCHAM′. *Here* LCHAM′ *is obtained from* LCHAM *by replacing the Cut-rule and the Times-rule with the following four rules:*

$$\frac{}{\vdash_l ;} \quad \frac{\vdash_l \Theta;\ \Gamma \quad \vdash_l \Xi;\ \Delta}{\vdash_l \Theta, \Xi;\ \Gamma, \Delta}\,Mix \quad \frac{\vdash_l \Theta;\ \Gamma, t : A, u : A^\perp}{\vdash_l \Theta, t \perp u;\ \Gamma} \quad \frac{\vdash_l \Theta;\ \Gamma, t : A, u : B}{\vdash_l \Theta;\ \Gamma, t \otimes u : A \otimes B}$$

Intuitively, the derivation of $\vdash_l \Theta;\ \bar{t} : \Gamma$ corresponds to a *proof structure* [9] with conclusions Γ. It is easy to see the following:

Proposition 8. *1. If $\vdash_l \Theta;\ \Gamma, u_1 \otimes u_2 : C$, then C must be of the form $A \otimes B$ and $\vdash_l \Theta;\ \Gamma, u_1 : A, u_2 : B$.*
2. If $\vdash_l \Theta, u_1 \perp u_2;\ \Gamma$, then there is an A such that $\vdash_l \Theta;\ \Gamma, u_1 : A^\perp, u_2 : A$.

Proposition 9. *If $\vdash_l \Theta;\ \Gamma$ and $\Theta \supseteq \Theta'$, $\Gamma \supseteq \Gamma'$, then $\vdash_l \Theta';\ \Gamma'$; provided that $\Theta'; \Gamma' = \Theta'; \bar{t}' : \Gamma''$ for some linear* PEXP *$\Theta'; \bar{t}'$.*

Proof. By induction on the deduction of $\vdash_l \Theta;\ \Gamma$.

Theorem 3. *For a typable* PEXP $\Theta;\ \bar{t}$, $\vdash \Theta;\ \bar{t} : \Gamma$ *if and only if* $\vdash_l \Theta;\ \bar{t} : \Gamma$.

Proof. The only-if part. Note that for each rule $\dfrac{\vdash Q_1 \cdots \vdash Q_n}{\vdash P}$ of the system $\vdash$, if in the system $\vdash_l$ we assume $\vdash_l Q_1 \cdots \vdash_l Q_n$ as axioms, we can infer $\vdash_l P$ by using the Mix-rule. Therefore, we are done. *The if part.* Because $\Theta;\ \bar{t}$ is typable, there is some Δ such that $\vdash \Theta;\ \bar{t} : \Delta$. The proof is by induction on the height of this derivation.

If the last rule is the Times-rule, the derivation ends with

$$\frac{\vdash \Theta_i;\ \bar{t_i} : \Delta_i,\ u_i : A_i\ (i = 1,2)}{\vdash \Theta_1,\ \Theta_2;\ \bar{t_1} : \Delta_1,\ \bar{t_2} : \Delta_2,\ u_1 \otimes u_2 : A_1 \otimes A_2}\ .$$

By Proposition 8, $\vdash_l \Theta;\ \bar{t} : \Gamma$ must be of the form $\vdash_l \Theta;\ \bar{t_1} : \Gamma_1,\ \bar{t_2} : \Gamma_2,\ u_1 \otimes u_2 : A'_1 \otimes A'_2$. Moreover, $\vdash_l \Theta;\ \bar{t_1} : \Gamma_1,\ \bar{t_2} : \Gamma_2,\ u_1 : A'_1,\ u_2 : A'_2$. We note that each $\Theta_i; \bar{t_i}, u_i$ is linear. Hence, by Proposition 9, $\vdash_l \Theta_i;\ \bar{t_i} : \Gamma_i, u_i : A'_i$. By induction hypotheses, $\vdash \Theta_1;\ \bar{t_1} : \Gamma_1, u_1 : A'_1$ and $\vdash \Theta_2;\ \bar{t_2} : \Gamma_2, u_2 : A'_2$. Then, by applying the Times-rule we can conclude $\vdash \Theta_1,\ \Theta_2;\ \bar{t_1} : \Gamma_1, \bar{t_2} : \Gamma_2, u_1 \otimes u_2 : A'_1 \otimes A'_2$, i.e. $\vdash \Theta;\ \bar{t} : \Gamma$. The other cases are easy and similar.

Proposition 10 (Generation Lemma, part 2).

1. *Suppose some deduction ends with* $\dfrac{\vdash \Theta_1;\ \bar{t_1} : \Gamma_1,\ u_1 : A_1 \quad \vdash \Theta_2;\ \bar{t_2} : \Gamma_2,\ u_2 : A_2}{\vdash \Theta;\ \bar{t} : \Gamma,\ u_1 \otimes u_2 : A_1 \otimes A_2}$. *Then if* $\vdash \Theta_1,\ \Theta_2;\ \bar{t_1} : \Delta_1,\ \bar{t_2} : \Delta_2,\ u_1 \otimes u_2 : C$, *then* C *is of the form* $B_1 \otimes B_2$ *and* $\vdash \Theta_i;\ \bar{t_i} : \Delta_i,\ u_i : B_i$ $(i = 1,2)$.
2. *Suppose some deduction ends with* $\dfrac{\vdash \Theta_1;\ \bar{t_1} : \Gamma_1,\ u_1 : A \quad \vdash \Theta_2;\ \bar{t_2} : \Gamma_2,\ u_2 : A^{\perp}}{\vdash \Theta,\ u_1 \perp u_2;\ \bar{t} : \Gamma}$. *If* $\vdash \Theta_1,\ \Theta_2,\ u_1 \perp u_2;\ \bar{t_1} : \Delta_1,\ \bar{t_2} : \Delta_2$, *there is a* B *such that* $\vdash \Theta_1;\ \bar{t_1} : \Delta_1,\ u_1 : B$ *and* $\vdash \Theta_2;\ \bar{t_2} : \Delta_2,\ u_2 : B^{\perp}$.

Proof. The premise implies through Theorem 3 that $\vdash_l \Theta_1, \Theta_2; \bar{t_1} : \Gamma_1, \bar{t_2} : \Gamma_2, u_1 \otimes u_2 : C$. By Proposition 8, $\vdash_l \Theta_1, \Theta_2; \bar{t_1} : \Gamma_1, \bar{t_2} : \Gamma_2, u_1 : B_1, u_2 : B_2$ and $C = B_1 \otimes B_2$ for some B_1 and B_2. Because the premise $(i) :\ \vdash \Theta_i; \bar{t_i} : \Gamma_i, u_i : A_i$ implies the linearity of $\Theta_i; \bar{t_i}, u_i$, Proposition 9 implies $\vdash_l \Theta_i;\ \bar{t_i} : \Delta_i,\ u_i : B_i$, and because of (i), Theorem 3 implies $\vdash \Theta_i;\ \bar{t_i} : \Delta_i,\ u_i : B_i$. The second claim can be proved in the same way as above.

Proposition 11 (Principal Type Inference, part 2). *The following are admissible inference rules.*

$$\frac{\Vdash \Theta_i;\ \bar{t_i} : \Gamma_i,\ u_i : A_i\ (i = 1,2)}{\Vdash \Theta_1,\ \Theta_2;\ \bar{t_1} : \Gamma_1,\ \bar{t_2} : \Gamma_2,\ u_1 \otimes u_2 : A_1 \otimes A_2}$$

$$\frac{\Vdash \Theta_i;\ \bar{t_i} : \Gamma_i, u_i : A_i\ (i = 1,2) \quad \mu = \mathrm{mgu}(A_1; A_2^{\perp})\ \textit{exists}}{\Vdash \Theta_1,\ \Theta_2,\ u_1 \perp u_2;\ \bar{t_1} : \Gamma_1\mu,\ \bar{t_2} : \Gamma_2\mu}$$

The premises of the form $\Vdash \cdots$ *share no variables.*

Proof. To prove the admissibility of the first inference rule, let $\vdash \Theta_1, \Theta_2; \bar{t_1} : \Delta_1, \bar{t_2} : \Delta_2, u_1 \otimes u_2 : C$. The premise of the rule implies the existence of a deduction ending with $\dfrac{\vdash \Theta_i;\ \bar{t_i} : \Gamma_i,\ u_i : A_i\ (i = 1, 2)}{\vdash \Theta_1, \Theta_2;\ \bar{t_1} : \Gamma_1, \bar{t_2} : \Gamma_2, u_1 \otimes u_2 : A_1 \otimes A_2}$. By Proposition 10, C is of the form $A_1' \otimes A_2'$ and $\vdash \Delta_i;\ \bar{t_i} : \Gamma_i, u_i : A_i'$ with $i = 1, 2$. Thus, for some σ_i, $\Delta_i = \Gamma_i\sigma_i$ and $A_i' = A_i\sigma_i$. Because of the side condition, we have $\Delta_i = \Gamma_i\sigma$ and $C\sigma = (A_1 \otimes A_2)\sigma$ for σ being defined below. If α occurs in Γ_i, A_i, then $\sigma(\alpha)$ is $\sigma_i(\alpha)$, or else it is α. Hence we are done. The admissibility of the second inference rule can be shown in the same way.

4.3 The Algorithm for Principal Type

To compute $\mathrm{pt}(P)$*, do the following:*

1. *If P is of the form $\Theta; \bar{t}, t_1 \parr t_2$, then: if* $\mathrm{pt}(\Theta; \bar{t}, t_1, t_2) = [\Gamma, A, B]$*, then* $\mathrm{pt}(P) = [\Gamma, A \parr B]$*, else failure.*
2. *If P is of the form $\Theta; \bar{t}, \mathsf{inl}(t)$, then: if* $\mathrm{pt}(\Theta; \bar{t}, t) = [\Gamma, A]$*, then* $\mathrm{pt}(P) = [\Gamma, A \oplus \alpha]$*, else failure where α is a fresh type variable.*
3. *If P is of the form $\Theta; \bar{t}, \mathsf{inr}(t)$, then: if* $\mathrm{pt}(\Theta; \bar{t}, t) = [\Gamma, A]$*, then* $\mathrm{pt}(P) = [\Gamma, \alpha \oplus A]$*, else failure where α is a fresh type variable.*
4. *If P is of the form $\Theta; \bar{t}, \odot$, then: if* $\mathrm{pt}(\Theta; \bar{t}) = [\Gamma]$*, then* $\mathrm{pt}(P) = [\Gamma, \bot]$*, else failure.*
5. *If P is of the form $\Theta; \bar{t}, \_$, then: if* $\mathrm{pt}(\Theta; \bar{t}) = [\Gamma]$*, then* $\mathrm{pt}(P) = [\Gamma, ?\alpha]$*, else failure. Here α is a fresh type variable.*
6. *If P is of the form $\Theta; \bar{t}, ?t$, then: if* $\mathrm{pt}(\Theta; \bar{t}, t) = [\Gamma, A]$*, then* $\mathrm{pt}(P) = [\Gamma, ?A]$*, else failure.*
7. *If P is of the form $\Theta; \bar{t}, t_1 @ t_2$, then: if* $\mathrm{pt}(\Theta; \bar{t}, t, u) = [\Gamma, A, B]$ *and both of* $\mu = \mathrm{mgu}(A; B)$ *and* $\nu = \mathrm{mgu}(?\alpha; A\mu)$ *exist, then* $\mathrm{pt}(P) = \Gamma\mu\nu, A\mu\nu$*, else failure.*
8. *If P is of the form $; *$, then* $\mathrm{pt}(P) = [\mathbf{1}]$.
9. *If P is of the form $; x, x$, then* $\mathrm{pt}(P) = [\alpha^\perp, \alpha]$*, where α is a fresh type variable.*
10. *If P is of the form $; \bar{x}, \bar{x}(Q)$, then: if* $\mathrm{pt}(Q) = [\Gamma, A]$*, and if* $\mu = \mathrm{mgu}(\Gamma; ?\bar{\alpha})$ *exists (where $\bar{\alpha}$ is a list of fresh names), then* $\mathrm{pt}(P) = [\Gamma\mu, !A\mu]$*. Otherwise, failure.*
11. *If P is of the form $; \bar{x}, \bar{x}(Q \parallel Q')$, then: if* $\mathrm{pt}(Q) = [\Gamma_1, A]$, $\mathrm{pt}(Q') = [\Gamma_2, B]$*, and* $\mu = \mathrm{mgu}(\Gamma_1; \Gamma_2)$ *exists, then* $\mathrm{pt}(P) = [\Gamma_1\mu, (A \,\&\, B)\mu]$*. Otherwise, failure.*
12. *Otherwise, let $P = \Theta; \bar{t}$. For every decomposition of the form $\Theta = \Theta_1, \Theta_2$ and $\bar{t} = \bar{t_1}, \bar{t_2}, u_1 \otimes u_2$, try to compute* $\mathrm{pt}(\Theta_1; \bar{t_1}, u_1)$ *and* $\mathrm{pt}(\Theta_2; \bar{t_2}, u_2)$*. If it fails for every decomposition, go to 13. If it succeeds for a decomposition, let the result be $[\Gamma, A]$ and $[\Delta, B]$. Then,* $\mathrm{pt}(P) = [\Gamma, \Delta, A \otimes B]$.
13. *For every decomposition of the form $\Theta = \Theta_1, \Theta_2, u_1 \perp u_2$ and $\bar{t} = \bar{t_1}, \bar{t_2}$, try to compute* $\mathrm{pt}(\Theta_1; \bar{t_1}, u_1)$ *and* $\mathrm{pt}(\Theta_2; \bar{t_2}, u_2)$*. If it succeeds for a decomposition, let the result be $[\Gamma, A]$ and $[\Delta, B]$. If* $\mathrm{mgu}(A; B^\perp)$ *exists, then* $\mathrm{pt}(P) = [\Gamma\mu, \Delta\mu]$*. Otherwise, failure.*

This algorithm terminates for any input, because the number of constructors in P decreases strictly at each step. Moreover, the correctness of each step is verified as follows: When P is typable, let π be a derivation of it. Then we can show that $\mathrm{pt}(P)$ is a principal type of P, by induction on π, by using Proposition 7 and Proposition 11. When P is not typable, it outputs "failure," because of Proposition 10 and Proposition 6. Thus, the proof of Theorem 2 is completed.

5 Concluding Remarks

Mackie [10] introduced a version of linear λ-calculus, a translation from the calculus to a proof structure, and studied efficient implementation of call-by-(name/value/need) evaluation of the λ-calculus.

By using LCHAM and the extension, we will analyze computation of linear λ-calculi neatly. Then we will study the (sub)computation can be encapsulated (and parallelized) in recent concurrent calculi.

References

1. S. Abramsky. Computational interpretations of linear logic. *TCS*, 111:3–57, 1993.
2. G. Bellin and P. Scott. On the π-calculus and linear logic. *TCS*, 135:11–65, 1994.
3. N. Benton, G. Bierman, J. Martin E. Hyland, and V. de Paiva. A term calculus for intuitionistic linear logic. In M. Bezem and J. F. Groote, eds., *Typed Lambda Calculi and Applications, Proceedings*, vol. 664 of *LNCS*, pp. 75–90. 1993.
4. N. Benton and P. Wadler. Linear logic, monads and the lambda calculus. In *Proceedings of the 11th LICS*, pp. 420–431, 1996.
5. G. Berry and G. Boudol. The chemical abstract machine. In *Conference Record of the 17th POPL*, pp. 81–94 1990.
6. L. Cardelli and A. D. Gordon. Mobile ambients. In M. Nivat, ed., *Foundations of Software Science and Computational Structures*, vol. 1378 of *LNCS*, pp. 140–155, 1998.
7. J. Chirimar, C. A. Gunter, and J. G. Riecke. Proving memory management invariants for a language based on linear logic. In *Proceedings of the 1992 ACM Conference on Lisp and Functional Programming*, pp. 139–150. 1992.
8. C. Fournet and G. Gonthier. The reflexive cham and the join-calculus. In *Conference Record of the 23rd POPL*, pp. 372–385, 1996.
9. J.-Y. Girard. Linear logic. *TCS*, 50:1–102, 1987.
10. I. Mackie, The Geometry of Implementation, Imperial College of Science, 1994.
11. I. Mackie. Lilac — a functional programming language based on linear logic. *JFP*, 4(4):395–433, 1994.
12. J. Maraist, M. Odersky, D. N. Turner, and P. Wadler. Call-by-name, call-by-value, call-by-need and the linear lambda calculus. *TCS, special issue on papers presented at MFPS'95.*
13. S. Mikami. A theory of a rewriting system based on proof-reduction of linear logic. Senior thesis, ftp://nicosia.is.s.u-tokyo.ac.jp:pub/staff/mikami/lcham.ps, 1996.
14. R. Milner. *Communication and Concurrency.* Prentice Hall, 1989.

Resource Interpretations, Bunched Implications and the $\alpha\lambda$-Calculus
(Preliminary Version)

Peter W. O'Hearn

Queen Mary & Westfield College

Abstract. We introduce the $\alpha\lambda$-calculus, a typed calculus that includes a multiplicative function type $-\!\!*$ alongside an additive function type $\rightarrow$. It arises proof-theoretically as a calculus of proof terms for the logic of bunched implications of O'Hearn and Pym, and semantically from doubly closed categories, where a single category possesses two closed structures. Typing contexts in $\alpha\lambda$ are bunches, i.e., trees built from two combining operations, one that admits the structural rules of Weakening and Contraction and another that does not. To illuminate the consequences of $\alpha\lambda$'s approach to the structural rules we define two resource interpretations, extracted from Reynolds's "sharing reading" of affine λ-calculus. Based on this we show how $\alpha\lambda$ enables syntactic control of interference and Idealized Algol, imperative languages based on affine and simply-typed λ-calculi, to be smoothly combined in one system.

1 Introduction

The logic **BI** of bunched implications has two implications, one additive ($\rightarrow$) and the other multiplicative ($-\!\!*$), which it accepts on an equal footing [18]. It may be viewed as a merging of intuitionistic logic (**IL**) and multiplicative, intuitionistic linear logic (**MILL**), where the two subsystems are combined by using contexts Γ in sequents $\Gamma \vdash A$ built from two combining operations, ";" and ",". Instead of lists, contexts are trees with internal nodes labelled by ";" or ",", or in brief, *bunches*. By allowing the two context-forming operators to nest arbitrarily deeply in a bunch the two subsystems intermix freely.

Here we consider **BI** from the point of view of types, by using its rules to typecheck terms in what we call the $\alpha\lambda$-calculus. Pym introduces $\alpha\lambda$ independently in a separate paper, as part of his account of the theory of propositional **BI** [20], and establishes some basic properties of the calculus, including completeness and strong normalization. Our focus here is more on the use of **BI** as a type system, and especially the semantic and computational implications of its approach to structural rules.

Bunches first arose in work on relevant logic in the seventies [9], where they were used to manage interactions between additive (or extensional, in the relevant terminology) and multiplicative (or intensional) connectives. (See [18] for an account of the relation of **BI** to relevant and other substructural logics.) The crucial point is that, with bunches, it is possible to control access to structural rules by allowing them for one form of combination but not another. For example, the rules of Weakening and Contraction for the ";" form of combination can be stated as follows:

$$\frac{\Gamma(\Delta) \vdash B}{\Gamma(\Delta; \Delta') \vdash B} \text{ Weakening} \qquad \frac{\Gamma(\Delta; \Delta) \vdash B}{\Gamma(\Delta) \vdash B} \text{ Contraction}$$

where notation of the form $\Gamma(\Delta)$ indicates a bunch with Δ appearing as a subtree. **BI** accepts Weakening and Contraction for ";" but not for ",".

Our main concern in this paper is with how this bunch-based approach to the structural rules impacts the meanings of function types. That is, the two implications should evidently correspond to function types $A \mathrel{-\!\!*} B$ and $A \to B$; but for what kinds of functions?

1.1 From Doubly Closed Categories to $\alpha\lambda$

To see how the $\alpha\lambda$-calculus arises semantically, consider that an introduction rule for a function type typically corresponds to an adjunction. That is, a typing rule

$$\frac{\Gamma, x : A \vdash M : B}{\Gamma \vdash \lambda x \,.\, M : A \Rightarrow B}$$

corresponds to an isomorphism of maps of the corresponding shape in a closed category

$$\frac{\Gamma \otimes A \longrightarrow B}{\Gamma \longrightarrow (A \Rightarrow B)} \,.$$

Now, suppose that we have a *doubly closed category*, i.e., a single category equipped with two monoidal closed structures instead of only one:

$$\frac{\Gamma \wedge A \longrightarrow B}{\Gamma \longrightarrow (A \to B)} \qquad \frac{\Gamma * A \longrightarrow B}{\Gamma \longrightarrow (A \mathrel{-\!\!*} B)} \,.$$

To match this situation, we extend the syntax of typing contexts with an additional combining operation, semi-colon, which allows us to formulate introduction rules corresponding to the two adjunctions:

$$\frac{\Gamma; x : A \vdash M : B}{\Gamma \vdash \alpha x \,.\, M : A \to B} \qquad \frac{\Gamma, x : A \vdash M : B}{\Gamma \vdash \lambda x \,.\, M : A \mathrel{-\!\!*} B} \,.$$

This leads directly to the use of bunches for typing contexts. The resulting calculus is named after its binders: α for the αdditive binder, and λ for multiplicative, or λinear, binder.

The language we consider admits Weakening and Contraction for ";" but not for ",", and both forms of combination will be commutative. But the same scheme can be used for other combinations, such as for non-symmetric monoidal structures, and even more than two.

The $\alpha\lambda$-calculus contains simply-typed λ-calculus and multiplicative, intuitionistic linear λ-calculus as subsystems. Various forms of linear λ-calculus also contain the two subsystems [1, 5, 24, 3], but $\alpha\lambda$'s approach is rather different. Where linear logic uses a modality "!" (or sometimes distinct zones in contexts) to control access to the structurals, in $\alpha\lambda$ access is governed by the two means of combination. The difference can be stated crisply in terms of categorical models. In models of linear logic two closed categories are involved, where one is often presented as a Kleisli category [5, 4, 3]. For instance, in the original coherence space model there are indeed two function types, but $\multimap$ is closed structure in the category of linear maps, while the additive $\rightarrow$, which can be represented as $!A \multimap B$, is closed for the category of stable maps. In contrast, in a doubly closed category the two closed structures must reside in one and the same category. (Again, we refer to [18] for a detailed account of the relation to linear logic.)

Categorical semantics makes the formal difference of **BI**'s, and $\alpha\lambda$'s, approach to the structural rules very clear, but the point of view on function types it offers is very abstract. We can investigate the implications of these structural properties further, and more concretely, by considering "resource interpretations" of the types.

1.2 The Sharing Interpretation

The sharing reading of the $\alpha\lambda$-calculus has two main sources of inspiration. One is Girard's vivid depiction of linear logic as "resource sensitive" [10, 11]. The key point, for us, is the focus on the significance of controlling Contraction; this has been explained in linear logic with the number-of-uses reading, where a linear function $f : A \multimap B$ is one that uses its argument exactly once.

The other main source is Reynolds's syntactic control of interference [21], which is based on a novel reading of the affine λ-calculus. We extrapolate from this reading to arrive at what we call the *sharing interpretation* of $\alpha\lambda$.

The background idea for the sharing interpretation is of functional programming data such as functions, pairs, etc, but with an additional, intensional, notion of resources that computational entities are allowed to access. The reading of function types is as follows.

$A \mathrel{-\!\!*} B$: functions that have access to disjoint resources from their arguments.
$A \rightarrow B$: functions that have access to the same resources as their arguments.

The bare statement of the interpretation is so direct that, at first glance, it may seem as if it must amount to the same thing as resource interpretations for other systems that control the structural rules. For, if we think of a context, roughly, as corresponding to a collection of resources, then use of separate contexts in an elimination rule for a multiplicative implication $\mathrel{-\!\!*}$ directly expresses the disjointness mentioned in the informal interpretation, and the use of a common context in a rule for the additive corresponds to the sameness.

$$\frac{\Gamma \vdash A \mathrel{-\!\!*} B \quad \Delta \vdash A}{\Gamma, \Delta \vdash B} \qquad \frac{\Gamma \vdash A \rightarrow B \quad \Gamma \vdash A}{\Gamma \vdash B}$$

However, there is an important point to notice: the reading places no constraint on how many times a $\mathrel{-\!\!*}$-typed function uses its argument, it just cannot be applied to arguments accessing the same resources. In fact, we will even show in Section 2 that the $\alpha\lambda$-calculus allows multiplicative functions that use their arguments many times, or not at all. This exemplifies the interactions between multiplicatives and additives permitted by the sharing interpretation. More generally, and speaking figuratively, we would suggest that this bunch-based control over structural rules can be understood as being about who has access to what, rather than how often a piece of data is used.

Strictly speaking, the sharing interpretation as stated above is for the *linear* version of the $\alpha\lambda$-calculus. The reading for the *affine* variant, which admits Weakening for the multiplicative combination ",", is obtained by changing the interpretation of the additive function type.

$A \multimap B$: functions that don't share resources with their arguments.
$A \rightarrow B$: functions that may share resources with their arguments.

The use of "may" here indicates that an additive function might share resources with its argument, but it does not have to. In the affine language this will be reflected in the fact that functions of type $A \rightarrow B$ can be converted to functions of type $A \multimap B$.

1.3 Syntactic Control of Interference

The affine interpretation just given is derived directly from syntactic control of interference and Idealized Algol, two imperative languages due to Reynolds [21, 22]. The key difference between functions in the two languages is that in Idealized Algol a function is allowed to interfere, by use of common storage, with its argument, while in syntactic control it is not. One might even say that the answer to the question of what kind of functions correspond to bunched implications preceded (in the affine case) the question.

In fact, our original interest in a calculus like $\alpha\lambda$ stemmed from an observation about a specific model that had been used for the two languages separately.

> "The semantic model presented here posesses two kinds of exponential, one for the monoidal closed structure, and another, adjoint to $\times$ for cartesian closed structure. This raises the question of whether interference control and uncontrolled Algol can coexist harmoniously in one system ... An interesting point to note is that here the two kinds of closed structure coexist in the *same* category, so there is no need to pass to a separate category, such as a Kleisli category, to interpret the intuitionistic (i.e., Algol's) function type. [13]"

In Sections 5 and 6 we show how the affine variant of the $\alpha\lambda$-calculus does indeed give rise to the requested enveloping language. There we give a brief introduction to syntactic control, emphasizing the unusual nature of its sharing interpretation. We also discuss limitations which motivate the question of an enveloping language containing both it and Idealized Algol.

The linear sharing interpretation will not be developed in detail in this preliminary paper, beyond a simple model in Section 3 intended to illustrate the basic ideas. At a later time we plan to show how the linear $\alpha\lambda$-calculus can also be used to control interference, but where banishing Weakening leads to additional positive properties that enable the dynamic extent of resources such as pointers to be controlled.

2 The $\alpha\lambda$-calculus

The definition of the $\alpha\lambda$-calculus is motivated by models as follows.

Definition 1. *A (symmetric)* doubly closed category, *or dcc in short, is a category equipped with two symmetric monoidal closed structures* $(I, *, -\!\!*)$ *and* $(1, \wedge, \rightarrow)$. *A dcc is called* cartesian *if one of the monoidal structures, say* $(1, \wedge)$, *is cartesian,* affine *if it is cartesian and the two units* 1 *and* I *are isomorphic, and* bicartesian *if it is cartesian and has finite coproducts.*

Models of the version of $\alpha\lambda$ here are given using cartesian dcc's. Full **BI**, which includes additive disjunction, uses bicartesian dcc's.

2.1 The Basic System

TYPES

$A ::=$	ρ	primitive types
	$\mid A * A$	multiplicative product
	$\mid A \wedge A$	additive product
	$\mid A -\!\!* A$	multiplicative exponent
	$\mid A \rightarrow A$	additive exponent

(We do not include types for the units of the products here; but these, and the additive disjunction of **BI**, pose no substantial difficulties for our purposes.)

BUNCHES

$\Gamma ::=$	$x : A$	identifier assumption
	$\mid I$	multiplicative unit
	$\mid \Gamma, \Gamma$	multiplicative combination
	$\mid 1$	additive unit
	$\mid \Gamma; \Gamma$	additive combination

The essence of the two forms of combination is that ";" admits Weakening and Contraction, whereas "," does not. Bunches are subject to the restriction that no identifier may occur twice in the tree. This restriction determines implicit side conditions on some of the rules below. We write $\Gamma(\Delta)$ to indicate a bunch in which Δ appears as a subtree, and $\Gamma(\Delta')$ for the similar tree where Δ' replaces Δ. $i(\Gamma)$ is the list of identifiers encountered one after the other in an inorder traversal of the tree Γ. $\Gamma \cong \Delta$ indicates that Γ and Δ are isomorphic as trees; i.e., one can be obtained from the other by a suitable renaming of identifiers. Isomorphism is used in the formulation of Contraction below.

We won't try to come up with a more compact representation of bunches using, say, sets or sequences instead of binary operators; the real point of bunches is to let us get the α- and λ-abstractions right. We use an equivalence on trees instead of worrying about representation.

COHERENT EQUIVALENCE: $\Gamma \equiv \Gamma'$.
$\equiv$ is the smallest equivalence relation on bunches satisfying

1 Commutative monoid equations for 1 and ;
2 Commutative monoid equations for I and ,
3 Congruence: if $\Delta \equiv \Delta'$ then $\Gamma(\Delta) \equiv \Gamma(\Delta')$

Note that ";" and "," do not distribute over one another.

TYPING JUDGEMENTS

These are of the form

$$\Gamma \vdash M : A$$

where the terms M are defined in the following rules.

IDENTITY AND STRUCTURE

$$\frac{}{x : A \vdash x : A}\ Id \qquad \frac{\Gamma \vdash M : A}{\Delta \vdash M : A} \equiv \text{(where } \Delta \equiv \Gamma\text{)}$$

$$\frac{\Gamma(\Delta) \vdash M : A}{\Gamma(\Delta; \Delta') \vdash M : A}\ W \qquad \frac{\Gamma(\Delta; \Delta') \vdash M : A}{\Gamma(\Delta) \vdash M[i(\Delta)/i(\Delta')] : A}\ C \text{ (where } \Delta \cong \Delta'\text{)}$$

ADDITIVES

$$\frac{\Gamma \vdash M : A \quad \Delta \vdash N : B}{\Gamma; \Delta \vdash \langle M, N\rangle : A \wedge B}\ \wedge I \qquad \frac{\Gamma \vdash M : A_1 \wedge A_2}{\Gamma \vdash \pi_i M : A_i}\ \wedge E \text{ (where } i \text{ is 1 or 2)}$$

$$\frac{\Gamma; x : A \vdash M : B}{\Gamma \vdash \alpha x \,.\, M : A \to B} \to I \qquad \frac{\Gamma \vdash M : A \to B \quad \Delta \vdash N : A}{\Gamma; \Delta \vdash MN : B} \to E$$

MULTIPLICATIVES

$$\frac{\Gamma \vdash M : A \quad \Delta \vdash N : B}{\Gamma, \Delta \vdash M * N : A * B}\ *I \qquad \frac{\Gamma(x : A, y : B) \vdash N : C \quad \Delta \vdash M : A * B}{\Gamma(\Delta) \vdash \mathbf{let}\,(x, y) = M \,\mathbf{in}\, N : C}\ *E$$

$$\frac{\Gamma, x : A \vdash M : B}{\Gamma \vdash \lambda x \,.\, M : A -\!\!* B} -\!\!* I \qquad \frac{\Gamma \vdash M : A -\!\!* B \quad \Delta \vdash N : A}{\Gamma, \Delta \vdash M@N : B} -\!\!* E$$

EQUATIONS

$$(\alpha x \,.\, M)N = M[N/x] \qquad (\alpha x \,.\, Mx) = M \ (x \notin free(M))$$

$$(\lambda x \,.\, M)@N = M[N/x] \qquad (\lambda x \,.\, M@x) = M \ (x \notin free(M))$$

$$\pi_1\langle M, N\rangle = M \qquad \langle \pi_1 M, \pi_2 M\rangle = M$$
$$\pi_2\langle M, N\rangle = N$$

$$(\mathbf{let}\,(x, y) = M_1 * M_2 \,\mathbf{in}\, N) = N[M_1/x, N_2/y] \qquad (\mathbf{let}\,(x, y) = M \,\mathbf{in}\, x * y) = M$$

The left and right columns contain β and η laws for each connective. A fuller treatment of **let** requires commutative conversions; these and the question of normalization are considered in [20].

Since ";" admits Weakening and Contraction, we can derive rules where the additive maintenance of premises is explicit.

Lemma 2. *The following are admissible rules.*

$$\frac{\Gamma \vdash M : A \rightarrow B \quad \Gamma \vdash N : A}{\Gamma \vdash MN : B} \qquad \frac{\Gamma \vdash M : A \quad \Gamma \vdash N : B}{\Gamma \vdash \langle M, N \rangle : A \wedge B}$$

The cut lemma, asserting that substitution preserves well-formedness, is formulated for identifiers appearing arbitrarily deeply in a bunch.

Lemma 3. *The following is an admissible rule.*

$$\frac{\Gamma(x : A) \vdash M : B \quad \Delta \vdash N : A}{\Gamma(\Delta) \vdash M[N/x] : B}$$

Using this lemma, we obtain the following admissible rules

$$\frac{\Gamma, x : A \vdash M : B \quad \Delta \vdash N : A}{\Gamma, \Delta \vdash M[N/x] : B} \qquad \frac{\Gamma; x : A \vdash M : B \quad \Gamma \vdash N : A}{\Gamma \vdash M[N/x] : B}$$

$$\frac{\Gamma(x : A, y : B) \vdash M : C \quad \Delta \vdash N : A \quad \Delta' \vdash N' : B}{\Gamma(\Delta, \Delta') \vdash M[N/x, N'/y] : B}$$

where the top right rule uses Contraction together with cut.

Lemma 4. *β reduction preserves typing.*
(where reductions are obtained by reading the equations left to right)

2.2 The Affine Variant

The affine variant extends the basic calculus as follows.

AFFINE COHERENT EQUIVALENCE adds

4 $I \equiv 1$

to Coherent Equivalence.

CONVERTABILITY OF "," TO ";"

$$\frac{\Gamma(\Delta; \Delta') \vdash M : A}{\Gamma(\Delta, \Delta') \vdash M : A} \; Conv$$

EQUATIONS. There are additional equations for projections for $*$.

$$(\mathbf{let}\,(x, y) = M * N\,\mathbf{in}\,x) \;=\; M \qquad (\mathbf{let}\,(x, y) = M * N\,\mathbf{in}\,y) \;=\; N$$

Lemma 5. *Weakening for "," is admissible in the affine variant.*

$$\frac{\Gamma(\Delta) \vdash M : A}{\Gamma(\Delta, \Delta') \vdash M : A} \; W,$$

2.3 Trivial Examples

Given a judgement $x : A \vdash x : A$ we cannot immediately use an introduction rule to type an identity function of type $A \mathrel{-\!\!*} A$ or $A \to A$, because to apply an introduction rule for a function type we must have a context of the form $\Gamma, x : A$ or $\Gamma; x : A$. So we need to use coherent equivalence first.

$$\dfrac{\dfrac{\dfrac{}{x : A \vdash x : A}}{1; x : A \vdash x : A}}{1 \vdash \alpha x\,.\,x : A \to A} \qquad \dfrac{\dfrac{\dfrac{}{x : A \vdash x : A}}{I, x : A \vdash x : A}}{I \vdash \lambda x\,.\,x : A \mathrel{-\!\!*} A}$$

Using coherent equivalence we can also mimic the isomorphisms

$$[1, A \to B] \cong [A, B] \cong [I, A \mathrel{-\!\!*} B]$$

of hom sets in a dcc.

$$\dfrac{\dfrac{x : A \vdash M : B}{1; x : A \vdash M : B}}{1 \vdash \alpha x\,.\,M : A \to B} \qquad \dfrac{\dfrac{x : A \vdash M : B}{I, x : A \vdash M : B}}{I \vdash \lambda x\,.\,M : A \mathrel{-\!\!*} B}$$

$$\dfrac{\dfrac{1 \vdash M : A \to B \quad \dfrac{}{x : A \vdash x : A}}{1; x : A \vdash Mx : B}}{x : A \vdash Mx : B} \qquad \dfrac{\dfrac{I \vdash M : A \mathrel{-\!\!*} B \quad \dfrac{}{x : A \vdash x : A}}{I, x : A \vdash M@x : B}}{x : A \vdash M@x : B}$$

$A \to B$ and $A \mathrel{-\!\!*} B$ are not convertable to one another in general, but in the affine variant we can go from the former to the latter.

$$\dfrac{\dfrac{\dfrac{\dfrac{\dfrac{}{f : A \to B \vdash f : A \to B}}{f : A \to B, x : A \vdash f : A \to B}\,W, \quad \dfrac{\dfrac{}{x' : A \vdash x' : A}}{f' : A \to B, x' : A \vdash x' : A}\,W,}{(f : A \to B, x : A); (f' : A \to B, x' : A) \vdash fx' : B}\,\to E}{f : A \to B, x : A \vdash fx : B}\,C}{f : A \to B \vdash \lambda x.\,fx : A \mathrel{-\!\!*} B}\,\mathrel{-\!\!*} I$$

2.4 Unusual Examples

In the $\alpha\lambda$-calculus we can have a multiplicative function that uses its argument many times. For example, in the following, a variable abstracted using λ, the multiplicative abstraction, appears multiple times in the body of the term.

$$\dfrac{\dfrac{\dfrac{\dfrac{\dfrac{\vdots}{x; f \vdash f\,x : A \to B} \qquad \dfrac{\vdots}{x; f \vdash x : A}}{x : A; f : A \to A \to B \vdash (f\,x)\,x : A \to B}\,C, \to E}{x : A \vdash \alpha f\,.\,(f\,x)\,x : ((A \to A \to B) \to B)}\,\to I}{I, x : A \vdash \alpha f\,.\,(f\,x)\,x : ((A \to A \to B) \to B)}\,\equiv}{I \vdash \lambda x\,.\,\alpha f\,.\,(f\,x)\,x : A \mathrel{-\!\!*} ((A \to A \to B) \to B)}\,\mathrel{-\!\!*} I$$

Here, in the key, top-pictured, step we use the admissible rule for $\rightarrow$ elimination (or equivalently we use $\rightarrow E$ followed by Contraction, with suitable renaming of premises).

This term seems unusual, or wrong, if one thinks of a number-of-uses reading. But it is justified by the sharing interpretation. To see why, consider that the subterm $f\,x$ is of type $A \rightarrow B$. According to the sharing interpretation, it is allowed to share with its argument, in this case x, which is why $(f\,x)\,x$ is reasonable. On the other hand, the sharing interpretation would not support an application $(f@x)@x$ where f had type $A \mathrel{-\!\!*} A \mathrel{-\!\!*} B$.

Similarly, we can have a multiplicative function that doesn't use its argument at all.

$$\dfrac{\dfrac{\dfrac{\dfrac{\overline{y : B \vdash y : B}}{x : A; y : B \vdash y : B}\ W}{x : A \vdash (\alpha y \,.\, y) : B \rightarrow B}\ \rightarrow I}{I, x : A \vdash (\alpha y \,.\, y) : B \rightarrow B}\ \equiv}{I \vdash \lambda x \,.\, (\alpha y .\, y) : A \mathrel{-\!\!*} (B \rightarrow B)}\ \mathrel{-\!\!*} I$$

It is instructive to compare the corresponding types in linear type theory. For the first example, the type would be $A \multimap\, !(!A \multimap\, !A \multimap B) \multimap B$. In trying to derive a term we could λ-abstract on $x : A$ and function parameter f. But then, to apply (the dereliction of) f to x, we would need to convert x to something of type $!A$, and we cannot do a conversion from A to $!A$ in general. Similarly, for the type $A \multimap\, !B \multimap B$ we can abstract on $x : A$ and $y : !B$, but we cannot throw x away.

These examples serve to illustrate that the idea that a multiplicative function uses its argument exactly once does not directly carry over to $\alpha\lambda$.

3 Two Models

In this section we give two simple models, which express some aspects of the informal sharing interpretation.

The definition of the $\alpha\lambda$-calculus is close enough to its models – it was, in fact, extracted from them – that the interpretation of (derivations of) typing judgements should be evident. We concentrate on models themselves here. (A thorough account of the relation between syntax and semantics, including coherence and a completeness result, may be found in [20].)

3.1 A Linear Model

Let $\mathcal{B}$ be the category of finite sets and bijections. The functor category $\mathbf{Set}^{\mathcal{B}}$ will be used as a model of the $\alpha\lambda$-calculus.

We think of $\mathcal{B}$ here as a category of possible worlds, where each world X determines a finite collection of resources. For a functor A and element $a \in AX$, we regard a as a computational entity of type A that has access to X.

The structure of the additives is determined pointwise; on objects it is

$$\begin{aligned} 1X &= \{*\} \\ (A \wedge B)X &= AX \times BX \\ (A \to B)X &= AX \Rightarrow BX. \end{aligned}$$

Here, $\Rightarrow$ is function space in **Set**, and $\times$ is cartesian product. The exponent $A \to B$ in a functor category is usually represented as

$$(A \to B)X = \mathbf{Set}^{\mathcal{B}}[\mathcal{B}[X,-] \wedge A, B]$$

but in the special case that all morphisms of $\mathcal{B}$ are isomorphisms this is equivalent to the pointwise representation.

Notice how the pointwise definition corresponds closely to the informal reading of $\to$ in the sharing interpretation, where an additive function and its argument access the same resources. The additive function type has a strongly local character, where the application of a function stays located at a given world.

The multiplicative function type, in contrast, explicitly refers to other worlds, which are set apart from X through the use of $+$.

$$(A \mathrel{-\!\!*} B)X = \mathbf{Set}^{\mathcal{B}}[A(-), B(X + -)]$$

Here, $+$ is the evident functor on $\mathcal{B}$ given by disjoint union of sets. The absence of X in $A(-)$ mirrors the informal description of multiplicative functions as disjoint from their arguments. An element $p \in (A \mathrel{-\!\!*} B)X$ accepts a world Y and element $a \in AY$ as arguments, and produces $p[Y]a \in B(X+Y)$. The "resources" for p are X, while those for a are Y, and these are separate in the result type by virtue of their positions in the combined world $X + Y$.

We give an example to illustrate the sharing aspect. Consider the inclusion functor $L : \mathcal{B} \longrightarrow \mathbf{Set}$. For each finite set X, we think of $LX = X$ as a set of names, or locations. Let N be the constant functor delivering the natural numbers, and define

$$S = L \to (1 \vee (N \wedge L))$$

where $\vee$ is the coproduct of functors (which is defined pointwise). Because of the pointwise definition of $\to$ we have that $SX = X \Rightarrow \{*\} + (N \times X)$. We regard an element $s \in SX$ as a representation of a portion of a computer store, where each $x \in X$ is a pointer to a linked list (possibly with loops).

Now consider any function $f \in ((S \wedge L) \to ((S \wedge L) \mathrel{-\!\!*} S))X$. f accepts $(s, x) \in SX \times X$ and $(s', y) \in SY \times Y$, for finite set Y, as arguments, and produces a state in $S(X+Y)$ as a final result. From the point of view of $S(X+Y)$, there is no overlap between x and y, or between the other pointers in the list pointed to by x and those pointed to by y. Thus, we can view f as a procedure that accepts two linked lists as arguments, with the proviso that the two input lists are defined using disjoint collections of pointers. This kind of proviso is often required in the statement of correctness of an algorithm that, say, removes the elements of one list that appear in the other.

On the other hand, consider the type $L \rightarrow (L \twoheadrightarrow (S \rightarrow S))$. A function of this type would accept two pointers to linked lists as arguments, and the two pointer arguments would again have to be distinct, but now they could point to lists that overlap in the store.

No particular practical significance is claimed for this example; it is offered just as a concrete illustration of how $\twoheadrightarrow$ and $\rightarrow$ can express sharing properties. But in a future paper we plan to show how a type system based on the linear $\alpha\lambda$-calculus can be used to control pointer aliasing in an imperative language.

Returning to the definition of the model, the multiplicative unit is I where $I(\{\}) = \{*\}$, and $I(X) = \{\}$ for all other X. We refer to Section 4 for the definition of $*$ (a concrete representation of it is given below). However, even prior to definition it is useful to observe that a multi-map characterization of maps out of $A * B$ is forced by the definition of $\twoheadrightarrow$. That is, if we are to have the isomorphism $\mathbf{Set}^{\mathcal{B}}[A * B, C] \cong \mathbf{Set}^{\mathcal{B}}[A, B \twoheadrightarrow C]$, then we must obtain the following [7].

Maps $p : A * B \longrightarrow C$ out of a tensor are in bijection with families of functions

$$\overline{p}[X][Y] : AX \times BY \longrightarrow C(X + Y) \ ,$$

natural in X and Y.

The idea in terms of sharing is that the components of $*$ are assigned different resources (this is in line with the form of semantics proposed by Reynolds for syntactic control of interference [15]).

Proposition 6. $\mathbf{Set}^{\mathcal{B}}$ *with this data is a cartesian dcc.*

Let us reconsider the first example from Section 2.4 in light of this model. The judgement

$$I \vdash \lambda x \,.\, \alpha f \,.\, (f\, x)\, x : A \twoheadrightarrow ((A \rightarrow A \rightarrow B) \rightarrow B)$$

determines an element $p \in A \twoheadrightarrow ((A \rightarrow A \rightarrow B) \rightarrow B)\{\}$ (where we indulge in a confusion between types and objects in $\mathbf{Set}^{\mathcal{B}}$). It accepts a world X and $a \in AX$, and produces (using the isomorphism $\{\} + X \cong X$) a function $p[X]a \in ((A \rightarrow A \rightarrow B) \rightarrow B)X$. By the pointwise definition of $\rightarrow$, this is a function of type $(AX \Rightarrow AX \Rightarrow BX) \Rightarrow BX$ in **Set**, and it is the expected function that maps f to $(fa)a$.

Remark 7 A concrete representation of the multiplicative product can be given as follows. If n and m are natural numbers let $[n|m]$ denote the set $\{n, \ldots, m-1\}$ and let $|X|$ denote the size of a finite set X. Then

$$(A * B)X \ \cong \ \{\langle n, m, a \in A[0|n], b \in B[n|m]\rangle \mid n + m = |X|\}.$$

Remark 8 It is important to see that there is no hidden Weakening or Contraction for "," lurking in the examples of terms that use their arguments two or zero times. In fact, we can see that these rules are absent in $\mathbf{Set}^{\mathcal{B}}$ in a very strong sense; there are not even any candidate maps of the required types to model them, let alone maps with the proper properties.

To model Contraction we would need maps of shape $A \longrightarrow A * A$. But there are no maps $L \longrightarrow L * L$, where L is the inclusion from $\mathcal{B}$ to **Set**. To see why, given $a \in L\{a\}$ we would have to produce an element in $(L * L)\{a\}$, but this set is empty. For, the representation of $*$ just given implies that an element would have to be of the form $\langle 0, 1, b \in L\{\}, b' \in L\{0\}\rangle$ or $\langle 1, 0, b \in L\{0\}, b' \in L\{\}\rangle$, but there are no such elements as $L\{\}$ is empty.

To model Weakening, we would need maps $A \longrightarrow I$, for all A. But there are no maps $1 \longrightarrow I$.

3.2 An Affine Model

Let $\mathcal{I}$ denote the category of finite sets and injective functions. The functor category $\mathbf{Set}^{\mathcal{I}}$ is cartesian closed, with finite products defined pointwise. The additive function type can be given a special representation, using the fact that any morphism in $\mathcal{I}$ factors into an injection $X \to X+Y$ into the left component of a disjoint union, followed by an isomorphism:

$$(A \to B)(X) \;=\; \mathbf{Set}^{\mathcal{I}}[A(X+-), B(X+-)].$$

This accurately reflects the informal reading from Section 1.2, in that the presence of X in the argument type $A(X+-)$ indicates how a function $p \in (A \to B)X$ may share access to X with its argument.

The multiplicative function type once again expresses disjointness of a function from its argument:

$$(A \mathrel{-\!\!*} B)X = \mathbf{Set}^{\mathcal{I}}[A, B(X+-)]$$

where $+$ is the functor on $\mathcal{I}$ given by disjoint union of finite sets.

In Section 2.3 we showed how to convert $\to$ to $\mathrel{-\!\!*}$ in the affine $\alpha\lambda$-calculus. In this model, the conversion takes a natural transformation $A(X+-) \longrightarrow B(X+-)$ and composes on the left with the map $A \longrightarrow A(X+-)$ that sends $a \in AY$ to $A(inr)a \in A(X+Y)$, where inr is the right injection. Here, an additive function in world X is applied to an argument $a \in AY$ that doesn't happen to depend on X.

Once again we refer to the following section for $*$, and simply state

Proposition 9. $\mathbf{Set}^{\mathcal{I}}$ *is an affine dcc.*

We can try to use the inclusion functor $L : \mathcal{I} \longrightarrow \mathbf{Set}$ as a variant on the functor used to illustrate the linear sharing interpretation, but it has something of a different character in the affine model. It would not be as reasonable to think of $s \in (L \to (1 \vee (N \wedge L)))X$ as a state, because s would have to accept other worlds Y, and potentially $y \in LY$, as arguments. So the development above, for the linear interpretation, does not carry through well to the affine case. However, a more thorough account of the sharing aspect of the affine model is given in Sections 5 and 6, where we study study syntactic control of interference.

Remark 10 When working with pullback preserving functors (which we always will), a concrete description of $*$ is possible. The basic intuition is that we can define the *support* of any computational value, as the smallest collection of resources upon which it depends. For $a \in AX$ define $supp(a)$ to be the smallest subset $Y \subseteq X$ such that $a \in range(Y \hookrightarrow X)$). Pullback preservation is enough to guarantee existence of such supports [12]. Then, for pullback preserving functors

$$(A * B)X \cong \{\langle a, b\rangle \in AX \times BX \mid supp(a) \cap supp(b) = \{\}\}.$$

Remark 11 There is no functor $! : \mathbf{Set}^{\mathcal{I}} \to \mathbf{Set}^{\mathcal{I}}$ admitting an isomorphism $!A \twoheadrightarrow B \cong A \to B$. To see why, consider the constant functor 2 which delivers the two element set $\{t, f\}$. Then

$$(A \twoheadrightarrow 2)X = \mathbf{Set}^{\mathcal{I}}[A, 2(X + -)] = \mathbf{Set}^{\mathcal{I}}[A, 2]$$

is independent of X, and so $A \twoheadrightarrow 2$ is a constant functor. On the other hand, $(A \to 2)X = \mathbf{Set}^{\mathcal{I}}[A(X + -), 2]$ depends on X, and is not necessarily (isomorphic to) a constant functor. For instance, if L is the inclusion functor from $\mathcal{I}$ into **Set**, then $(L \to 2)\{\}$ has two elements, corresponding to the two constant functions into $\{t, f\}$. On the other hand, $(L \to 2)\{a, b\}$ has elements that are not in the range of $(L \to 2)(f : \{\} \hookrightarrow \{a, b\})$. One such maps a to t and b to f (and all other inputs to, say, f). Therefore, no matter what "!" we try to pick, $!L \twoheadrightarrow 2$ will be a constant functor, while $L \to 2$ is not, so they cannot be isomorphic. This indicates that a dcc is not simply a model of linear logic in disguise.

4 Day's Construction

The material in the previous section can be regarded as two worked examples, of specific instances of a general construction due to Brian Day [7]. He shows that any (small) monoidal category $(\mathcal{C}, *, I)$ induces a monoidal closed structure on $\mathbf{Set}^{\mathcal{C}^{op}}$, and that when $(\mathcal{C}, *, I)$ is symmetric monoidal so is $\mathbf{Set}^{\mathcal{C}^{op}}$. This, combined with the standard fact that $\mathbf{Set}^{\mathcal{C}^{op}}$ is bicartesian closed, yields a bicartesian dcc.

We have already seen $\twoheadrightarrow$: given functors A and B,

$$(A \twoheadrightarrow B)Z = \mathbf{Set}^{\mathcal{C}^{op}}[A, B(Z * -)].$$

The formula for the tensor product is written using a coend:

$$(A * B)Z = \int^{X,Y} AX \times BY \times \mathcal{C}[Z, X * Y].$$

It is sometimes possible to give an explicit description of $*$ without using coends, as we did in remarks in the previous section. The unit I of the monoidal structure is $\mathcal{C}[-, I]$. The formulas for $(A * B)Z$ and $(A \twoheadrightarrow B)Z$ are both contravariant in Z, giving the morphism parts of the functors.

Although the models given in the last section are instances of this structure, their connection to the sharing interpretation does not fall out from it. Put another way, not all instances of Day's construction would be consistent with the informal sharing reading, such as $\mathbf{Set}^{\mathcal{C}^{op}}$ where the tensor in $\mathcal{C}$ admits Contraction. We wonder if there are abstract properties of $\mathcal{C}$, together with an accompanying analysis, that could provide an axiomatic understanding of the essence of the "resource" aspect of $\alpha\lambda$.

Besides giving us a host of models, Day's construction enables us to make remarks about full and faithful embeddings. Faithfulness is the semantic counterpart of a syntactic conservativity result, while fullness says that adding such structure does not cause any new maps to added, when we focus on just ccc or smcc types being embedded.

We can embed a ccc in a dcc in a trivial way, by regarding it as a dcc in which the two closed structures coincide. This implies the conservativity of the equality on $\alpha\lambda$-terms given by cartesian dcc's over that for simply-typed λ-calculus. For smcc's we refer to a result of [8], which says that the Yoneda embedding takes symmetric monoidal closed structure on a small category $\mathcal{C}$ to that structure just described on $\mathbf{Set}^{\mathcal{C}^{op}}$.

These embeddings raise the question of a "purely functional" understanding of $\alpha\lambda$. For example, we could formulate a model consisting of (certain, [19,14]) functors from the category of coherence spaces and linear functions to a category of cpo's (with bottom) and continuous functions. This gives us a model of $\alpha\lambda$ which, when restricted to multiplicative types, agrees with the coherence space model of linear logic. But for terms that mix multiplicatives and additives there would be strange behaviour, from the point of view of coherence spaces, as the examples from Section 2.4 show. So, although it is possible to define such a model, the proper meaning to attach to it is not clear.

5 Interference Control and Affine λ-calculus

In this and the next section we develop the affine model from Section 3.2, and show how the $\alpha\lambda$-calculus can be used to extend syntactic control of interference (SCI). We begin with an introduction to SCI, focusing on the sharing interpretation for it, and properties desired of an extension.

The central statement of imperative programming is the assignment $x := e$, which overwrites the contents of a cell, or location, denoted by x. Imperative languages give rise to the phenomenon of *interference* [21], where executing one statement can affect another when they share access to the same cells. In particular, there can be *covert* interference, where seemingly unrelated statements, such as $x := y$ and $z := w$, can affect one another; this can happen when the identifiers x and z are aliases (denote the same cell).

Contraction is a source of aliasing in imperative programming. In

$$((\lambda y \lambda z . \cdots y := 2 \cdots \mathbf{if}\ z = 3\ \mathbf{then} \cdots) x) x$$

if x denotes a cell c, then that same cell will be passed to both x and y. A result of this is that, in the body of the λ-expression, the statement $y := 2$ will set the contents of c to 2. This, in turn, will affect the truth of the condition $z = 3$, because that condition checks the contents of cell c, to see if it is 3.

To enable this passing of c to both y and z we have to have Contraction, either explicitly or as an admissible rule, in order to get two occurrences of x in an application $(Mx)x$. SCI rejects Contraction by using affine λ-calculus as its type system.

5.1 Basic SCI

We work with a version of SCI whose types are as follows.

$$\rho ::= \mathbf{exp} \mid \mathbf{cell} \mid \mathbf{comm} \qquad \text{primitive types}$$
$$\theta ::= \rho \mid \theta \wedge \theta' \mid \theta \mathrel{-\!\!*} \theta' \qquad \text{types}$$

The primitive type **exp** is the type of natural number-valued expressions, **comm** is the type of commands, and **cell** is the type of storage cells, or locations.

We have used $\mathrel{-\!\!*}$ to emphasize that functions in SCI are multiplicative.

AFFINE λ-CALCULUS

$$\frac{}{x:\theta \vdash x:\theta}\ Id \qquad \frac{\Gamma \vdash M:\theta}{\Delta \vdash M:\theta}\ Ex \text{ (where } \Delta \text{ is a permutation of } \Gamma)$$

$$\frac{\Gamma \vdash M:\theta'}{\Gamma, x:\theta \vdash M:\theta'}\ W$$

$$\frac{\Gamma, x:\theta \vdash M:\theta'}{\Gamma \vdash \lambda x:\theta\,.\,M : \theta \mathrel{-\!\!*} \theta'}\ \mathrel{-\!\!*} I \qquad \frac{\Gamma \vdash M:\theta \mathrel{-\!\!*} \theta' \quad \Delta \vdash N:\theta}{\Gamma, \Delta \vdash MN:\theta'}\ \mathrel{-\!\!*} E$$

$$\frac{\Gamma \vdash M:\theta \quad \Gamma \vdash N:\theta'}{\Gamma \vdash \langle M, N\rangle : \theta \wedge \theta'}\ \wedge I \qquad \frac{\Gamma \vdash M:\theta_1 \wedge \theta_2}{\Gamma \vdash \pi_i M:\theta_i}\ \wedge E \text{ (where } i \text{ is 1 or 2)}$$

A typing context Γ here is a list of assumptions $x : \theta$ pairing identifiers with types, with the proviso that no identifier appears twice.

SELECTED SCI-SPECIFIC RULES

$$\frac{\Gamma \vdash N:\mathbf{exp}}{\Gamma \vdash \mathbf{succ}\ N:\mathbf{exp}} \qquad \frac{\Gamma \vdash N:\mathbf{exp}}{\Gamma \vdash \mathbf{pred}\ N:\mathbf{exp}}$$

$$\frac{}{\Gamma \vdash 0:\mathbf{exp}} \qquad \frac{\Gamma \vdash N_1:\mathbf{exp} \quad \Gamma \vdash N_i:\mathbf{comm}\ ,\ i=2,3}{\Gamma \vdash \mathbf{if}\ N_1 = 0\ \mathbf{then}\ N_2\ \mathbf{else}\ N_3 : \mathbf{comm}}$$

$$\frac{x:\theta \vdash M:\theta}{\vdash \mathbf{rec}\,x\,.\,M:\theta} \qquad \frac{\Gamma, x:\mathbf{cell} \vdash M:\mathbf{cell}}{\Gamma \vdash \mathbf{new}\,x\,.\,M\ :\mathbf{comm}}$$

$$\frac{}{\Gamma \vdash \mathbf{skip}:\mathbf{comm}} \qquad \frac{\Gamma \vdash M:\mathbf{comm} \quad \Gamma \vdash N:\mathbf{comm}}{\Gamma \vdash M;N:\mathbf{comm}}$$

$$\frac{\Gamma \vdash M:\mathbf{cell}}{\Gamma \vdash\, !M:\mathbf{exp}} \qquad \frac{\Gamma \vdash M:\mathbf{cell} \quad \Gamma \vdash N:\mathbf{exp}}{\Gamma \vdash M := N:\mathbf{comm}}$$

Of these constructs, $!M$ is the operation that reads the contents of a cell, and **new** allocates a fresh cell (which is put on the runtime stack).

5.2 The Sharing Interpretation of SCI

We saw above how eliminating Contraction could rule out one instance of aliasing. More generally, the absence of aliasing is subsumed under the

DISJOINTNESS POLICY: distinct identifiers never interfere.

The SCI sharing interpretation of types is as follows.

$A \multimap B$: functions that don't interfere with their arguments.
$A \wedge B$: pairs that may interfere with one another.

If we substitute "share resources" for interfere, then the reading of $\multimap$ is just the one we gave for the affine case in Section 1.2.

It is important to realize how this is an unusual reading of the affine λ-calculus. Often, the idea in the affine calculus is that a function uses its argument at most once, so that for instance in a function of type $A \wedge B \multimap C$ either the A or the B component may be used, but not both. But according to SCI's reading, it is perfectly reasonable for a function p of such a type to use either or both components of a pair $\langle a, b\rangle$ supplied to it as an argument, and either of these elements could be used many times. The only constraint is that p doesn't interfere with $\langle a, b\rangle$.

For example, in SCI we can write a function

$$(\lambda c : \mathbf{comm} \wedge \mathbf{comm} \,.\, \pi_1 c\,;\, \pi_2 c\,;\, \pi_1 c) : \mathbf{comm} \wedge \mathbf{comm} \to \mathbf{comm}$$

that uses the first component of a pair twice and the second component once.

The sharing reading also helps to understand the typing of **if**. In the number-of-uses reading, in $\mathbf{if}\, N_1 = 0 \,\mathbf{then}\, N_2 \,\mathbf{else}\, N_3$ one would expect to use one context for N_1, and a separate context for N_2 and N_3. But the conditional essentially corresponds to a constant of type $\mathbf{exp} \wedge \mathbf{comm} \wedge \mathbf{comm} \multimap \mathbf{comm}$ in SCI and there is no inconsistency if all the N_i's share the same context. And in imperative programming this sharing is often wanted, so that information can pass from the condition into the branches.

Now the affine calculus certainly does not *force* the sharing reading. But it is consistent with it. The pure affine calculus is actually too small for this "many uses" aspect to be seen; the additional features of SCI are where it comes out. The pure $\alpha\lambda$-calculus, in contrast, already admits multiplicative functions that use their arguments many times, as we saw in Section 2.4. This is why $\alpha\lambda$ is consistent with the sharing reading but not, as far as we are aware, with a straightforward adaptation of the number-of-uses reading.

5.3 A Model

We can describe a model of SCI using the category $\mathcal{I}$ of finite sets and injections from Section 3.2. To cope with recursion, we use $\mathbf{Predom}^{\mathcal{I}}$ in place of $\mathbf{Set}^{\mathcal{I}}$, where **Predom** is the category of predomains (ω-complete posets and continuous maps). The definition of the dcc structure is as in Section 3.2, with small adjustments to account for order.

For X a finite set we define

$$[\![\mathbf{comm}]\!]X = SX \Rightarrow SX_\bot$$

$$[\![\mathbf{exp}]\!]X = SX \Rightarrow N_\bot$$

$$[\![\mathbf{cell}]\!]X = X_\bot$$

Here, $SX = X \Rightarrow N$ is the set of states at world X, and N is the set of natural numbers. The action of each primitive type on morphisms f in $\mathcal{I}$ is defined by renaming cells according to f, and ignoring cells not in the range of f.

Recursion is interpreted as follows. If $[\![M]\!] : [\![A]\!] \longrightarrow [\![A]\!]$, then we require a map $[\![\mathbf{rec}\, x.\, M]\!] : 1 \longrightarrow A$. The definition is that $[\![\mathbf{rec}\, x.\, M]\!]X*$ is the least fixed point of the function $[\![M]\!]X : AX \Rightarrow AX$. For existence, this definition requires the observation that each $[\![A]\!]X$ has a least element, and for naturality that all maps $[\![A]\!]f : AX \to AY$ are strict as well as continuous [19].

We will not give the detailed semantics of other terms, but we comment on the sense in which the semantics of $-\!\!*$ faithfully reflects the sharing interpretation. Consider the type $\mathbf{cell} -\!\!* \mathbf{cell} -\!\!* \mathbf{comm}$. Semantically, an element $p \in [\![\mathbf{cell} -\!\!* \mathbf{cell} -\!\!* \mathbf{comm}]\!]\{\}$ accepts

two worlds Y and Z,
cells $c \in Y_\bot$ and $e \in Z_\bot$

and produces (using $\{\} + Y + Z \cong Y + Z$)

$$p[Y]c[Z]e : S(Y + Z) \Rightarrow S(Y + Z)_\bot$$

It is evident from this that the arguments c and e cannot be aliases, as (presuming neither is $\bot$) they live in disjoint portions of the store at world $Y + Z$.

This model of SCI uses the multiplicatives $-\!\!*$, I, and $*$ (which is used to interpret typing contexts), along with the additive $\wedge$. However, the model also contains the additive function type $\to$, which can separately be used to model Idealized Algol, a language based on the simply-typed λ-calculus [22, 16]. This observation leads to the question of whether there is a semantically natural enveloping language, that contains both SCI and Idealized Algol.

Before describing how the $\alpha\lambda$-calculus can be used to answer this question, we discuss why we might want to do so.

5.4 Limitations

There are two specific cases when the disjointness policy of SCI appears overly restrictive [21, 23]. First, notice that the rule for recursion in SCI is restricted so that $\mathbf{rec}\, x\,.M$ is a closed term. The reason for this restriction is that, if an assignment to a free identifier y were to occur, as in $(\mathbf{rec}\, x.\ \cdots y := e \cdots x \cdots)$, then y would interfere with x, violating the disjointness policy. The problem can also be seen with a fixed-point operator $\mathbf{Y}$: an unwinding $\mathbf{Y}F \rhd F(\mathbf{Y}F)$ would violate the requirement of affine typing that the free identifiers of a function and argument be disjoint, unless F is closed.

The second limitation is jumps. To see the difficulty, consider a label declaration block **escape** x **in** M. This declares a new label which, when jumped to from within M, results in a transfer of control to the end of the block. From the point of view of continuation semantics, it binds x to the current continuation, which is a function from states to final answers that describes computation that will take place after the block is finished. This means that, if the computation associated with the current continuation changes any storage variable then x will interfere with that storage variable. So, in $(\mathbf{escape}\ x\ \mathbf{in}\ M);\ z := 4$ the identifiers z and x interfere, if z occurs within M. Thus, from the point of view of continuation semantics, the **escape** statement violates the requirement that distinct identifiers never interfere (unless we put rather draconian conditions on identifiers appearing in or following an **escape** block).

A solution is to relax the disjointness requirement of SCI by using the $\alpha\lambda$-calculus. A bunch Γ, Δ will indicate that identifiers appearing in Γ do not interfere with any identifiers in Δ, while the combination $\Gamma; \Delta$ will allow interference to occur. Then, when a recursively defined x has an assignment to y in its body, the typing rule for recursion must ensure that x and y are separated by ";" during the typing of the body, indicating that interference might occur . We will not give the solution for jumps in this preliminary paper, but the idea is similar: when typing an **escape** block we require a declared label to sit in additive combination with other identifiers appearing freely in its body.

6 An Enveloping Language

The enveloping language, SCI+, uses the affine $\alpha\lambda$-calculus as its type system. The primitive types are the same as those given for SCI in Section 5.1, as are all of the language-specific rules, with the exception of recursion.

The SCI+ rule for recursion allows for free identifiers, as long as they are in additive combination with x.

$$\frac{\Gamma; x : \theta \vdash M : \theta}{\Gamma \vdash \mathbf{rec}\, x\,.\,M : \theta}\ \text{SCI+}\,\mathbf{rec}$$

The sense in which SCI+ allows detection of interference is that, whenever we see a sequence $\alpha x\,\lambda y$ or $\lambda x\,\lambda y$, we know that x and y don't interfere. So, non-interference can be inferred (in a fail-safe manner) from a simple inspection of a context. The one difference is that in Basic SCI this determination is context free. It is context sensitive in SCI+ because when we see $\alpha x\, \alpha y$ or $\lambda x \alpha y$ we don't know if x and y interfere or not.

Thus, the combination of additive and multiplicative features using bunches gives rise to a flexible form of interference control, where it is possible to switch between interference and non-interference. It allows programs that violate the disjointness policy of SCI to be accepted in a local context, but then embedded in larger contexts where the policy remains in effect. An example of this is given at the end of the section.

6.1 Mappings

Idealized Algol is similar to SCI, except that it has a more general rule for recursion, and it uses the full simply-typed λ-calculus as its type system. Formally, define IA to be the language in Section 5.1, with the addition of the two rules

$$\frac{\Gamma, x:\theta', y:\theta' \vdash M:\theta}{\Gamma, y:\theta' \vdash M[y/x]:\theta}\ \mathrm{IA}\,C \qquad \frac{\Gamma, x:\theta \vdash M:\theta}{\Gamma \vdash \mathbf{rec}\, x\,.\, M:\theta}\ \mathrm{IA\,rec}$$

and with the symbol $\rightarrow$ replacing $-\!\!*$ everywhere.

Proposition 12. *1. SCI+ has IA as a sublanguage. That is, if*

$$x_1 : A_1, ..., x_n : A_n \vdash M : B$$

in IA then

$$x_1 : A_1; ...; x_n : A_n \vdash M^\star : B$$

in SCI+, where $(\cdot)^\star$ maps λ to α, and everything else (inductively) to itself.
2. SCI+ has SCI as a sublanguage. That is, if

$$x_1 : A_1, ..., x_n : A_n \vdash B$$

in SCI then

$$x_1 : A_1, ..., x_n : A_n \vdash M^\circ : B$$

where $(\cdot)^\circ$ maps MN to $M^\circ @ N^\circ$ and everything else (inductively) to itself.

Most rules translate directly; the only exceptions are the SCI rule for recursion and the IA rule for **new**. SCI+ has the SCI recursion as a special case, using $\Gamma \equiv 1$ and a coherent equivalence. The IA version of **new** translates to

$$\frac{\Gamma; x:\mathbf{cell} \vdash M:\mathbf{comm}}{\Gamma \vdash \mathbf{new}\, x\,.\, M:\mathbf{comm}}$$

We can derive this at once using the SCI+ rule for **new** and the *Conv* rule of affine $\alpha\lambda$.

6.2 Semantics

The category described in Section 5.3 can be used to interpret the types and the $\alpha\lambda$ typing rules, and semantic valuations for all the SCI+-specific terms can follow the standard route taken in functor category semantics [16,19,22]. We just indicate the treatment of recursion. If $[\![M]\!] : [\![\Gamma]\!] \times [\![A]\!] \longrightarrow [\![A]\!]$, then we require a map $[\![\mathbf{rec}\, x.\, M]\!] : [\![\Gamma]\!] \longrightarrow [\![A]\!]$. The definition is that $[\![\mathbf{rec}\, x.\, M]\!]Xu$ is the least fixed point of $f : [\![A]\!]X \Rightarrow [\![A]\!]X$, where $f = \lambda a\,.\,[\![M]\!]X\langle u, a\rangle$. This definition is tantamount to giving a fixed-point combinator of type $(A \to A) \to A$, using the additive function type.

The presence of u is the difference from the SCI case. If we were to have attempted to parameterize the definition there, we would have had to contend with $[\![\Gamma]\!] * [\![A]\!]$ instead of $[\![\Gamma]\!] \times [\![A]\!]$. Then, for a fixed u, we could not have considered arbitrary $a \in AX$ as arguments, because of the disjointness requirement of $*$. Furthermore, an explicit attempt to "iterate from $\perp$" to define the recursion would run into iterates that interfere with u which, again because of $*$, would disable the use of $[\![M]\!]$ to iterate further.

Using this model, it is possible to show a sense in which IA and SCI are *semantic* sublanguages of SCI+, adding to Proposition 12.

6.3 An Example

We give an example (with sugar) that violates the disjointness policy of SCI: the Towers of Hanoi program, where disks are moved between pegs.

$$
\begin{array}{l}
\mathit{moveone} : \mathbf{exp} \to \mathbf{exp} \to \mathbf{comm} \\
\vdash\ \mathbf{rec}\, \mathit{movemany}\,.\,\alpha k\, a\, b\, c : \mathbf{exp} \\
\qquad\qquad\qquad\quad \mathbf{if}\ k > 0\ \mathbf{then} \\
\qquad\qquad\qquad\qquad\quad \mathit{movemany}(k-1, a, c, b); \\
\qquad\qquad\qquad\qquad\quad \mathit{moveone}(a, b); \\
\qquad\qquad\qquad\qquad\quad \mathit{movemany}(k-1, c, b, a) \\
: \mathbf{exp} \to \mathbf{exp} \to \mathbf{exp} \to \mathbf{exp} \to \mathbf{comm}
\end{array}
$$

The procedure *moveone* can work by printing a message to the screen, or by recording a move in a global data structure. The point is that *moveone* and *movemany* interfere in the body of the procedure.

To type this using the rule for recursion the crucial point is that, during the typing of the body, we turn interference control off by using the bunch

$$
\begin{array}{l}
\ \ \mathit{moveone} : \mathbf{exp} \to \mathbf{exp} \to \mathbf{comm} \\
;\ \mathit{movemany} : \mathbf{exp} \to \mathbf{exp} \to \mathbf{exp} \to \mathbf{exp} \to \mathbf{comm}
\end{array}
$$

which indicates that *moveone* and *movemany* might interfere. But more globally we can turn interference control back on. For instance, in

$$
\begin{array}{l}
\mathit{moveone} : \mathbf{exp} \to \mathbf{exp} \to \mathbf{comm}\,,\ c : \mathbf{comm} \\
\vdash\ ((\mathbf{rec}\, \mathit{movemany}\,.\,\cdots)7\,1\,2\,3)\,;\,c : \mathbf{comm}
\end{array}
$$

we will know that the two sequentially-composed commands don't interfere. They could, therefore, be permuted without affecting the final result, or even run in parallel.

Acknowledgements

I am especially grateful to David Pym and John Reynolds, both for their original ideas and for numerous discussions, which provided much of the impetus for the work in this paper. Thanks to Martin Hofmann and Guy McCusker for pointing out glitches in an earlier version. This research was supported by the EPSRC and, during a one month visit to Carnegie Mellon in July, 1997, by the NSF.

References

1. S. Abramsky. Computational interpretations of linear logic. *Theoretical Computer Science*, 111(1–2):3–57, April 12 1993.
2. A.R. Anderson and N.D. Belnap. *Entailment: the Logic of Relevance and Necessity, volume I.* Princeton University Press, 1975.
3. A. Barber and G.D. Plotkin. Dual intuitionistic linear logic. Submitted, October 1997.
4. P. N. Benton. A mixed linear and non-linear logic: proofs, terms and models. Proceedings of *Computer Science Logic '94*, Kazimierz, Poland. Springer-Verlag LNCS 933, 1995.
5. P.N. Benton, G.M. Bierman, V.C.V. de Paiva, and J.M.E. Hyland. Linear λ-calculus and categorical models revisited. In E. Börger et al., editors, *Proceedings of the Sixth Workshop on Computer Science Logic*, volume 702 of *Lecture Notes in Computer Science*, pages 61–84. Springer-Verlag, Berlin, 1992.
6. S. Brookes, M. Main, A. Melton, and M. Mislove, editors. *Mathematical Foundations of Programming Semantics, Eleventh Annual Conference*, volume 1 of *Electronic Notes in Theoretical Computer Science*, Tulane University, New Orleans, Louisiana, March 29–April 1 1995. Elsevier Science.
7. B. J. Day. On closed categories of functors. In S. Mac Lane, editor, *Reports of the Midwest Category Seminar*, volume 137 of *Lecture Notes in Mathematics*, pages 1–38. Springer-Verlag, Berlin-New York, 1970.
8. B. J. Day. An embedding theorem for closed categories. In G.M. Kelly, editor, *Category Seminar, Sydney*, volume 420 of *Lecture Notes in Mathematics*, pages 55–64. Springer-Verlag, Berlin-New York, 1974.
9. J.M. Dunn. Consecution formulation of positive R with co-tenability and t. In [2], pp381–391.
10. J.-Y. Girard. Linear logic. *Theoretical Computer Science*, pages 1–102, 1987.
11. J.-Y Girard. Towards a geometry of interaction. *Contemporary Mathematics 92: Categories in Computer Science and Logic*, 69–108, 1989.
12. P. W. O'Hearn. A model for syntactic control of interference. *Mathematical Structures in Computer Science*, 3(4):435–465, 1993.
13. P. W. O'Hearn, A. J. Power, M. Takeyama, and R. D. Tennent. Syntactic control of interference revisited. *Theoretical Computer Science*, 1999. To appear. Earlier version in [6] and in [17].

14. P. W. O'Hearn and U. S. Reddy. Objects, interference and the Yoneda embedding. *Theoretical Computer Science*, to appear. Preliminary version in [6], 1999.
15. P. W. O'Hearn and J. C. Reynolds. From Algol to polymorphic linear lambda-calculus. *J.ACM*, to appear, 1999.
16. P. W. O'Hearn and R. D. Tennent. Semantics of local variables. In M. P. Fourman, P. T. Johnstone, and A. M. Pitts, editors, *Applications of Categories in Computer Science*, volume 177 of *London Mathematical Society Lecture Note Series*, pages 217–238. Cambridge University Press, Cambridge, England, 1992.
17. P. W. O'Hearn and R. D. Tennent, editors. *Algol-like Languages*. Birkhauser, Boston, 1997. Two volumes.
18. P.W. O'Hearn and D.J. Pym. The logic of bunched implications. 1998. Submitted.
19. F. J. Oles. *A Category-Theoretic Approach to the Semantics of Programming Languages*. Ph.D. thesis, Syracuse University, Syracuse, N.Y., 1982.
20. D.J. Pym. The semantics and proof theory of the logic of bunched implications, I: Propositional **BI**. 1998. Available on the web at http://www.dcs.qmw.ac.uk/ ~ pym.
21. J. C. Reynolds. Syntactic control of interference. In *Conference Record of the Fifth Annual ACM Symposium on Principles of Programming Languages*, pages 39–46, Tucson, Arizona, January 1978. ACM, New York. Also in [17], vol 1.
22. J. C. Reynolds. The essence of Algol. In J. W. de Bakker and J. C. van Vliet, editors, *Algorithmic Languages*, pages 345–372, Amsterdam, October 1981. North-Holland, Amsterdam. Also in [17], vol 1.
23. J. C. Reynolds. Syntactic control of interference, part 2. In G. Ausiello, M. Dezani-Ciancaglini, and S. Ronchi Della Rocca, editors, *Automata, Languages and Programming, 16th International Colloquium*, volume 372 of *Lecture Notes in Computer Science*, pages 704–722, Stresa, Italy, July 1989. Springer-Verlag, Berlin.
24. P. Wadler. A syntax for linear logic. In S. Brookes et al., editors, *Mathematical Foundations of Programming Semantics*, volume 802 of *Lecture Notes in Computer Science*, pages 513–529, New Orleans, 1993. Springer-Verlag, Berlin.

A Curry-Howard Isomorphism for Compilation and Program Execution
(Extended Abstract)

Atsushi Ohori*

Research Institute for Mathematical Sciences
Kyoto University, Kyoto 606-8502, Japan
ohori@kurims.kyoto-u.ac.jp

Abstract. This paper establishes a Curry-Howard isomorphism for compilation and program execution by showing the following facts. (1) The set of *A-normal forms*, which is often used as an intermediate language for compilation, corresponds to a subsystem of Kleene's contraction-free variant of Gentzen's intuitionistic sequent calculus. (2) Compiling the lambda terms to the set of A-normal forms corresponds to proof transformation from the natural deduction to the sequent calculus followed by proof normalization. (3) Execution of an A-normal form corresponds to a special proof reduction in the sequent calculus. Different from cut elimination, this process eliminates left rules by converting them to cuts of proofs corresponding to closed values. The evaluation of an entire program is the process of inductively applying this process followed by constructing data structures.

1 Introduction

Curry-Howard isomorphism [3, 11] is one of most influential concepts in design and analysis of programming languages. It reveals the exact correspondence between the typed lambda calculus and the natural deduction proof system: typing derivations correspond to proofs and β reduction corresponds to proof normalization. This notion is, however, not entirely appropriate for an actual programming language because of the apparent mismatch between β reduction and language implementation. In actual programming languages (except for some interpreted languages) a program is not β reduced but instead is compiled to a low-level code and then executed by an (abstract) machine. Because of this mismatch, the profound correspondence between β reduction and proof normalization does not have much significance in language implementation. If Curry-Howard isomorphism is extended to implementation process, then research on compilation and implementation would be greatly benefited through high-level logical analysis made available by the extended isomorphism. This would be particularly

* This work was partly supported by the Japanese Ministry of Education Grant-in-Aid for Scientific Research on Priority Area no. 275 "Advanced databases," and by the Parallel and Distributed Processing Research Consortium, Japan.

useful for recent active researches on types in compilation where compilation is directed by typing derivation. The goal of this paper is to establish a Curry-Howard isomorphism for compilation and program execution.

There are several formalisms for compilation and program execution. Here we base our development on the work by Flanagan et. al. [6] for a call-by-value functional language using an intermediate language called *A-normal forms*, which is equivalent to the language obtained from CPS terms by "un-CPS" transformation [5, 19] that eliminates continuation. In this formalism, compilation is modeled by transformation from lambda terms into A-normal forms, and program execution is defined by an abstract machine for A-normal forms. As forcefully argued by Flanagan et al, compiling into A-normal forms can be regarded as "the essence" of compiling a functional language, and the execution model for A-normal forms closely reflects an actual implementation of a functional language using environments. They give a simple linear time compilation algorithm and demonstrate that it can be used as a basis for an efficient practical compiler through their experimentation. Because of these facts, we also believe that compiling with A-normal forms can serve as an realistic model for efficient implementation of functional languages.

Our specific goal is therefore to develop logical foundations for compiling the lambda terms to the set of A-normal forms and for evaluation of A-normal forms. We achieve this goal by establishing the following facts.

1. A logic that corresponds to a language for mechanical execution in a conventional computer system is a Gentzen-style sequent calculus, which represents finer notion of computation than the natural deduction system. Instead of performing general substitution, it decomposes a computation on a data type into smaller structures by the corresponding left rule. In particular, Kleene's [13] contraction-free sequent calculus, denoted here by $\mathcal{GK}$, serves as a logic for an implementation language. The set of A-normal forms is identified with a subsystem $\mathcal{GKA}$ whose proofs are those of $\mathcal{GK}$ in a certain normal form.
2. A compilation algorithm from lambda terms to A-normal forms in the style of [6] corresponds to the composition of a proof transformation from the natural deduction system (denoted here by $\mathcal{N}$) to $\mathcal{GK}$ and a proof normalization from $\mathcal{GK}$ to $\mathcal{GKA}$.
3. Execution of an A-normal form corresponds to a special proof reduction process in $\mathcal{GKA}$. Different from cut elimination, this process eliminates left rules by converting them to cuts of proofs corresponding to closed values. The evaluation of an entire program is the process of inductively applying this process followed by constructing data structures. This process exactly corresponds to execution of a program using environments.

These results establish a *Curry-Howard isomorphism* for compilation and program execution. The summary of the correspondence is shown in Fig. 1.

Intuitively, an A-normal compiler performs two types of transformations: (1) it identifies all the redexes by naming the intermediate results of reductions, and (2) it flattens and linearizes redexes by extending the scope of intermediate

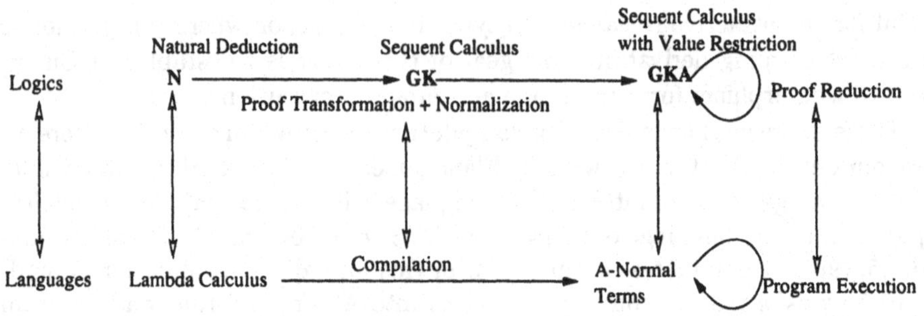

Fig. 1. Curry-Howard isomorphism for compilation and program execution

bindings. As a simple example, consider the source term $(f\ (g\ x))$. The first type of transformation converts this term into the following:

$$\mathbf{app}\ (f\ (\mathbf{app}\ (g\ x)\ \mathbf{is}\ z\ \mathbf{in}\ z))\ \mathbf{is}\ w\ \mathbf{in}\ w$$

where **app** $(x\ M)$ **is** y **in** N is our syntax in $\mathcal{GK}$ for applying function x to M and naming the result as y in N. This is exactly the transformation of a natural deduction proof to a sequent calculus proof. The second type of transformation converts this into the following A-normal form:

$$\mathbf{app}\ (g\ x)\ \mathbf{is}\ z\ \mathbf{in}\ (\mathbf{app}\ (f\ z)\ \mathbf{is}\ w\ \mathbf{in}\ w)$$

This process is the proof normalization from $\mathcal{GK}$ to $\mathcal{GKA}$. Execution of an A-normal form is also tightly modeled in $\mathcal{GKA}$: an operational semantics of A-normal forms exactly corresponds to a proof reduction in $\mathcal{GKA}$.

We believe that those logical correspondences worked out in this paper will contribute to design, analysis and optimization of compilation in a higher-order functional language. As an example of one of such benefits, A-normal compilation of [6] is immediately extended to products and sums by using the corresponding logical principles, as seen in this paper.

Related work. Before giving the technical development, we compare the results presented in this paper with related works. The use of a Gentzen-style sequent calculus as a model of computation is not new. Abramsky [1] has given a term calculus for linear logic. Breaze-Tannen et. al. [2] have given a typed pattern calculus where the underlying logic is a sequent calculus. In [4, 10] a sequent calculus is regarded as a model of computation. In particular, Herbelin [10] has argued that a sequent calculus can be a basis for computation and presented a term calculus. Based on a similar observation, Ogata [16] has shown that the term calculus presented in [4] corresponds to CPS terms under Griffin's [9] interpretation of CPS terms. In a tutorial article, Gallier [7] has given a term calculus for a Gentzen sequent calculus and suggested that a Gentzen-style sequent calculus represents finer notion of computation than β reduction. In general perspective, all those term calculi have the similarity to ours in the sense that they represent refined notion of computation, and they have been source of inspiration of the present paper. However, to the author's knowledgement, the connection to compilation and execution of compiled code has not been investigated.

In establishing this connection in the present paper, we use a well known result stating that any natural deduction proof can be transformed into a proof in the sequent calculus. Zucker [21] and Pottinger [17] conducted extensive studies on the relationship between the two proof systems. As we will show, however, this relationship alone does not provide the desired interpretation for compilation and program execution, and significant new results are needed to extend Curry-Howard isomorphism to them. In the existing works on Gentzen's sequent calculus and computation, the advocated thesis is that "cut elimination corresponds to computation." Our analysis shows, however, that this commonly believed thesis does not apply to actual implementation of a (call-by-value) functional language. In the usual cut elimination, cut rule is inductively moved upward to smaller proofs. A somewhat surprising result of our work is that program evaluation in conventional implementation pushes cut downwards, and corresponds a quite difference proof normalization process.

Paper Organization. Section 2 defines the typed lambda calculus. Section 3 defines $\mathcal{GK}$, $\mathcal{GKA}$, and a proof normalization from $\mathcal{GK}$ to $\mathcal{GKA}$. Section 4 shows that a compilation algorithm from lambda terms to A-normal forms is the combination of a proof transformation from the natural deduction to $\mathcal{GK}$ and a proof normalization from $\mathcal{GK}$ to $\mathcal{GKA}$. Section 5 shows that the operational semantics of A-normal forms is a proof reduction in $\mathcal{GKA}$. Section 6 concludes the paper.

Limitations of space make it difficult to cover the technical development fully; the author intends to present a more detailed description elsewhere.

Acknowledgments. The author would like to thank Yasuhiko Minamide and Ichiro Ogata for useful discussions on A-normal forms and sequent calculi. He also thanks Susumu Nishimura for helpful comments on a draft of this paper.

2 Typed Lambda Calculus

To make the relationship to logic explicit, we use the following logical notations for types (ranger over by τ):

$$\tau ::= b \mid \tau \supset \tau \mid \tau \wedge \tau \mid \tau \vee \tau$$

where b stands for a given set of atomic types. A *type assignment* Γ is a function from a finite set of variables to types. We write $\{x_1 : \tau_1, \ldots, x_n : \tau_n\}$ for the function that maps each x_i to τ_i $(1 \leq i \leq n)$. If f is a function, we write $f, x : \tau$ for the function f' such that $dom(f') = dom(f) \cup \{x\}$ and $f'(x) = \tau$, $f'(y) = f(y)$ if $y \neq x$. The set of terms is given by the following syntax:

$$\begin{aligned} M ::= {} & c^b \mid x \mid \lambda x : \tau.M \mid M\ M \mid (M, M) \mid M.1 \mid M.2 \mid \\ & \mathbf{inl}(M : \tau) \mid \mathbf{in2}(M : \tau) \mid \mathbf{case}\ M\ \mathbf{of}\ x.M,\ x.M \end{aligned}$$

c^b stands for atomic constants of type b. x stands for a given set of variables. $M.1$ and $M.2$ are first and second projection, respectively. $\mathbf{in1}(M : \tau), \mathbf{in2}(M : \tau)$ are left and right injection to a variant, respectively. The type annotations in $\lambda x : \tau.M$, $\mathbf{in1}(M : \tau)$, and $\mathbf{in2}(M : \tau)$ are necessary to achieve the uniqueness of typing derivation (uniqueness of a term representation of a proof). In what follows, however, we make those type annotations implicit.

$$\text{(axiom)}\ \Gamma \rhd c^b : b \qquad \text{(taut)}\ \Gamma, x : \tau \rhd x : \tau \qquad (\supset\text{:I})\ \frac{\Gamma, x : \tau_1 \rhd M : \tau_1}{\Gamma \rhd \lambda x : \tau.M : \tau_1 \supset \tau_2}$$

$$(\supset\text{:E})\ \frac{\Gamma \rhd M_1 : \tau_1 \supset \tau_2 \quad \Gamma \rhd M_2 : \tau_1}{\Gamma \rhd M_1\ M_2 : \tau_2} \qquad (\wedge\text{:I})\ \frac{\Gamma \rhd M_1 : \tau_1 \quad \Gamma \rhd M_2 : \tau_2}{\Gamma \rhd (M_1, M_2) : \tau_1 \wedge \tau_2}$$

$$(\wedge\text{:Ei})\ \frac{\Gamma \rhd M : \tau_1 \wedge \tau_2}{\Gamma \rhd M.i : \tau_i}\ i \in \{1,2\} \qquad (\vee\text{:Ii})\ \frac{\Gamma \rhd M : \tau_i}{\Gamma \rhd \mathbf{in}i(M : \tau_1 \vee \tau_2) : \tau_1 \vee \tau_2}\ i \in \{1,2\}$$

$$(\vee\text{:E})\ \frac{\Gamma \rhd M_1 : \tau_1 \vee \tau_2 \quad \Gamma, x : \tau_1 \rhd M_2 : \tau_3 \quad \Gamma, y : \tau_2 \rhd M_3 : \tau_3}{\Gamma \rhd \mathbf{case}\ M_1\ \mathbf{of}\ x.M_2, y.M_3 : \tau_3}$$

Fig. 2. Typed Lambda Calculus with Products and Sums

The proof system for the typed lambda terms is given in Fig. 2. The following properties are well known as *Curry-Howard isomorphism.*

- If we erase M from $\Gamma \rhd M : \tau$ and replace Γ with the multi-set obtained by erasing the variables, then we obtain the natural deduction system [18] (with additional axioms for atomic propositions), which is denoted here by $\mathcal{N}$.
- If $\vdash \Gamma \rhd M : \tau$ then the term M uniquely represents a proof of $\vdash \Gamma \rhd M : \tau$ in $\mathcal{N}$.
- The β reduction on lambda terms corresponds to proof normalization in $\mathcal{N}$.

We write $\mathcal{N} \vdash \Gamma \rhd M : \tau$ if $\Gamma \rhd M : \tau$ is provable in this proof system. Our aim is to extend this logical correspondence to compilation and program execution using a Gentzen-style sequent calculus.

3 Intuitionistic Sequent Calculus : $\mathcal{GK}$

We choose a contraction-free variant of the Gentzen's intuitionistic sequent calculus due to Kleene [13, Ch.XV,§80], which is particularly suitable for establishing the exact correspondence between program execution and proof reduction.

The set of types is the same as that of $\mathcal{N}$. The set of terms is given by the following syntax.

$$M ::= c^b \mid x \mid \lambda x.M \mid \mathbf{app}\ (x\ M)\ \mathbf{is}\ y\ \mathbf{in}\ M \mid (M, M) \mid \mathbf{proj}\ x\ \mathbf{on}\ (y, z)\ \mathbf{in}\ M$$
$$\mid \mathbf{in1}(M) \mid \mathbf{in2}(M) \mid \mathbf{case}\ x\ \mathbf{of}\ y.M, z.M \mid \mathbf{let}\ x = M\ \mathbf{in}\ M$$

We have explained **app** $(x\ M_1)$ **is** y **in** M_2. **proj** x **on** (y, z) **in** M binds y to the first component of x, and binds z to the second component of x in M. **case** x **of** $y.M_1, z.M_2$ performs case analysis on x and if it is of the form $\mathbf{in1}(v)$ then binds x to v in M_1 otherwise x is of the form $\mathbf{in2}(v)$ and it binds z to v in M_2. The proof system is given in Fig. 3.

$$(\text{axiom})\ \Gamma \rhd c^b : b \qquad (\text{taut})\ \Gamma, x : \tau \rhd x : \tau \qquad (\supset{:}\text{R})\ \frac{\Gamma, x : \tau_1 \rhd M : \tau_2}{\Gamma \rhd \lambda x.M : \tau_1 \supset \tau_2}$$

$$(\supset{:}\text{L})\ \frac{\Gamma, x : \tau_1 \supset \tau_2 \rhd M_1 : \tau_1 \quad \Gamma, x : \tau_1 \supset \tau_2, y : \tau_2 \rhd M_2 : \tau_3}{\Gamma, x : \tau_1 \supset \tau_2 \rhd \mathbf{app}\ (x\ M_1)\ \mathbf{is}\ y\ \mathbf{in}\ M_2 : \tau_3}$$

$$(\wedge{:}\text{R})\ \frac{\Gamma \rhd M_1 : \tau_1 \quad \Gamma \rhd M_2 : \tau_2}{\Gamma \rhd (M_1, M_2) : \tau_1 \wedge \tau_2} \qquad (\vee{:}\text{R}i)\ \frac{\Gamma \rhd M : \tau_2}{\Gamma \rhd \mathbf{in}i(M) : \tau_1 \vee \tau_2} \quad (i \in \{1, 2\})$$

$$(\wedge{:}\text{L})\ \frac{\Gamma, x : \tau_1 \wedge \tau_2, y : \tau_1, z : \tau_2 \rhd M : \tau_3}{\Gamma, x : \tau_1 \wedge \tau_2 \rhd \mathbf{proj}\ x\ \mathbf{on}\ (y, z)\ \mathbf{in}\ M : \tau_3}$$

$$(\vee{:}\text{L})\ \frac{\Gamma, x : \tau_1 \vee \tau_2, y : \tau_1 \rhd M_1 : \tau_3 \quad \Gamma, x : \tau_1 \vee \tau_2, z : \tau_2 \rhd M_2 : \tau_3}{\Gamma, x : \tau_1 \vee \tau_2 \rhd \mathbf{case}\ x\ \mathbf{of}\ y.M_1, z.M_2 : \tau_3}$$

$$(\text{cut})\ \frac{\Gamma \rhd M_1 : \tau_1 \quad \Gamma, x : \tau_1 \rhd M_2 : \tau_2}{\Gamma \rhd \mathbf{let}\ x = M_1\ \mathbf{in}\ M_2 : \tau_2}$$

Fig. 3. Gentzen-style Intuitionistic Sequent Calculus $\mathcal{GK}$

For our calculus, the notion of bound and free variables are defined on both terms and proofs, and we can show that α-equivalence hold in this calculus. In the following development, we assume the "bound variable convention", i.e. all bound variables are distinct and are different from any free variables. It should be noted, however, that α equivalence is not entirely obvious for sequent calculi. For example, if we adopt the Gentzen's original proof system where each left rule introduces a new assumption, then some extra machinery will be needed to obtain α equivalence.

3.1 A-Normal Forms and Proof Normalization

We define a subsystem $\mathcal{GKA}$ of $\mathcal{GK}$ whose proofs correspond to the set of A-normal forms. We say that a premise is an *argument premise* if it is a premise of one of right rules except ($\supset$:R), or it is the left premise of ($\supset$:L) or (cut). $\mathcal{GKA}$ is obtained from $\mathcal{GK}$ by distinguishing those proofs that correspond to "values", and restricting argument premises to be value proofs. The set of values (ranged over by V) and the set of A-normal forms are given as follows.

$$\begin{aligned} V &::= c^b \mid x \mid \lambda x.M \mid (V, V) \mid \mathbf{in1}(V) \mid \mathbf{in2}(V) \\ M &::= V \mid \mathbf{app}\ (x\ V)\ \mathbf{is}\ y\ \mathbf{in}\ M \mid \mathbf{proj}\ x\ \mathbf{on}\ (y, z)\ \mathbf{in}\ M \mid \\ &\qquad \mathbf{case}\ x\ \mathbf{of}\ y.M, z.M \mid \mathbf{let}\ x = V\ \mathbf{in}\ M \end{aligned}$$

The proof system $\mathcal{GKA}$ is given in Fig. 4, where the use of a meta variable V indicates that it must be a value.

Values.

$$\text{(axiom)}\ \Gamma \rhd c^b : b \qquad \text{(taut)}\ \Gamma, x:\tau \rhd x : \tau \qquad (\wedge\text{:R})\ \frac{\Gamma \rhd V_1 : \tau_1 \quad \Gamma \rhd V_2 : \tau_2}{\Gamma \rhd (V_1, V_2) : \tau_1 \wedge \tau_2}$$

$$(\vee\text{:Ri})\ \frac{\Gamma \rhd V : \tau_i}{\Gamma \rhd \mathbf{in}i(V) : \tau_1 \vee \tau_2}\ (i \in \{1,2\}) \qquad (\supset\text{:R})\ \frac{\Gamma, x:\tau_1 \rhd M : \tau_2}{\Gamma \rhd \lambda x.M : \tau_1 \supset \tau_2}$$

General A-normal forms

$$(\supset\text{:L})\ \frac{\Gamma, x:\tau_1 \supset \tau_2 \rhd V : \tau_1 \quad \Gamma, x:\tau_1 \supset \tau_2, y:\tau_2 \rhd M : \tau_3}{\Gamma, x:\tau_1 \supset \tau_2 \rhd \mathbf{app}\ (x\ V)\ \mathbf{is}\ y\ \mathbf{in}\ M : \tau_3}$$

$$(\wedge\text{:L})\ \frac{\Gamma, x:\tau_1 \wedge \tau_2, y:\tau_1, z:\tau_2 \rhd M : \tau_3}{\Gamma, x:\tau_1 \wedge \tau_2 \rhd \mathbf{proj}\ x\ \mathbf{on}\ (y,z)\ \mathbf{in}\ M : \tau_3}$$

$$(\vee\text{:L})\ \frac{\Gamma, x:\tau_1 \vee \tau_2, y:\tau_1 \rhd M_1 : \tau_3 \quad \Gamma, x:\tau_1 \vee \tau_2, z:\tau_2 \rhd M_2 : \tau_3}{\Gamma, x:\tau_1 \vee \tau_2 \rhd \mathbf{case}\ x\ \mathbf{of}\ y.M_1, z.M_2 : \tau_3}$$

$$\text{(cut)}\ \frac{\Gamma \rhd V : \tau_1 \quad \Gamma, x:\tau_1 \rhd M : \tau_2}{\Gamma \rhd \mathbf{let}\ x = V\ \mathbf{in}\ M : \tau_2}$$

Fig. 4. Proof system $\mathcal{GKA}$ for A-normal forms

We define the set $\mathcal{S}$ of proof transformations from $\mathcal{GK}$ to $\mathcal{GKA}$. Each transformation pushes a cut rule or a left rule appearing in an argument premise downward. For each of {(cut), ($\supset$:L), ($\wedge$:L), ($\vee$:L)} there are 6 transformation rules corresponding to the 6 different argument premises.

The sets of transformations for {($\supset$:L), ($\wedge$:L), (cut)} are similar to one another. Here we only show the two cases where (cut) appears in an argument premise as follows:

$$\cfrac{\cfrac{\begin{array}{cc}\Delta_1 & \Delta_2 \\ (\Gamma \rhd M_1 : \tau_1) & (\Gamma, x:\tau_1 \rhd M_2 : \tau_2)\end{array}}{\Gamma \rhd \mathbf{let}\ x = M_1\ \mathbf{in}\ M_2 : \tau_2}\text{(cut)} \quad \cfrac{\Delta_3}{\Gamma \rhd M_3 : \tau_3}}{\Gamma \rhd (\mathbf{let}\ x = M_1\ \mathbf{in}\ M_2, M_3) : \tau_2 \wedge \tau_3}(\wedge\text{:R})$$

$$\Longrightarrow \cfrac{\begin{array}{c}\Delta_1 \\ (\Gamma \rhd M_1 : \tau_1)\end{array} \quad \cfrac{\begin{array}{cc}\Delta_2 & \Delta_3 + \{x:\tau_1\} \\ (\Gamma, x:\tau_1 \rhd M_2 : \tau_2) & (\Gamma, x:\tau_1 \rhd M_3 : \tau_3)\end{array}}{\Gamma, x:\tau_1 \rhd (M_2, M_3) : \tau_2 \wedge \tau_3}(\wedge\text{:R})}{\Gamma \rhd \mathbf{let}\ x = M_1\ \mathbf{in}\ (M_2, M_3) : \tau_2 \wedge \tau_3}\text{(cut)}$$

$$\cfrac{\cfrac{\begin{array}{cc}\Delta_1 & \Delta_2 \\ (\Gamma \rhd M_1 : \tau_1) & (\Gamma, x:\tau_1 \rhd M_2 : \tau_2)\end{array}}{\Gamma \rhd \mathbf{let}\ x = M_1\ \mathbf{in}\ M_2 : \tau_2}\text{(cut)} \quad \begin{array}{c}\Delta_3 \\ (\Gamma, y:\tau_2 \rhd M_3 : \tau_3)\end{array}}{\Gamma \rhd \mathbf{let}\ y = (\mathbf{let}\ x = M_1\ \mathbf{in}\ M_2)\ \mathbf{in}\ M_3 : \tau_3}\text{(cut)}$$

$$\Longrightarrow \cfrac{\begin{array}{c}\Delta_1 \\ (\Gamma \rhd M_1 : \tau_1)\end{array} \quad \cfrac{\begin{array}{cc}\Delta_2 & \Delta_3 + \{x:\tau_1\} \\ (\Gamma, x:\tau_1 \rhd M_2 : \tau_2) & (\Gamma, x:\tau_1, y:\tau_2 \rhd M_3 : \tau_3)\end{array}}{\Gamma, x:\tau_1 \rhd \mathbf{let}\ y = M_2\ \mathbf{in}\ M_3 : \tau_3}\text{(cut)}}{\Gamma \rhd \mathbf{let}\ x = M_1\ \mathbf{in}\ \mathbf{let}\ y = M_2\ \mathbf{in}\ M_3 : \tau_3}\text{(cut)}$$

where $\overset{\Delta}{(\Gamma \rhd M : \tau)}$ is a proof of the sequent $\Gamma \rhd M : \tau$, and $\Delta + \{x : \tau\}$ is the proof obtained from the proof Δ by adding $\{x : \tau\}$ to the assumption of each sequent in Δ.

If we project the set $\mathcal{S}$ of proof transformations on untyped term structures, then they become the following set of reduction rules.

$$\begin{aligned} C[\textbf{app } (x\ M_1) \textbf{ is } y \textbf{ in } M_2] &\Longrightarrow \textbf{app } (x\ M_1) \textbf{ is } y \textbf{ in } C[M_2] \\ C[\textbf{proj } z \textbf{ on } (w,v) \textbf{ in } M_1] &\Longrightarrow \textbf{proj } z \textbf{ on } (w,v) \textbf{ in } C[M_1] \\ C[\textbf{let } x = M_1 \textbf{ in } M_2] &\Longrightarrow \textbf{let } x = M_1 \textbf{ in } C[M_2] \\ C[\textbf{case } x \textbf{ of } y.M_1, z.M_2] &\Longrightarrow \textbf{case } x \textbf{ of } y.C[M_1], z.C[M_2] \end{aligned}$$

where $C[\,]$ denotes any one of the following contexts:

$$C[\,] ::= ([\,], M) \mid (M, [\,]) \mid \textbf{in1}([\,]) \mid \textbf{in2}([\,]) \mid \textbf{app } (x\ [\,]) \textbf{ is } y \textbf{ in } M \mid \textbf{let } x = [\,] \textbf{ in } M$$

This set of rules can be regarded as a "one-step version" of some of A-reductions defined in [6]. (The other A-reduction rules corresponds to proof transformation from $\mathcal{N}$ to $\mathcal{GK}$ for function application.)

Next we consider transformations for ($\vee$:L). The structures of the transformations are similar to the previous cases except that part of derivation is copied. Suppose ($\vee$:L) appears in an argument premise of a rule R. There are two types of transformations depending on whether R is a right rule or not. Here we only show the case where where ($\vee$:L) appears in the left argument premise of ($\wedge$:R).

$$\dfrac{\dfrac{\Delta_1 \quad \Delta_2}{\Gamma, x : \tau_1 \vee \tau_2 \rhd \textbf{case } x \textbf{ of } y.M_1, z.M_2 \ :\ \tau_3}(\vee{:}\mathrm{L}) \qquad \Delta_3}{\Gamma, x : \tau_1 \vee \tau_2 \rhd ((\textbf{case } x \textbf{ of } y.M_1, z.M_2),\ M_3) \ :\ \tau_3 \wedge \tau_4}(\wedge{:}\mathrm{R}) \quad \Longrightarrow$$

$$\dfrac{\dfrac{\Delta_1 \quad \Delta_3 + \{y : \tau_1\}}{\Gamma, x : \tau_1 \vee \tau_2, y : \tau_1 \rhd (M_1, M_3) \ :\ \tau_3 \wedge \tau_4}(\wedge{:}\mathrm{R}) \qquad \dfrac{\Delta_2 \quad \Delta_3 + \{z : \tau_2\}}{\Gamma, x : \tau_1 \vee \tau_2, z : \tau_2 \rhd (M_2, M_3) \ :\ \tau_3 \wedge \tau_4}(\wedge{:}\mathrm{R})}{\Gamma, x : \tau_1 \vee \tau_2 \rhd \textbf{case } x \textbf{ of } y.(M_1, M_3), z.(M_2, M_3) \ :\ \tau_3 \wedge \tau_4}(\vee{:}\mathrm{L})$$

In this rule, the derivation Δ_3 is duplicated. The same phenomenon occurs in the transformation of a conditional statement in [6]. It is not hard to modify the rule for ($\vee : L$) to avoid copying a part of derivation by introducing additional assumption for holding the intermediate result of the case analysis.

We write

$$\mathcal{S} \vdash \Gamma \rhd M \overset{*}{\Longrightarrow} M' \ :\ \tau$$

if the proof of $\Gamma \rhd M \ :\ \tau$ can be transformed to that of $\Gamma \rhd M' \ :\ \tau$ by repeated application of some of the transformation rules $\mathcal{S}$. Since each rule in $\mathcal{S}$ is a valid proof transformation, it is immediate that if $\mathcal{GK} \vdash \Gamma \rhd M \ :\ \tau$ and $\mathcal{S} \vdash \Gamma \rhd M \overset{*}{\Longrightarrow} M' \ :\ \tau$ then $\mathcal{GK} \vdash \Gamma \rhd M' \ :\ \tau$. Moreover, we have the following.

Theorem 1. *If $\mathcal{GK} \vdash \Gamma \rhd M \ :\ \tau$ then there is some M' such that $\mathcal{S} \vdash \Gamma \rhd M \overset{*}{\Longrightarrow} M' \ :\ \tau$ and $\mathcal{GKA} \vdash \Gamma \rhd M' \ :\ \tau$*

This is proved by a routine induction on derivation of M. By these results, A-normal forms can be regarded as a form of normal proofs in $\mathcal{GK}$ identified by the subsystem $\mathcal{GKA}$.

This transformation is the first half of the proof normalization that corresponds to the computation of a program in a conventional implementation. A distinguishing property of this normalization process is that cuts are moved downwards in the compound proofs of products and sums, which is the opposite of the usual cut elimination procedure. As we shall show later, program execution does not corresponds to cut elimination either.

Our choice of the contraction-free variant of $\mathcal{GK}$ is suitable for the normalization transformation. If we adopted the original Gentzen's sequent calculus, then additional machinery would have been needed. To see the difficulty, consider the term (**proj** z **on** (x, y) **in** $(x, y), z)$. This is provable in the Gentzen's sequent calculus, but the corresponding proof of A-normal form **proj** z **on** (x, y) **in** $((x, y), z)$ is not directly provable in the subsystem corresponding to the Gentzen's sequent calculus.

4 A-Normal Compilation as Proof Transformation

Our first main result is that the compilation from the set of typed lambda terms into the set of A-normal forms is characterized as the combination of a proof transformation from $\mathcal{N}$ to $\mathcal{GK}$ and a proof normalization from $\mathcal{GK}$ to $\mathcal{GKA}$.

We first state the well known result in proof theory.

Theorem 2 ([8, 18, 21, 17]). *There is an algorithm, denoted here by $\mathcal{NG}$, that transforms any $\mathcal{N}$ proof to a $\mathcal{GK}$ proof.*

The main idea behind $\mathcal{NG}$ is to decompose an elimination rule into the combination of a left rule and a cut rule. Since this result will be used in the following development, we include some important cases of the algorithm $\mathcal{NG}$ in Fig. 5.

By combining Theorem 1 and Theorem 2, we have the following.

Corollary 3. *Every proof in $\mathcal{N}$ is transformed to a proof in $\mathcal{GKA}$.*

Moreover, compiling a lambda term to an A-normal form is exactly this transformation, which we prove below.

Flanagan et. al. [6] have given a linear time compilation algorithm from lambda terms to A-normal forms in Scheme using the two-level programming technique for CPS algorithms by Danvy and Fillinski [5]. To establish the desired result, it is essential to reason about the meta-level language as well. For this purpose, we re-state their algorithm using a simply typed first-order language for manipulating sequent proofs. To define the language, we extend the proof system with *proof variables* (ranged over by X) typed with a logical sequent $\Gamma \rhd \tau$. We also extend the set of terms with the same set of variables. We use σ as a meta variable ranging over logical sequent $\Gamma \rhd \tau$ regarded as a type. Let Ω be a set of type assignment for proof variables, which is a mapping from a finite set of proof variables to types (logical sequents), and write $\{X_1 : \sigma, \ldots, X_n : \sigma_n\}$ for a type assignment that assigns σ_i to X_i. Let $\mathcal{GK}(\Omega)$ be the proof system obtained from $\mathcal{GK}$ by adding $\Gamma \rhd X : \tau$ as an axiom for each $X : \Gamma \rhd \tau$ in Ω, and also by adding the set of variables appearing in Ω as new term variables.

$$\mathcal{NG}\left(\dfrac{\begin{array}{c}\Pi_1\\(\Gamma \rhd M_1 : (\tau_1 \supset \tau_2))\end{array} \quad \begin{array}{c}\Pi_2\\(\Gamma \rhd M_2 : \tau_1)\end{array}}{\Gamma \rhd M_1\ M_2 : \tau_2}\ \supset\text{:E}\right)$$

$$= \dfrac{\begin{array}{c}\mathcal{NG}(\Pi_1)\\(\Gamma \rhd M_1' : \tau_1 \supset \tau_2)\end{array} \quad \dfrac{\begin{array}{c}\mathcal{NG}(\Pi_2) + \{x : \tau_1 \supset \tau_2\}\\(\Gamma, x : \tau_1 \supset \tau_2 \rhd M_2' : \tau_1)\end{array} \quad \dfrac{}{\Gamma, x : \tau_1 \supset \tau_2, y : \tau_2 \rhd y : \tau_2}\ \text{(taut)}}{\Gamma, x : \tau_1 \supset \tau_2 \rhd \mathbf{app}\ (x\ M_2')\ \mathbf{is}\ y\ \mathbf{in}\ y : \tau_2}\ (\supset\text{:L})}{\Gamma \rhd \mathbf{let}\ x = M_1'\ \mathbf{in}\ \mathbf{app}\ (x\ M_2')\ \mathbf{is}\ y\ \mathbf{in}\ y : \tau_2}\ \text{(cut)}$$

$$\mathcal{NG}\left(\dfrac{\begin{array}{c}\Pi_1\\(\Gamma \rhd M : \tau_1 \wedge \tau_2)\end{array}}{\Gamma \rhd M.i : \tau_i}\ \wedge\text{:E}i\right)$$

$$= \dfrac{\begin{array}{c}\mathcal{NG}(\Pi_1)\\(\Gamma \rhd M' : \tau_1 \wedge \tau_2)\end{array} \quad \dfrac{\dfrac{}{\Gamma, y : \tau_1 \wedge \tau_2, x_1 : \tau_1, x_2 : \tau_2 \rhd x_i : \tau_i}\ \text{(taut)}}{\Gamma, y : \tau_1 \wedge \tau_2 \rhd \mathbf{proj}\ y\ \mathbf{on}\ (x_1, x_2)\ \mathbf{in}\ x_i : \tau_i}\ (\wedge\text{:L})}{\Gamma \rhd \mathbf{let}\ y = M'\ \mathbf{in}\ \mathbf{proj}\ y\ \mathbf{on}\ (x_1, x_2)\ \mathbf{in}\ x_i : \tau_i}\ \text{(cut)} \qquad (i \in \{1, 2\})$$

$$\mathcal{NG}\left(\dfrac{\begin{array}{c}\Pi_1\\(\Gamma \rhd M_1 : \tau_1 \vee \tau_2)\end{array} \quad \begin{array}{c}\Pi_2\\(\Gamma, x_1 : \tau_1 \rhd M_2 : \tau_2)\end{array} \quad \begin{array}{c}\Pi_3\\(\Gamma, x_2 : \tau_2 \rhd M_3 : \tau_2)\end{array}}{\Gamma \rhd \mathbf{case}\ M_1\ \mathbf{of}\ x_1.M_1,\ x_2.M_2 : \tau_2}\ (\vee\text{:E})\right)$$

$$= \dfrac{\begin{array}{c}\mathcal{NG}(\Pi_1)\\(\Gamma \rhd M_1' : \tau_1 \vee \tau_2)\end{array} \quad \dfrac{\begin{array}{c}\mathcal{NG}(\Pi_2) + \{y : \tau_1 \vee \tau_2\}\\(\Gamma, y : \tau_1 \vee \tau_2, x_1 : \tau_1 \rhd M_2' : \tau_2)\end{array} \quad \begin{array}{c}\mathcal{NG}(\Pi_3) + \{y : \tau_1 \vee \tau_2\}\\(\Gamma, y : \tau_1 \vee \tau_2, x_2 : \tau_2 \rhd M_3' : \tau_3)\end{array}}{\Gamma, y : \tau_1 \vee \tau_2 \rhd \mathbf{case}\ y\ \mathbf{of}\ x_1.M_2', x_2.M_3' : \tau_3}\ (\vee\text{:L})}{\Gamma \rhd \mathbf{let}\ y = M_1'\ \mathbf{in}\ \mathbf{case}\ y\ \mathbf{of}\ x_1.M_2', x_2.M_3' : \tau_3}\ \text{(cut)}$$

Fig. 5. Some of Proof Translation Rules form $\mathcal{N}$ to $\mathcal{GK}$

We write $\mathcal{GK}(\Omega) \vdash \Gamma \rhd M : \tau$ if $\Gamma \rhd M : \tau$ is provable in $\mathcal{GK}(\Omega)$. If Δ_1 is a proof containing an axiom for X of type σ and Δ_2 is a proof of type σ then we write $[\Delta_2/X]\Delta_1$ for the proof obtained from Δ_1 by replacing each occurrence of axiom for X with Δ_2 and the variable occurrences of X in the terms of Δ_1 by M_2. The following substitution property holds.

Proposition 4. *1. If Δ_1 is a proof of σ_1 in $\mathcal{GK}(\Omega, X : \sigma_2)$ and Δ_2 is a proof of σ_2 in $\mathcal{GK}(\Omega)$ then $[\Delta_2/X]\Delta_1$ is a proof of σ_1 in $\mathcal{GK}(\Omega)$.*
2. If $\mathcal{GK}(\Omega, X : (\Gamma_2 \rhd \tau_2)) \vdash \Gamma_1 \rhd M_1 : \tau_1$ and $\mathcal{GK}(\Omega) \vdash \Gamma_2 \rhd M_2 : \tau_2$ then $\mathcal{GK}(\Omega) \vdash \Gamma_1 \rhd [M_2/X]M_1 : \tau_1$.

The set of typings of the first-order language (whose terms are ranged over by D) is defined by the following rules to derive a typing of the form $\Omega \vdash D : \sigma$ denoting the fact that D is a well typed term under Ω.

- $\Omega \vdash D : (\Gamma \rhd \tau)$ if $\mathcal{GK}(\Omega) \vdash \Gamma \rhd D : \tau$.
- $\Omega \vdash \delta X : \sigma_1.D : \sigma_1 \to \sigma$ if $\Omega, X : \sigma_1 \vdash D : \sigma_2$.
- $\Omega \vdash D_1 \odot D_2 : \sigma$ if $\Omega \vdash D_1 : \sigma_1 \to \sigma$ and $\Omega \vdash D_2 : \sigma_1$.

The reduction relation on this language is defined by the following rule

$$(\delta X : \sigma_1.D_1) \odot D_2 \longrightarrow [D_2/X]D_1$$

In the following we omit the type annotation in $\delta X : \sigma_1.D$ if it does not cause any confusion.

It is easily seen that the reduction is confluent and terminating. We regard terms of this language as those modulo the equality induced by this reduction relation. Also, X in $\delta X.D$ is a bound variable, and we regard terms module bound variable renaming. From this and Proposition 4, the following properties are immediate.

Proposition 5. *1. If $\Omega \vdash D : \sigma$ then D determines a proof of σ in $\mathcal{GK}(\Omega)$.*
2. If $\Omega \vdash D : \sigma_1 \to \sigma_2$ then D is a term of the form $\delta X : \sigma_1.D'$ such that $\Omega, X : \sigma_1 \vdash D' : \sigma_2$.

We can therefore regard a typed term D such that $\Omega \vdash D : (\Gamma \rhd \tau_1) \to (\Gamma' \rhd \tau_2)$ as a "context," i.e. a proof of $\Gamma' \rhd \tau_2$ in $\mathcal{GK}(\Omega)$ containing a "hole" to be filled with a proof of $\Gamma \rhd \tau_1$ in $\mathcal{GK}(\Omega)$.

Suppose $\Omega \vdash D : \sigma$. We write $\mathcal{S}(\Omega) \vdash D \overset{*}{\Longrightarrow} D' : \sigma$ if the proof determined by D is reduced to the one determined by D' in $\mathcal{GK}(\Omega)$ using the set $\mathcal{S}$ of proof reduction rules defined earlier. Suppose $\Omega \vdash D : (\Gamma \rhd \tau_1) \to (\Gamma' \rhd \tau_2)$. We also write $\mathcal{S}(\Omega) \vdash D \overset{*}{\Longrightarrow} D' : \sigma_1 \to \sigma_2$ if $D = \delta X : \sigma_1.D_0$, $D' = \delta X : \sigma_1.D_1$ and $\mathcal{S}(\Omega, X : \sigma_1) \vdash D_0 \overset{*}{\Longrightarrow} D_1 : \sigma_2$.

The following two lemmas can then be shown by the properties of proofs in $\mathcal{GK}$ using Proposition 5.

Lemma 6. *If $\Omega \vdash D : (\Gamma_1 \rhd \tau_1) \to (\Gamma_2 \rhd \tau_2)$, $x \notin dom(\Gamma_1) \cup dom(\Gamma_2)$ and $\Omega \subseteq \Omega'$ then $\Omega' \vdash D : (\Gamma_1, x : \tau_3 \rhd \tau_1) \to (\Gamma_2, x : \tau_3 \rhd \tau_2)$.*

Lemma 7. *If $\mathcal{S}(\Omega) \vdash D_1 \overset{*}{\Longrightarrow} D_1' : \sigma \to \sigma'$ and $\Omega \vdash D_2 \overset{*}{\Longrightarrow} D_2' : \sigma$ then $\mathcal{S}(\Omega) \vdash D_1 \odot D_2 \overset{*}{\Longrightarrow} D_1' \odot D_2' : \sigma'$.*

Using this first-order language, A-normal translation algorithm is given as a function $[\![\_]\!]\_$ that takes a terms D such that $\Omega \vdash D : \Gamma_1 \rhd \tau_1$ and a function term k such that $\Omega \vdash k : (\Gamma_1 \rhd \tau_1) \to (\Gamma_2 \rhd \tau_2)$, and return a term D' such that $\Omega \vdash D' : \Gamma_2 \rhd \tau_2$. For the notational reason, we give the algorithm as an algorithm to transformation untyped terms in Fig. 6. (Note that the first-order language does not contain variables of function type; k used in this definition is a meta variable denoting a term of the form $\delta X.D$.) It is straightforward to construct the complete algorithm from this description. This algorithm, when regarded as one on untyped lambda terms, is a generalization of the A-normalization algorithm given in [6].

Under these preparations, we can now establish the following desired result.

Theorem 8. *If $\mathcal{N} \vdash \Gamma \rhd M : \tau_1$ and $\Omega \vdash k : (\Gamma' \rhd \tau_1) \to (\Gamma \rhd \tau_2)$ such that $\Gamma \subseteq \Gamma'$ then $\Omega \vdash [\![M]\!]k : \Gamma \rhd \tau_2$ and $\mathcal{S}(\Omega) \vdash k \odot \mathcal{NG}(M) \overset{*}{\Longrightarrow} [\![M]\!]k : (\Gamma \rhd \tau_2)$.*

As a special case of Theorem 8 where k is $\delta X.X$, we have the following.

Corollary 9. *If $\mathcal{N} \vdash \Gamma \rhd M : \tau$ then $\mathcal{GKA} \vdash \Gamma \rhd [\![M]\!]\delta X.X : \tau$ and $\mathcal{S} \vdash \Gamma \rhd \mathcal{NG}(M) \overset{*}{\Longrightarrow} [\![M]\!]\delta X.X : \tau$.*

This establishes that A-normal compilation corresponds to proof transformation.

$$
\begin{aligned}
[\![c^b]\!]k &= k \odot c^b \\
[\![x]\!]k &= k \odot x \\
[\![\lambda x.M]\!]k &= k \odot (\lambda x.[\![M]\!](\delta X.X)) \\
[\![(M\ N)]\!]k &= [\![M]\!](\delta X.[\![N]\!](\delta Y.\textbf{let } x = X \textbf{ in app } (x\ Y) \textbf{ is } z \textbf{ in } k \odot z)) \\
[\![(M, N)]\!]k &= [\![M]\!](\delta X.[\![N]\!](\delta Y.k \odot (X, Y))) \\
[\![M.i]\!]k &= [\![M]\!](\delta X.\textbf{let } x = X \textbf{ in proj } x \textbf{ on } (x_1, x_2) \textbf{ in } k \odot x_i) \\
[\![\textbf{in}i(M)]\!]k &= [\![M]\!](\delta X.k \odot \textbf{in}i(X)) \\
[\![case\ M\ \lambda x.N, \lambda y.L]\!]k &= [\![M]\!](\delta X.(\textbf{let } z = X \textbf{ in case } z \textbf{ of } x.[\![N]\!]k, y.[\![L]\!]k))
\end{aligned}
$$

Fig. 6. A-normal compilation algorithm $[\![\_]\!]\_$

5 Program Execution as Proof Reduction

We move to the second half of our Curry-Howard isomorphism and establish that the execution of the compiled program by an abstract machine corresponds to proof reduction process in $\mathcal{GKA}$. For the set of A-normal forms, Flanagan et.al. [6] have defined an abstract machine called C_aEK. Here we define an equivalent operational semantics in the style of natural semantics [12], which makes the correspondence to logic more evident. The set of *runtime values* (ranged over by r) is given by the following syntax:

$$r ::= c^b \mid cls(E, \lambda x.M) \mid (r, r) \mid \textbf{in1}(r) \mid \textbf{in2}(r)$$

$cls(E, \lambda x.M)$ is a *closure* representing a function, where E is a *runtime environment* which is a mapping form a finite set of variables to runtime values. Fig. 7 define the operational semantic as a set of rules to derive the relation $E \vdash M \Downarrow r$ indicating the fact that M is evaluated to r under E.

Computation Rules:

$$\frac{E(x) = cls(E_1, \lambda z.M_1) \quad \gamma(E, V) = r_1 \quad E_1, z : r_1 \vdash M_1 \Downarrow r_2 \quad E, y : r_2 \vdash M \Downarrow r}{E \vdash \textbf{app } (x\ V) \textbf{ is } y \textbf{ in } M \Downarrow r}$$

$$\frac{E(x) = (r_1, r_2) \quad E, y : r_1, z : r_2 \vdash M \Downarrow r}{E \vdash \textbf{proj } x \textbf{ on } (y, z) \textbf{ in } M \Downarrow r} \qquad \frac{E(x) = \textbf{in1}(r_1) \quad E, y : r_1 \vdash M_1 \Downarrow r}{E \vdash \textbf{case } x \textbf{ of } y.M_1, z.M_2 \Downarrow r}$$

$$\frac{E(x) = \textbf{in2}(r_1) \quad E, z : r_1 \vdash M_2 \Downarrow r}{E \vdash \textbf{case } x \textbf{ of } y.M_1, z.M_2 \Downarrow r} \qquad \frac{\gamma(E, V) = r_1 \quad E, x : r_1 \vdash M \Downarrow r}{E \vdash \textbf{let } x = V \textbf{ in } M \Downarrow r}$$

Value Construction rules:

$$\gamma(E, c^b) = c^b \qquad \gamma(E, x) = E(x) \qquad \gamma(E, \lambda x.M) = cls(E, \lambda x.M)$$
$$\gamma(E, (V_1, V_2)) = (\gamma(E, V_1), \gamma(E, V_2)) \quad \gamma(E, \textbf{in}i(V)) = \textbf{in}i(\gamma(E, V))$$

Fig. 7. Operational semantics for A-normal forms

Lambda Derivations

$$(\supset:\text{R})\ \frac{\Gamma, x:\tau_1 \rhd M : \tau_1}{\Gamma \rhd_\lambda \lambda x.M : \tau_1 \supset \tau_2} \qquad (\text{cut})\ \frac{\Gamma \rhd_v M_1 : \tau_1 \quad \Gamma, x:\tau_1 \rhd_\lambda M_2 : \tau_2}{\Gamma \rhd_\lambda \textbf{let } x = M_1 \textbf{ in } r_2 : \tau_2}$$

Value derivations

$$(\text{axiom})\ \Gamma \rhd_v c^b : b \qquad (\text{closure})\ \frac{\Gamma \rhd_\lambda M : \tau \quad FV(M) = \emptyset}{\Gamma \rhd_v M : \tau}$$

$$(\wedge:\text{R})\ \frac{\Gamma \rhd_v M_1 : \tau_1 \quad \Gamma \rhd_v M_2 : \tau_2}{\Gamma \rhd_v (M_1, M_2) : \tau_1 \wedge \tau_2} \qquad (\vee:\text{Ri})\ \frac{\Gamma \rhd_v M : \tau_i}{\Gamma \rhd_v \textbf{in}i(M) : \tau_1 \vee \tau_2}$$

Fig. 8. Runtime Value Derivations

We show that this operational semantics corresponds to proof reduction in $\mathcal{GKA}$. We first define a restriction of $\mathcal{GKA}$ in Fig. 8 that corresponds to a set of runtime values defined above. In this system, a judgment of the form $\Gamma \rhd_v M : \tau$ corresponds to a runtime value, and one of the form $\Gamma \rhd_\lambda M : \tau$ is an auxiliary judgment used to derive a closure. If $\Gamma \rhd_v M : \tau$ is derivable, then one of M is of the form

$$\textbf{let } x_1 = M_1 \textbf{ in } \ldots \textbf{let } x_n = M_n \textbf{ in } \lambda x.M$$

By the definition of the restricted proof system, each M_i is closed, and the order of the cuts is irrelevant. We can therefore consider the series of cuts as a mapping from variables to closed terms of the form $\{x_1 = M_1, \ldots, x_n = M_n\}$ and consider the term modulo the equivalence induced by reordering of cuts and write **let** $\{x_1 = M_1, \ldots, x_n = M_n\}$ **in** $\lambda x.M$. Let E be a mapping from variables to closed terms. We write $E : \Gamma$ if $dom(E) = dom(\Gamma)$ and for each $x \in dom(E)$, $\emptyset \rhd_v E(x) : \Gamma(x)$ is provable. If $E : \Gamma$, then the sequence of cuts corresponding to E is abbreviated as follows.

$$\frac{E : \Gamma_1 \quad \Gamma_2; \Gamma_1 \rhd M : \tau_2}{\Gamma_2 \rhd \textbf{let } E \textbf{ in } M : \tau_2}\ \text{cut}^*$$

Under this interpretation, if $\Gamma \rhd_v M : \tau$ is provable by the proof rules in Fig. 8 then M is isomorphic to some runtime value r defined above, and therefore the typing rules can be regarded as a type system of runtime values. In what follows, we identify runtime values with the corresponding terms and write $\Gamma \rhd_v r : \tau$ if a term corresponding to r is derivable.

Our plan now is to interpret the evaluation relation $E \vdash M \Downarrow r$ as a proof reduction that transforms the proof represented by **let** E **in** M to the one represented by r. The second major result of this paper is to establish that this is indeed the case, as shown in the following.

Theorem 10. *There is an algorithm taking a proof of* $\emptyset \rhd$ **let** E **in** $M : \tau$, *producing a runtime value* r *and a proof of* $\emptyset \rhd r : \tau$.

Proof (Outline). This is proved by defining a proof reduction algorithm, denoted as $\emptyset \rhd$ **let** E **in** $M \Downarrow r : \tau$, and showing its correctness. Due to the space limitation, we can only explain the main idea behind the proof.

The proof reduction algorithm first transforms a proof (represented by a term) of the form **let** E **in** M to a proof of the form **let** E' **in** V. This is done by inductively applying the algorithm to each argument proof to obtain a runtime value proof, and converting left rules in M to cuts of those runtime value proofs. The algorithm then converts **let** E' **in** V to a runtime value proof. The correctness of the algorithm is shown using the idea of logical relation and reducibility [20]. We first define a family of predicates $P(\tau)$ indexed by types. $r \in P(\tau)$ if one of the following holds.

- if $\tau \equiv b$ then $r \equiv c^b$.
- if $\tau \equiv \tau_1 \supset \tau_2$ then $r \equiv$ **let** E **in** $\lambda x.M$ such that $\forall r_1 \in P(\tau_1).\exists r_2.\emptyset \rhd$ **let** $E\{x : r_1\}M$ **in** $\Downarrow r_2 : \tau_2$ and $r_2 \in P(\tau_2)$.
- if $\tau \equiv \tau_1 \wedge \tau_2$ then $r \equiv (r_1, r_2)$ such that $r_1 \in P(\tau_1)$ and $r_2 \in P(\tau_2)$.
- if $\tau \equiv \tau_1 \vee \tau_2$ then either $r \equiv$ **inl**(r_1) such that $r_1 \in P(\tau_1)$, or **in2**(r_1) such that $r_1 \in P(\tau_2)$.

We then show the following property

if $\Gamma \rhd_v M : \tau$ and for each $x \in dom(E)$, $E(x) \in P(\Gamma(x))$ then $\emptyset \rhd$ **let** E **in** $M \Downarrow r : \tau$ for some r such that $r \in P(\tau)$

by induction on the derivation of M. □

This proof reduction algorithm, when projected on untyped terms, is the the operational semantics for A-normal forms given in Fig. 7. We have:

$$\emptyset \rhd \textbf{let } E \textbf{ in } M \Downarrow r : \tau$$
$$\iff \text{there is some } \Gamma \text{ such that } E : \Gamma,\ \Gamma \rhd M : \tau, \text{ and } E \vdash M \Downarrow r$$

A distinguishing characteristic of the algorithm is that it is not based on the usual cut elimination procedure. Instead of inductively eliminating cuts, it converts left rules and cut rules to those cuts whose proofs correspond to runtime values and keeps them until the final result is obtained. This process reveals the correspondence: cut rule corresponds to building (extending) a runtime environment, and left rule corresponds to computation on a data constructor.

6 Conclusions

We have developed a logical foundation for compilation and program execution by showing that compilation of lambda terms to A-normal forms corresponds to a proof transformation from the natural deduction system to a Gentzen-style sequent calculus followed by a proof normalization in the sequent calculus, and that evaluation of an A-normal form corresponds to a special proof reduction process in the sequent calculus. These results extend *Curry-Howard* isomorphism to compilation and program execution. There are a number of topics that merit further investigation. An interesting topic is to extend the formalism to second-order logic. Such extension wold provide a logical basis for type in compilation paradigm where a second-order type system is used to optimize programs.

A-normal forms also appear to be related to various other computational interpretation of lambda calculi. In particular, it would be beneficial to compare the logical correspondence we have worked out with Moggi's [15] computational lambda calculus and Kobayashi's work on modal logic [14].

References

1. S. Abramsky. Computational interpretation of linear logic. *Theoretical Computer Science*, 3(57), 1993.
2. B. Breazu-Tannen, D. Kesner, and L. Puel. A typed pattern calculus. In *Proc. IEEE Symposium on Logic in Computer Science*, pages 262–274, 1993.
3. H. B. Curry and R. Feys. *Combinatory Logic*, volume 1. North-Holland, Amsterdam, 1968.
4. V. Danos, J-B. Jointe, and H. Schellinx. A new deconstructive logic: linear logic. *Journal of Symbolic Logic*, 63(3):755–807, 1997.
5. O. Danvy. Back to direct style. In *Proc. European Symposium on Programming*, volume 582 of *Lecture Notes in Computer Science*, pages 130–150, 1992.
6. C. Flanagan, A. Sabry, B.F. Duba, and M. Felleisen. The essence of compiling with continuation. In *Proc. ACM PLDI Conference*, pages 237–247, 1993.
7. J. Gallier. Constructive logics part I: A tutorial on proof systems and typed λ-calculi. *Theoretical Computer Science*, 110:249–339, 1993.
8. G. Gentzen. Investigation into logical deduction. In M.E. Szabo, editor, *The Collected Papers of Gerhard Gentzen*. North-Holland, 1969.
9. T. Griffin. A formulae-as-types notion of control. In *Conference Record of the Seventeenth Annual ACM Symposium on Principles of Programming Languages*, pages 47–58, 1990.
10. H. Herbelin. A λ-calculus structure isomorphic to Gentzen-style sequent calculus structure. In *Proc. European Association for Computer Science Logic*, Lecture Notes in Computer Science 933, pages 61–74, 1994.
11. W. Howard. The formulae-as-types notion of construction. In *To H. B. Curry: Essays on Combinatory Logic, Lambda Calculus and Formalism*, pages 476–490. Academic Press, 1980.
12. G. Kahn. Natural semantics. In *Proc. Symposium on Theoretical Aspects of Computer Science*, pages 22–39. Springer Verlag, 1987.
13. S. Kleene. *Introduction to Metamathematics*. North-Holland, 1952. 7th edition.
14. S. Kobayashi. Monads as modality. *Theoretical Computer Science*, 175(1):29–74, 1997.
15. E. Moggi. Computational lambda-calculus and monads. In *Proceedings of the Symposium on Logic in Computer Science*, 1989.
16. I. Ogata. Cut elimination for classical proofs as continuation passing style computation. In *Proc. ASIAN Computing Science Conference, LNCS 1538*, 1998.
17. G. Pottinger. Normalization as a homomorphic image of cut-elimination. *Ann. Math. Logic*, 12:323–357, 1977.
18. D. Prawitz. *Natural Deduction*. Almqvist & Wiksell, 1965.
19. A. Sabry and M. Felleisen. Reasoning about programs in continuation-passing style. *J. Lisp and Symbolic Computation*, 6(3):287–358, 1993.
20. W. Tait. Intensional interpretations of functionals of finite type i. *Journal of Symbolic Logic*, 32(2), 1966.
21. J. Zucker. The correspondence between cut-elimination and normalization. *Ann. Math. Logic*, 7:1–112, 1974.

Natural Deduction for Intuitionistic Non-commutative Linear Logic

Jeff Polakow* and Frank Pfenning**

Department of Computer Science
Carnegie Mellon University
Pittsburgh, PA 15213, USA
jpolakow@cs.cmu.edu, fp@cs.cmu.edu

Abstract. We present a system of natural deduction and associated term calculus for intuitionistic non-commutative linear logic (INCLL) as a conservative extension of intuitionistic linear logic. We prove subject reduction and the existence of canonical forms in the implicational fragment.

1 Introduction

Intuitionistic logic captures pure functional computation in a logical way, as can be seen from the Curry-Howard isomorphism between constructive proofs and functional programs. However, there are many structural properties of programs that are not captured within the intuitionistic framework, such as resource usage, computational complexity, and sequentiality.

Intuitionistic linear logic [Gir87,Abr93,Bar97] can be thought of as a refinement of intuitionistic logic in which resource consumption properties of functions can be expressed internally. Here, we refine it further to allow the expression of sequencing of computations. We achieve this by controlling the use of the structural rule of exchange to arrive at *intuitionistic non-commutative linear logic* (INCLL). Much research in non-commutative linear logic has been focused on simply removing the exchange rule from the underlying logic and only allowing exchange to be used in tandem with other structural rules on modal formulas. As an alternative we propose a system which distinguishes among unrestricted, linear, and ordered hypotheses.

Our presentation of INCLL is in the form of natural deduction with proof terms, thereby departing from previous formulations based on the sequent calculus [BG91,Abr90,Rue97]. This establishes the connection to functional computation by an extension of the Curry-Howard isomorphism. INCLL is a conservative extension of dual intuitionistic linear logic [Bar97] which means that we strictly increase its expressive power.

We have several motivating applications for this logic, although space does not permit their detailed analysis in this paper. One direct application is a

* Partially supported by the National Science Foundation under grant CCR-9804014.
** Partially supported by the National Science Foundation under grant CCR-9619584.

logical explanation for ordering properties of terms in continuation-passing style investigated by Danvy and the second author in [DP95]. The ordering inherent in non-commutative function arguments can be used to internalize stackability properties of program evaluation in a fragment of INCLL, which is large enough to capture the case of terms resulting from the standard CPS transformation.

Furthermore, our system integrates the Lambek calculus [Lam58] into a functional framework which also permits ordinary and linear functions in a consistent manner. With the coexistence of linear and ordered functions, we can logically describe more natural language phenomena than with either one by itself; for example, pied-piping and unbounded filler-gap dependencies [Par89,Hod94]. Related approaches to similar problems from computational linguistics are pursued, for example, by Kurtonina and Moortgat [KM96].

We show that our calculus permits canonical (that is, long $\beta\eta$-normal) forms, which means that it is a candidate for a foundation of a logical framework and logic programming language along the lines of Lolli [HM94] and linear LF [CP96]. In related work on a sequent calculus formulation of INCLL [PP99], we have developed an efficient proof search mechanism suitable for logic programming and applied it to algorithms for natural language parsing, sorting, and execution of abstract machines [PP98].

We begin in Section 2 by introducing the implicational fragment of INCLL which is characterized by four implications: intuitionistic ($\rightarrow$), linear ($\multimap$), left ordered ($\rightarrowtail$), and right ordered ($\twoheadrightarrow$). From a functional point of view, this corresponds to having four different types of functions—those which have no restrictions placed upon the use of their arguments; those which must use all their arguments once in any order; and those which which must use all of their arguments once in a specified order. We prove that this fragment satisfies subject reduction thereby validating the introduction and elimination rules. Strong normalization and the Church-Rosser property also hold, but are elided in this extended abstract.

In Section 3 we prove that every well-typed term has an equivalent canonical form, which is important for applications to logic programming and logical frameworks. The proof of this property employs logical relations and we develop the necessary machinery of substitutions. Then we introduce further logical connectives in Section 4 which include a modal operator for mobility (¡) and the usual connectives of linear logic. While subject reduction continues to hold, the existence of commutative conversions destroys the canonical form property.

2 The Implicational Fragment

We define intuitionistic non-commutative linear logic (INCLL) via a judgment

$$\Gamma; \Delta; \Omega \vdash M : A$$

where Γ is a context of unrestricted hypotheses (allowing exchange, weakening, and contraction), Δ is a context of linear hypotheses (allowing only exchange), Ω is a context of ordered hypotheses, M is a proof term, and A is a formula.

Associativity is assumed implicitly for all three contexts. In general, we use "formula" and "type" interchangeably, which is justified by the Curry-Howard isomorphism.

If we reflect the three kinds of hypotheses as connectives in the language of types, we obtain the familiar intuitionistic ($\rightarrow$) and linear ($\multimap$) implications, and two forms of ordered implication, depending on whether hypotheses are taken from the left ($\rightarrowtail$) or the right ($\twoheadrightarrow$) end of the ordered context. In the Lambek calculus [Lam58], the left ordered implication $A \rightarrowtail B$ is written as $A\backslash B$, while the right ordered implication $A \twoheadrightarrow B$ is written as B/A.

Types	$A ::= P$		atomic types
	$\mid$	$A_1 \rightarrow A_2$	intuitionistic implication
	$\mid$	$A_1 \multimap A_2$	linear implication
	$\mid$	$A_1 \twoheadrightarrow A_2$	ordered right implication
	$\mid$	$A_1 \rightarrowtail A_2$	ordered left implication

Proof terms are drawn from a λ-calculus in the style of Church, that is, each valid term has a unique type, which seems essential for the logical framework applications we have in mind. We distinguish between intuitionistic (x), linear (y), and ordered (z) variables and write v if a variable might be declared in any of the three contexts.

Terms	$M ::= x \mid y \mid z$		variables
	$\mid$	$\lambda x{:}A.\ M \mid M_1\, M_2$	intuitionistic functions ($A \rightarrow B$)
	$\mid$	$\hat{\lambda} y{:}A.\ M \mid M_1 \,\hat{}\, M_2$	linear functions ($A \multimap B$)
	$\mid$	$\overset{>}{\lambda} z{:}A.\ M \mid M_1 \,{}^{>}\, M_2$	right ordered functions ($A \twoheadrightarrow B$)
	$\mid$	$\overset{<}{\lambda} z{:}A.\ M \mid M_1 \,{}^{<}\, M_2$	left ordered functions ($A \rightarrowtail B$)

Contexts Γ, Δ, and Ω are simply lists of assumptions, $v{:}A$, where all variables v are distinct but of the same category (intuitionistic, linear, or ordered). We use "$\cdot$" to stand for the empty context, but we often omit it at the beginning of a context. We allow bound variables to be renamed tacitly.

In order to describe the inference rules, we need some auxiliary operations on contexts, *context concatenation* Ω, Ω' and *context merge* $\Delta \bowtie \Delta'$. Concatenation preserves the order of the assumptions, while the non-deterministic merge allows any interleaving of assumptions.

When viewing a natural deduction bottom-up, we think of context concatenation Ω_1, Ω_2 as *ordered context split* and context merge $\Delta_1 \bowtie \Delta_2$ as *context split*. Both of these are non-deterministic when read in this way, that is, there may be many ways to split a context $\Omega = \Omega_1, \Omega_2$ or $\Delta = \Delta_1 \bowtie \Delta_2$.

We now present the introduction and elimination rules for each implicational connective in turn. Other connectives are treated in Section 4. Generally, we use Γ, Δ and Ω to stand for contexts declaring intuitionistic, linear, and ordered variables, respectively.

Intuitionistic Functions $A \rightarrow B$. Since neither the linear nor the ordered context admit weakening, the rule for unrestricted variables requires them to be empty.

In the introduction rule, new variables are added at the right of Γ, but they could just as well be added on the left since the intuitionistic context admits exchange (see Lemma 1). In the elimination rule we cannot allow the derivation of the minor premise to depend on linear or ordered assumptions, since the use of A in the proof of $A \rightarrow B$ is unrestricted and subject reduction would fail. The intuitionistic context must be the same in both premises, which indicates that the rules are biased towards a bottom-up reading, where we distribute the hypotheses Γ to both premises, relying on the validity of contraction for the intuitionistic context.

$$\frac{}{(\Gamma_1, x{:}A, \Gamma_2); \cdot; \cdot \vdash x : A}\,\mathbf{ivar} \qquad \frac{(\Gamma, x{:}A); \Delta; \Omega \vdash M : B}{\Gamma; \Delta; \Omega \vdash \lambda x{:}A.\ M : A \rightarrow B}\rightarrow I$$

$$\frac{\Gamma; \Delta; \Omega \vdash M : A \rightarrow B \qquad \Gamma; \cdot; \cdot \vdash N : A}{\Gamma; \Delta; \Omega \vdash M\,N : B}\rightarrow E$$

Linear Functions $A \multimap B$. The rules for linear functions exhibit the new phenomenon that the linear contexts from the premises of the elimination rules are interleaved to form the linear context of the conclusion, which expresses the linearity condition concisely.

$$\frac{}{\Gamma; y{:}A; \cdot \vdash y : A}\,\mathbf{lvar} \qquad \frac{\Gamma; (\Delta, y{:}A); \Omega \vdash M : B}{\Gamma; \Delta; \Omega \vdash \hat{\lambda} y{:}A.\ M : A \multimap B}\multimap I$$

$$\frac{\Gamma; \Delta_1; \Omega \vdash M : A \multimap B \qquad \Gamma; \Delta_2; \cdot \vdash N : A}{\Gamma; (\Delta_1 \bowtie \Delta_2); \Omega \vdash M\,\hat{}\,N : B}\multimap E$$

Ordered Variables. Ordered variables must be the only ones in the hypothesis rule, which expresses that ordered variables must also be linear. In other words, order is seen as a further restriction on linearity, rather than as an independent property (which is also conceivable).

$$\frac{}{\Gamma; \cdot; z{:}A \vdash z : A}\,\mathbf{ovar}$$

Right Ordered Functions $A \twoheadrightarrow B$. In the introduction rule for right ordered functions the variable z must be new (by our general convention that variables in context are unique) and appear at the right end of the ordered context. In the matching elimination rule, the ordered contexts of the premises are concatenated in order to form the ordered context of the conclusion. The linear context is still interleaving, so as not to violate linearity.

$$\frac{\Gamma; \Delta; (\Omega, z{:}A) \vdash M : B}{\Gamma; \Delta; \Omega \vdash \overset{>}{\lambda} z{:}A.\ M : A \twoheadrightarrow B}\twoheadrightarrow I$$

$$\frac{\Gamma; \Delta_1; \Omega_1 \vdash M : A \twoheadrightarrow B \qquad \Gamma; \Delta_2; \Omega_2 \vdash N : A}{\Gamma; (\Delta_1 \bowtie \Delta_2); (\Omega_1, \Omega_2) \vdash M \overset{>}{}\, N : B}\twoheadrightarrow E$$

Left Ordered Functions $A \rightarrowtail B$. The rules for left ordered implication are symmetric to right ordered implication: the assumption $z{:}A$ appears at the left end of the ordered context in the introduction rule, and the contexts are concatenated in reverse order in the elimination rule. The fact that these rules are consistent is demonstrated by the subject reduction theorem 1.

$$\frac{\Gamma; \Delta; (z{:}A, \Omega) \vdash M : B}{\Gamma; \Delta; \Omega \vdash \overset{<}{\lambda} z{:}A.\ M : A \rightarrowtail B} \rightarrowtail I$$

$$\frac{\Gamma; \Delta_2; \Omega_2 \vdash M : A \rightarrowtail B \qquad \Gamma; \Delta_1; \Omega_1 \vdash N : A}{\Gamma; (\Delta_1 \bowtie \Delta_2); (\Omega_1, \Omega_2) \vdash M^{<} N : B} \rightarrowtail E$$

To give more intuition to our formulation, we now reconsider the rules as they would be used in the bottom-up construction of a proof.

In the three variable rules **ivar**, **lvar**, and **ovar**, the linear and ordered contexts must either be empty or contain only the subject variable, while the intuitionistic context is unrestricted. This forces linear and ordered assumptions to appear at least once in a term.

In the $\multimap E$, $\twoheadrightarrow E$, and $\rightarrowtail E$ rules, the linear context is split into two disjoint parts (when reading from the bottom up), which means that each assumption can be used at most once. In the $\rightarrow E$ rules, all linear assumptions propagate to the left premise. These observations together show that each linear variable is used at most once. Since it is also used at least once by the observation made about the variable rules, linear assumptions occur exactly once.

In the $\twoheadrightarrow E$ rule, the ordered context is split in an order-preserving way, with the leftmost assumptions Ω_1 going to the left premise and the rightmost assumptions Ω_2 going to the right premise. The converse applies to the $\rightarrowtail E$ rule. In the $\multimap E$ and $\rightarrow E$ rules the whole ordered context Ω goes to the left premise. These observations, together with the observation on the variable rules, show that ordered assumptions occur exactly once and in the order they were made.

As we will see, the emptiness restrictions on the linear and ordered contexts in the $\multimap E$ and $\rightarrow E$ rules are necessary to guarantee subject reduction. The reduction rules are simply β-reduction for all three kinds of functions. We will later also consider η-expansion.

Reduction Rules.

$$(\lambda x{:}A.\ M)\, N \Longrightarrow [N/x]M \qquad (\hat{\lambda} y{:}A.\ M)\,\hat{}\, N \Longrightarrow [N/y]M$$

$$(\overset{>}{\lambda} z{:}A.\ M)^{>} N \Longrightarrow [N/z]M \qquad (\overset{<}{\lambda} z{:}A.\ M)^{<} N \Longrightarrow [N/z]M$$

In order to prove subject reduction we proceed to establish the expected structural properties for contexts and substitution lemmas.

Lemma 1 (Structural Properties).

1. *If* $(\Gamma_1, x{:}A, x'{:}A', \Gamma_2); \Delta; \Omega \vdash M : B$ *then* $(\Gamma_1, x'{:}A', x{:}A, \Gamma_2); \Delta; \Omega \vdash M : B$.
2. *If* $(\Gamma_1, \Gamma_2); \Delta; \Omega \vdash M : B$ *then* $(\Gamma_1, x{:}A, \Gamma_2); \Delta; \Omega \vdash M : B$.
3. *If* $(\Gamma_1, x{:}A, x'{:}A, \Gamma_2); \Delta; \Omega \vdash M : B$ *then* $(\Gamma_1, x{:}A, \Gamma_2); \Delta; \Omega \vdash [x/x']M : B$.
4. *If* $\Gamma; (\Delta_1, y{:}A, y'{:}A', \Delta_2); \Omega \vdash M : B$ *then* $\Gamma; (\Delta_1, y'{:}A', y{:}A, \Delta_2); \Omega \vdash M : B$.

Proof: By induction on the structure of the given derivations. □

Lemma 2 (Substitution Properties).

1. *If* $(\Gamma_1, x{:}A, \Gamma_2); \Delta; \Omega \vdash M : B$ *and* $\Gamma_1; \cdot; \cdot \vdash N : A$
   then $(\Gamma_1, \Gamma_2); \Delta; \Omega \vdash [N/x]M : B$.
2. *If* $\Gamma; (\Delta_1, y{:}A, \Delta_2); \Omega \vdash M : B$ *and* $\Gamma; \Delta'; \cdot \vdash N : A$
   then $\Gamma; (\Delta_1, \Delta', \Delta_2); \Omega \vdash [N/y]M : B$.
3. *If* $\Gamma; \Delta; (\Omega_1, z{:}A, \Omega_2) \vdash M : B$ *and* $\Gamma; \Delta'; \Omega' \vdash N : A$
   then $\Gamma; (\Delta \bowtie \Delta'); (\Omega_1, \Omega', \Omega_2) \vdash [N/z]M : B$.

Proof: By induction over the structure of the given typing derivation for M in each case, using Lemma 1. □

Subject reduction now follows immediately.

Theorem 1 (Subject Reduction).
If $M \Longrightarrow M'$ *and* $\Gamma; \Delta; \Omega \vdash M : A$ *then* $\Gamma; \Delta; \Omega \vdash M' : A$.

Proof: For each reduction, we apply inversion to the given typing derivation and then use the substitution lemma 2 to obtain the typing derivation for the conclusion. □

Subject reduction demonstrates that an introduction rule immediately followed by an elimination rule for the same connective can be reduced. This is a form of a *local soundness* theorem expressing that the elimination rules are not too strong. The corresponding global soundness property states that every derivation can be normalized entirely. This is easy to establish via a standard forgetful interpretation into the simply-typed λ-calculus. The normal form is also unique, which is a direct consequence of confluence. We will not formally state these theorems here, since they are besides the main interest of this paper. The proof of confluence is also completely standard (either developing a theory of residuals or using the Tait/Martin-Löf method of parallel reduction).

Local soundness (expressed as subject reduction) guarantees that, for each connective, the elimination rules are not too strong. To check that they are not too weak, we need to show that there is a way to apply elimination rules so that the original judgment can be recovered by introduction rules. This property of local completeness is expressed on proof terms as *subject expansion*, where "expansion" refers to η-expansion.

Theorem 2 (Subject Expansion).

1. *If* $\Gamma; \Delta; \Omega \vdash M : A \rightarrow B$ *then* $\Gamma; \Delta; \Omega \vdash \lambda x{:}A.\ M\, x : A \rightarrow B$.
2. *If* $\Gamma; \Delta; \Omega \vdash M : A \multimap B$ *then* $\Gamma; \Delta; \Omega \vdash \hat{\lambda} y{:}A.\ M\,\hat{}\, y : A \multimap B$.
3. *If* $\Gamma; \Delta; \Omega \vdash M : A \twoheadrightarrow B$ *then* $\Gamma; \Delta; \Omega \vdash \overset{>}{\lambda} z{:}A.\ M^{>} z : A \twoheadrightarrow B$.
4. *If* $\Gamma; \Delta; \Omega \vdash M : A \rightarrowtail B$ *then* $\Gamma; \Delta; \Omega \vdash \overset{<}{\lambda} z{:}A.\ M^{<} z : A \rightarrowtail B$.

Proof: By a direct derivation in each case, using weakening (lemma 1(2)) in part 1. □

A corresponding global property is the existence of long normal forms. This is the subject of the next section.

3 Canonical Forms

The existence of *canonical* (or long $\beta\eta$-normal) forms is critical in logical framework applications of our calculus, since it is the canonical forms which are in bijective correspondence with the objects to be represented. This property is inherited both from the logical framework LF [HHP93] and its linear refinement LLF [CP96]. For the intuitionistic case, both syntactic and semantic proofs exist (see, for example, [Gha97]). Here we pursue a proof by logical relations, whose development also sheds light on the nature of substitutions in our calculus.

We first formalize the property that a term can be converted to canonical form via a deductive system which can easily be related to the usual notion of long $\beta\eta$-normal form. This deductive system can also be read as an algorithm for converting a term to canonical form.

We then prove that any well-typed term can indeed be converted to canonical form. Our proof will be an argument by Kripke logical relations (also called Tait's method) consisting of two parts: (1) If M is a well-typed term of type A then M is in the logical relation represented by A, and (2) if M is in the logical relation represented by A then there is some canonical term N convertible to M. Our reduction strategy is based on weak head reduction defined below.

$$\frac{}{(\lambda x{:}A.\ M)\, N \overset{\mathbf{whr}}{\longrightarrow} [N/x]M}\,\beta_{\rightarrow} \qquad \frac{M \overset{\mathbf{whr}}{\longrightarrow} M'}{M\, N \overset{\mathbf{whr}}{\longrightarrow} M'\, N}\,\mathbf{whr}_{\rightarrow}$$

$$\frac{}{(\hat{\lambda} y{:}A.\ M)\,\hat{}\, N \overset{\mathbf{whr}}{\longrightarrow} [N/y]M}\,\beta_{\multimap} \qquad \frac{M \overset{\mathbf{whr}}{\longrightarrow} M'}{M\,\hat{}\, N \overset{\mathbf{whr}}{\longrightarrow} M'\,\hat{}\, N}\,\mathbf{whr}_{\multimap}$$

$$\frac{}{(\overset{<}{\lambda} z{:}A.\ M)^{<} N \overset{\mathbf{whr}}{\longrightarrow} [N/z]M}\,\beta_{\rightarrowtail} \qquad \frac{M \overset{\mathbf{whr}}{\longrightarrow} M'}{M^{<} N \overset{\mathbf{whr}}{\longrightarrow} M'^{<} N}\,\mathbf{whr}_{\rightarrowtail}$$

$$\frac{}{(\overset{>}{\lambda} z{:}A.\ M)^{>} N \overset{\mathbf{whr}}{\longrightarrow} [N/z]M}\,\beta_{\twoheadrightarrow} \qquad \frac{M \overset{\mathbf{whr}}{\longrightarrow} M'}{M^{>} N \overset{\mathbf{whr}}{\longrightarrow} M'^{>} N}\,\mathbf{whr}_{\twoheadrightarrow}$$

Intuitively, canonical terms are atomic terms of atomic type or λ-abstractions of canonical terms. Atomic terms are variables or applications of atomic terms to canonical terms. This is formalized in the judgments $\Gamma; \Delta; \Omega \vdash M \Uparrow M' : A$, which denotes that M has canonical form M' at type A, and $\Gamma; \Delta; \Omega \vdash M \downarrow M' : A$, which denotes that M has atomic form M' at type A.

Atomic Types.

$$\frac{\Gamma; \Delta; \Omega \vdash M \downarrow M' : P}{\Gamma; \Delta; \Omega \vdash M \Uparrow M' : P}\,\textbf{coercion}$$

$$\frac{M \xrightarrow{\textbf{whr}} M' \qquad \Gamma; \Delta; \Omega \vdash M' \Uparrow M'' : P}{\Gamma; \Delta; \Omega \vdash M \Uparrow M'' : P}\,\textbf{reduction}$$

Intuitionistic Functions.

$$\frac{}{(\Gamma_1, x{:}A, \Gamma_2); \cdot; \cdot \vdash x \downarrow x : A}\,\textbf{ivar} \qquad \frac{(\Gamma, x{:}A); \Delta; \Omega \vdash M\,x \Uparrow M' : B}{\Gamma; \Delta; \Omega \vdash M \Uparrow \lambda x{:}A.\, M' : A \to B}\,{\to}I$$

$$\frac{\Gamma; \Delta; \Omega \vdash M \downarrow M' : A \to B \qquad \Gamma; \cdot; \cdot \vdash N \Uparrow N' : A}{\Gamma; \Delta; \Omega \vdash M\,N \downarrow M'\,N' : B}\,{\to}E$$

Linear Functions.

$$\frac{}{\Gamma; y{:}A; \cdot \vdash y \downarrow y : A}\,\textbf{lvar} \qquad \frac{\Gamma; (\Delta, y{:}A); \Omega \vdash M\,\hat{}\,y \Uparrow M' : B}{\Gamma; \Delta; \Omega \vdash M \Uparrow \hat{\lambda} y{:}A.\, M' : A \multimap B}\,{\multimap}I$$

$$\frac{\Gamma; \Delta; \Omega \vdash M \downarrow M' : A \multimap B \qquad \Gamma; \Delta_A; \cdot \vdash N \Uparrow N' : A}{\Gamma; (\Delta \bowtie \Delta_A); \Omega \vdash M\,\hat{}\,N \downarrow M'\,\hat{}\,N' : B}\,{\multimap}E$$

Ordered Functions.

$$\frac{}{\Gamma; \cdot; z{:}A \vdash z \downarrow z : A}\,\textbf{ovar}$$

$$\frac{\Gamma; \Delta; (\Omega, z{:}A) \vdash M^{>} z \Uparrow M' : B}{\Gamma; \Delta; \Omega \vdash M \Uparrow \overset{>}{\lambda} z{:}A.\, M' : A \twoheadrightarrow B}\,{\twoheadrightarrow}I$$

$$\frac{\Gamma; \Delta; \Omega \vdash M \downarrow M' : A \twoheadrightarrow B \qquad \Gamma; \Delta_A; \Omega_A \vdash N \Uparrow N' : A}{\Gamma; (\Delta \bowtie \Delta_A); (\Omega, \Omega_A) \vdash M^{>} N \downarrow M'^{>} N' : B}\,{\twoheadrightarrow}E$$

$$\frac{\Gamma; \Delta; (z{:}A, \Omega) \vdash M^{<} z \Uparrow M' : B}{\Gamma; \Delta; \Omega \vdash M \Uparrow \overset{<}{\lambda} z{:}A.\, M' : A \rightarrowtail B}\,{\rightarrowtail}I$$

$$\frac{\Gamma; \Delta; \Omega \vdash M \downarrow M' : A \rightarrowtail B \qquad \Gamma; \Delta_A; \Omega_A \vdash N \Uparrow N' : A}{\Gamma; (\Delta \bowtie \Delta_A); (\Omega_A, \Omega) \vdash M^{<} N \downarrow M'^{<} N' : B}\,{\rightarrowtail}E$$

We remark that the expected structural properties of the intuitionistic and linear contexts also hold for this system. Furthermore, if $\Gamma; \Delta; \Omega \vdash M \Uparrow M' : A$ then $\Gamma; \Delta; \Omega \vdash M' : A$ and M' is in long $\beta\eta$-normal form. These properties follow by immediate structural inductions.

The following unary Kripke logical relation is the crux of our argument. It is defined by induction on the type A. Note how the structural properties of intuitionistic, linear, and ordered contexts are captured in this definition.

$\Gamma; \Delta; \Omega \vdash M \in [\![P]\!]$ iff $\Gamma; \Omega; \Delta \vdash M \Uparrow N : P$ for some N.

$\Gamma; \Delta; \Omega \vdash M \in [\![A_1 \to A_2]\!]$ iff for all Γ_N and N,
 if $\Gamma, \Gamma_N; \cdot; \cdot \vdash N \in [\![A_1]\!]$ then $\Gamma, \Gamma_N; \Omega; \Delta \vdash M\,N \in [\![A_2]\!]$.

$\Gamma; \Delta; \Omega \vdash M \in [\![A_1 \multimap A_2]\!]$ iff for all Δ_N and N,
 if $\Gamma; \Delta_N; \cdot \vdash N \in [\![A_1]\!]$ then $\Gamma; \Delta \bowtie \Delta_N; \Omega \vdash M\hat{\ }N \in [\![A_2]\!]$.

$\Gamma; \Delta; \Omega \vdash M \in [\![A_1 \twoheadrightarrow A_2]\!]$ iff for all Δ_N, Ω_N and N,
 if $\Gamma; \Delta_N; \Omega_N \vdash N \in [\![A_1]\!]$ then $\Gamma; \Delta \bowtie \Delta_N; \Omega, \Omega_N \vdash M^{>}N \in [\![A_2]\!]$.

$\Gamma; \Delta; \Omega \vdash M \in [\![A_1 \rightarrowtail A_2]\!]$ iff for all Δ_N, Ω_N and N,
 if $\Gamma; \Delta_N; \Omega_N \vdash N \in [\![A_1]\!]$ then $\Gamma; \Delta \bowtie \Delta_N; \Omega_N, \Omega \vdash M^{<}N \in [\![A_2]\!]$.

We can now formally state and prove the second part of our proof— that well-typed terms in the logical relation at all types have canonical forms. We can prove this only simultaneously with the reverse statement for terms with an atomic form.

Lemma 3 (Logical Relations and Canonical Forms).

1. *If $\Gamma; \Delta; \Omega \vdash M \in [\![A]\!]$ then $\Gamma; \Delta; \Omega \vdash M \Uparrow N : A$ for some N.*
2. *If $\Gamma; \Delta; \Omega \vdash M \downarrow N : A$ then $\Gamma; \Delta; \Omega \vdash M \in [\![A]\!]$.*

Proof: By induction on A using structural properties of contexts. □

Lemma 4 (Closure Under Head Expansion).
If $M \overset{\text{whr}}{\longrightarrow} M'$ and $\Gamma; \Delta; \Omega \vdash M' \in [\![A]\!]$ then $\Gamma; \Delta; \Omega \vdash M \in [\![A]\!]$.

Proof: By induction on A making use of lemma 3. □

In order to show $\Gamma; \Delta; \Omega \vdash M : A$ implies $\Gamma; \Delta; \Omega \vdash M \in [\![A]\!]$, we need to explicitly manipulate substitutions. We shall define a substitution to be a triple, $(\gamma; \delta; \omega)$, where each component is a list of term/variable pairs.

$$(\gamma; \delta; \omega) = (\cdot; \cdot; \cdot) \mid (\gamma, M/x; \delta; \omega) \mid (\gamma; \delta, M/y; \omega) \mid (\gamma; \delta; \omega, M/z)$$

We assume no variable is defined more than once in $(\gamma; \delta; \omega)$ and we write $(\gamma; \delta; \omega)(v) = M$ if M/v occurs in $(\gamma; \delta; \omega)$. We define well-typed substitutions with the judgment $\Gamma'; \Delta'; \Omega' \vdash (\gamma; \delta; \omega) : \Gamma; \Delta; \Omega$ which means that $\gamma; \delta; \omega$ supply appropriate terms for the variables declared in $\Gamma; \Delta; \Omega$, respectively.

$$\frac{}{\Gamma'; \Delta'; \Omega' \vdash (\cdot; \cdot; \cdot) : \cdot; \cdot; \cdot}$$

$$\frac{\Gamma'; \Delta'; \Omega' \vdash (\gamma; \delta; \omega) : \Gamma; \Delta; \Omega \qquad \Gamma'; \cdot; \cdot \vdash M : A}{\Gamma'; \Delta'; \Omega' \vdash (\gamma, M/x; \delta; \omega) : \Gamma, x{:}A; \Delta; \Omega}$$

$$\frac{\Gamma'; \Delta'_1; \Omega' \vdash (\gamma; \delta; \omega) : \Gamma; \Delta; \Omega \qquad \Gamma'; \Delta'_2; \cdot \vdash M : A}{\Gamma'; \Delta'_1 \bowtie \Delta'_2; \Omega' \vdash (\gamma; \delta, M/y; \omega) : \Gamma; \Delta, y{:}A; \Omega}$$

$$\frac{\Gamma'; \Delta'_1; \Omega'_1 \vdash (\gamma; \delta; \omega) : \Gamma; \Delta; \Omega \qquad \Gamma'; \Delta'_2; \Omega'_2 \vdash M : A}{\Gamma'; \Delta'_1 \bowtie \Delta'_2; \Omega'_1, \Omega'_2 \vdash (\gamma; \delta; \omega, M/z) : \Gamma; \Delta; \Omega, z{:}A}$$

Note the restrictions which prohibit, for example, that the substitution term for a linear variable depends on an ordered variable. Such a dependence would falsify Theorem 5.

When computing the result of applying a substitution to a term, we would like to maintain the invariant that the substitution matches the contexts in which the term is well-formed. This means we have to split the substitution at applications. Thus, we define the application of a substitution to a term as follows:

$$\begin{aligned}
&[(\gamma; \delta; \omega)]v = (\gamma; \delta; \omega)(v) \\
&[(\gamma; \delta; \omega)](\lambda x{:}A.\ M) = \lambda x{:}A.\ [(\gamma, x/x; \delta; \omega)]M \\
&[(\gamma; \delta; \omega)](M\,N) = ([(\gamma; \delta; \omega)]M)([\gamma; \cdot; \cdot]N) \\
&[(\gamma; \delta; \omega)](\hat{\lambda} y{:}A.\ M) = \hat{\lambda} y{:}A.\ [(\gamma; \delta, y/y; \omega)]M \\
&[(\gamma; \delta_1 \bowtie \delta_2; \omega)](M\hat{\ }N) = ([(\gamma; \delta_1; \omega)]M)\hat{\ }([\gamma; \delta_2; \cdot]N) \\
&[(\gamma; \delta; \omega)](\overset{>}{\lambda} z{:}A.\ M) = \overset{>}{\lambda} z{:}A.\ [(\gamma; \delta; \omega, z/z)]M \\
&[(\gamma; \delta_1 \bowtie \delta_2; \omega_1, \omega_2)](M^{>}N) = ([(\gamma; \delta_1; \omega_1)]M)^{>}([\gamma; \delta_2; \omega_2]N) \\
&[(\gamma; \delta; \omega)](\overset{<}{\lambda} z{:}A.\ M) = \overset{<}{\lambda} z{:}A.\ [(\gamma; \delta; z/z, \omega)]M \\
&[(\gamma; \delta_1 \bowtie \delta_2; \omega_2, \omega_1)](M^{<}N) = ([(\gamma; \delta_1; \omega_1)]M)^{<}([\gamma; \delta_2; \omega_2]N)
\end{aligned}$$

At first glance the substitution splitting may seem non-deterministic. However, the proper split can be easily determined from the typing derivation of the term we substitute into. Since typing derivations are unique, there is no ambiguity. We rely on this in the proof of the fundamental theorem of logical relations (Lemma 7).

Lemma 5 (Typing and Substitutions). *If* $\Gamma; \Delta; \Omega \vdash M : A$ *and* $\Gamma'; \Delta'; \Omega' \vdash (\gamma; \delta; \omega) : \Gamma; \Delta; \Omega$ *then* $\Gamma'; \Delta'; \Omega' \vdash [(\gamma; \delta; \omega)]M : A$.

Proof: By induction on the structure of the derivation of $\Gamma; \Delta; \Omega \vdash M : A$. □

Substitutions compose in the obvious way, although we do not investigate properties of substitutions further here. We write $id_{\Gamma;\Delta;\Omega}$ for the identity substitution on the variables declared in Γ, Δ, and Ω. We define logical relations on substitutions by induction on the structure of contexts.

$$\Gamma'; \cdot; \cdot \vdash \cdot \in [\![\cdot; \cdot; \cdot]\!]$$

$$\Gamma'; \Delta'; \Omega' \vdash (\gamma, M/x; \delta; \omega) \in [\![\Gamma, x{:}A; \Delta; \Omega]\!] \text{ iff}$$
$$\quad \Gamma'; \Delta'; \Omega' \vdash (\gamma; \delta; \omega) \in [\![\Gamma; \Delta; \Omega]\!] \text{ and } \Gamma'; \cdot; \cdot \vdash M \in [\![A]\!]$$

$$\Gamma'; \Delta'_1 \bowtie \Delta'_2; \Omega' \vdash (\gamma; \delta, M/y; \delta; \omega) \in [\![\Gamma; \Delta, y{:}A; \Omega]\!] \text{ iff}$$
$$\quad \Gamma'; \Delta'_1; \Omega' \vdash (\gamma; \delta; \omega) \in [\![\Gamma; \Delta; \Omega]\!] \text{ and } \Gamma'; \Delta'_2; \cdot \vdash M \in [\![A]\!]$$

$$\Gamma'; \Delta'_1 \bowtie \Delta'_2; \Omega'_1 \Omega'_2 \vdash (\gamma; \delta; \omega, M/z) \in [\![\Gamma; \Delta; \Omega, z{:}A]\!] \text{ iff}$$
$$\quad \Gamma'; \Delta'_1; \Omega'_1 \vdash (\gamma; \delta; \omega) \in [\![\Gamma; \Delta; \Omega]\!] \text{ and } \Gamma'; \Delta'_2; \Omega'_2 \vdash M \in [\![A]\!]$$

Lemma 6 (Identity). $\Gamma; \Delta; \Omega \vdash id_{\Gamma;\Delta;\Omega} \in [\![\Gamma; \Delta; \Omega]\!]$

Proof: Immediate by definition and lemma 3. □

Lemma 7 (Typing and Logical Relations). *If $\Gamma; \Delta; \Omega \vdash M : A$ then for any $\Gamma'; \Delta'; \Omega' \vdash (\gamma; \delta; \omega) \in [\![\Gamma; \Delta; \Omega]\!]$ we have $\Gamma'; \Delta'; \Omega' \vdash [(\gamma; \delta; \omega)]M \in [\![A]\!]$.*

Proof: By induction on the structure of the given derivation using lemma 4. □

Theorem 3 (Canonical Forms).
If $\Gamma; \Delta; \Omega \vdash M : A$ then for some N, $\Gamma; \Delta; \Omega \vdash M \Uparrow N : A$.

Proof: Immediate from lemmas 7, 3, and 6. □

4 Other Logical Connectives

Before considering the other standard connectives from linear logic, we note further structural properties.

Theorem 4 (Demotion).

1. *If $\Gamma; (\Delta_1, y{:}A, \Delta_2); \Omega \vdash M : B$ then $(\Gamma, x{:}A); (\Delta_1, \Delta_2); \Omega \vdash [x/y]M : B$.*
2. *If $\Gamma; \Delta; (\Omega_1, z{:}A, \Omega_2) \vdash M : B$ then $\Gamma; (\Delta, y{:}A); (\Omega_1, \Omega_2) \vdash [y/z]M : B$.*

Proof: In both cases by induction on the structure of the given derivation. □

When considering the typing rules for the new connectives, we shall take care that the preceding property continues to hold. The subject reduction and strong normalization theorems also continue to hold, with straightforward extensions of the proofs mentioned in Section 2.

Some of the new connectives, namely an ordered conjunction (•), multiplicative unit (1), disjunction (⊕), falsehood (0), mobility (¡) and exponential (!) introduce commutative conversions into the proof term calculus. Unique canonical forms no longer exist, even though each connective remains locally sound and complete. This means that these connectives must be ruled out or restricted in logic programming or logical frameworks applications of INCLL. Fortunately, this does not seem to be a serious drawback in practice [PP98].

Ordered Conjunction $A \bullet B$.

$$\frac{\Gamma; \Delta_1; \Omega_1 \vdash M{:}A \qquad \Gamma; \Delta_2; \Omega_2 \vdash N{:}B}{\Gamma; (\Delta_1 \bowtie \Delta_2); (\Omega_1, \Omega_2) \vdash M \bullet N : A \bullet B} \bullet I$$

$$\frac{\Gamma; \Delta_2; \Omega_2 \vdash M : A \bullet B \qquad \Gamma; \Delta_1; (\Omega_1, z{:}A, z'{:}B, \Omega_3) \vdash N : C}{\Gamma; (\Delta_1 \bowtie \Delta_2); (\Omega_1, \Omega_2, \Omega_3) \vdash \mathbf{let}\ z \bullet z' = M\ \mathbf{in}\ N : C} \bullet E$$

We have the following reduction rule:

$$\mathbf{let}\ z \bullet z' = M \bullet M'\ \mathbf{in}\ N \Longrightarrow [M/z, M'/z']N$$

Multiplicative Unit 1. This is the right and left unit element for the ordered conjunction connective. We have $1 \twoheadrightarrow C$ iff C iff $1 \rightarrowtail C$, and $A \bullet 1$ iff A iff $1 \bullet A$. The introduction rule shows why there is only one multiplicative unit.

$$\frac{}{\Gamma; \cdot; \cdot \vdash \star : 1} 1I$$

$$\frac{\Gamma; \Delta_2; \Omega_2 \vdash M : 1 \qquad \Gamma; \Delta_1; (\Omega_1, \Omega_3) \vdash N : C}{\Gamma; (\Delta_1 \bowtie \Delta_2); (\Omega_1, \Omega_2, \Omega_3) \vdash \mathbf{let}\ \star = M\ \mathbf{in}\ N : C} 1E$$

We have the following reduction rule:

$$\mathbf{let}\ \star = \star\ \mathbf{in}\ N \Longrightarrow N$$

Additive Conjunction $A \,\&\, B$. This is additive on both the linear and ordered contexts, in order to preserve Theorem 4.

$$\frac{\Gamma; \Delta; \Omega \vdash M : A \qquad \Gamma; \Delta; \Omega \vdash N : B}{\Gamma; \Delta; \Omega \vdash \langle M, N \rangle : A \,\&\, B} \&I$$

$$\frac{\Gamma; \Delta; \Omega \vdash M : A \,\&\, B}{\Gamma; \Delta; \Omega \vdash \mathbf{fst}\, M : A} \&E_1 \qquad \frac{\Gamma; \Delta; \Omega \vdash M : A \,\&\, B}{\Gamma; \Delta; \Omega \vdash \mathbf{snd}\, M : B} \&E_2$$

We have the following reduction rules:

$$\mathbf{fst}\, \langle M, N \rangle \Longrightarrow M$$
$$\mathbf{snd}\, \langle M, N \rangle \Longrightarrow N$$

Additive Unit $\top$. Because it is additive, the left and right units for & coincide.

$$\frac{}{\Gamma; \Delta; \Omega \vdash \langle\rangle : \top} \top I$$

Since there is no elimination rule, there are no reductions for the additive unit.

Disjunction $\oplus$. The disjunction is additive and therefore does not split into left and right versions.

$$\frac{\Gamma; \Delta; \Omega \vdash M : A}{\Gamma; \Delta; \Omega \vdash \mathbf{inl}^B\, M : A \oplus B}\oplus I_1 \qquad \frac{\Gamma; \Delta; \Omega \vdash M : B}{\Gamma; \Delta; \Omega \vdash \mathbf{inr}^A\, M : A \oplus B}\oplus I_2$$

$$\frac{\begin{array}{l}\Gamma; \Delta_2; \Omega_2 \vdash M : A \oplus B\\ \Gamma; \Delta_1; (\Omega_1, z{:}A, \Omega_3) \vdash N : C\\ \Gamma; \Delta_1; (\Omega_1, z'{:}B, \Omega_3) \vdash N' : C\end{array}}{\Gamma; (\Delta_1 \bowtie \Delta_2); (\Omega_1, \Omega_2, \Omega_3) \vdash \mathbf{case}\, M\ \mathbf{of}\ \mathbf{inl}\, z \Rightarrow N \mid \mathbf{inr}\, z' \Rightarrow N' : C}\oplus E$$

We have the following reduction rules:

$$\begin{array}{c}\mathbf{case}\,\mathbf{inl}^B\, M\ \mathbf{of}\ \mathbf{inl}\, z \Rightarrow N \mid \mathbf{inr}\, z' \Rightarrow N' \Longrightarrow [M/z]N\\ \mathbf{case}\,\mathbf{inr}^A\, M'\ \mathbf{of}\ \mathbf{inl}\, z \Rightarrow N \mid \mathbf{inr}\, z' \Rightarrow N' \Longrightarrow [M'/z']N'\end{array}$$

Additive Falsehood 0. This is the unit for disjunction.

$$\frac{\Gamma; \Delta_2; \Omega_2 \vdash M : 0}{\Gamma; (\Delta_1 \bowtie \Delta_2); (\Omega_1, \Omega_2, \Omega_3) \vdash \mathbf{abort}^C\, M : C}0E$$

Since there is no introduction rule for 0, there are no new reductions.

Linear Exponential $!A$.

$$\frac{\Gamma; \cdot; \cdot \vdash M : A}{\Gamma; \cdot; \cdot \vdash\, !M : !A}!I$$

$$\frac{\Gamma; \Delta_2; \Omega_2 \vdash M : !A \qquad (\Gamma, x{:}A); \Delta_1; (\Omega_1, \Omega_3) \vdash N : C}{\Gamma; (\Delta_1 \bowtie \Delta_2); (\Omega_1, \Omega_2, \Omega_3) \vdash \mathbf{let}\, !x = M\ \mathbf{in}\ N : C}!E$$

We have the following reduction rule:

$$\mathbf{let}\, !x = !M\ \mathbf{in}\ N \Longrightarrow [M/x]N$$

Mobility Modal $\text{¡}A$. We may also consider a modality not present in linear logic which allows an ordered hypothesis to be used out of order. In analogy with !, we wish to have $\text{¡}A \rightarrowtail B \equiv \text{¡}A \twoheadrightarrow B \equiv A \multimap B$.

$$\frac{\Gamma; \Delta; \cdot \vdash M : A}{\Gamma; \Delta; \cdot \vdash \text{¡}M : \text{¡}A}\text{¡}I$$

$$\frac{\Gamma; \Delta_2; \Omega_2 \vdash M : \text{¡}A \qquad \Gamma; (\Delta_1, y{:}A); (\Omega_1, \Omega_3) \vdash N : C}{\Gamma; (\Delta_1 \bowtie \Delta_2); (\Omega_1, \Omega_2, \Omega_3) \vdash \mathbf{let}\, \text{¡}y = M\ \mathbf{in}\ N : C}\text{¡}E$$

We have the following reduction rule:

$$\mathbf{let}\, \text{¡}y = \text{¡}M\ \mathbf{in}\ N \Longrightarrow [M/y]N$$

5 Conclusion and Future Work

We have presented a natural deduction version of intuitionistic non-commutative linear logic which conservatively extends intuitionistic linear logic. We have shown that the proof term calculus satisfies subject reduction and strong normalization, and that canonical forms exist for the implicational fragment. In [PP99] we present a sequent calculus for INCLL, prove cut-elimination and show that it closely corresponds to the natural deduction system presented here.

Applications lie in the areas of logical frameworks, functional programming, logic programming, and natural language processing. These applications are sketched in the introduction and are the subject of current research. At present, for example, we have shown that the ordering properties of functional programs which result from CPS conversion discovered by Danvy [Dan94] can be captured completely internally in the INCLL term calculus. We have also shown that uniform derivations are sound and complete with respect to our calculus, which means that the implicational fragment of INCLL can be considered an *abstract logic programming language* [MNPS91]. A prototype implementation using advanced resource management strategies analogous to Lolli [Hod94] has been used for the concise expression of various algorithms for sorting, natural language parsing, and the execution of abstract machines. The systems and examples may be found in [PP98].

We have also given an operational semantics to an extension of the functional core presented here and are investigating the connection between stackability of intermediate values and ordered function arguments.

References

[Abr90] V. Michele Abrusci. Non-commutative intuitionistic linear propositional logic. *Zeitschrift für Mathematische Logik und Grundlagen der Mathematik*, 36:297–318, 1990.

[Abr93] Samson Abramsky. Computational interpretations of linear logic. *Theoretical Computer Science*, 111:3–57, 1993.

[Bar97] Andrew Barber. *Linear Type Theories, Semantics and Action Calculi.* PhD thesis, Department of Computer Science, University of Edinburgh, 1997.

[BG91] C. Brown and D. Gurr. Relations and non-commutative linear logic. Technical Report DAIMI PB-372, Computer Science Department, Aarhus University, November 1991.

[CP96] Iliano Cervesato and Frank Pfenning. A linear logical framework. In E. Clarke, editor, *Proceedings of the Eleventh Annual Symposium on Logic in Computer Science — LICS'96*, pages 264–275, New Brunswick, New Jersey, 27–30 July 1996. IEEE Computer Society Press. This work also appeared as Preprint 1834 of the Department of Mathematics of Technical University of Darmstadt, Germany.

[Dan94] Olivier Danvy. Back to direct style. *Science of Computer Programming*, 22(3):183–195, 1994.

[DP95] Olivier Danvy and Frank Pfenning. The occurrence of continuation parameters in CPS terms. Technical Report CMU-CS-95-121, Department of Computer Science, Carnegie Mellon University, February 1995.

[Gha97] Neil Ghani. Eta-expansion in dependent type theory — the Calculus of Constructions. In Philippe de Groote and J. Roger Hindley, editors, *Proceedings of the 3rd International Conference on Typed Lambda Calculi and Applications (TLCA'97)*, pages 164–180, Nancy, France, April 1997. Springer-Verlag LNCS 1210.

[Gir87] Jean-Yves Girard. Linear logic. *Theoretical Computer Science*, 50:1–102, 1987.

[HHP93] Robert Harper, Furio Honsell, and Gordon Plotkin. A framework for defining logics. *Journal of the Association for Computing Machinery*, 40(1):143–184, January 1993.

[HM94] Joshua S. Hodas and Dale Miller. Logic programming in a fragment of intuitionistic linear logic. *Information and Computation*, 110(2):327–365, 1994. Extended abstract in the Proceedings of the Sixth Annual Symposium on Logic in Computer Science, Amsterdam, July 15–18, 1991.

[Hod94] Joshua S. Hodas. *Logic Programming in Intuitionistic Linear Logic: Theory, Design and Implementation.* PhD thesis, University of Pennsylvania, Department of Computer and Information Science, 1994.

[KM96] Natasha Kurtonina and Michael Moortgat. Structural control. In P. Blackburn and M. de Rijke, editors, *Specifying Syntactic Structures.* CSLI Publications, 1996.

[Lam58] Joachim Lambek. The mathematics of sentence structure. *American Mathematical Monthly*, 65:363–386, 1958.

[MNPS91] Dale Miller, Gopalan Nadathur, Frank Pfenning, and Andre Scedrov. Uniform proofs as a foundation for logic programming. *Annals of Pure and Applied Logic*, 51:125–157, 1991.

[Par89] Remo Pareschi. *Type-Driven Natural Language Analysis.* PhD thesis, University of Edinburgh, Edinburgh, Scotland, July 1989. Available as technical report MS-CIS-89-45, Department of Computer and Information Sciences, University of Pennsylvania.

[PP98] Jeff Polakow and Frank Pfenning. Ordered linear logic programming. Technical Report CMU-CS-98-183, Department of Computer Science, Carnegie Mellon University, December 1998.

[PP99] Jeff Polakow and Frank Pfenning. Relating natural deduction and sequent calculus for intuitionistic non-commutative linear logic. In Andre Scedrov and Achim Jung, editors, *Proceedings of the 15th Conference on Mathematical Foundations of Programming Semantics*, New Orleans, Louisiana, April 1999. To appear.

[Rue97] Paul Ruet. *Non-Commutative Logic and Concurrent Constraint Programming.* PhD thesis, Université Denis Diderot, Paris 7, 1997.

A Logic for Abstract Data Types as Existential Types

Erik Poll[1] and Jan Zwanenburg[2]

[1] Computing Lab, University of Kent at Canterbury, England
E.Poll@ukc.ac.uk
[2] Eindhoven University of Technology, The Netherlands
janz@win.tue.nl

Abstract. The second-order lambda calculus allows an elegant formalisation of abstract data types (ADT's) using existential types. Plotkin and Abadi's logic for parametricity [PA93] then provides the useful proof principle of *simulation* for ADT's, which can be used to show equivalence of data representations. However, we show that this logic is not sufficient for reasoning about specifications of ADT's, and we present an extension of the logic that does provide the proof principles for ADT's that we want.

1 Introduction

The second-order lambda calculus allows an elegant formalisation of abstract data types (ADT's), as shown in [MP88], using existential types. This description of ADT's provides a useful basis to investigate properties of ADT's. In particular, it has been successfully used to investigate a notion of equivalence of implementations of ADT's. [Mit91] considers a semantic notion of equivalence of data representations, which suggests a method for proving the equivalence of data representations, namely by showing that there exists a simulation relation between the representations. We will refer to this proof principle as **simulation**. Plotkin and Abadi's logic for parametricity [PA93] is a logic for reasoning about the second order lambda calculus (system F). It formalises the notion of *parametricity*, and for the existential types this logic does indeed provide the proof principle of simulation envisaged in [Mit91].

Unfortunately, it turns out that this proof principle of simulation for existential types is not enough for reasoning about specifications of ADT's, in particular specifications that use equality. We propose an extension of the logic of [PA93] (with axioms stating the existence of quotients, to be precise) that does provide all the proof principles one would like for reasoning about ADT's. The same PER model used in [PA93] as a semantics for their logic immediately justifies these additional axioms. (Indeed, in the PER model all types are "quotient types".)

The remainder of this introduction discusses one of the proof principles we want for ADT's. It is a very natural one, that immediately arises whenever an implementation of an ADT allows different concrete representations of the same abstract value. This example will be treated in more detail later in Section 4.

Suppose we implement an ADT for bags using lists to represents bags. Then there will be many different lists that represent the same bag: any two lists that are permutations represent the same bag. As a consequence, there are *different notions of equality* in play: equality of lists, equality of bags, and the relation $\sim_{perm}$ on lists that relates lists representing the same bag (i.e. that are permutations). A programmer implementing an ADT has to be aware of the fact that there are these different notions of equality. But a programmer using an ADT should only have to deal with equality of bags, and not have to know anything about an underlying relation $\sim_{perm}$ on lists. Indeed, this is precisely the *abstraction* that an *abstract* data type is supposed to provide. A consequence of all is that the programmer implementing an ADT and the programmer using an ADT may want to use a slightly different specification: the former in terms of the relation $\sim_{perm}$ on the concrete data type of lists, the latter in terms of equality on the abstract data type of bags. For instance, the programmer using the ADT might require that

$$\forall m, n : Nat, s : Bag.\ add(m, add(n, s)) = add(n, add(m, s)) \qquad \text{(i)}$$

and to meet this specification, the programmer implementing the ADT must ensure that

$$\forall m, n : Nat, s : List.\ cons(m, cons(n, s)) \sim_{perm} cons(n, cons(m, s)) \qquad \text{(ii)}$$

if *add* is implemented as *cons*. In a logic for reasoning with (specifications of) ADT's we should be able to relate statements such as (i) and (ii). In particular, here one would want to be able to prove that (ii) implies (i). We will refer to a proof principle that would allow us to deduce (i) from (ii) as **abstraction**.

The logic for parametricity of [PA93] does not quite provide this proof principle of abstraction for arbitrary ADT's and specifications. But extending the logic with axioms stating the existence of quotients solves this problem: we will show that then the proof principle of abstraction can be obtained from the proof principle of simulation, which is provided by the logic for parametricity of [PA93]. (For this particular example, we would want the existence of lists quotiented by $\sim_{perm}$.)

The organisation of this paper is as follows. Section 2 defines our notation for the second-order lambda calculus and gives a quick recap on how existential types can be used for ADT's. Section 3 discusses the logic for parametricity of [Tak97], which is a slightly different formulation of the logic as first introduced in [PA93]; in particular, we discuss the proof principle of simulation for proving equivalence of data representations that this logic provides. Section 4 then considers a simple example of a specification of an ADT for bags and illustrates the problem with reasoning about ADT's hinted at above. Section 5 then present our extension of the logic that does provide the power we want.

2 The second-order lambda calculus

We first give the definition of the second-order lambda calculus, and then illustrate how the existential types can be used for ADT's.

The *terms* t and *types* T of the second-order lambda calculus are given by

$$t ::= x \mid \lambda x{:}T.\,t \mid tt \mid (t,t) \mid t.i \mid \lambda X.\,t \mid tT \mid \mathsf{pack}\ \langle T,t\rangle\ \mathsf{to}\ T \mid \mathsf{open}\ t\ \mathsf{as}\ \langle T,t\rangle\ \mathsf{in}\ t$$
$$T ::= X \mid T\times T \mid T\to T \mid \forall X.\,T \mid \exists X.\,T$$

Here x ranges over *term-variables*, X over *type-variables*. Free and bound variables are defined as usual. Terms and types equal up to the names of bound variables and permutation of fields are identified.

We use the following convention for our meta-variables: x, y, z range over term variables, X, Y, Z range over type variables, a, b, c, f range over terms (or programs), A, B, C range over types.

We include products and existentials as primitives here because they play an important role later, but of course they can be regarded as syntactic sugar for their usual encodings. (In fact, we will not even need the universal types in this paper.) Later on we will also use some base types, namely a type *Nat* of natural numbers and a type *List* of lists of natural numbers. These can be encoded in the usual way, too.

The typing rules for judgements of the form $\Gamma \vdash t : T$, where Γ is a sequence of declarations $x_1 : T_1, \ldots, x_n : T_n$, are

$$\frac{}{\Gamma, x : A, \Gamma' \vdash x : A}$$

$$\frac{\Gamma, x : A \vdash b : B}{\Gamma \vdash \lambda x{:}A.\, b : A \to B} \qquad \frac{\Gamma \vdash f : A \to B \quad \Gamma \vdash a : A}{\Gamma \vdash fa : B}$$

$$\frac{\Gamma \vdash a_1 : A_1 \quad \Gamma \vdash a_2 : A_2}{\Gamma \vdash (a_1, a_2) : A_1 \times A_2} \qquad \frac{\Gamma \vdash a : A_1 \times A_2}{\Gamma \vdash a.i : A_i}\ i = 1,2$$

$$\frac{\Gamma \vdash b : B}{\Gamma \vdash \lambda X.\, b : \forall X.\, B}\ X \text{ not free in } \Gamma \qquad \frac{\Gamma \vdash f : \forall X.\, B}{\Gamma \vdash fA : B[A/X]}$$

$$\frac{\Gamma \vdash c : A[C/X]}{\Gamma \vdash (\mathsf{pack}\ \langle C, c\rangle\ \mathsf{to}\ \exists X.\, A) : \exists X.\, A}\ X \text{ not free in } \Gamma$$

$$\frac{\Gamma, x : A \vdash b : B \quad \Gamma \vdash s : \exists X.\, A}{\Gamma \vdash (\mathsf{open}\ s\ \mathsf{as}\ \langle X, x\rangle\ \mathsf{in}\ b) : B}\ X \text{ not free in } B \text{ or } \Gamma$$

The reduction rules are $(\lambda x{:}A.\, b)a \rhd_\beta b[a/x]$, $(\lambda X.\, a)A \rhd_\beta a[A/X]$, $(a_1, a_2).i \rhd_\beta a_i$, and $\mathsf{open}\ (\mathsf{pack}\ \langle C, c\rangle\ \mathsf{to}\ \exists X.\, A)\ \mathsf{as}\ \langle X, x\rangle\ \mathsf{in}\ b \rhd_\beta b[C/X, c/x]$.

Notation. The notation for pairs is extended to n-tuples, which are simply nested pairs. E.g. we write $A \times B \times C$ for $A \times (B \times C)$ and (a, b, c) for $(a, (b, c))$. We typically omit the second type parameter of **pack**, writing $\mathsf{pack}\ \langle C, a\rangle$ for

$(\mathsf{pack}\ \langle C, a\rangle\ \mathsf{to}\ \exists X.\, A)$, whenever this type is clear from the context. Finally, we will sometimes use a "pattern-matching" style notation for tuples, e.g. writing $\lambda(y, z){:}A \times B.\, c$ instead of $\lambda x{:}A \times B.\, c[x.1/y, x.2/z]$.

Abstract Data Types as Existential Types

Existential types allow an elegant formalisation of abstract data types (ADT's), as shown in [MP88]. This formalisation provides a clean separation between using an ADT on the one hand and implementing an ADT on the other hand. Moreover, as is often the case with descriptions of notions from programming languages in terms of typed lambda calculus, this formalisation provides a more powerful notion than exists in most existing programming languages: existential types provide implementations of ADT's as "first-class citizens", i.e. as values that can be passed as parameters to functions or returned as results like any other value. This also means that we can talk about equality of implementations of ADT's just like we can talk about equality of other values. (This will be useful later, in Section 3, when we consider proof rules for ADT's.)

The remainder of this section briefly explains the use of existential types for ADT's (for a more extensive discussion see [MP88]), and introduces our running example of bags.

Our running example will be an ADT of bags, which provides a type *Bag* with three operations: the operation of adding an element to a bag, an operation to inspect how often a given element occurs in a bag, and the empty bag:

$$\begin{aligned} \mathit{empty} &: \mathit{Bag}, \\ \mathit{add} &: \mathit{Nat} \times \mathit{Bag} \to \mathit{Bag}, \\ \mathit{card} &: \mathit{Nat} \times \mathit{Bag} \to \mathit{Nat}. \end{aligned}$$

Tupling the three operations yields

$$(\mathit{empty}, \mathit{add}, \mathit{card}) : \mathit{Bag} \times (\mathit{Nat} \times \mathit{Bag} \to \mathit{Bag}) \times (\mathit{Nat} \times \mathit{Bag} \to \mathit{Nat}),$$

so the signature of the ADT can be given as

$$\mathit{BagSig}(X) \mathrel{\hat{=}} X \times (\mathit{Nat} \times X \to X) \times (\mathit{Nat} \times X \to \mathit{Nat}).$$

The existential type *BagImp*, $\mathit{BagImp} \mathrel{\hat{=}} \exists X.\, \mathit{BagSig}(X)$, can be used as type of implementations of the ADT of bags, as we will now explain.

To implement the ADT of bags, we have to come up with some type *Rep* which will be used as representations of bags, and a 3-tuple of functions of type *BagSig*(*Rep*) that implement the bag-operations for this representation. An obvious way to represent bags is to use lists. In this case *empty* can be implemented as the empty list $\mathit{nil} : \mathit{List}$, *add* as the operation $\mathit{cons} : \mathit{Nat} \times \mathit{List} \to \mathit{List}$ on lists, and *card* as a function $\mathit{count} : \mathit{Nat} \times \mathit{List} \to \mathit{List}$ that counts how often a given natural number occurs in a given list of natural numbers. These three operations have the right types, since

$$(\mathit{nil}, \mathit{cons}, \mathit{count}) : \mathit{BagSig}(\mathit{List}).$$

The introduction rule for existential types can be used to construct an element of type *BagImp* from the type *List* and the triple (*nil*, *cons*, *count*):

$$impl1 \mathrel{\widehat{=}} (\mathsf{pack}\ \langle List, (nil, cons, count)\rangle\ \mathsf{to}\ BagImp) : BagImp.$$

Now suppose we want to define some program b that uses the ADT of bags. Then in b we want to use the abstract operations *empty*, *add*, and *card*, and b has to be well-typed under the assumption that these three abstract operations have their correct types:

$$empty : Bag, add : Nat \times Bag \to Bag, card : Nat \times Bag \to Nat \vdash b : B$$

Here *Bag* is a type variable. The elimination rule for existential types tells us how to combine this program b with the implementation $impl1 : BagImp$ defined above:

$$\mathsf{open}\ impl1\ \mathsf{as}\ \langle Bag, (empty, add, card)\rangle\ \mathsf{in}\ b : B$$

It is easy to verify that this program behaves as expected:

$$\begin{array}{l} \quad \mathsf{open}\ impl1\ \mathsf{as}\ \langle Bag, (empty, add, card)\rangle\ \mathsf{in}\ b \\ \rhd_\beta \\ \quad b[List/Bag, nil/empty, cons/add, count/card]. \end{array}$$

So the concrete representation *List* gets substituted for the abstract type *Bag*, and the concrete implementations of the operations on *List*'s get substituted for the abstract operations on *Bag*'s.

The typing rules play a crucial role in hiding the concrete implementation of the ADT (using *List*'s) from the main program b. It is not possible to apply list operations to bags in b, because this would not be well-typed. The program b has to be typed under the assumptions that

$$empty : Bag, add : Nat \times Bag \to Bag, card : Nat \times Bag \to Nat,$$

where *Bag* is a type variable.

3 The logic for parametricity

Plotkin and Ababi's logic for parametric polymorphism [PA93] is a logic for reasoning about the second-order lambda calculus that exploits the notion of parametricity. We will use the somewhat different presentation of the logic given by Takeuti [Tak97].

We only describe the fragment of the logic that is of interest to us. This makes the description much simpler and this paper much easier to digest. (In particular, Definition 3 only deals with the type constructors $\to$ and $\times$, not $\forall$ and $\exists$ – which are more complex – and considers the parametricity property only for existential types $\exists X. T$ where T is a "first-order" signature built using $\times$ and $\to$. The small price we pay for this is that we can only consider ADT's with such signatures, but this covers most examples.)

Takeuti defines the logic for parametricity in two stages: first a base logic **L** which provides the standard logical connectives and their rules, and then a logic **Par** which extends **L** with axioms expressing parametricity.

3.1 The base logic L

L is a second-order predicate logic over the second-order lambda calculus, i.e. it provides predicates on the types of the second-order lambda calculus. **L** is a *typed* logic, with predicates – and also propositions – having types. The type of propositions is denoted by $*_p$. Predicates can be viewed as functions that return propositions, so $T \to *_p$ is the type of predicates over type T. Relations are binary predicates, so $T \to T \to *_p$ is the type of binary predicates – or relations – on T. So the types of propositions and predicates are given by

$$\mathbb{P} ::= *_p \mid T \to \mathbb{P}.$$

The propositions and predicates are given by

$$P ::= P \Rightarrow Q \mid \forall x{:}T.\, P \mid \forall X.\, P \mid \forall P{:}\mathbb{P}.\, Q \mid \lambda x{:}T.\, P \mid P\, t.$$

The first four constructions provide ways to built propositions: namely implication $P \Rightarrow Q$, and three kinds of universal quantification, universal quantification over all elements of a type $\forall x{:}T.\, P$, universal quantification over all types $\forall X.\, P$, and (second-order) universal quantification over propositions and predicates $\forall P{:}\mathbb{P}.\, Q$. The last two constructs allow the definition of predicates $\lambda x{:}T.\, P$ and the application of predicates to terms $P\, t$.

Judgements in the logic **L** are of the form $\Gamma, \Delta \vdash P$ where Γ is a sequence of declarations $x_1 : T_1, \ldots, x_n : T_n$ as before, Δ is a sequence of assumptions $P_1, \ldots, P_m$, and P is a proposition. We have the standard structural rules, and the standard elimination and introduction for the logical connective $\Rightarrow$ and the quantifiers $\forall$ (for details see [Tak97]).

The second-order universal quantification over propositions and predicates enables the definition of the logical connectives $\vee$, $\wedge$ and $\exists$ in the usual way. It also enables *Leibniz' equality* for datatypes T to be defined in the standard way:

Definition 1 (Leibniz' equality). *For any type T, Leibniz' equality of type T, $=_T\colon T \to T \to *_p$, is defined by*

$$=_T \;\hat{=}\; \lambda x, y{:}T.\;\; \forall P{:}(T \to *_p).\, (Px) \Rightarrow (Py).$$

The subscript of $=_T$ will sometimes be omitted when it is clear from the context. Leibniz' equality will be written infix. Other relations will sometimes also be written infix, and sometimes "postfix", i.e. $(t_1, t_2) \in P$ for Pt_1t_2. □

Remark 1. For readers familiar with *Pure Type Systems* (PTS's) [Bar92], we note that the logic **L** of Takeuti can be concisely described as a PTS, namely the PTS $(\mathsf{S}, \mathsf{A}, \mathsf{R})$ with $\mathsf{S} = \{*_s, \Box_s, *_p, \Box_p\}$, $\mathsf{A} = \{(*_s : \Box_s)\;,\; (*_p : \Box_p)\}$ and

$$\begin{aligned}\mathsf{R} = \{\, &(\Box_s, *_s), (*_s, *_s),\\ &\qquad\qquad (*_s, \Box_p),\\ &(\Box_s, *_p), (*_s, *_p),\; (\Box_p, *_p), (*_p, *_p)\}\end{aligned}$$

Here $*_s$ is the type of all datatypes, just like $*_p$ is the type of all propositions. The fact that **L** is a PTS is the main reason why we chose Takeuti's presentation of the logic rather than Plotkin & Abadi's; it enabled us to verify some examples using the theorem prover Yarrow [Zwa97] which implements arbitrary PTS's.

L is a subsystem of the logic $\lambda\omega_L$ introduced in [Pol94] as a logic for reasoning about the higher-order typed lambda calculus (system F^ω). $\lambda\omega_L$ includes a few more PTS rules, so that it includes the higher-order rather than the second order lambda calculus as "programming language" and allows more powerful abstractions in the logic (such as polymorphic predicates). □

3.2 The logic for parametricity

The logic **Par** extends **L** with an axiom for every type T which states that all elements of T satisfy a certain parametricity property. Since we are only interested in certain properties of existential types in **Par** – viz. the simulation principles - we simply introduce these properties as axioms here.

First, the constructions $\rightarrow$ and $\times$ for building types have to be "lifted" to constructions for building relations on types.

Definition 2. *Let R_1 and R_2 be relations (i.e. binary predicates), with $R_i : A_i \rightarrow A'_i \rightarrow *_p$. Then the relations $R_1 \rightarrow R_2 : (A_1 \rightarrow A_2) \rightarrow (A'_1 \rightarrow A'_2) \rightarrow *_p$ and $R_1 \times R_2 : (A_1 \times A_2) \rightarrow (A'_1 \times A'_2) \rightarrow *_p$ are defined as follows*

$$f(R_1 \rightarrow R_2)f' \mathrel{\hat{=}} \forall x : A_1, x' : A'_1 .\, xR_1x' \Rightarrow (fx)R_2(f'x')$$
$$f(R_1 \times R_2)f' \mathrel{\hat{=}} (f.1)R_1(f'.1) \wedge (f.2)R_2(f'.2)$$

□

Now we lift the type expressions $A(X)$ to relations:

Definition 3. *Let $A(X)$ be a type expression built using $\rightarrow$ and $\times$ from X and closed type expressions. We write $A(B)$ for $A[B/X]$.*

For any relation $\sim: B_1 \rightarrow B_2 \rightarrow *_p$ the relation $A(\sim) : A(B_1) \rightarrow A(B_2) \rightarrow *_p$ is defined by induction on the structure of A, as follows:

$$\begin{array}{lll} A(\sim) \mathrel{\hat{=}} A_1(\sim) \rightarrow A_2(\sim) & , \textit{if } A(X) \equiv A_1(X) \rightarrow A_2(X) \\ A(\sim) \mathrel{\hat{=}} A_1(\sim) \times A_2(\sim) & , \textit{if } A(X) \equiv A_1(X) \times A_2(X) \\ A(\sim) \mathrel{\hat{=}} \sim & , \textit{if } A(X) \equiv X \\ A(\sim) \mathrel{\hat{=}} =_C & , \textit{otherwise, i.e. } A(X) \equiv C \textit{ and } X \notin FV(C) \end{array}$$

In the right-hand sides $\rightarrow$ and $\times$ denote the construction on relations defined in Definition 2, and $=_C$ is Leibniz' equality as defined in Definition 1. □

As an example, consider the interface of the ADT for bags. Suppose $\sim : B_1 \rightarrow B_2 \rightarrow *_p$. Then $\mathit{BagSig}(\sim) : \mathit{BagSig}(B_1) \rightarrow \mathit{BagSig}(B_2) \rightarrow *_p$ is the following relation on 3-tuples:

$$((\mathit{empty}_1, \mathit{add}_1, \mathit{card}_1), (\mathit{empty}_2, \mathit{add}_2, \mathit{card}_2)) \in \mathit{BagSig}(\sim)$$

$$\iff$$

$$\begin{array}{l} \mathit{empty}_1 \sim \mathit{empty}_2 \wedge \\ \forall n : \mathit{Nat}, b_1 : B_1, b_2 : B_2 .\, b_1 \sim b_2 \Rightarrow \mathit{add}_1(n, b_1) \sim \mathit{add}_2(n, b_2) \wedge \\ \forall n : \mathit{Nat}, b_1 : B_1, b_2 : B_2 .\, b_1 \sim b_2 \Rightarrow \mathit{card}_1(n, b_1) =_{\mathit{Nat}} \mathit{card}_2(n, b_2) \end{array}$$

Definition 4 (Par). *The logic* **Par** *is the extension of* **L** *with the axioms*

$$\begin{array}{l} \forall u_1, u_2{:}\exists X.A(X). \\ \quad u_1 = u_2 \\ \iff (\exists X_1, X_2.\ \exists x_1{:}A(X_1), x_2{:}A(X_2).\ \exists \sim{:}X_1 \to X_2 \to *_p. \\ \qquad u_1 = \mathsf{pack}\ \langle X_1, x_1\rangle \wedge u_2 = \mathsf{pack}\ \langle X_2, x_2\rangle \wedge (x_1, x_2) \in A(\sim)) \end{array}$$

for all type expressions $A(X)$ built using $\to$ and $\times$ from X and closed type expressions. □

This axiom allows us to prove equivalence of different implementations of an ADT by showing there exists a simulation relation $\sim$ between them. We will refer to this proof principle as *simulation*.

Example: Equality of bag implementations.

We briefly illustrate how we can prove equivalence of different data representations in **Par**.

Recall the implementation *impl* : *BagImp*. Now consider another implementation of the ADT for bags, where we implement the *add*-operation not as the *cons*-operation on *List*'s, but as the *snoc*-operation on *List*'s, which adds a element to the end rather than the front of a list:

$$imp2 \mathrel{\widehat{=}} \mathsf{pack}\ \langle List, (nil, snoc, count)\rangle : BagImp.$$

Intuitively, this should not make any difference, because the order of the list representing a bag is irrelevant. In **Par** we can prove $impl =_{BagImp} imp2$, namely by proving $((nil, cons, count), (nil, snoc, count)) \in BagSig(\sim_{perm})$, where $\sim_{perm}$: $List \to List \to *_p$ relates all lists that are permutations.

Of course, *impl* and *imp2* use the same datatype to represent bags. But we can also prove equivalence of implementations that use different representation types. For example, consider the implementation *imp3* below, which represents bags as functions of type $Nat \to Nat$:

$$imp3 \mathrel{\widehat{=}} \mathsf{pack}\ \langle Nat \to Nat, (const_0, addimp, app)\rangle : BagImp$$

where

$$\begin{aligned} const_0 &= \lambda n{:}Nat.\ 0 \\ addimp &= \lambda(n,f){:}(Nat \times (Nat \to Nat)).\ \lambda m{:}Nat.\ \begin{cases} 1 + (f\,m) & \text{if } m = n \\ f\,m & \text{otherwise} \end{cases} \\ app &= \lambda(n,f){:}(Nat \times (Nat \to Nat)).\ fn \end{aligned}$$

The principle of simulation can be used to prove $impl =_{BagImp} imp3$, namely by showing that $((nil, cons, count), (const_0, addimp, app)) \in BagSig(\sim)$, where $\sim$: $List \to (Nat \to Nat) \to *_p$ relates $l : List$ and $f : Nat \to Nat$ iff $\forall n.\ fn = count(n, l)$.

4 Insufficiency of Par

We will show that the principle of simulation that **Par** provides is not sufficient for reasoning over ADT's. To illustrate this, we consider a specification for the ADT of bags.

Naive Specification

A possible specification for the operations *empty*, *add*, and *card* could be:

$$\begin{aligned}
\forall n : Nat.\ & card(n, empty) =_{Nat} 0 \ \wedge \\
\forall m : Nat, s : Bag.\ & card(m, add(m, s)) =_{Nat} 1 + card(m, s) \ \wedge \\
\forall m, n : Nat, s : Bag.\ & m \neq_{Nat} n \Rightarrow card(m, add(n, s)) =_{Nat} card(m, s) \ \wedge \\
\forall m, n : Nat, s : Bag.\ & add(m, add(n, s)) =_{Bag} add(n, add(m, s))
\end{aligned}$$

We will consider a simple specification *Spec* giving only the last conjunct. This is the most interesting part of the specification, as it uses equality of bags. For any type *Bag* and any triple $(empty, add, card) : BagSig(Bag)$ we define

$$\begin{aligned}
& Spec(Bag, (empty, add, card)) \\
& \widehat{=} \ \forall m, n : Nat, s : Bag.\ add(m, add(n, s)) =_{Bag} add(n, add(m, s)).
\end{aligned}$$

Spec can be turned into a predicate on *BagImp* as follows

$$\begin{aligned}
& Spec^{\exists} : BagImp \rightarrow *_p \\
& \widehat{=} \ \lambda imp{:}BagImp.\ \exists Rep, ops.\ imp =_{BagImp} \mathsf{pack}\ \langle Rep, ops\rangle \ \wedge \ Spec(Rep, ops)
\end{aligned}$$

Note that here $Spec(Rep, ops)$ uses Leibniz' equality on type *Rep*, i.e. $=_{Rep}$.

Clearly

$$Spec(Rep, ops) \ \Rightarrow \ Spec^{\exists}(\mathsf{pack}\ \langle Rep, ops\rangle).$$

(But beware that the reverse implication does not always hold. In fact, this would be inconsistent with parametricity, following the example given in Remark 3.)

Remark 2. It is tempting to extend the "open as $\langle\ \rangle$ in " construction that we have for programs to predicates, c.f. the inductive types proposed in [CP90]. This so-called "strong" elimination principle is included in Coq [PM93]. It would mean having the rule

$$\frac{\Gamma, x : A \vdash P : *_p \qquad \Gamma \vdash s : \exists X.\, A}{\Gamma \vdash (\mathsf{open}\ s\ \mathsf{as}\ \langle X, x\rangle\ \mathsf{in}\ P) : *_p} \ X \notin \mathsf{FV}(\Gamma)$$

With this rule the specification *Spec* could be turned into a predicate on *BagImp* in a much more direct way:

$$Spec^{\exists}(imp) \ \widehat{=} \ \mathsf{open}\ imp\ \mathsf{as}\ \langle Bag, ops\rangle\ \mathsf{in}\ Spec(Bag, ops)$$

and $Spec^{\exists}(\mathsf{pack}\ \langle List, (nil, cons, count)\rangle)$ would then simply β-reduce to $Spec(List, (nil, cons, count))$, so these two propositions would be equivalent. Unfortunately, this is inconsistent with parametricity, as shown in Remark 3. □

The problem with the naive specification

The specification $Spec^{\exists}$ might be what the user of the ADT wants, but it may be a problem for the implementor of the ADT to meet this specification. As an example we take the implementation $impl1$,

$$impl1 \mathrel{\widehat{=}} \mathsf{pack}\ \langle List, (nil, cons, count)\rangle : BagImp,$$

and consider the following question: *Can we prove* $Spec^{\exists}(impl1)$?

We could prove $Spec^{\exists}(impl1)$ by proving $Spec(List, (nil, cons, count))$, i.e. by proving

$$\forall m, n : Nat, s : List.\ cons(m, cons(n, s)) =_{List} cons(n, cons(m, s)).$$

But this is clearly not true! Note that the proposition above uses Leibniz' equality of lists, $=_{List}$, since *Spec* uses Leibniz' equality. The equality above makes sense for bags, but not for lists. We could only prove the proposition above for a weaker notion of equality for lists than $=_{List}$, e.g. $\sim_{perm}$.

We now discuss two ways to solve (or avoid) the problem above. Neither of these is really acceptable, which is why we then propose an extension of the logic **Par** to solve the problem in a more satisfactory way.

Solution 1: Finding another implementation

Recall that by the definition of $Spec^{\exists}$

$$Spec^{\exists}(impl1) \iff \exists Rep, ops.\ impl1 =_{BagImp} \mathsf{pack}\ \langle Rep, ops\rangle \ \wedge\ Spec(Rep, ops).$$

So we can prove $Spec^{\exists}(impl1)$ by finding another implementation $\mathsf{pack}\ \langle Rep, ops\rangle$ of the ADT such that $impl1 =_{BagImp} \mathsf{pack}\ \langle Rep, ops\rangle$ for which we *can* prove $Spec(Rep, ops)$.

It turns out that such an implementation exists, namely the implementation which represents bags as *sorted* lists. Let

$$imp_{sort} \mathrel{\widehat{=}} \mathsf{pack}\ \langle List, (nil, insert, count)\rangle,$$

where $insert : Nat \times List \rightarrow List$ inserts a natural number in a list and returns the list sorted. For this implementation we can prove it meets *Spec*, since

$$\forall m, n : Nat, s : List.\ insert(m, insert(n, s)) =_{List} insert(n, insert(m, s)). \qquad \text{(i)}$$

The reason we can prove *Spec* for this implementation is due to the fact that for this particular representation – bags are represented as sorted lists – equality of the concrete representation type, i.e. equality of lists, coincides with equality of the abstract type, i.e. equality of bags.

Using parametricity we can prove

$$impl1 =_{BagImp} imp_{sort}, \qquad \text{(ii)}$$

namely by showing that $\sim_{perm}$ is a simulation relation between the two implementations. Now $Spec^{\exists}(impl)$ follows from (i) – i.e. $Spec(List, (nil, insert, count))$ – and (ii).

There are obvious drawbacks to this way of proving $Spec^{\exists}(impl)$. Firstly, it is not acceptable that to prove correctness of our original implementation $impl$ we have to come up with a second implementation imp_{sort}. Moreover, it may not always be possible to find a second implementation that does meet the specification, i.e. for which concrete and abstract equality coincide! For example, for a generic datatype $Bag(X)$ of bags over an arbitrary type X we would have a problem; there is no way to extend the implementation using sorted lists of natural numbers to lists of an arbitrary type, since there is no generic sorting algorithm for arbitrary types.

Remark 3. We can use imp_{sort} to show the inconsistency of the elimination scheme discussed in Remark 2. If $Spec^{\exists}$ were defined with this scheme, then $Spec^{\exists}(\mathsf{pack}\ \langle Rep, ops\rangle)$ would be β-equivalent with $Spec(Rep, ops)$, so then

$$Spec^{\exists}(impl) \iff Spec(List, (nil, cons, count))$$
$$Spec^{\exists}(imp_{sort}) \iff Spec(List, (nil, insert, count))$$

But $Spec(List, (nil, cons, count))$ is false, (because $cons$ is not "commutative"), whereas $Spec(List, (nil, insert, count))$ is true, (because $insert$ is "commutative"). And by parametricity $impl = imp_{sort}$, so $Spec^{\exists}(impl) \iff Spec^{\exists}(imp_{sort})$, and we have a contradiction. □

Solution 2: Using a weaker specification

The best we could prove for $impl$ is that

$$\forall m, n : Nat, s : List.\ cons(m, cons(n, s)) \sim_{perm} cons(n, cons(m, s)).$$

Note that $\sim_{perm}$ is a bisimulation for the implementation, i.e.

$$((nil, cons, count), (nil, cons, count)) \in BagSig(\sim_{perm}), \qquad (*)$$

since $nil \sim_{perm} nil$, $\forall n : Nat, l, l' : List.\ l \sim_{perm} l' \Rightarrow cons(n, l) \sim_{perm} cons(n, l')$, and $\forall n : Nat, l, l' : List.\ l \sim_{perm} l' \Rightarrow count(n, l) =_{Nat} count(n, l')$. Intuitively, $(*)$ says that lists in the relation $\sim_{perm}$ cannot be distinguished using the bag-operations, so that lists in the relation $\sim_{perm}$ represent the same bag. With this in mind, one could propose a weaker specification for bags. First, we abstract the specification $Spec$ over a notion of equality for bags, to get the following "generic" specification $GenSpec$:

$$GenSpec(Bag, (empty, add, card), \sim)$$
$$\mathrel{\hat{=}} \forall m, n : Nat, s : Bag.\ add(m, add(n, s)) \sim add(n, add(m, s)).$$

(So $Spec(Bag, ops) = GenSpec(Bag, ops, =_{Bag})$.)

We can now consider the following weaker specification

$$\begin{array}{l} \mathit{WeakSpec}(\mathit{Bag}, \mathit{ops}) \\ \mathrel{\widehat{=}} \exists \sim : \mathit{Bag} \to \mathit{Bag} \to *_p. \\ \quad \mathit{GenSpec}(\mathit{Bag}, \mathit{ops}, \sim) \;\wedge\; (\mathit{ops}, \mathit{ops}) \in \mathit{BagSig}(\sim) \;\wedge\; \mathit{Equiv}(\sim), \end{array}$$

where $\mathit{Equiv}(\sim)$ says that $\sim$ is an equivalence relation.

Turning *WeakSpec* into a predicate $\mathit{WeakSpec}^{\exists}$ on *BagImp* we get

$$\begin{array}{l} \mathit{WeakSpec}^{\exists} : \mathit{BagImp} \to *_p \\ \mathrel{\widehat{=}} \lambda \mathit{imp}{:}\mathit{BagImp}. \\ \quad \exists \mathit{Rep}, \mathit{ops}.\ \mathit{imp} =_{\mathit{BagImp}} (\mathsf{pack}\ \langle \mathit{Rep}, \mathit{ops}\rangle) \;\wedge\; \mathit{WeakSpec}(\mathit{Rep}, \mathit{ops}). \end{array}$$

The implementor of the ADT will be happy with this weaker specification, as it is possible to prove $\mathit{WeakSpec}^{\exists}(\mathit{imp}1)$, simply by proving $\mathit{WeakSpec}(\mathit{List}, (\mathit{nil}, \mathit{cons}, \mathit{count}))$, taking $\sim_{perm}$ for $\sim$.

The user of the ADT on the other hand will be less happy with $\mathit{WeakSpec}^{\exists}$: rather than using the standard Leibniz' equality of bags, the user has to reason about bags using some bisimulation $\sim$ as notion of equality for bags. This seems an unnecessary complication: there is no reason why the user shouldn't use Leibniz' equality instead of $\sim$. Indeed, this is precisely the *abstraction* that the *abstract* data type is supposed to provide.

5 Our Solution: Extending the logic

Given that the two solutions discussed above are not really satisfactory, we now consider an extension of the logic **Par** that provides a satisfactory solution of the problem.

What we really want is a way to relate the two specifications, $\mathit{WeakSpec}^{\exists}$ and $\mathit{Spec}^{\exists}$, by proving

$$\forall \mathit{imp} : \mathit{BagImp}.\ \mathit{WeakSpec}^{\exists}(\mathit{imp}) \Rightarrow \mathit{Spec}^{\exists}(\mathit{imp}). \qquad (*)$$

Then the implementor of the ADT would only have to establish $\mathit{WeakSpec}^{\exists}$ – i.e. prove the specification up to some bisimulation $\sim$ – and the user of the ADT could assume the stronger specification $\mathit{Spec}^{\exists}$ – i.e. assume the specification with (Leibniz') equality –. Intuitively the property $(*)$ seems OK. (Indeed, it is true in the PER model.)

It turns out that if we have *quotient types* then $(*)$ could be proved. Quotient types are available in some type theories, e.g. Nuprl [Con86], and have been proposed as extensions of other type theories, see e.g. [Hof95] [BG96].

We will first give the general idea of how quotient types could be used to prove the property above. Suppose $\mathit{WeakSpec}^{\exists}(\mathit{imp})$, i.e.

$$\mathit{GenSpec}(\mathit{Rep}, \mathit{ops}, \sim) \wedge (\mathit{ops}, \mathit{ops}) \in \mathit{BagSig}(\sim) \wedge \mathit{Equiv}(\sim)$$

for some pack $\langle Rep, ops\rangle =_{BagImp}$ *imp* and some $\sim$. The trick to proving (∗) is to consider the quotient type $Rep/\!\sim$, i.e. the type with $\sim$-equivalence classes of *Rep* as elements.

$$(ops, ops) \in BagSig(\sim)$$

says that *ops* respects $\sim$-equivalence classes, so *ops* induces a related function $ops/\!\sim$ on $\sim$-equivalence classes, $ops/\!\sim\, : BagSig(Rep/\!\sim)$. And by the principle of simulation it follows that

$$\mathsf{pack}\ \langle Rep, ops\rangle = \mathsf{pack}\ \langle Rep/\!\sim, ops/\!\sim\rangle.$$

The interesting thing about $ops/\!\sim$ is that is satisfies the specification *up to Leibniz' equality*: it follows from $GenSpec(Rep, ops, \sim)$ that

$$GenSpec(Rep/\!\sim, ops/\!\sim, =_{Rep/\sim}),$$

i.e. $Spec(Rep/\!\sim, ops/\!\sim)$!

Note that the argument above goes along the lines as indicated in Solution 1. But the use of quotient types means that the additional work of finding another implementation of ADT is avoided, as this implementation is constructed as a quotient. (So we avoid the drawbacks mentioned on page 320.)

We could consider adding quotient types to the syntax of the second-order lambda calculus. But we do not actually have to do this: it suffices if we add axioms to the logic stating that quotients exist:

Definition 5 (ParQuot). *The logic* **ParQuot** *is the extension of* **Par** *with the axioms*

$$\begin{array}{l} \forall X.\, \forall \sim : X \to X \to *_p. \\ \quad Equiv(\sim) \Rightarrow \\ \quad \exists Q.\quad \forall opsX : A(X).\, (opsX, opsX) \in A(\sim) \Rightarrow \\ \qquad\qquad\quad \exists opsQ{:}A(Q).\, isQuot(X, opsX, \sim, Q, opsQ) \end{array}$$

where

$$\begin{array}{ll} & isQuot(X, opsX, \sim, Q, opsQ) \\ \hat{=} & \exists inj{:}X \to Q.\ \forall r, r'{:}X.\, r \sim r' \iff (inj\ r) =_Q (inj\ r') \wedge \\ & \qquad\qquad \forall q{:}Q.\, \exists r{:}X.\, q =_Q (inj\ r) \wedge \\ & \qquad\qquad (opsX, opsQ) \in A(\lambda r{:}X, q{:}Q.\, q =_Q (inj\ r)) \end{array}$$

for all type expressions $A(X)$ built using $\to$ and $\times$ from X and closed type expressions. □

The same PER model used in [PA93] as a semantics for their logic, viz. [BFSS90], quite trivially justifies these additional axioms. Indeed, in a PER model all types are "quotient types"!

Theorem 4. *In the logic* **ParQuot** *it can be proved that*

$$\forall imp : BagImp.\ WeakSpec^{\exists}(imp) \Rightarrow Spec^{\exists}(imp).$$

Proof. Assume that $WeakSpec^{\exists}(imp)$ holds. Then there exists a type *Rep* with $ops : BagSig(Rep)$ such that

$$imp =_{BagImp} \mathsf{pack}\ \langle Rep, ops\rangle$$

for which $GenSpec(Rep, ops, \sim) \wedge (ops, ops) \in BagSig(\sim) \wedge Equiv(\sim)$ for some $\sim$: $Rep \to Rep \to *_p$.

By $(ops, ops) \in BagSig(\sim)$ and $Equiv(\sim)$ there then exists a type Q with $opsQ : BagSig(Q)$ and $inj{:}Rep \to Q$ such that

$$\forall r, r'{:}Rep.\ r \sim r' \iff (inj\ r) =_Q (inj\ r') \quad \text{(i)}$$

$$\forall q{:}Q.\ \exists r{:}Rep.\ q =_Q (inj\ r) \quad \text{(ii)}$$

$$(ops, opsQ) \in A(\lambda r{:}Rep, q{:}Q.\ q =_Q (inj\ r)) \quad \text{(iii)}$$

It follows from (iii) that $\mathsf{pack}\ \langle Q, opsQ\rangle =_{BagImp} \mathsf{pack}\ \langle Rep, ops\rangle$. Using the definition of *GenSpec*, we can prove

$$GenSpec(Q, opsQ, =_Q) \quad \text{(iv)}$$

using $GenSpec(Rep, ops, \sim)$ and (i), (ii), and (iii).

And (iv) is equivalent with $Spec(Q, opsQ)$, and since $\mathsf{pack}\ \langle Q, opsQ\rangle =_{BagImp} \mathsf{pack}\ \langle Rep, ops\rangle =_{BagImp} imp$ it then follows that $Spec^{\exists}(imp)$. □

Similar theorems can be proved for other ADT's and other (equational) specifications: For any other ADT and specification for it, a weak version of the specification using some relation $\sim$ (similar to $WeakSpec^{\exists}$) and the strong version using Leibniz' equality (similar to $Spec^{\exists}$) can be related in exactly the same way as in the theorem above.

6 Conclusion

In this paper we have explored the gap between the formal notion of parametricity of [PA93] and the important "folk" reasoning principle about ADT's, which we have called *abstraction.*

Roughly, this principle of abstraction says that elements of the concrete representation type of an ADT can be considered equal if they are not distinguishable using the ADT-operations. For example, if we implement bags as lists, then lists that are permutations cannot be distinguished using the bag-operations – they represent the same bag – and can hence be considered equal. To prove that such an implementation of bags satisfies an equational specification we may therefore use permutation of lists as the notion of equality. This principle of abstraction is a well-known reasoning principle for ADT's.

Parametricity provides the proof principle of *simulation* for existential types [Mit91] [PA93]. This is a useful proof principle if existential types are used for abstract data types: it provides a method to prove that different implementations of an ADT are equivalent, namely by showing that there exists a simulation relation between them.

However, we have shown that this principle of simulation alone is not enough to reason about ADT's, since in general it does not provide the proof principle of abstraction that we want. This observation is new, as far as we know. However, extending the logic for parametricity of [PA93] with axioms stating the existence of quotients is enough to solve this problem. Like the original logic for parametricity of [PA93] these additional axioms can be justified by a PER model.

Proofs for the example of the specification for bags have all been verified using the interactive theorem prover Yarrow [Zwa97]. Indeed, it was only in the course of formalising specifications for ADT's in Yarrow that we noticed that more was needed than just the proof principle of simulation to reason about specifications of ADT's.

References

[Bar92] H.P. Barendregt. Lambda calculi with types. In D.M. Gabbai, S. Abramsky, and T.S.E. Maibaum, editors, *Handbook of Logic in Computer Science*, volume 1. Oxford University Press, 1992.

[BFSS90] E.S. Bainbridge, P.J. Freyd, A. Scedrov, and P.J. Scott. Functorial polymorphism. *Theoretical Computer Science*, 70(1):35–64, 1990.

[BG96] G. Barthe and J.H. Geuvers. Congruence types. In *CSL'95*, volume 1092 of *Lecture Notes in Computer Science*, pages 36–51. Springer, 1996.

[Con86] R.L. Constable et al. *Implementing Mathematics in the Nuprl proof development system.* Prentice-Hall, 1986.

[CP90] Thierry Coquand and Christine Paulin. Inductively Defined Types. In P. Martin-Löf and G. Mints, editors, *COLOG-88*, volume 417 of *Lecture Notes in Computer Science*, pages 50–66. Springer, 1990.

[GM93] M. J. Gordon and T. F. Melham. *Introduction to HOL.* Cambridge, 1993.

[Hof95] Martin Hofmann. A simple model for quotient types. In *TLCA'95*, volume 902 of *Lecture Notes in Computer Science*, pages 216–234, 1995.

[Mit91] John C. Mitchell. On the equivalence of data representations. In *Artificial Intelligence and Mathematical Theory of Computation*, pages 305–330. Academic Press, 1991.

[MP88] John C. Mitchell and Gordon D. Plotkin. Abstract types have existential type. *ACM Trans. on Prog. Lang. and Syst.*, 10(3):470–502, 1988.

[PA93] Gordon Plotkin and Martin Abadi. A logic for parametric polymorphism. In *TLCA'93*, volume 664 of *Lecture Notes in Computer Science*, pages 361–375, 1993.

[PM93] Christine Paulin-Mohring. Inductive definitions in the system Coq. In *TLCA'93*, volume 664 of *Lecture Notes in Computer Science*, pages 328–345. Springer, 1993.

[Pol94] Erik Poll. *A Programming Logic based on Type Theory.* PhD thesis, Technische Universiteit Eindhoven, 1994.

[Tak97] Izumi Takeuti. An axiomatic system of parametricity. In *TLCA'97*, volume 1130 of *Lecture Notes in Computer Science*, pages 354–372, 1997.

[Zwa97] Jan Zwanenburg. The proof assistant Yarrow. Submitted for publication. See also http://www.win.tue.nl/cs/pa/janz/yarrow/, 1997.

Characterising Explicit Substitutions which Preserve Termination
(Extended Abstract)

Eike Ritter *

School of Computer Science, University of Birmingham
http://www.cs.bham.ac.uk/~exr

Abstract. Contrary to all expectations, the $\lambda\sigma$-calculus, the canonical simply-typed lambda-calculus with explicit substitutions, is not strongly normalising. This result has led to a proliferation of calculi with explicit substitutions. This paper shows that the reducibility method provides a general criterion when a calculus of explicit substitution is strongly normalising for all untyped lambda-terms that are strongly normalising. This result is general enough to imply preservation of strong normalisation of the calculi considered in the literature. We also propose a version of the $\lambda\sigma$-calculus with explicit substitutions which is strongly normalising for strongly normalising λ-terms.

1 Introduction

The essence of the λ-calculus is the β-reduction rule $(\lambda x.M)N \rightsquigarrow M[N/x]$. It uses substitution, which is a meta-operation and *not* part of the calculus. This is unsatisfactory for implementations because the handling of substitutions is the difficult part and has to be done in several steps and not in one, as the β-rule might suggest. Take for example environment machines for functional languages like OCAML, ML or Haskell: an environment is just a list of outstanding substitutions, and replacing M for x is turned into accessing the component in the environment corresponding to x. As a consequence the correspondence between λ-calculus and the implementations becomes highly nontrivial. This complicates reasoning about implementations significantly.

As a way of making this reasoning easier, Abadi et al. [1] define the $\lambda\sigma$-calculus, which incorporates substitutions explicitly into the calculus. For this purpose they introduce an extra syntactic category of substitutions, which consists essentially of a list of pairs $\langle M_i/x_i\rangle$, where x_i is a variable and M_i a term. To distinguish the explicit substitution from the meta-operation, we use a different symbol to denote explicit substitution, e.g. the term $M[N/x]$ in the λ-calculus becomes $M\langle N/x\rangle$ in the $\lambda\sigma$-calculus. The reduction rules of the $\lambda\sigma$-calculus are the beta-rule, which creates an explicit substitution, and rules for carrying out substitutions, which formalise the standard inductive definition of substitution.

* Research supported by EPSRC-grant GR/L28296 under the title "The eXplicit Substitution Linear Abstract Machine".

Because this change has only turned substitution from an implicit operation into an explicit one it is natural to expect the meta-theory of the $\lambda\sigma$-calculus and the λ-calculus to coincide. More precisely, one expects the following properties of the $\lambda\sigma$-calculus:

1. it is confluent, possibly even confluent when meta-variables are added (the meta-variables are useful for applications of explicit substitutions in theorem proving);
2. the normal forms of terms are normal λ-terms;
3. each reduction step in the λ-calculus gives rise to possibly many reduction steps in the $\lambda\sigma$-calculus and each reduction step in the $\lambda\sigma$-calculus corresponds to some number of β-reductions in the λ-term which is obtained by eliminating explicit substitutions;
4. the $\lambda\sigma$-calculus preserves strong normalisation, *i.e.*, any β-strongly normalising λ-term is also a strongly normalising $\lambda\sigma$-term.

The last property will be abbreviated by PSN in the sequel, and we will write also SN for strongly normalising.

Abadi et al. [1] show the confluence without meta-variables and the second and third property for the $\lambda\sigma$-calculus. Curien et al. [5] show that the $\lambda\sigma$-calculus is not confluent if meta-variables are added. (Terms with meta-variables are often called open terms, and hence confluence a calculus with meta-variables is called confluence on open terms.) They also introduce additional syntax for special substitutions in the $\lambda\sigma$-calculus. This yields a calculus which is confluent on open terms, the so-called $\lambda\sigma_{\Uparrow}$-calculus.

To everyone's surprise the fourth property fails spectacularly. Mellies [12] gives a strongly normalising λ-term which reduces to the identity $\lambda x.x$ but nevertheless admits an infinite reduction sequence in the $\lambda\sigma$-calculus as well as in the $\lambda\sigma_{\Uparrow}$-calculus. Typing does not provide a solution: the counterexample is a well-formed term of the simply-typed λ-calculus.

Fixing this problem and finding a λ-calculus with explicit substitutions that has all desired meta-theoretic properties turned out to be rather difficult. Mellies' counterexample enforces severe restrictions on the possible reduction sequences. Since he presented this example various ways of capturing these restrictions syntactically have been designed. Firstly, the use of nested substitutions has been severely limited. This restriction is motivated by the fact that environment machines can be modelled without nested substitutions. In this way we obtain PSN because the nested substitutions are the main reason for the failure of PSN. Examples of this approach are the λx-calculus [4], the $\lambda\upsilon$-calculus [2] and the $\lambda\zeta$-calculus [13], the last being also confluent on open terms. In the second approach, composition has been retained but the use of environments (*i.e.*, substitutions which are lists of terms) has been severely curtailed. Examples of this approach are a $\lambda\sigma_{\Uparrow}$-calculus without environments [7], the λs_e-calculus [10] and the λxci-calculus [11]. The second and third calculus are confluent on open terms, and the first and third preserve strong normalisation. Except from one proof of strong normalisation for a typed λx-calculus using a mapping of the λx-calculus into proof nets [6] all other proofs use term rewriting techniques. Recursive path orderings provide a good way of showing preservation of strong normalisation for these calculi [3].

This paper pursues a different line of reasoning and uses the reducibility method to show preservation of strong normalisation. In [15] we used this method to show that all reduction strategies that first reduce an expression to weak head normal form and then to a normal form terminate for the typed $\lambda\sigma$-calculus. This paper generalises this argument to give a criterion when a λ-calculus with explicit substitution preserves strong normalisation. This criterion ensures that all possible contracta of a β-redex correspond to the stepwise execution of the corresponding implicit substitutions in the λ-calculus. The criterion is a generalisation of the restriction in [11] which ensures preservation of strong normalisation.

The reducibility method is powerful enough to show that preservation of strong normalisation follows from this condition. This method is sufficiently general to be applicable with minor modifications to all the calculi with PSN mentioned above. The modifications arise from the fact that not all calculi are subcalculi of the $\lambda\sigma$-calculus, and hence the proof has to be suitably adapted.

Because the underlying λ-calculus is untyped, we cannot use the standard reducibility method, which works only for typed calculi. We use here an adaptation of the reducibility method to untyped calculi which replaces induction over the structure of types by induction over the length of the longest reduction sequence of a strongly normalising term of the untyped λ-calculus. The standard structure of a reducibility proof is preserved by this change.

We do not consider meta-variables in this paper as the $\lambda\sigma$-calculus is not confluent on open terms. The criterion can also be stated for the $\lambda\sigma_{\Uparrow}$-calculus, and it should be possible to extend the results of this paper to this calculus as well.

The paper is structured as follows. In section 2 we review our version of the $\lambda\sigma$-calculus. The core of the paper is section 3, where we define the criterion for preservation of strong normalisation and use the reducibility method to show that the criterion is valid. In section 4 we present a restriction of the $\lambda\sigma$-calculus which preserves strong normalisation. We finish by showing preservation of strong normalisation by applying the criterion for all the calculi mentioned above.

2 A version $\lambda\sigma$-calculus with names

In this section we review our version of the $\lambda\sigma$-calculus. Abadi et al. [1] use mainly a version with de Bruijn-numbers but mention also briefly versions with names without proving all meta-theoretic properties. We use here a version of the $\lambda\sigma$-calculus with names and explicit weakening, which we call $\lambda\sigma_n$-calculus. The use of names improves the readability significantly, and the weakening is necessary to state the condition for preservation of strong normalisation. We present the version with names in this section and relate it in the appendix to the original presentation with de Bruijn-numbers. The results shown in this paper hold also for a calculus with de Bruijn-numbers.

The raw expressions of the (untyped) $\lambda\sigma_n$-calculus are given by the following grammar:

$$M ::= x \mid \lambda x.M \mid MM \mid M\langle f\rangle$$
$$f ::= \mathtt{weak}(X) \mid M/x \cdot f \mid f \circ f$$

where x is a variable and X is a set of variables. We call expressions of the first kind *terms* and expressions of the second kind *substitutions.* Moreover, we write $M_n/x_n \cdots M_1/x_1 \cdot f$ for $M_n/x_n \cdot (M_{n-1}/x_{n-1} \cdot \cdots \cdot (M_1/x_1 \cdot f) \cdots)$ and write Id for $\mathtt{weak}(\emptyset)$ whenever convenient. We also write $\mathbf{ext}_x(f)$ for $x/x \cdot (f \circ \mathtt{weak}(x))$ and $\mathtt{weak}(X, x)$ for $\mathtt{weak}(X \cup \{x\})$.

We identify terms which are identical up to change of bound variables. We have three binding operations: λ-abstraction, composition of substitution $\circ$ and application of a substitution $\_\langle\_\rangle$. In the sequel we will always use Barendregt's variable convention: the names of bound and free variables are different for any expression in any context. In particular, we identify the terms $x\langle N/x \cdot f\rangle$ and $y\langle N/y \cdot f\rangle$. For details see [14, 16].

To rule out ill-formed terms like $N/x \cdot (M/x \cdot f)$ we introduce typing judgements $\Gamma \vdash M{:}\,\Omega$ and $\Gamma \vdash f{:}\,\Delta$, where Γ and Δ are lists of variables with no variable occurring twice. The idea is that Γ contains the free variables of M and f, and Δ contains the variables for which the substitution f provides terms to be substituted for. Because we have an untyped calculus, we use Ω as the universal type. These judgements are as follows:

(i) On terms:

$$\frac{}{\Gamma, x, \Gamma' \vdash x{:}\,\Omega} \qquad \frac{\Gamma, x \vdash M{:}\,\Omega}{\Gamma \vdash \lambda x.M{:}\,\Omega}$$

$$\frac{\Gamma \vdash M{:}\,\Omega \qquad \Gamma \vdash N{:}\,\Omega}{\Gamma \vdash MN{:}\,\Omega} \qquad \frac{\Gamma \vdash f{:}\,\Delta \qquad \Delta \vdash M{:}\,\Omega}{\Gamma \vdash M\langle f\rangle{:}\,\Omega}$$

(ii) On substitutions (In the first rule, Γ' is the list Γ with all variables in X deleted, and all variables in X occur also in Γ):

$$\frac{}{\Gamma \vdash \mathtt{weak}(X){:}\,\Gamma'} \qquad \frac{\Gamma \vdash f{:}\,\Delta \qquad \Gamma \vdash M{:}\,\Omega}{\Gamma \vdash M/x \cdot f{:}\,\Delta, x}\ (\text{if } x \notin \Delta)$$

$$\frac{\Gamma \vdash f{:}\,\Gamma' \qquad \Gamma' \vdash g{:}\,\Gamma''}{\Gamma \vdash g \circ f{:}\,\Gamma''}$$

In the sequel we consider only well-formed terms and substitutions.

The syntax of the $\lambda\sigma_n$-calculus is best explained by relating the terms with explicit substitutions to terms with the usual implicit substitution of the simply-typed λ-calculus. The basic idea is that a substitution f in the $\lambda\sigma_n$-calculus corresponds to a list of terms $\boldsymbol{M} = (M_1, \ldots, M_n)$ in the λ-calculus. The operation $\_\langle\_\rangle$ in the $\lambda\sigma_n$-calculus models the explicit substitution: a term $M\langle f\rangle$ in the $\lambda\sigma_n$-calculus corresponds to a term $M[\boldsymbol{M}/\boldsymbol{x}]$ [1] in the λ-calculus.

[1] We write $M[\boldsymbol{M}/\boldsymbol{x}]$ for $M[M_1/M_1, \ldots, M_n/M_n]$, and will use the vector notation in a similar way in the future.

The substitution $\mathtt{weak}(X)$ models weakening: The term $M\langle \mathtt{weak}(x)\rangle$ is only well-formed if x is not free in M. The operation $\circ$ models nesting of substitutions: the substitution $(M/y \cdot \mathsf{Id}) \circ (N/x \cdot \mathsf{Id})$ corresponds to the substitution operator $[M/y][N/x]$ in the λ-calculus. We have two kinds of reduction rules: firstly, a β-reduction rule $(\lambda x.M)N \leadsto M\langle s/x \cdot \mathsf{Id}\rangle$, and secondly rules which formalise the inductive definition of implicit substitutions. We call the latter rules σ-rules. The reduction rules are given in Figure 1. We denote by $\leadsto^+$ the transitive closure of the relation $\leadsto$, and by $\leadsto^*$ the reflexive and transitive closure of $\leadsto$.

$$
\begin{array}{rlrl}
(\lambda x.M)N & \leadsto M\langle N/x \cdot \mathsf{Id}\rangle & x\langle M/x \cdot f\rangle & \leadsto M \\
y\langle M/x \cdot f\rangle & \leadsto y\langle f\rangle \text{ if } x \neq y & x\langle \mathtt{weak}(X)\rangle & \leadsto x \\
(\lambda x.M)\langle f\rangle & \leadsto \lambda x.M\langle \mathtt{ext}_x(f)\rangle & (MN)\langle f\rangle & \leadsto (M\langle f\rangle)(N\langle f\rangle) \\
\mathtt{weak}(X)\circ(M/x \cdot f) & \leadsto M/x \cdot (\mathtt{weak}(X)\circ f) \text{ if } x \notin X & f \circ \mathsf{Id} & \leadsto f \\
\mathtt{weak}(X,x)\circ(M/x \cdot f) & \leadsto \mathtt{weak}(X) \circ f & \mathsf{Id} \circ f & \leadsto f \\
\mathtt{weak}(X) \circ \mathtt{weak}(Y) & \leadsto \mathtt{weak}(X,Y) & M\langle \mathsf{Id}\rangle & \leadsto M \\
(M \cdot f) \circ g & \leadsto (M\langle g\rangle) \cdot (f \circ g) & (f \circ g) \circ h & \leadsto f \circ (g \circ h) \\
M\langle f\rangle\langle g\rangle & \leadsto M\langle f \circ g\rangle & &
\end{array}
$$

Fig. 1. Reduction rules for the $\lambda\sigma_n$-calculus

This notion of reduction satisfies all desired properties except preservation of strong normalisation. The proof of this proposition uses the interpretation method [9].

Proposition 1. *This notion of reduction is ground confluent (i.e., confluent on expressions without meta-variables), and the normal form of terms are normal λ-terms. If $M \leadsto M'$, then also $M^e \leadsto^* M'^e$, where M^e is the λ-term obtained by applying all σ-rules and hence executing all explicit substitutions. Moreover, if the term M reduces to M' in the λ-calculus, then $M \leadsto^+ M'$ in the $\lambda\sigma_n$-calculus.*

A substitution in normal form is an environment $\boldsymbol{M}/\boldsymbol{x} \cdot \mathtt{weak}(X)$, where the terms in $\boldsymbol{M}$ are normal λ-terms.

3 Preservation of Strong Normalisation

Mellies [12] gives a counterexample to preservation of strong normalisation for the $\lambda\sigma$-calculus. This counterexample applies to many λ-calculi with explicit substitution, in particular to the $\lambda\sigma_n$-calculus used in this paper. It is based on a nasty interaction between explicit substitution and β-reduction. Mellies gives a term of the simply-typed λ-calculus which reduces to the identity but which admits an infinite reduction sequence. He exploits that in the term $((\lambda x.M)\langle f\rangle)N$ the term N can interact with the substitution f by the reduction sequence

$$
\begin{aligned}
((\lambda x.M)\langle f\rangle)N &\leadsto (\lambda x.M\langle \mathtt{ext}_x(f)\rangle)N \leadsto^+ M\langle \mathtt{ext}_x(f) \circ (N/x \cdot \mathsf{Id})\rangle \\
&\leadsto^+ M\langle N/x \cdot (f \circ \mathtt{weak}(x) \circ (N/x \cdot \mathsf{Id}))\rangle \qquad (1)
\end{aligned}
$$

Although x does not occur in f, Mellies now pushes the substitution $\mathtt{weak}(x) \circ (N/x \cdot \mathsf{Id})$ inside f and manages thereby to create a reduction sequence $M \rightsquigarrow^+ M'$, where M is a proper subterm of M'. This continuation is counterintuitive: the variable x does not occur in f, so the only meaningful reduction sequence is $f \circ \mathtt{weak}(x) \circ (N/x \cdot \mathsf{Id}) \rightsquigarrow^+ f$.

In this section we show that any restriction of the reduction relation for which the substitution $f \circ \mathtt{weak}(x) \circ (N/x \cdot \mathsf{Id})$ is SN if f and N are SN preserves strong normalisation. We adapt the reducibility method to show this claim, and give a reduction relation for the $\lambda\sigma_n$-calculus with this property in the next section. The idea of the reducibility method to show SN for the simply-typed λ-calculus is to define a subset of λ-terms, the so-called *reducible* terms, which are SN and in addition satisfy strong closure properties. Now proving strong normalisation amounts to showing that every term is reducible. The key condition is the definition of reducibility for terms of type $A \to B$: A term M of type $A \to B$ is reducible if for all reducible terms N of type A, the term MN is reducible of type B. Then one shows by an induction over the structure of the term M that any term $M[\boldsymbol{N}/\boldsymbol{x}]$ is reducible if all terms in $\boldsymbol{N}$ are reducible and $\boldsymbol{x}$ is the set of free variables of M. The critical case in this proof is the case of a λ-abstraction $\lambda x.M$. One has to show that any term $((\lambda x.M)[\boldsymbol{N}/\boldsymbol{x}])N$ is reducible for any reducible term N. It turns out that it is enough to show that this term reduces to reducible terms only. This is easy to see, as by induction hypothesis the term $M[\boldsymbol{N}/\boldsymbol{x}, N/x]$ is reducible.

If we transfer this approach to a calculus with explicit substitutions, the definition of reducibility stays unchanged, and one shows by induction over the structure of M that for any reducible substitution f the term $M\langle f\rangle$ is reducible. Again, the interesting case is the case of a λ-abstraction $(\lambda x.M)\langle f\rangle$. The reduction sequence in (1) is exactly the one which causes this proof to fail: there is no way to derive reducibility of $f \circ \mathtt{weak}(x) \circ (M/x \cdot \mathsf{Id})$ from the reducibility of f and M. This is the only place where the reducibility proof fails. We show in this paper that the reducibility proof goes through if we require that $f \circ \mathtt{weak}(x) \circ (M/x \cdot \mathsf{Id})$ is SN if f and M are.

Because we consider an untyped λ-calculus in this paper, we cannot use an induction over types as in the standard reducibility proof. The induction over the type structure is replaced by an induction over $\nu^\beta(M)$, where $\nu^\beta(M)$ is the length of the longest β-reduction sequence of the λ-term which is obtained from M by executing all explicit substitutions. If this λ-term is not strongly normalising, then set $\nu^\beta(M) = \infty$. In the typed case, the reducibility of a term of function type is tested by applying it to reducible arguments; in the untyped setting the reducibility of a term is tested by applying it to reducible terms M_i such that $\nu^\beta(M_i)$ is smaller than $\nu^\beta(M)$.

The definition of reducibility has to consider also finite expansions of a term M and a substitution f by the associativity rules $(f \circ g) \circ h \rightsquigarrow f \circ (g \circ h)$, $M\langle g\rangle\langle f\rangle \rightsquigarrow M\langle g \circ f\rangle$. We call the congruence relation generated by these reduction rules R. From now on until the end of this section we will consider only the equivalence classes of terms modulo R. It is easy to see that if this equivalence class is SN without these two rules, any term of this equivalence class is SN even with these rules.

Now we turn to the precise statement and proof of the criterion. To state the criterion we need to identify all substitutions which could arise as a result of the reduction sequence (1).

Definition 2. *(i) Let $f = f_n \circ f_{n-1} \circ \cdots \circ f_m \circ f_{m-1} \circ \cdots \circ f_1$ be any substitution with $m \leq n$. We call the substitution $f_n \circ \cdots \circ f_{m+1} \circ \mathtt{weak}(x) \circ \mathtt{ext}_x(f_m) \circ \cdots \circ \mathtt{ext}_x(f_1)$ a λ-extension of f with extension variable x. The special case $m = n$ denotes the substitution $\mathtt{ext}_x(f_m) \circ \cdots \circ \mathtt{ext}_x(f_1)$.*

(ii) We call g an i-fold *λ-extension of f with extension variables X_i if there are substitutions $f = f_0, f_1, \ldots, f_i = g$ such that f_j is an extension of f_{j-1} with extension variable x_j and $X_j = X_{j-1} \cup \{x_j\}$ for all $1 \leq j \leq i$.*

(iii) Let $f_n \circ f_{n-1} \circ \cdots \circ f_m \circ \cdots \circ f_1$ be any substitution. For any $k \leq m \leq n$ and any term M we call the substitution $f_n \circ \cdots \circ f_{m+1} \circ \mathtt{weak}(x) \circ \mathtt{ext}_x(f_m) \circ \cdots \circ \mathtt{ext}_x(f_{k+1}) \circ (M/x \cdot \mathsf{Id}) \circ f_k \circ \cdots \circ f_1$ a β-extension of f with extension variable x. We call the term $M\langle f_k \circ \cdots \circ f_1 \rangle$ the extending term. *The special case $m = n$ denotes the substitution $\mathtt{ext}_x(f_m) \circ \cdots \circ \mathtt{ext}_x(f_{k+1}) \circ (M/x \cdot \mathsf{Id}) \circ f_k \circ \cdots \circ f_1$.*

(iv) We call g a $\lambda\beta$-extension with extension variables $X \cup \{x\}$ and extending term M of f if there exists a β-extension f' of f with extension variable x and extending term M such that g is a k-fold λ-extension of f' with extension variables X.

(v) The term $M\langle g \rangle$ is called a k-fold λ-($\lambda\beta$-)extension of $M\langle f \rangle$ if g is a λ-($\lambda\beta$)-extension of f such that the extension variables of g are not free in M. A 0-fold $\lambda\beta$-extension of $M\langle f \rangle$ is an i-fold λ-extension of $M\langle f \rangle$ for some i.

Now we can state the criterion.

Definition 3. *A reduction relation $\leadsto$ on expressions of the $\lambda\sigma_n$-calculus is called* strong normalisation-preserving *if the following two conditions are satisfied:*

(i) Let $\leadsto_o$ be the reduction relation of Figure 1. If $M \leadsto M'$, then also $M \leadsto_o^+ M'$ and if $f \leadsto f'$, then $f \leadsto_o^+ f'$;

(ii) For any substitution f and any λ-extension h of f and any β-extension g of f with extending term M, g and h are SN if both f and M are.

Note that for any $\lambda\beta$-extension f' of f and any variable $y \neq x$, where x is the extension variable of the β-extension of f, we have that $\nu^\beta(y\langle f' \rangle) = \nu^\beta(y\langle f \rangle)$. Similarly, any $\lambda\beta$-extension M' of M satisfies $\nu^\beta(M') = \nu^\beta(M)$.

Now we can define reducible expressions. The standard definition defines reducible terms of ground types to be exactly the strongly normalisable terms and reducible terms of function type to be those terms which when applied to a reducible term yield a reducible result. As already mentioned, we have to replace the induction over the structure of types by an induction over $\nu^\beta(M)$. This exploits the fact that $\nu^\beta(M\langle N/x \cdot \mathsf{Id} \rangle) < \nu^\beta((\lambda x.M)N)$. Hence we define not reducible terms, but reducible terms of grade n with $n \geq 0$. We also add closure properties for preservation of strong normalisation by extending substitutions.

Definition 4. *i Call a sequence of terms $N_1, \ldots, N_m$ with $m \geq 1$ a test-sequence for M of degree n if $N_i \in Red(n_i)$ with $\nu^\beta(MN_1 \cdots N_m) - 1 \leq n_i < n$ for all $1 \leq i \leq m$.*

ii Define the set of reducible terms of degree n for any $n \geq 0$, written $Red(n)$, inductively as follows:

- *A term M is an element of $Red(0)$ if $\nu^\beta(M) = 0$ implies that M and any λ-extension of M are SN.*
- *M is reducible of degree $n > 0$ if $\nu^\beta(M) > n$ or $\nu^\beta(M) \leq n$ and for every j-fold extension $M'\langle h'\rangle$ with $0 \leq j < n$ such that all extending terms are SN and reducible of degrees $n > k_1 > \cdots > k_j \geq \nu^\beta(M'\langle h'\rangle)$, the following conditions are satisfied:*
  - *$M'\langle h'\rangle$ is SN;*
  - *For all terms M'' and substitutions h such that that $M'\langle h'\rangle \rightsquigarrow^* (\lambda x.M'')\langle h\rangle$ and for all test-sequences $N_1, \ldots, N_m$ for $M'\langle h'\rangle$ of degree n such that*

$$\nu^\beta(M'\langle h'\rangle N_1 \cdots N_m) \leq k_j \qquad (\text{if } j = 0, \text{then} \nu^\beta(M'\langle h'\rangle N_1 \cdots N_m) \leq n),$$

there exists a term $(\lambda x.P)\langle f\rangle$ such that firstly $(\lambda x.P)\langle f\rangle \rightsquigarrow^* (\lambda x.M'')\langle h\rangle$ and secondly, for any $\lambda\beta$-extension

$$f' = \mathbf{ext}_x(f_m) \circ \cdots \circ \mathbf{ext}_x(f_{k+1}) \circ (N'/x \cdot \mathsf{Id}) \circ f_k \circ \cdots \circ f_1$$

of $f = f_m \circ \cdots \circ f_{k+1} \circ f_k \circ \cdots \circ f_1$ with extending term N_1 we have $x_i\langle f'\rangle \in Red(d_i)$ and

$$\nu^\beta((P\langle f'\rangle)N_2 \cdots N_m) \leq d_i < n$$

for all free variables x_i of P, and also $P\langle f'\rangle \in Red(d)$ with

$$\nu^\beta((P\langle f'\rangle)N_2 \cdots N_m) \leq d < n\,.$$

The last clause is the appropriate generalisation of the standard reducibility condition: a term of function type is reducible if it is SN and whenever it reduces to a λ-abstraction, the result of applying this function to a reducible term is reducible again.

Note that reducibility of degree n does not imply strong normalisation. This implication is only guaranteed for terms M with $\nu^\beta(M) \leq n$. This is the major difference to the standard reducibility proof. The next lemma states two basic properties of reducible expressions.

Lemma 5. *(i) If M is reducible of degree n and $\nu^\beta(M) \leq n$, then M is SN. (ii) $Red(n) \subseteq Red(n-1)$ for all $n > 0$.*

The next lemma states that reducibility is preserved by extensions of substitutions.

Lemma 6. *Let f be a substitution such that f is SN and $y\langle f\rangle \in Red(n)$ for some variable y.*

(i) *If $Q \in Red(k)$ with $k < n$ and Q is SN, then for any $\lambda\beta$-extension g of f with extending term Q and extension variable x both terms $x\langle g\rangle$ and $y\langle g\rangle$ are reducible of degree k.*

(ii) *For any λ-extension g of f, if $x\langle f\rangle$ is reducible of degree n, so is $x\langle g\rangle$ for any variable x.*

Proof. (i) Definition 3 implies that g is SN. By definition of reducibility, $y\langle g\rangle$ is reducible of degree k for any variable $y \neq x$. For the variable x, consider any j-fold $\lambda\beta$-extension M of $x\langle g\rangle$. Now one shows that M is SN if all j-fold $\lambda\beta$-extensions of Q are SN and that $M \leadsto^* (\lambda z.N)\langle h\rangle$ implies that for some j-fold $\lambda\beta$-extension Q' of Q, $Q' \leadsto^* (\lambda z.N)\langle h\rangle$. It is then easy to see that $x\langle g\rangle$ is reducible.

(ii) Similar argument.

Now we can show that every expression is reducible.

Theorem 7. *Consider any λ-term M. Let f be a substitution such that f is SN and $x_i\langle f\rangle$ is reducible of degree n for all free variables x_i of M. Then the expression $M\langle f\rangle$ is reducible of degree n as well.*

Proof. We use induction over n, and for each n an induction over the structure of M.

x : Assumption.

$\lambda x.M$: Let g be a j-fold $\lambda\beta$-extension of f. By Lemma 6, for each free variable y of $\lambda x.M$ the substitution $y\langle g'\rangle$ is reducible of degree k_j with $n > k_j \geq \nu^\beta((\lambda x.M)\langle g\rangle)$, where g' is the λ-extension obtained by pushing g under the λ-abstraction. Hence by induction hypothesis, $M\langle g'\rangle$ is SN, and hence also $(\lambda x.M)\langle g\rangle$ is SN.

For the second condition, let g be any j-fold $\lambda\beta$-extension of f. Again by Lemma 6, for the $\lambda\beta$-extension g' of g with extending term N_1, the term $x\langle g'\rangle$ is reducible of degree k with $k \geq \nu^\beta((M\langle g'\rangle)N_2\cdots N_m)$ for all free variables x of M. Hence by induction hypothesis, $M\langle g'\rangle \in Red(k)$.

MN : If there are no M'' and f' such that $M\langle f\rangle \leadsto^* (\lambda x.M'')\langle f'\rangle$, the second condition is vacuously true, and the first condition holds by induction hypothesis.

If there are such M'' and f', then consider any j-fold $\lambda\beta$-extension g of f. Now consider terms $N_1 \ldots, N_m$ with $m \geq 0$ such that $N_i \in Red(n_i)$ and $n_i \geq \nu^\beta((MN)\langle g\rangle N_1\cdots N_m)$ for all $1 \leq i \leq m$. By induction hypothesis, whenever $M\langle g\rangle \leadsto^* (\lambda x.P)\langle h\rangle$, then there exists a term $(\lambda y.Q)\langle h'\rangle$ such that $M\langle f\rangle \leadsto^* (\lambda y.Q)\langle h'\rangle$ and $(\lambda y.Q)\langle h'\rangle \leadsto^* (\lambda x.P)\langle h\rangle$ and for the $\lambda\beta$-extension h'' of h' with extending term $N\langle g\rangle$, we have $\nu^\beta((Q\langle h''\rangle)N_1\cdots N_m) \leq d_i < n$, where d_i is the degree of $x_i\langle h''\rangle$ for the free variables x_i of Q. Hence by induction hypothesis $Q\langle h''\rangle \in Red(d)$, where $\nu^\beta((Q\langle h''\rangle)N_1\cdots N_m) \leq d < n$. Now it is easy to see that $MN\langle g\rangle$ is reducible of degree n.

Preservation of strong normalisation follows now as an easy corollary by applying the previous theorem to the empty substitution.

Corolloray 8. *Assume $\rightsquigarrow$ satisfies the condition of Definition 3. Let M be any strongly normalising λ-term. Then M is a strongly normalising term of the $\lambda\sigma_n$-calculus.*

Proof. Using Lemma 6, one shows that $x\langle \mathsf{Id}\rangle \in Red(n)$ for all n. By the previous theorem, $M\langle \mathsf{Id}\rangle \in Red(\nu^\beta(M))$. Hence by Lemma 5 M is SN.

Note that both the theorem and the corollary do not say anything about substitutions. The reason is that it is difficult to state the preservation of strong normalisation for substitutions. For a typed calculus, this proof can be extended to show that a typed $\lambda\sigma_n$-calculus with this restriction is strongly normalising. This includes the substitutions. Using Ghani's techniques [8] we can extend this proof also to the η-rules.

4 A strong normalisation-preserving restriction of the $\lambda\sigma_n$-calculus

The previous section gave a condition when a reduction relation on the $\lambda\sigma_n$-calculus preserves strong normalisation. In this section we give a concrete example of such a relation for the $\lambda\sigma_n$-calculus. It suffices to restrict the reduction rule which carries out nested substitutions, namely

$$(\langle\rangle - nat)\ (M/x \cdot g) \circ f \rightsquigarrow M\langle f\rangle/x \cdot (g \circ f)$$

in such a way that weakening operations are carried out as early as possible. More precisely, the reduction rules are the rules for the $\lambda\sigma_n$-calculus with the exception of this rule, which is replaced by the two rules

$$\begin{array}{c}(M/x \cdot g) \circ ((N/y \cdot f) \circ h) \rightsquigarrow (M\langle N/y \cdot f\rangle/x \cdot (g \circ (N/y \cdot f))) \circ h \\ (M/x \cdot g) \circ (N/y \cdot f) \rightsquigarrow M\langle N/y \cdot f\rangle/x \cdot (g \circ (N/y \cdot f))\end{array}$$

Note that the explicit weakening makes it possible to formulate these two rules as unconditional rewrite rules, *i.e.*, without a side condition on free variables as in [11]. All reduction rules are local in the sense that their applicability can be decided by inspecting the top of the syntax tree only. This means these rules are directly suitable for an abstract machine.

To achieve confluence we add the rules

$$\begin{array}{r}x\langle (M/x \cdot g) \circ f\rangle \rightsquigarrow M\langle f\rangle \\ y\langle (M/x \cdot g) \circ f\rangle \rightsquigarrow y\langle g \circ f\rangle \text{ if } x \neq y \\ \mathtt{weak}(X, x) \circ ((M/x \cdot f) \circ g) \rightsquigarrow \mathtt{weak}(X) \circ (f \circ g) \\ \mathtt{weak}(X) \circ ((M/x \cdot f) \circ g) \rightsquigarrow M/x \cdot (\mathtt{weak}(X) \circ (f \circ g)) \text{ if } x \notin X\end{array}$$

These rules are derivable in the original calculus but no longer in the restricted one. We denote this restricted reduction relation by $\rightsquigarrow_p$.

If we apply this restriction to the reduction sequence (1) at the beginning of section 3, we see that with this restriction there is no way that the substitution

$N/x \cdot \mathsf{Id}$ can interact with the substitution f: if f is a substitution $M/y \cdot g$, then the reduction sequence

$$(M/y \cdot g) \circ \mathtt{weak}(x) \circ (N/x \cdot \mathsf{Id}) \rightsquigarrow M\langle \mathtt{weak}(x) \circ (N/x \cdot \mathsf{Id})\rangle \cdot (g \circ \mathtt{weak}(x) \circ (N/x \cdot \mathsf{Id}))$$

is not allowed. The only possible reduction sequence (apart from reductions inside M, N and g) is $(M/y \cdot g) \circ \mathtt{weak}(x) \circ (N/x \cdot \mathsf{Id}) \rightsquigarrow_p (M/y \cdot g) \circ \mathsf{Id}$ which eliminates any possible interaction.

This restriction is still confluent and refines the β-reduction of the λ-calculus.

Proposition 9. *The notion of reduction $\rightsquigarrow_p$ is ground confluent, and the normal form of terms are normal λ-terms. If $M \rightsquigarrow_p M'$, then also $M^e \rightsquigarrow^* M'^e$, where M^e is the λ-term obtained by applying all σ-rules and hence executing all explicit substitutions. Moreover, if the term M reduces to M' in the λ-calculus, then $M \rightsquigarrow_p^+ M'$ in the $\lambda\sigma_n$-calculus.*

Next we show that this notion of reduction is SN. We use the criterion of Definition 3. A key step in the proof is the following Lemma, which uses the restriction of the rule ($\langle\rangle$-nat) in an essential way.

Lemma 10. *If f is SN, so is $f \circ \mathtt{weak}(X)$ for any set X.*

Now we can prove the main theorem.

Theorem 11. *Let M be any strongly normalising λ-term. Then M is a strongly normalising term of the $\lambda\sigma_n$-calculus with respect to the reduction relation $\rightsquigarrow_p$.*

Proof. By Corollary 8 it suffices to show that any λ-and β-extension of any substitution f is SN provided f and the extending term are SN. We show here only the case of the β-extension; the case of the λ-extension is similar. So we have to show that $g \circ \mathtt{weak}(x) \circ \mathtt{ext}_x(g_n) \circ \cdots \circ \mathtt{ext}_x(g_1) \circ (N/x \cdot \mathsf{Id}) \circ f$ is SN provided $g \circ g_n \circ \cdots \circ g_1 \circ f$ and $N\langle f\rangle$ are SN. Now consider the inductively defined substitutions $h_0 = (N/x \cdot \mathsf{Id}) \circ f$ and $h_{k+1} = x\langle h_k\rangle/x \cdot (g_{k+1} \circ \mathtt{weak}(x) \circ h_k)$ which arise by applying the rules replacing $\langle\rangle$-nat to the substitution $g \circ \mathtt{weak}(x) \circ \mathtt{ext}_x(g_n) \circ \cdots \circ \mathtt{ext}_x(g_1) \circ (N/x \cdot \mathsf{Id}) \circ f$. Lemma 10 shows that it suffices to prove that $x\langle h_k\rangle/x \cdot (g \circ g_n \circ \cdots \circ g_{k+1} \circ \mathtt{weak}(x) \circ h_k)$ is SN. For this, we show by induction over the lexicographically ordered pair $(m+k, \nu(f_1)+\nu(f_2)+\nu(f_3))$ that $x\langle f_1\rangle/x \cdot (f_2 \circ \mathtt{weak}(x) \circ f_3)$ is SN whenever $h_m \rightsquigarrow_p^* f_1$, $g \circ g_n \circ \cdots \circ g_{k+1} \rightsquigarrow_p^* f_2$ and $h_k \rightsquigarrow_p^* f_3$. The restriction of the rule ($\langle\rangle$-nat) implies that this substitution can be reduced in one step only to

- $x\langle f_1\rangle/x \cdot (f_2 \circ g'_k \circ \mathtt{weak}(x) \circ f')$ with $g_k \rightsquigarrow_p^* g'_k$ and and also $h_{k-1} \rightsquigarrow_p^* f'$;
- $x\langle f_1\rangle/x \cdot (f_2 \circ \mathtt{weak}(x) \circ f')$ with $f_3 \rightsquigarrow_p f'$;
- $x\langle f_1\rangle/x \cdot (g' \circ \mathtt{weak}(x) \circ f_3)$, where $f_2 \rightsquigarrow_p g'$;
- $x\langle f'\rangle/x \cdot (f_2 \circ \mathtt{weak}(x) \circ f_3)$ with $h_{m-1} \rightsquigarrow_p^* f'$;
- $x\langle f'\rangle/x \cdot (f_2 \circ \mathtt{weak}(x) \circ f_3)$ with $f_1 \rightsquigarrow_p f'$.

All these substitutions are SN by induction hypothesis, hence $x\langle f_1\rangle/x \cdot (f_2 \circ \mathtt{weak}(x) \circ f_3)$ is SN.

5 Preservation of strong normalisation for other calculi

The criterion in definition 3 also yields the preservation of strong normalisation for other calculi in the literature, *e.g.*, the $\lambda\upsilon$-calculus [2], the $\lambda\zeta$-calculus [13], a $\lambda\sigma_{\Uparrow}$-calculus without environments [7] and the λxci-calculus [11]. As an example for the argument, we consider the λx-calculus [4]. This calculus has the raw expressions $M ::= x \mid \lambda x.M \mid MM \mid M\langle x := M\rangle$. with no separate syntactic category of substitutions. The reduction rules are

$$\begin{aligned}(\lambda x.M)N &\rightsquigarrow M\langle x := N\rangle\\ (\lambda x.M)\langle y := N\rangle &\rightsquigarrow \lambda x.M\langle y := N\rangle\\ (MN)\langle x := P\rangle &\rightsquigarrow (M\langle x := P\rangle)(N\langle x := P\rangle)\\ x\langle x := M\rangle &\rightsquigarrow M\\ M\langle x := N\rangle &\rightsquigarrow M \text{ if } x \notin FV(M)\end{aligned}$$

For the proof it is convenient to re-introduce this distinction and present the syntax with two categories, namely terms and substitutions:

$$\begin{aligned}M &::= x \mid \lambda x.M \mid MM \mid Mf\\ f &::= \langle x := M\rangle \mid f\langle x := M\rangle\end{aligned}$$

We identify the terms $(Mf)g$ and Mh where $f = \langle x_1 := M_1\rangle \cdots \langle x_n := M_n\rangle$, $g = \langle y_1 := N_1\rangle \cdots \langle y_m := N_m\rangle$ and $h = \langle x_1 := M_1\rangle \cdots \langle x_n := M_n\rangle\langle y_1 := N_1\rangle \cdots \langle y_m := N_m\rangle$. To state the criterion for preservation of strong normalisation, we have to analyse the reduction sequence (1) at the beginning of section 3. This sequence is now $((\lambda x.M)f)N \rightsquigarrow^+ (\lambda x.Mf)N \rightsquigarrow Mf\langle x := N\rangle$ Because there is no rule for composition of substitution it is trivial that $f\langle x := N\rangle$ is SN if f and N are. Hence the criterion for PSN is satisfied. An extension of a substitution f is now a substitution $f\langle x := M\rangle$, where x does not occur freely in f, and an extension of Mf is a term $Mf\langle x := N\rangle$ where x does not occur freely in M nor f. Now the proof of section 3 goes through. Because a reduction in the λx-calculus cannot be mapped directly to a sequence of reductions in the $\lambda\sigma_n$-calculus it does not suffice to check the criterion and then to appeal to PSN for the $\lambda\sigma_n$-calculus.

6 Conclusions

This paper presents a criterion for the preservation of strong normalisation in calculi with explicit substitutions. The proof uses a novel adaptation of the reducibility method to the untyped λ-calculus. The criterion can be easily checked and yields the preservation of strong normalisation of various common calculi with explicit substitution. This method applies to typed λ-calculi as well, where the standard reducibility method can be used.

The transfer of the reducibility method to the untyped λ-calculus could show the way for the transfer of reasoning principles of the typed λ-calculus to the untyped λ-calculus. In particular it might be possible to transfer logical relations to the untyped λ-calculus and replace the induction over the type structure by a computation induction, *i.e.*, an induction over the number of β-reductions in a computation.

Acknowledgements

I would like to thank Paul-Andre Mellies for helpful comments on an earlier draft of this paper.

References

1. Martin Abadi, Luca Cardelli, Pierre-Louis Curien, and Jean-Jaques Lévy. Explicit substitutions. *Journal of Functional Programming*, 1(4):375–416, 1991.
2. Z.-E.A. Benaissa, D. Briaud, P. Lescanne, and J. Rouyer-Degli. $\lambda\upsilon$, a calculus of explicit substitutions which preserves strong normalisation. *Journal of Functional Programming*, 6(5), 1996.
3. R. Bloo and H. Geuvers. Explicit substitutions: on the edge of strong normalization. *To appear in Theoretical Computer Science*, 1998.
4. R. Bloo and K.H. Rose. Preservation of strong normalisation in named lambda calculi with explicit substitution and garbage collection. In *Proc. CSN'95—Computer Science in the Netherlands*, pages 62–72, 1995.
5. P.-L. Curien, Th. Hardin, and J.-J.Lévy. Confluence properties of weak and strong calculi of explicit substitutions. *Journal of the ACM*, 43:362–397, March 1996.
6. R. Di Cosmo and D. Kesner. Strong normalization of explicit substitutions via cut-elimination in proof nets. In *Proc. of 12th IEEE Symp. on Logic in Computer Science*, pages 35–47, 1997.
7. M.C. Ferreira, D. Kesner, and L. Puel. λ-calculi with explicit substitutions and compositions which preserve β-strong normalization (extended abstract). In *Proc. of 5th Int. Conf. on Algebraic and Logic Programming (ALP)*, volume 1139 of *Lecture Notes in Computer Science*, pages 248–298, 1996.
8. N. Ghani. *Adjoint Rewriting*. PhD thesis, University of Edinburgh, 1995.
9. Thérèse Hardin. Confluence results for the pure strong categorical logic CCL. λ-calculi as subsystems of CCL. *Theoretical Computer Science*, 65:291–342, 1989.
10. F. Kamareddine and A. Rios. Extending a lambda-calculus with explicit substitutions which preserves strong normalisation into a confluent calculus on open terms. *Journal of Functional Programming*, 7(4):395–420, 1997.
11. F. Lang and K.H. Rose. Two equivalent calculi of explicit substitution with confluence on meta-terms and preservation of strong normalisation (one with names and one first order). In *Proc. of Westapp'98*, 1998.
12. P.-A. Mellies. Typed λ-calculi with explicit substitution may not terminate. In *Proc. of TLCA'95*, pages 328–334. LNCS 902, 1995.
13. C. Munoz. Confluence and preservation of strong normalisation in an explicit substitution calculus. In *Proc. of 11th IEEE Symp. on Logic in Computer Science*, pages 440–447, 1996.
14. E. Ritter and V. de Paiva. On explicit substitution and names (extended abstract). In *Proc. of ICALP'97*, LNCS 1256, pages 248–258, 1997.
15. Eike Ritter. Normalization for typed lambda calculi with explicit substitution. In *Proc. of CSL'93*, pages 295–304. LNCS 832, 1994.
16. K.H. Rose. Explicit substitution—tutorial & survey. Lecture series LS-96-3, BRICS, Department of Computer Science, University of Aarhus, 1996.

Appendix

The relation between the $\lambda\sigma$-calculus and the $\lambda\sigma_n$-calculus

This appendix sketches the relation between the $\lambda\sigma$-calculus and the $\lambda\sigma_n$-calculus. names and with de Bruijn-numbers. The raw expressions of the $\lambda\sigma$-calculus are

$$\begin{aligned} M &::= n \mid \lambda M \mid MM \mid M\langle f\rangle \\ f &::= \mathsf{Id} \mid \mathsf{Fst} \mid M\cdot f \mid f\circ f \end{aligned}$$

where n is an integer such that $n \geq 1$, and the reduction rules are

$$\begin{array}{rlrl} (\lambda M)N &\leadsto M\langle N\cdot \mathsf{Id}\rangle & 1\langle M\cdot f\rangle &\leadsto M \\ n+1\langle M\cdot f\rangle &\leadsto n\langle f\rangle & n\langle \mathsf{Fst}\rangle &\leadsto n+1 \\ (\lambda M)\langle f\rangle &\leadsto \lambda M\langle 1\cdot (f\circ \mathsf{Fst})\rangle & (MN)\langle f\rangle &\leadsto (M\langle f\rangle)(N\langle f\rangle) \\ f\circ \mathsf{Id} &\leadsto f & \mathsf{Id}\circ f &\leadsto f \\ \mathsf{Fst}\circ (M\cdot f) &\leadsto f & M\langle \mathsf{Id}\rangle &\leadsto M \\ (M\cdot f)\circ g &\leadsto (M\langle g\rangle)\cdot (f\circ g) & (f\circ g)\circ h &\leadsto f\circ (g\circ h) \\ M\langle f\rangle\langle g\rangle &\leadsto M\langle f\circ g\rangle & & \end{array}$$

The relation between the version with names and the version with de Bruijn-numbers is investigated in [14]. Here we only give translations between the two versions and state their properties. The translation from the calculus with names to the calculus with de-Bruijn numbers depends on a context which lists all free variables. If Γ is a list $x_n, \ldots x_1$ which includes all free variables of the expressions with names M and f, we define expressions with de Bruijn-numbers $(\Gamma, M)^{dB}$ and $(\Gamma, f)^{dB}$ by induction over the structure of M and f. For the definition of the translation of f we need to determine the list of variables for which f provides a term to be substituted for.

$$\begin{aligned} ((x_n,\ldots,x_1),x_k)^{dB} &= k \\ ((\Gamma,\lambda x.M))^{dB} &= \lambda(((\Gamma,x),M)^{dB}) \\ ((\Gamma,MN))^{dB} &= (\Gamma,M)^{dB}(\Gamma,N)^{dB} \\ (\Gamma,M\langle f\rangle)^{dB} &= (\Delta,M)^{dB} \text{ if } (\Gamma,f)^{dB} = (g,\Delta) \\ (\Gamma,\mathsf{Id})^{dB} &= (\mathsf{Id},\Gamma) \\ ((x_n,\ldots,x_1),\mathtt{weak}(X))^{dB} &= \begin{cases} (1\cdot(g\circ\mathsf{Fst}),(\Delta,x_1) & if x_1 \notin X and \\ & ((x_n,\ldots,x_2), \\ & \mathtt{weak}(X))^{dB} = (g,\Delta) \\ (g\circ\mathsf{Fst},\Delta) & if x_1 \in X and \\ & ((x_n,\ldots,x_2), \\ & \mathtt{weak}(X\setminus\{x_1\}))^{dB} = (g,\Delta) \end{cases} \\ (\Gamma,g\circ f)^{dB} &= (h_2\circ h_1,\Gamma_2) \text{ if } (\Gamma,f)^{dB} \\ &= (h_1,\Gamma_1) \text{ and } (\Gamma_1,g)^{dB} = (h_2,\Gamma_2) \end{aligned}$$

The translation in the other direction again depends on an association to names to the free variables of M and f. Given a context $x_n, \ldots, x_1$ and an expression with de Bruijn-numbers, we define an expression with names by

$$((x_n,\ldots,x_1),k)^N = \begin{cases} x_k & \text{if } 1 \le k \le n \\ \text{undefined otherwise} \end{cases}$$

$$\begin{aligned}
(\Gamma,\lambda M)^N &= \lambda x.((\Gamma,x),M)^N \text{ (if } x \text{ not contained in } \Gamma)\\
(\Gamma,MN)^N &= (\Gamma,M)^N(\Gamma,N)^N\\
(\Gamma,M\langle f\rangle)^N &= (\Delta,M)^N\langle g\rangle \text{ if } (\Gamma,f)^N = (g,\Delta)\\
(\Gamma,\mathsf{Id})^N &= (\Gamma,\mathsf{Id})\\
((\Gamma,x),\mathsf{Fst})^N &= (\mathtt{weak}(x),\Gamma)\\
(\Gamma,M\cdot f)^N &= ((\Gamma,M)^N\cdot g,(\Delta,x)) \text{ if } x\notin\Delta \text{ and } (\Gamma,f)^N=(g,\Delta)\\
(g\circ f)^N &= (h_2\circ h_1,\Gamma_2) \text{ if } (\Gamma,f)^N=(h_1,\Gamma_1) \text{ and } (\Gamma_1,g)^N=(h_2,\Gamma_2)
\end{aligned}$$

If one translates a de-Bruijn-expression into an expression with names and back, one obtains the same expression. This is not true if one translates an expression with names into a de Bruijn-expression and back: the reason is that the substitution $\mathtt{weak}(x)$ is more general than the weakening operator Fst in the de Bruijn-calculus. Both translations preserve reduction. The details are as follows.

Proposition 12. *(i) Let M and f be expressions with names and let Γ be a context which contains all free variables of M and f. If $M \rightsquigarrow M'$ and $f \rightsquigarrow f'$, then also $(\Gamma,M)^{dB} \rightsquigarrow^+ (\Gamma,M')^{dB}$ and $g \rightsquigarrow^+ g'$ where $(\Gamma,f)^{dB}=(g,\Delta)$ and $(\Gamma,f')^{dB}=(g',\Delta)$.*

(ii) Let M and f be expressions with de Bruijn-numbers and Γ a context such that $(\Gamma,M)^N$ and $(\Gamma,f)^N$ are defined. Then $(\Gamma,(\Gamma,M)^N)^{dB}=M$ and if $(\Gamma,f)^N=(g,\Delta)$, then $(\Gamma,g)^{dB}=(f,\Delta)$.

(iii) Let M and f be expressions with de Bruijn-numbers and let Γ be a context such that $(\Gamma,M)^N$ and $(\Gamma,f)^N$ are defined. If $M \rightsquigarrow M'$ and $f \rightsquigarrow f'$, then also $(\Gamma,M)^N \rightsquigarrow^+ (\Gamma,M')^N$ and $g \rightsquigarrow^+ g'$ where $(\Gamma,f)^N=(g,\Delta)$ and $(\Gamma,f')^N=(g',\Delta)$.

This proposition shows the close connection between the version with names and the version with Bruijn-numbers: there is a one-to-one correspondence between the two calculi except for the generalised weakening operation in the calculus with names. The proposition also shows that if preservation of strong normalisation holds in one calculus it holds also in the other calculus.

Explicit Environments
(Extended Abstract)

Masahiko Sato[1], Takafumi Sakurai[2], and Rod Burstall[3]

[1] Graduate School of Informatics, Kyoto University
masahiko@kuis.kyoto-u.ac.jp
[2] Department of Mathematics and Informatics, Chiba University
sakurai@math.s.chiba-u.ac.jp
[3] Department of Computer Science, University of Edinburgh
rb@dcs.ed.ac.uk

Abstract. We introduce $\lambda\varepsilon$, a simply typed calculus with environments as first class values. As well as the usual constructs of λ and application, we have $e[\![a]\!]$ which evaluates term a in an environment e. Our environments are a set of variable-value pairs, but environments can also be computed by function application and evaluation in some other environments. The notion of environments here is a generalization of explicit substitutions and records. We show that the calculus has desirable properties such as subject reduction, confluence, conservativity over the simply typed $\lambda\beta$-calculus and strong normalizability.

1 Introduction

In this paper, we solve the problem of designing a *pure* functional language that has *explicit environments*. We understand that a functional language is *pure* if (i) it is a conservative extension of the untyped or simply typed $\lambda\beta$-calculus, (ii) confluent and (iii) strongly normalizing (SN) if the language is typed and has preservation of strong normalization (PSN) property if the language is untyped. The conservative extension property guarantees that the language is logically well-behaved and the confluence property and SN or PSN would guarantee that the language is computationally well-behaved.

What do we mean by explicit environments? An *explicit environment* is a set of variable-value pairs representing a finite function from variables to values

(i) which is equipped with an operation $\cdot[\![\cdot]\!]$ such that, $e[\![a]\!]$ is the evaluation of a term a in the environment e, and
(ii) which can be the argument or result of a function. (It is a "first class value".)

Explicit environments can be regarded as a generalization of explicit substitutions and records:

- An explicit substitution has property (i) above but not (ii).
- A record has property (ii) above but not (i).

Without the purity conditions, there are already several languages with explicit environments. For example, several versions of the programming language Scheme have explicit environments. The Pebble language of Lampson and

Burstall [8] also treated explicit environments (bindings). It used dependent types and no confluence result was obtained. Nishizaki [9, 10] also attempted to treat explicit environments. But, his system does not satisfy the conservative extension condition because he avoided the problem of defining the set of free variables for a term of his language. As we will soon see, giving a correct definition of free variables in a term becomes a difficult problem if the term in question contains variables whose values are environments. These languages are, however, exceptional, and in most programming languages, environments are *implicit* in the sense that they are used at meta-level as a device for giving formal semantics of these languages or they are used when implementing these languages, but they do not appear as syntactic entities of these languages.

On the other hand, there are quite a few typed or untyped calculi of explicit substitutions including [1–3, 7], and some of these are pure in our sense. However, to the best of our knowledge, there are no calculi of explicit substitutions in which substitutions are first class objects. So, we believe that our language is the first pure language that has substitutions as its first class objects, since as we explain below we may regard environments as substitutions.

Our use of 'environment' derives from LISP which has explicit environments in our sense. In what follows we will generally just say 'environment' for 'explicit environment' where the context makes this clear. A concept closely related to explicit environment is **let** declaration (SML) or local definition. In SML, declarations are not first class values, but they do permit some combination operations, such as ;.

Let us compare these language features, showing notations and their expressivity. When we give the formation rules for our calculus with environments we can be more precise.

- *Explicit environment* $\{a/x, b/y\}[\![c]\!]$: Evaluate term c in an environment which binds x to a and y to b.
- *Explicit substitution* $[x := a, y := b]c$: Substitute a for x and b for y in term c.
- *Record* $[x = a, y = b].x$: Extract the x field of the record $[x = a, y = b]$; x is a field name.
- *Let* **let** $x = a$ **and** $y = b$ **in** c: Declare local variables x with value a, y with value b. Use these to evaluate c.

The syntax for these features show that our environments are the most general allowing terms for both e and a in $e[\![a]\!]$.

Explicit environment	term[[term]]
Explicit substitution	[var := term, ..., var := term]term
Record	term.fieldName
Let	**let** declaration **in** term

In the above table, declaration (d) is defined by the grammar:

$$d ::= \mathsf{var} = \mathsf{term} \mid d; d \mid d \ \textbf{and} \ d \mid \textbf{local} \ d \ \textbf{in} \ d.$$

In section 2, we introduce a typed language $\lambda\varepsilon$ by introducing typing rules for terms that will determine the (typable) terms of the language. In section 3, we give reduction rules of the $\lambda\varepsilon$-calculus and prove the subject reduction theorem for the calculus. In section 4, we prove the confluence of $\lambda\varepsilon$ and prove that the $\lambda\varepsilon$-calculus is a conservative extension of the simply typed $\lambda\beta$-calculus. In section 5, we prove the strong normalizability of $\lambda\varepsilon$. By the results obtained in sections 3-6, we can see that $\lambda\varepsilon$ provides a language we wanted to design. In section 6, we give concluding remarks. Due to lack of space, we have omitted some lemmas and almost all proofs. A full version of this paper with proofs is accessible at `http://www.sato.kuis.kyoto-u.ac.jp/~masahiko/index-e.html`.
Acknowledgements We thank Takayasu Ito and Yukiyoshi Kameyama for helpful comments on earlier versions of the paper.

2 The Type System

In this section we introduce the $\lambda\varepsilon$-calculus[1] as an extension of the simply typed $\lambda\beta$-calculus. We assume that we have an infinite set of variables which is a disjoint union of an infinite set of *bindable variables* and an infinite set of *unbindable variables*. We will design our syntax in such a way that an unbindable variable never gets bound. The distinction of these two kinds of variables will become important only in the proof of strong normalizability of our calculus. So, until then, the reader may read the paper assuming that all the variables are bindable.

The untyped $\lambda\varepsilon$-*terms* are defined by the following grammar, where z ranges over variables (bindable or unbindable), x over bindable variables and $x_1, \ldots, x_n$ $(n \geq 0)$ are distinct bindable variables.

$$a, b, e ::= z \mid \lambda x.b \mid ba \mid \{a_1/x_1, \ldots, a_n/x_n\} \mid e[\![a]\!].$$

For each term a we associate a finite set $\Pi(a)$ of strings over 1, 2, ..., which we call *positions*, as follows. We call this set the *position set* of a. We use Λ to denote the empty string and π, σ, τ etc. to denote positions.

1. $\Pi(x) := \{\Lambda\}$.
2. $\Pi(\lambda x.a) := \{\Lambda\} \cup 1\Pi(a)$.
3. $\Pi(ba) := \{\Lambda\} \cup 1\Pi(b) \cup 2\Pi(a)$.
4. $\Pi(\{a_1/x_1, \ldots, a_n/x_n\}) := \{\Lambda\} \cup 1\Pi(a_1) \cup \cdots \cup n\Pi(a_n)$.
5. $\Pi(e[\![a]\!]) := \{\Lambda\} \cup 1\Pi(e) \cup 2\Pi(a)$.

For each $\pi \in \Pi(a)$, we associate a term a/π as follows.

1. $a/\Lambda := a$.
2. $\lambda x.a/1\pi := a/\pi$.
3. $ba/1\pi := b/\pi$, $ba/2\pi := a/\pi$.
4. $\{a_1/x_1, \ldots, a_n/x_n\}/i\pi := a_i/\pi$ $(1 \leq i \leq n)$.
5. $e[\![a]\!]/1\pi := e/\pi$, $e[\![a]\!]/2\pi := a/\pi$.

---
[1] ε is for environment.

If $b \equiv a/\pi$, then we say that b *occurs in* a *at position* π. Let π and σ be positions. We write $\pi \leq \sigma$ if π is an initial substring of σ, that is, $\sigma \equiv \pi\pi'$ for some π'. We write $\pi < \sigma$ if $\pi' \not\equiv \Lambda$.

We define *types* (A, B) and *environment types* (E) simultaneously as follows, where K ranges over atomic types and in the definition of E, $n \geq 0$, x_i must be bindable variables and $x_i^{A_i}$ must be distinct.

$$A, B ::= K \mid E \mid A \Rightarrow B$$
$$E ::= \{x_1^{A_1}, \ldots, x_n^{A_n}\}$$

If $E \equiv \{x_1^{A_1}, \ldots, x_n^{A_n}\}$ we say that $x_i^{A_i}$ $(1 \leq i \leq n)$ are *in* the environment type E. If a term e has type E by the typing rules we introduce below, then e is an environment and it is a first-class value of the calculus.

A *declaration* is an expression of the form x^A where x is a variable and A is a type. A *context* is a sequence $x_1^{A_1}, \ldots, x_n^{A_n}$ of declarations. We use Γ, Δ etc. as meta variables for contexts. As notational conventions, we will write Γ, Δ for the concatenation of the contexts Γ and Δ. Also, $\Gamma - E$ will denote the context obtained from Γ by removing all the elements in E from Γ.

A *typing judgment* is an expression of the form $\Gamma \vdash a : A$ where Γ is a context, a is a (typed) $\lambda\varepsilon$-term and A is a type. We have the following typing rules that are used to derive typing judgments, where rules whose names end with 'I' ('E') introduce (eliminate, respectively) the types mentioned in the rule names.

$$\frac{}{x^A \vdash x^A : A} \ (\mathsf{assume})$$

$$\frac{\Gamma \vdash b : B}{\Gamma - \{x^A\} \vdash \lambda x^A.b : A \Rightarrow B} \ (\Rightarrow I)$$

$$\frac{\Gamma \vdash b : A \Rightarrow B \quad \Delta \vdash a : A}{\Gamma, \Delta \vdash ba : B} \ (\Rightarrow E)$$

$$\frac{\Gamma_1 \vdash a_1 : A_1 \quad \cdots \quad \Gamma_n \vdash a_n : A_n}{\Gamma_1, \ldots, \Gamma_n \vdash \{a_1/x_1^{A_1}, \ldots, a_n/x_n^{A_n}\} : \{x_1^{A_1}, \ldots, x_n^{A_n}\}} \ (\mathsf{env}I)$$

$$\frac{\Gamma \vdash e : E \quad \Delta \vdash a : A}{\Gamma, (\Delta - E) \vdash e[\![a]\!] : A} \ (\mathsf{env}E).$$

In $(\Rightarrow I)$, x must be a bindable variable. We see by the (env$I$) rule that the environment type $\{x_1^{A_1}, \ldots, x_n^{A_n}\}$ for $n = 0$ becomes the *unit type* $\{\}$ whose unique element is the *empty environment* $\{\}$.

A $\lambda\varepsilon$-term is *canonical* if it is of the form $\lambda x^A.b$ or $\{a_1/x_1^{A_1}, \ldots, a_n/x_n^{A_n}\}$, that is, if it is obtained by one of the introduction rules. A $\lambda\varepsilon$-term is *neutral* if it is not canonical. If $\Gamma \vdash a : A$ is derivable and Γ does not contain bindable variables (that is, all variables in Γ are unbindable variables), then a is said to be *bindable variable free*. A $\lambda\varepsilon$-term is said to be an *environment term* if its type is an environment type.

An untyped $\lambda\varepsilon$-term a' is said to be *typable* if we can derive $\Gamma \vdash a : A$ for some Γ, a and A by using the above typing rules and a' is obtained from a by erasing all the types in a (that is, by replacing each declaration x^A in a by x). In this case a' is called the *type-free form* of a. Henceforth, we will often use the type-free form of $\lambda\varepsilon$-terms for the sake of notational simplicity.

It is easy to see that if $\Gamma \vdash a : A$ is derivable, then we can completely recover the entire derivation tree uniquely by inspecting the typed term a[2]. In this case, we write $\mathrm{TY}(a)$ (type of a) for A. Note that if $e \equiv \{a_1/x_1, \ldots, a_n/x_n\}$, then $\mathrm{TY}(e)$ is $\{x_1, \ldots, x_n\}$. We will say that these variables are *bound by* e.

We can now use $\mathrm{TY}(e)$ to tell if a given occurrence of a variable in a term is free. Consider the expression $(\lambda x.fxy)y$. There are two free occurrences of y. We will describe an occurrence by a term with a hole ($\Box$) in it, and we say y is free at $(\lambda x.fx\Box)y$ and y is free at $(\lambda x.fxy)\Box$. The syntax of *occurrences* is:

$$\begin{aligned}\alpha, \beta, \varepsilon ::= {} & \Box \mid \lambda x.\beta \mid \beta a \mid b\alpha \\ & \mid \{\ldots, a_{i-1}/x_{i-1}, \alpha/x_i, a_{i+1}/x_{i+1}, \ldots\} \\ & \mid \varepsilon[\![a]\!] \mid e[\![\alpha]\!].\end{aligned}$$

The inductive definition of '$\mathsf{variable\ is\ free\ at\ occurrence}$' follows the syntax of terms.

$$\frac{}{x \text{ is free at } \Box} \qquad \frac{x \text{ is free at } \beta}{x \text{ is free at } \lambda y.\beta}\ (\text{if } x \not\equiv y)$$

$$\frac{x \text{ is free at } \beta}{x \text{ is free at } \beta a} \qquad \frac{x \text{ is free at } \alpha}{x \text{ is free at } b\alpha}$$

$$\frac{x \text{ is free at } \alpha}{x \text{ is free at } \{\ldots, a_{i-1}/x_{i-1}, \alpha/x_i, a_{i+1}/x_{i+1}, \ldots\}}$$

$$\frac{x \text{ is free at } \varepsilon}{x \text{ is free at } \varepsilon[\![a]\!]} \qquad \frac{x \text{ is free at } \alpha}{x \text{ is free at } e[\![\alpha]\!]}\ (\text{if } x \notin \mathrm{TY}(e)) .$$

We define $\mathrm{FV}(a)$ as the set of variables occurring free in a. It can be easily verified that if $\Gamma \vdash a : A$ is derivable, then Γ, considered as a set, is equal to $\mathrm{FV}(a)$. In particular, a is *closed*, i.e., $\mathrm{FV}(a) = \emptyset$, iff $\vdash a : A$ is derivable.

We can easily define replacement of a variable in an occurrence to get a term, just textual replacement of the $\Box$.

α-congruence. Let a and b be terms and x, y be bindable variables of the same type such that y is fresh for a (that is, y is neither free nor bound in a). Then we will identify $\lambda x.a$ with $\lambda y.c$ where c is obtained from a by replacing each free occurrence of x by y. Here, the α-congruence means changing the names of λ bound variables but not those appearing in environment expressions as x in $\{a/x\}$.

[2] It is, in general, not possible to recover the typed form when its type-free form is given. Therefore, even if we write terms in type-free forms, it is only for the sake of simplicity, and we are in fact dealing with fully decorated typed terms.

We will write $a : A$ if $A \equiv \mathrm{TY}(a)$. An example gives a typing derivation for the term $\{y/x\}[\![x]\!] \equiv \{y^A/x^A\}[\![x^A]\!]$:

$$\frac{\dfrac{\overline{y^A \vdash y : A}}{y^A \vdash \{y/x\} : \{x^A\}} \quad \overline{x^A \vdash x : A}}{y^A \vdash \{y/x\}[\![x]\!] : A} \quad .$$

By the above typing, we see that the two occurrences of x in $\{y/x\}[\![x]\!]$ are both bound while the only one occurrence of y in this term is free.

The following example reveals the subtle point which is related to the fact that in the $\lambda\varepsilon$-calculus we can pass an environment as an argument to a function.

$$\frac{\dfrac{\overline{z^{\{x^{A\Rightarrow B},y^A\}} \vdash z : \{x^{A\Rightarrow B},y^A\}} \quad \dfrac{\overline{x^{A\Rightarrow B} \vdash x : A \Rightarrow B} \quad \overline{y^A \vdash y : A}}{x^{A\Rightarrow B},y^A \vdash xy : B}}{z^{\{x^{A\Rightarrow B},y^A\}} \vdash z[\![xy]\!] : B}}{\vdash \lambda z.(z[\![xy]\!]) : \{x^{A\Rightarrow B},y^A\} \Rightarrow B}$$

3 Reduction Rules

In this section we define a reduction relation $\to_{\lambda\varepsilon}$ on $\lambda\varepsilon$-terms. Then, in this and the following sections, we show that $\to_{\lambda\varepsilon}$ enjoys the subject reduction property, $\to_{\lambda\varepsilon}$ is confluent and strongly normalizing, and the $\lambda\varepsilon$-calculus is a conservative extension of the simply typed $\lambda\beta$-calculus. In this way, we can show that the $\lambda\varepsilon$-calculus solves the problem we posed in section 1.

We first define $\mapsto_{\lambda\varepsilon}$ as the union of the two relations $\mapsto_\lambda$ and $\mapsto_\varepsilon$, where the relation $\mapsto_\lambda$ is defined by the following single rule:

(λ) $(\lambda x.b)a \mapsto_\lambda \{a/x\}[\![b]\!]$,

and the relation $\mapsto_\varepsilon$ is defined by the following 6 conversion rules. These 6 rules will be called ε-rules. They evaluate expressions of the form $e[\![a]\!]$.

(gc) $e[\![a]\!] \mapsto_\varepsilon a$, if $\mathrm{TY}(e) \cap \mathrm{FV}(a) = \emptyset$.
(var) $\{a_1/x_1, \ldots, a_n/x_n\}[\![x_i]\!] \mapsto_\varepsilon a_i$ $(1 \le i \le n)$.
(abs) $e[\![\lambda x.b]\!] \mapsto_\varepsilon \lambda x.e[\![b]\!]$, if $x \notin \mathrm{TY}(e) \cup \mathrm{FV}(e)$.
(app) $e[\![ba]\!] \mapsto_\varepsilon e[\![b]\!]e[\![a]\!]$.
(env) $e[\![\{a_1/x_1, \ldots, a_n/x_n\}]\!] \mapsto_\varepsilon \{e[\![a_1]\!]/x_1, \ldots, e[\![a_n]\!]/x_n\}$.
(eval) $e[\![f[\![x]\!]]\!] \mapsto_\varepsilon e[\![f]\!][\![x]\!]$, if $x \in \mathrm{TY}(f)$.

The *garbage collection* rule (gc) collects e as a garbage. We can see the correctness of this rule intuitively, because $\mathrm{TY}(e) \cap \mathrm{FV}(a) = \emptyset$ means that variables in $\mathrm{FV}(a)$ are not bound by e. The (abs) rule pushes the environment e through the λx binder. Thanks to the typing information about e, we could precisely state the condition under which we may push the environment e through the λx binder. We note that the condition $x \notin \mathrm{TY}(e) \cup \mathrm{FV}(e)$ can always be met by taking a suitable α-congruent term.

We introduced the last rule (eval) to take care of the nested evaluation. It may seem that one needs a rule which reduces a term of the form $e[\![f[\![a]\!]]\!]$ for any term a which may or may not be a variable. But, such a rule is not necessary, since if a is not a variable, then a is either of the form $g[\![b]\!]$, or otherwise we can reduce $f[\![a]\!]$ by one of the rules (abs), (app) or (env). If a is of the form $g[\![b]\!]$, then we can repeatedly apply this argument to $f[\![g[\![b]\!]]\!]$. So, we only need a rule that reduces $e[\![f[\![x]\!]]\!]$. Now, if $x \notin \mathrm{TY}(f)$, then we can convert $f[\![x]\!]$ to x by the (gc) rule. In this way, we arrived at the rule (eval). Theorem 1 below shows that these reduction rules comprise a sufficiently rich set of reduction rules. We will discuss further about our choice of the reduction rules at the end of subsection 5.2.

If $\pi \in \Pi(a)$ and b is a term which is of the same type as a/π, then $a_\pi[b]$ stands for the term which is obtained from a by replacing its subterm a/π at π with b. Then the reduction relation $\rightarrow$ is defined by stipulating that $a \rightarrow a_\pi[b]$ if and only if $a/\pi \mapsto_{\lambda\varepsilon} b$. We write $\stackrel{*}{\rightarrow}$ for the reflexive and transitive closure of the relation $\rightarrow$, and $\stackrel{+}{\rightarrow}$ for the transitive closure of $\rightarrow$.

A $\lambda\varepsilon$ term a is said to be *strongly normalizing* (SN, for short) if there are no infinite reduction sequences starting from a.

The following theorem shows that we have a sufficiently rich set of reduction rules.

Theorem 1 (Closed Normal Term is Canonical). *If $c : C$ is closed and normal, then c is canonical.*

Let Γ and Γ' be contexts. Then we say that Γ' is a subcontext of Γ if each x^A in Γ' is also in Γ. In other words, Γ' is a subcontext of Γ if the set consisting of the members of Γ' is a subset of the set consisting of the members of Γ. We will write $\Gamma' \subseteq \Gamma$ if Γ' is a subcontext of Γ.

Our reduction relation $\rightarrow_{\lambda\varepsilon}$ enjoys the subject reduction property.

Theorem 2 (Subject Reduction). *If $\Gamma \vdash a : A$ and $a \rightarrow_{\lambda\varepsilon} b$, then $\Delta \vdash b : A$ for some subcontext Δ of Γ.*

This theorem says not only that the type is preserved by reduction, but also that the reduction never introduces new free variables. Thus, if we start from a closed term, then we always get closed terms by reductions of the given term.

We conclude this section by giving simple examples of $\lambda\varepsilon$ programs (i.e., closed $\lambda\varepsilon$-terms) which show the expressivity of the language. We assume that we have integer type and the successor function s.

$$\{\lambda f.\lambda x.f(f(x))/\mathtt{double}\}[\![$$
$$\qquad \{\mathtt{double}\ \mathtt{s}/\mathtt{add2}\}[\![\{\mathtt{double}\ \mathtt{add2}/\mathtt{add4}\}[\![\mathtt{add2}(\mathtt{add4}(0))]\!]]\!]$$
$$]\!] \stackrel{*}{\rightarrow}_{\lambda\varepsilon} 6.$$

$$\{\lambda f.\lambda e.\{f(e[\![x]\!])/x, f(e[\![y]\!])/y\}/\mathtt{pointwise}\}[\![\mathtt{pointwise}\ \mathtt{s}\ \{1/x, 2/y\}]\!]$$
$$\stackrel{*}{\rightarrow}_{\lambda\varepsilon} \{2/x, 3/y\}.$$

4 Confluence and Conservativity

In this section, we prove the confluence property of the $\lambda\varepsilon$-calculus by combining Hardin's interpretation method [5] (which is a standard method used to prove the confluence of calculi of explicit substitutions [1–3, 7]) with Takahashi's parallel reduction method [11].

Lemma 1. *$\to_\varepsilon$ on $\lambda\varepsilon$-terms is noetherian and confluent.*

A $\lambda\varepsilon$-term a is said to be *ε-normal* if $a \to_\varepsilon b$ holds for no b. By the above lemma, we see that for any $\lambda\varepsilon$-term a there uniquely exist an ε-normal term b such that $a \overset{*}{\to}_\varepsilon b$. We will write $\varepsilon(a)$ for this b. ε-normal terms are characterized by the following grammar, where u ranges over ε-normal terms and v over ε-normal terms such that $x \in \mathrm{TY}(v)$ and which are not canonical, that is, not of the form $\{a_1/x_1, \ldots, a_n/x_n\}$.

$$u ::= x \mid \lambda x.u \mid uu \mid \{u/x, \ldots, u/x\} \mid v[\![x]\!]$$

We now define the parallel reduction relation $\Rightarrow$ on ε-normal $\lambda\varepsilon$-terms as follows.

1. $x \Rightarrow x$.
2. If $a \Rightarrow b$, then $\lambda x.a \Rightarrow \lambda x.b$.
3. If $a \Rightarrow c$ and $b \Rightarrow d$, then $(\lambda x.a)b \Rightarrow \varepsilon(\{d/x\}[\![c]\!])$.
4. If $a \Rightarrow c$ and $b \Rightarrow d$, then $ab \Rightarrow cd$.
5. If $a_i \Rightarrow b_i$, then $\{a_1/x_1, \ldots, a_n/x_n\} \Rightarrow \{b_1/x_1, \ldots, b_n/x_n\}$.
6. If $e \Rightarrow f$, then $e[\![x]\!] \Rightarrow \varepsilon(f[\![x]\!])$.

Next, with each ε-normal term a, we associate an ε-normal term a^* as follows.

1. $x^* := x$.
2. $(\lambda x.a)^* := \lambda x.a^*$.
3. $((\lambda x.a)b)^* := \varepsilon(\{b^*/x\}[\![a^*]\!])$.
4. $(ab)^* := a^* b^*$, if a is not an abstraction.
5. $\{a_1/x_1, \ldots, a_n/x_n\}^* := \{a_1^*/x_1, \ldots, a_n^*/x_n\}$.
6. $(e[\![x]\!])^* := \varepsilon(e^*[\![x]\!])$.

It is easy to see that $a \Rightarrow a^*$ for any ε-normal term a. In this section, we work only in the $\lambda\varepsilon$-calculus. So, we will write $a \to b$ ($a \overset{*}{\to} b$) for $a \to_{\lambda\varepsilon} b$ ($a \overset{*}{\to}_{\lambda\varepsilon} b$), respectively. We have the following key Lemmas 2-5.

Lemma 2. *If $a \Rightarrow b$, then $a \overset{*}{\to} b$.*

Lemma 3. *If $\varepsilon(a) \Rightarrow \varepsilon(a')$ and $\varepsilon(e) \Rightarrow \varepsilon(e')$, then $\varepsilon(e[\![a]\!]) \Rightarrow \varepsilon(e'[\![a']\!])$.*

Lemma 4. *If $a \to_\lambda a'$, then $\varepsilon(a) \Rightarrow \varepsilon(a')$.*

Lemma 5. *$\Rightarrow$ on ε-normal $\lambda\varepsilon$-terms is confluent.*

Theorem 3 (Confluence). *$\to_{\lambda\varepsilon}$ on $\lambda\varepsilon$-terms is confluent.*

Next, we show the conservativity of $\lambda\varepsilon$ over the simply typed $\lambda\beta$-calculus, where by the simply typed $\lambda\beta$-calculus, we mean a typed calculus whose typing rules are (assume), ($\Rightarrow I$) and ($\Rightarrow E$) and whose only reduction rule is (β). We can state the conservativity theorem as follows.

Theorem 4 (Conservativity). *Let a and b be typed $\lambda\beta$-terms. Then $a \stackrel{*}{\rightarrow}_\beta b$ if and only if $a \stackrel{*}{\rightarrow}_{\lambda\varepsilon} b$.*

In order to prove this theorem, we define simple $\lambda\varepsilon$-terms as follows. A $\lambda\varepsilon$-term is *simple*, if its type A is of the following form:

$$A, B ::= K \mid A \Rightarrow B$$

and its untyped form a can by constructed by the following grammar:

$$a, b ::= x \mid \lambda x.b \mid ba \mid e[\![a]\!]$$
$$e, f ::= \{a/x\} \mid e[\![f]\!]$$

We note that any $\lambda\beta$-term is a simple $\lambda\varepsilon$-term. If $e[\![a]\!]$ is a simple $\lambda\varepsilon$-term, then we say that e is a *simple environment term.*

Before we state Lemma 6 we define a syntactic translation Φ which translates each simple $\lambda\varepsilon$-term to a $\lambda\beta$-term as follows. We also define auxiliary translation functions Ψ_1 and Ψ_2 at the same time.

1. $\Phi(x) \equiv x$.
2. $\Phi(\lambda x.b) \equiv \lambda x.\Phi(b)$.
3. $\Phi(ba) \equiv \Phi(b)\Phi(a)$.
4. $\Phi(e[\![a]\!]) \equiv \Phi(a)[\Psi_1(e) := \Psi_2(e)]$.
5. $\Psi_1(\{a/x\}) \equiv x$.
6. $\Psi_1(e[\![f]\!]) \equiv \Psi_1(f)$.
7. $\Psi_2(\{a/x\}) \equiv \Phi(a)$.
8. $\Psi_2(e[\![f]\!]) \equiv \Psi_2(f)[\Psi_1(e) := \Psi_2(e)]$.

Lemma 6. *Let a be a simple λs-term. If $a \rightarrow_{\lambda\varepsilon} b$, then b is also simple and $\Phi(a) \stackrel{*}{\rightarrow}_\beta \Phi(b)$.*

Proof of Theorem 4. Only if part is trivial and if part follows from Lemma 6 by noting that Φ is identity on $\lambda\beta$-terms. □

5 Strong Normalizability

In this section we prove the strong normalizability of the $\lambda\varepsilon$-calculus using the reducibility argument. (See, e.g., [4].) So, for each type A, we define a set $[A]$ of *reducible terms* of type A as follows.

1. If $a : A$ and A is atomic or the empty environment type $\{\}$, then $a \in [A]$ iff a is SN.
2. If $b : A \Rightarrow B$, then $b \in [A \Rightarrow B]$ iff $ba \in [B]$ for all $a \in [A]$.
3. If $e : E$ and E is a non-empty environment type, then $e \in [E]$ iff $e[\![x^A]\!] \in [A]$ for all $x^A \in E$.

We can prove the following fundamental proposition as in [4].

Proposition 1.
(CR1) *If $a \in [A]$, then a is SN.*
(CR2) *If $a \in [A]$ and $a \rightarrow a'$, then $a' \in [A]$.*
(CR3) *If $a : A$ is neutral and $a' \in [A]$ for all a' such that $a \rightarrow a'$, then $a \in [A]$.*

5.1 Decoration Trees and Decorated Terms

Although we can define reducibility in a standard way, we cannot prove the SN of $\lambda\varepsilon$ in a similar way as in the case of simply typed lambda calculus. The reason is that while substitution is carried out by a single step (β)-rule in the simply typed lambda calculus, in $\lambda\varepsilon$ we have to compute the substitution internally by moving around environments that carry information about substitution. To cope with this situation, we introduce the notion of a decoration tree which is useful in keeping track of the movements of the environments during the reduction steps.

We define *decoration tree* (δ) and its type ($\mathrm{TY}(\delta)$) inductively as follows.

1. If $\delta_1, \ldots, \delta_n$ ($n \geq 0$) are decoration trees, then $\delta \equiv (\delta_1, \ldots, \delta_n)$ is a decoration tree and $\mathrm{TY}(\delta) := \mathrm{TY}(\delta_1) \cup \cdots \cup \mathrm{TY}(\delta_n)$.
2. If δ is a decoration tree and e is a bindable variable free environment term, then (e, δ) is a decoration tree and $\mathrm{TY}((e, \delta)) := \mathrm{TY}(e) \cup \mathrm{TY}(\delta)$.

Note that each leaf of a decoration tree is $()$ or a bindable variable free environment term. A decoration tree is *trivial* if its leaves are always $()$. We will use δ, γ, ρ etc. to denote decoration trees.

Let a and $\tilde{a}$ be terms of type A and δ be a decoration tree. We define a ternary relation '$\tilde{a}$ is a *decoration* of a by δ' inductively as follows. In the following definition, δ provides the information about the positions in a which are to be decorated as well as which environments are used to decorate these positions. We will write $\delta : a \mapsto \tilde{a}$ for this relation.

$$\frac{\delta : a \mapsto \tilde{a}}{(e, \delta) : a \mapsto e[\![\tilde{a}]\!]} \qquad \frac{}{() : x \mapsto x}$$

$$\frac{\delta : b \mapsto \tilde{b}}{(\delta) : \lambda x.b \mapsto \lambda x.\tilde{b}}\ (*) \qquad \frac{\delta : b \mapsto \tilde{b} \quad \delta' : a \mapsto \tilde{a}}{(\delta, \delta') : ba \mapsto \tilde{b}\tilde{a}}$$

$$\frac{\delta_1 : a_1 \mapsto \tilde{a_1} \quad \cdots \quad \delta_n : a_n \mapsto \tilde{a_n}}{(\delta_1, \ldots, \delta_n) : \{a_1/x_1, \ldots, a_n/x_n\} \mapsto \{\tilde{a_1}/x_1, \ldots, \tilde{a_n}/x_n\}}$$

$$\frac{\delta : e \mapsto \tilde{e} \quad \delta' : a \mapsto \tilde{a}}{(\delta, \delta') : e[\![a]\!] \mapsto \tilde{e}[\![\tilde{a}]\!]}\ (**)$$

The rule marked by $(*)$ may be applied only under the condition that $x \notin \mathrm{TY}(\delta)$ and the rule marked by $(**)$ may be applied only when $\mathrm{TY}(e) \cap \mathrm{TY}(\delta') = \emptyset$.

If $\delta : c \mapsto a$ holds, we will call the triple (a, c, δ) a *decorated term over* c, and will simply write a for the decorated term, if c and δ can be inferred from the context.

It is easy to see that if $\delta : c \mapsto a$ and $\delta : c \mapsto b$, then $a \equiv b$. So, we will write $\delta(c)$ for this a. It is also easy to see that for each term a, there is a unique trivial δ such that $\delta : a \mapsto a$. We will write ι_a for this δ.

If $\overline{e} \equiv e_1, \ldots, e_n$, then we will write $(\overline{e}, \delta)$ for $(e_1, (e_2, \cdots (e_n, \delta) \cdots))$ and $\overline{e}[\![a]\!]$ for $e_1[\![\cdots e_n[\![a]\!] \cdots]\!]$. Suppose that $\delta : c \mapsto a$. Then δ can be written uniquely in the form $(\overline{e}, \gamma)$ where γ is of the form $(\delta_1, \ldots, \delta_n)$ $(n \geq 0)$. In this case, we have $a \equiv \overline{e}[\![\gamma(c)]\!]$.

Let $\delta : c \mapsto \tilde{c}$ and $\pi \in \Pi(\tilde{c})$. We say that π *is internal, marginal or external in* δ if it can be seen so by the following inductive clauses. At the same time, we also define δ/π which is either a decoration tree or a subterm of some environment term in δ.

1. Λ is internal in $(\delta_1, \ldots, \delta_n)$ $(n \geq 0)$ and $(\delta_1, \ldots, \delta_n)/\Lambda := (\delta_1, \ldots, \delta_n)$.
2. Λ is marginal in (e, δ) and $(e, \delta)/\Lambda := (e, \delta)$.
3. If $\pi \in \Pi(e)$, then 1π is external in (e, δ) and $(e, \delta)/1\pi := e/\pi$.
4. If π is internal, marginal or external in δ, then 2π is internal, marginal or external, respectively, in (e, δ) and $(e, \delta)/2\pi := \delta/\pi$.
5. If π is internal, marginal or external in δ_i, then $i\pi$ is internal, marginal or external, respectively, in $(\delta_1, \ldots, \delta_n)$ $(n \geq 1)$ and $(\delta_1, \ldots, \delta_n)/i\pi := \delta_i/\pi$.

Suppose that $\delta : c \mapsto \tilde{c}$ and $\pi \in \Pi(\tilde{c})$. The occurrence of $\tilde{a}/\pi$ in $\tilde{a}$ is said to be an *internal, marginal* or *external occurrence with respect to* δ if π is internal, marginal or external in δ, respectively. It is easily verified that for any $\pi \in \Pi(\tilde{c})$, π is either internal, marginal or external in δ and only one of these is the case. This classification can be characterized as follows. π is external if δ/π is a $\lambda\varepsilon$-term, π is marginal if $\delta/\pi 1$ is external and π is internal otherwise.

Suppose that $\delta : c \mapsto \tilde{c}$ and σ is external in δ. Then we can find a unique π such that π is marginal in δ and $\pi 1 \leq \sigma$. In this case, we say that π is the *root* of σ in δ.

Let $\delta : c \mapsto \tilde{c}$. Then, this decoration naturally induces a mapping from $\Pi(c)$ to $\Pi(\tilde{c})$ as follows. We use δ to denote this mapping. So, for each $\pi \in \Pi(c)$ we define $\delta(\pi) \in \Pi(\tilde{c})$ inductively as follows.

1. $(\delta_1, \ldots, \delta_n)(\Lambda) := \Lambda$.
2. $(\delta_1, \ldots, \delta_n)(i\pi) := i\delta_i(\pi)$ $(1 \leq i \leq n)$.
3. $(e, \delta)(\pi) := 2\delta(\pi)$.

We can easily check that the image of the mapping $\delta : \Pi(c) \to \Pi(\tilde{c})$ is exactly the set of internal positions in δ. If $\tilde{\pi} \equiv \delta(\pi)$, then we write $\delta^{-1}(\tilde{\pi})$ for π.

Let $\delta : c \mapsto \tilde{c}$, so that $\tilde{c}$ is of the form $\overline{e}[\![\gamma(c)]\!]$ where $\gamma \equiv (\delta_1, \ldots, \delta_n)$. Then, for each $\pi \in \Pi(c)$, we define a decoration tree $\delta|\pi$ as follows. If $\pi \equiv \Lambda$, we put $\delta|\pi := \gamma$. If $\pi \equiv i\sigma$, we put $\delta|\pi := \delta_i|\sigma$. We can then see that for each $\pi \in \Pi(c)$, we have $\delta|\pi : c/\pi \mapsto \tilde{c}/\tilde{\pi}$ where $\tilde{\pi} \equiv \delta(\pi)$. We will call $\tilde{c}/\tilde{\pi}$ *the image of* c/π *under* δ.

A decoration tree δ is *strongly normalizing* (SN) if each environment term e in δ is SN.

5.2 Orthogonal Decorations

In this subsection we introduce orthogonal decorations and show their fundamental properties. We will say that $\delta : c \mapsto \tilde{c}$ is *orthogonal* if for each marginal subterm $e[\![a]\!]$ of $\tilde{c}$, $\mathrm{TY}(e) \cap \mathrm{FV}(a) = \emptyset$ holds. Then we have the following proposition which is used in the proof of Proposition 4.

Proposition 2. *If $\gamma : c \mapsto \tilde{c}$ is orthogonal, γ is SN and $c \in [C]$, then $\tilde{c} \in [C]$.*

Discussions on the (eval) rule. Here, we would like to remark that the condition $x \in \mathrm{TY}(f)$ is essential in proving Proposition 2. Now, let us consider the following rule:

(eval′) $e[\![f[\![x]\!]]\!] \mapsto_\varepsilon e[\![f]\!][\![x]\!]$, if $x \notin \mathrm{TY}(e)$ or $x \in \mathrm{TY}(f)$.

This rule is a semantically correct and more liberal rule than our (eval) rule. So, we might be tempted to make our calculus more liberal by adopting the (eval′) rule in place of the (eval) rule. However, if we do so, the calculus will not be strongly normalizing. In fact, we have the following counter-example to SN due to Bloo et. al. [3]. Let us put $f := \{(\lambda x.z)z/x\}$ and $e := \{f[\![z]\!]/x\}$ where z is an arbitrary unbindable variable and z and x are of the same type, say, A. Then we can construct an infinite reduction sequence starting from $e[\![f]\!]$. We also note that both e and f are SN and $(e, \iota_f) : f \mapsto e[\![f]\!]$ is an orthogonal decoration since f is bindable variable free. Hence Proposition 2 no longer holds for the extended calculus. Thus we see that the condition $x \in \mathrm{TY}(f)$ is indispensable in our proof of Proposition 2.

Next, consider the rule:

(eval″) $e[\![f[\![a]\!]]\!] \mapsto_\varepsilon e[\![f]\!][\![a]\!]$, if $\mathrm{TY}(e) \cap (\mathrm{FV}(a) - \mathrm{TY}(f)) = \emptyset$.

This rule is even more liberal than the (eval′) rule, and if we adopt this rule instead of our (eval) rule, we would still have the confluence and the conservativity properties, but we would no longer have SN since SN fails for a less liberal system as we saw above. These are the reasons we have chosen the (eval) rule as it is now.

5.3 Partial Orders on Decorated Terms

In this subsection, for each term c, we define a partial order $\preceq_c$ on decorated terms over c. These partial orders play an essential role in our proof of SN.

Let $\delta : c \mapsto \tilde{c}$ be a decoration, and suppose that $\pi \in \Pi(c)$ is such that c/π is a variable. Then we define $\mathsf{env}(c, \pi, \delta)$ inductively below as a sequence of pairs of the form $\langle e, \sigma \rangle$ where e is an environment term in δ and $\sigma \in \Pi(c)$. In the following definition, if $\overline{e} \equiv \langle e_1, \sigma_1 \rangle, \ldots, \langle e_n, \sigma_n \rangle$, then $k\overline{e}$ denotes the sequence $\langle e_1, k\sigma_1 \rangle, \ldots, \langle e_n, k\sigma_n \rangle$.

1. $\mathsf{env}(x, \Lambda, ()) := \Lambda$ (empty sequence).
2. $\mathsf{env}(c, k\pi, (\delta_1, \ldots, \delta_n)) := k\mathsf{env}(c/k, \pi, \delta_k)$ $(1 \le k \le n)$.
3. $\mathsf{env}(c, \pi, (e, \delta)) := \mathsf{env}(c, \pi, \delta), \langle e, \Lambda \rangle$.

If $\delta : c \mapsto \tilde{c}$ and $\langle e, \sigma \rangle$ is in the sequence $\mathsf{env}(c, \pi, \delta)$, then we will say that the variable c/π at π in c is *decorated by e at σ in c and at $\delta(\sigma)$ in $\tilde{c}$*. We note that, here, $\delta(\sigma)$ is marginal in δ and $\tilde{c}/\delta(\sigma) \equiv e[\![a]\!]$ for some a. If $\langle e, \sigma \rangle$ appears before $\langle e', \sigma' \rangle$ in $\mathsf{env}(c, \pi, \delta)$, then we have $\sigma \ge \sigma'$ and $\delta(\sigma) > \delta(\sigma')$. Hence, we see that $\mathsf{env}(c, \pi, \delta)$ gives a sorted sequence of environments decorating c/π where the environment closest to c/π is the first element of the sequence.

Here is a simple example. Let $\delta \equiv (f, ((e, \iota_y), \iota_x))$. For the decoration $\delta : yx \mapsto f[\![e[\![y]\!][\![x]\!]]\!]$, we have $\mathsf{env}(yx, 1, \delta) \equiv \langle e, 1\rangle, \langle f, \Lambda\rangle$ and $\mathsf{env}(yx, 2, \delta) \equiv \langle f, \Lambda\rangle$.

Let $\overline{e} \equiv \langle e_1, \sigma_1\rangle, \ldots, \langle e_m, \sigma_m\rangle$ and $\overline{e'} \equiv \langle e'_1, \sigma'_1\rangle, \ldots, \langle e'_n, \sigma'_n\rangle$. We write $\overline{e} \le \overline{e'}$ if (1) $m \le n$ and (2) for each k such that $1 \le k \le m$, $e_k \equiv e'_k$ and $\sigma_k \le \sigma'_k$ hold. Let a and b be two decorated terms such that $\gamma : c \mapsto a$ and $\delta : c \mapsto b$. We write $a \preceq_c b$ if, for each occurrence of a variable $x \equiv c/\pi$ in c, we have $\mathsf{env}(c, \pi, \gamma) \le \mathsf{env}(c, \pi, \delta)$. It is easy to see that $\preceq_c$ determines a partial order on decorated terms over c and $c \equiv (c, c, \iota_c) \preceq_c a$ holds for any decorated term a over c.

Proposition 3. *If $c_1 \preceq_c c_2$ and $c_2 \in [C]$, then $c_1 \in [C]$.*

5.4 Reducibility Theorem

We say that a decoration tree δ is *reducible*, if each environment term in δ is reducible. In this subsection, we prove the following theorem as the final result of our paper.

Theorem 5 (Reducibility). *If $c : C$, $(\overline{e}, \iota_c) : c \mapsto \overline{e}[\![c]\!]$ and $(\overline{e}, \iota_c)$ is reducible, then $\overline{e}[\![c]\!] \in [C]$.*

We note that by using an empty sequence as $\overline{e}$ in the theorem, we can conclude that any $\lambda\varepsilon$-term is reducible and hence SN. We have to show some more propositions before we can prove the reducibility theorem.

Proposition 4. *If $\delta : x^A \mapsto \overline{e}[\![x]\!]$ and δ is reducible, then $\overline{e}[\![x]\!] \in [A]$.*

Proposition 5. *If $a \in [A]$ and $\{a/x\}[\![b]\!] \in [B]$, then $(\lambda x.b)a \in [B]$.*

Proposition 6. *If $a_1 \in [A_1], \ldots, a_n \in [A_n]$, then*
$$\{a_1/x_1, \ldots, a_n/x_n\} \in [\{x_1^{A_1}, \ldots, x_n^{A_n}\}].$$

Let c be a term, $\pi_1, \ldots, \pi_n \in \Pi(c)$ be such that $x_i \equiv c/\pi_i$ $(1 \le i \le n)$ are bindable variables free at π_i in c and $y_1, \ldots, y_n$ be unbindable variables such that, for each i $(1 \le i \le n)$, x_i and y_i are of the same type. Then we say that $c' \equiv c_{\pi_1,\ldots,\pi_n}[y_1, \ldots, y_n]$ is a *variant* of c. If, moreover, c' is bindable variable free, then we say that c' is a *bindable variable free variant* of c. Note that we can always find a bindable variable free variant of c for any c.

Proposition 7. *If c' is a variant of $c : C$, then $c \in [C]$ iff $c' \in [C]$.*

Now we can prove the reducibility theorem as follows.

Proof. By induction on the size of the derivation of $c : C$. We write $\tilde{c}$ for $\overline{e}[\![c]\!]$. We classify cases according to the last rule applied to derive $c : C$. We treat only two cases.

1. $(\Rightarrow I)$: In this case, $C \equiv A \Rightarrow B$, $c \equiv \lambda x.b$ and $\tilde{c} \equiv \overline{e}[\![\lambda x.b]\!]$. Here, by α-conversion, we may assume that $x \notin \mathrm{TY}(e)$. We put $\tilde{b} := \overline{e}[\![b]\!]$. We first show that $\lambda x.\tilde{b} \in [C]$. To show this, we take an arbitrary $a \in [A]$ and we let a' be a bindable variable free variant of a. Then, by Proposition 7, we have that $a' \in [A]$. Hence, by Proposition 6, we have $\gamma : b \mapsto \{a'/x\}[\![\tilde{b}]\!]$

where $\gamma := (\{a'/x\}, (\bar{e}, \iota_b))$ is reducible and bindable variable free. (Note that $(\{a/x\}, (\bar{e}, \iota_b))$ may not be bindable variable free since $\mathrm{FV}(a)$ may contain bindable variables. So, we consider a' in place of a.) So, by IH, we have $\{a'/x\}[\![\tilde{b}]\!] \in [B]$. Hence by Proposition 5, we have $(\lambda x.\tilde{b})a' \in [B]$. So, by Proposition 7, we have $(\lambda x.\tilde{b})a \in [B]$. Therefore, we have $\lambda x.\tilde{b} \in [C]$. Since $\tilde{c} \preceq_{\lambda x.b} \lambda x.\tilde{b}$, we have $\tilde{c} \in [C]$ by Proposition 3.

2. (env$E$): In this case $c \equiv e[\![a]\!]$ where $e : E$ and $a : C$. We show that $\bar{e}[\![e[\![a]\!]]\!] \in [C]$. We let e' be a variant of e such that all free occurrences of bindable variables in e that are not bound by $\bar{e}$ are renamed. Then, we see that $\bar{e}[\![e']\!]$ is bindable variable free. Now, consider the decoration: $(\bar{e}, \iota_{e'}) : e' \mapsto \bar{e}[\![e']\!]$. Since the size of the derivation of $e' : E$ is equal to the size of the derivation of $e : E$, we may appeal to IH for e'. Then, we have $\bar{e}[\![e']\!] \in [E]$. So, we have the decoration: $(\bar{e}, (\bar{e}[\![e']\!], \iota_a)) : a \mapsto \bar{e}[\![\bar{e}[\![e']\!][\![a]\!]]\!]$, where $(\bar{e}, (\bar{e}[\![e']\!], \iota_a))$ is reducible. Hence, by IH for a, we have $\bar{e}[\![\bar{e}[\![e']\!][\![a]\!]]\!] \in [C]$. Next, we consider the following two decorations.
$$(\bar{e}, ((\bar{e}, \iota_{e'}), \iota_a)) : e'[\![a]\!] \mapsto \bar{e}[\![\bar{e}[\![e']\!][\![a]\!]]\!],$$
$$(\bar{e}, (\iota_{e'}, \iota_a)) : e'[\![a]\!] \mapsto \bar{e}[\![e'[\![a]\!]]\!].$$
From these decorations, we see that $\bar{e}[\![e'[\![a]\!]]\!] \preceq_{e'[a]} \bar{e}[\![\bar{e}[\![e']\!][\![a]\!]]\!]$. Hence, by Proposition 3, we have $\bar{e}[\![e'[\![a]\!]]\!] \in [C]$. Therefore, by Proposition 7 we have $\bar{e}[\![e[\![a]\!]]\!] \in [C]$. □

6 Conclusion

We have defined a notion of explicit environments which generalizes explicit substitutions and records, and given a calculus for it which is confluent, SN and conservative extension of the simply typed $\lambda\beta$-calculus. The calculus we have presented here is the first such calculus that is conservative over the simply typed $\lambda\beta$-calculus.

We have shown how definition of free and bound variables can be achieved by a suitable type system. This is a form of static analysis.

Due to lack of space, we have not been able to explain how our calculus contains records. We only note here that a canonical record $[x_1 = a_1, \ldots, x_n = a_n]$ may be represented by the environment $\{a_1/x_1, \ldots, a_n/x_n\}$, and that accessing to the x field of a record r may be achieved by $r[\![x]\!]$. This representation is made possible owing to the fact that $\lambda\varepsilon$ is a system with variable names. It is therefore critical that our calculus has variables with names, since we insist that an explicit environment must generalize both a record and a substitution.

There are both named and nameless calculi of explicit substitutions. However, in the case of explicit environments, we have considered only a named calculus for the reason we explained above. Thus, viewed as a calculus of explicit substitution, $\lambda\varepsilon$ is a named calculus of explicit substitution. We also think it worthwhile to design a nameless version of $\lambda\varepsilon$ (although such a calculus would no longer contain a calculus of records but instead contain a calculus of tuples), since such a calculus would become a nameless calculus of explicit substitutions that has substitutions as first class values.

It seems possible to design an untyped version of our calculus, which is conservative over the untyped $\lambda\beta$-calculus and preserves SN. The syntax of such a system, however, would have to be an extension of the syntax of untyped $\lambda\varepsilon$-terms we have given, since, otherwise, we would not be able to determine free variables correctly.

Also, there are recently growing interests in the calculi of contexts. Among these calculi, a typed calculus of context introduced by Hashimoto and Ohori [6] uses type information to determine the set of free variables for a given term. We feel that we should be able to design a language which has both environments and contexts as first class values.

Other future research directions would be to extend the calculus to a calculus that supports dependent types. We are also considering how to use environments to mimic assignment in imperative programs and hope to do further work in this direction.

References

1. Abadi, M., Cardelli, L., Curien, P.-L. and Levy, J.-J., Explicit Substitutions, pp. 375-416, *Journal of Functional Programming*, **1**, 1991.
2. Benaissa, Z.E.A., Briand, D., Lescanne, P. and Maibaum, T.S.E., λv, a calculus of explicit substitutions which preserves strong normalization pp. 699-722, *Journal of Functional Programming*, **6**, 1996.
3. Bloo, R. and Rose, K.H., Preservation of Strong Normalization in Named Lambda Calculi with Explicit Substitution and Garbage Collection, *Proceedings of CSN'95 (Computer Science in Netherlands)*, van Vliet J.C. (ed.), 1995. (`ftp://ftp.diku.dk/diku/semantics/papers/D-246.ps`)
4. Girard, J.-Y., Lafont, Y. and Taylor, P., Proofs and Types, Cambridge University Press, 1989.
5. Hardin, T., Confluence results for the pure strong categorical combinatory logic CCL: λ-calculi as subsystems of CCL, pp. 305-312, *Theoretical Computer Science*, **46**, 1986.
6. Hashimoto, M. and Ohori, A., A typed context calculus, Preprint RIMS-1098, Res. Inst. for Math. Sci., Kyoto Univ., 1996. Available at: `http://www.kurims.kyoto-u.ac.jp/~ohori/list.html`.
7. Kamareddine, F. and Ríos, A., Extending a λ-calculus with explicit substitution which preserves strong normalization into a confluent calculus on open terms, pp. 395-420, *Journal of Functional Programming*, **7**, 1997.
8. Lampson, B. and Burstall, R., Pebble, a Kernel Language for Modules and Abstract Data Types, pp. 278-346, *Information and Computation*, **76**, 1988.
9. Nishizaki, S., Simply Typed Lambda Calculus with First-Class Environments, *Publications of the Research Institute for Mathematical Sciences, Kyoto University*, Vol. 30, No. 6, 1994.
10. Nishizaki, S., ML with First-Class Environments and its Type Inference Algorithm, pp. 95-116, *Logic, Language and Computation, Festschrift in Honor of Satoru Takasu*, Lecture Notes in Computer Science **792**, Jones, N. D., Hagiya, M. and Sato, M. (eds.), Springer-Verlag, 1994.
11. Takahashi, M., Parallel Reductions in λ-calculus, *J. Symbolic Computation*, **7**, pp. 113-123, 1989.

Consequences of Jacopini's Theorem: Consistent Equalities and Equations

Rick Statman*

1 Introduction

In this note we consider the problem of whether a combinator P can consistently (in most cases with beta conversion) be assumed to satisfy the functional equation $Mx = Nx$. Much of the literature in this area concerns easy terms first discovered by Jacopini. These are combinators P which can consistently be assumed to be solutions to the equation $x = Q$ for any Q. Here we shall prove several results which might be viewed as unexpected; although given Jacopini's result the unexpected should be expected in this topic in lambda calculus.

We shall construct an identity $M = N$ which is not a beta conversion but which is consistent with any consistent set of combinator equations. By a simpler construction we shall build a functional equation $Mx = Nx$ for which there is no solution modulo beta conversion but such that for each consistent set S of combinator equations there exists a combinator P with $S\ U\ \{MP = NP\}$ consistent. Next we consider the problem of which sets of combinators are "consistency sets" i.e., sets of the form $\{P : MP = NP$ is consistent $\}$. Each such set is closed under beta conversion and pi-zero-one ("co-Visseral" in [5]). We produce such a co-Visseral set which is not a consistency set, in contrast to the case for first order arithmetic. Finally, we consider some questions involving compactness. We give several examples of sets of functional equations $Mx = Nx$ such that

(*) for each finite subset there is a combinator which can be consistently assumed to be a solution

but there is no single combinator which can consistently be assumed to be a solution of the whole set. However, we show that if the condition (*) is made effective then no such examples are possible. This is in contrast to the familiar event of the effectivization of a classical theorem being false.

2 Preliminaries

We adopt for the most part the notation and terminology of [1]. A combinator is a closed term. The following are the usual combinators

* Research Supported by NSF CCR-9624681

$B \quad = \lambda xyz.\ x(yz)$
$C^* \quad = \lambda xy.\ yz$
$K \quad = \lambda xy.\ x$
$K^* \quad = \lambda xy.\ y$
$Y \quad = \lambda x.\ (\lambda y.\ x(yy))(\lambda y.\ x(yy))$
$O \quad = \lambda xy.\ y(xy)$
Omega $= (\lambda x.\ xx)(\lambda x.\ xx)$

but we reserve the symbol S for sets of combinators or combinator equations. We take the simple expediant of identifying the natural numbers with their Curch numerals. In this way we avoid the quotation device of [1]. In addition, with this understanding, any uniform sequence of combinators can, up to beta conversion, be denoted $P0, P1, \ldots, Pi, \ldots$ instead of the usual subscript notation. However, for readability we shall write such a sequence as $P[1], P[2], \ldots, P[i], \ldots$. We are interested in functional equations

$$U = V$$

in a single free variable x, which by abstraction can be put in the form

$$(\lambda x.U)x = (\lambda x.V)x.$$

Functional equations in more than one free variable can be reduced to one by pairing. For example, the equation

$$Mxy = Nxy$$

can be replaced by

$$M(zK)(zK^*) = N(zk)(zK^*)$$

with solutions $z = \langle x, y\rangle = \lambda a.\ axy$.

Similarly, several equations can be combined into one by pairing. If S is a set of combinator equations then, by the well known existence of free models ([1]), $M = N$ is inconsistent with S if and only if $S\ U\ \{M = N\} \vdash K = K^*$. Implicit in Jacopini's classic paper [3] is the following

Theorem (Jacopini): $M = N$ is inconsistent with S if and only if there exist combinators $P[1], \ldots, P[p]$ such that
$S \vdash K = P[1]M\ \&\ P[1]N = P[2]M\ \&\ldots\&\ P[p]N = K^*$.

Another way to state this theorem is to consider the graph whose points consist of the congruence classes of combinators modulo provable equivalence in S, and whose undirected edges join points of the form PM to those of the form PN. Then $M = N$ is inconsistent $\Leftrightarrow$ K and K^* are connected by a path $\Leftrightarrow$ the graph is connected.

Among the congruence classes of combinators modulo equivalence in S are some which contain no solvable terms such as the class of K – infinity $= YK$. We call the number of these classes the degree of S. For example, Barendregt's

H^* has degree 1 but the empty S (beta conversion) has infinite degree. Below, we shall observe that S's of each finite degree exist.

When it comes to functional equations $Mx = Nx$ it is possible that the equation $MA = NA$ is consistent with beta conversion for a new constant A but for any combinator L the equation $ML = NL$ is inconsistent with beta conversion. For example, in [4] we constructed a Plotkin term P such that for each combinator M, PM beta converts to P but P does not beta convert to KP. It is easy to see, by using Mitscheke's theorem [1] page 401, that the equation

$$PA = I$$

with a new constant A, is consistent with beta conversion but clearly it is not consistent for any combinator in place of A. It is also possible for a given combinator to be a consistent solution to each of several functional equations separately when the entire collection cannot have a solution. For example, Omega is a consistent solution to $x = K^*$ and to $x = Y\langle K, K^*\rangle$.

Definition: Suppose S is a set of combinator equations. The functional equation $Mx = Nx$ is said to be consistently solvable over S if there exists a combinator P such that $S\ U\{MP = NP\}$ is consistent. Such a P is called a consistent solution over S.

Remark: When S is empty we drop the phrase "over S".

Definition: The combinator equation $M = N$ is said to be inevitably consistent if M does not beta convert to N but for any consistent set S of combinator equations $S\ U\ \{M = N\}$ is consistent. The functional equation $Mx = Nx$ is said to be inevitably consistently solvable if there is no solution in the combinators modulo beta conversion but for any consistent set S of combinator equations there exists a combinator P such that $S\ U\ \{MP = NP\}$ is consistent.

Example: Y is a consistent solution to the equations

$$x = Ox \quad \text{and} \quad x = xO$$

since Y satisfies these equations in the Bohm tree model ([1]) but there is no solution to these equations modulo beta conversion (Intrigila, unpublished).

Example (generalization): We say that M is consistently solvable if there exists $N[1] \ldots N[n]$ such that $MN[1] \ldots N[n] = I$ is consistent with beta conversion. For each e construct a combinator $P[e]$ such that

$$P[e]n = \begin{cases} \lambda x.\ P[e](n+1) & \text{if the eth Turing machine} \\ & \text{converges on } n \text{ or} \\ \text{an order zero unsolvable} & \text{otherwise.} \end{cases}$$

This can be done directly or by the Visser fixed point theorem ([5]). Then $P[e]0$ is consistently solvable $\Leftrightarrow$ the eth Turing Machine is not total.

3 Inevitably Consistent and Consistently Solvable Equations

Theorem 1: Suppose that S is a set of combinator equations of finite degree. Then there exists a functional equation

$$Pxyz = Qxyz$$

such that for any combinator equation $M = N$.

$S\ U\ \{M = N\}$ is inconsistent $\Leftrightarrow PMNz = QMNz$ has a solution over S.

Proof: Suppose that S is given of degree n.

Consider the graph described after the statement of Jacopini's theorem above and all shortest paths joining the combinator class containing K and the class containing K^*. Now, for any of these paths, no intermediate point can contain a solvable combinator P. For if such a P exists it must have a distinct head normal form from either K or K^*. W.l.o.g. assume it is distinct from K and thus there exist $M[1]\ldots M[m]$ such that $KM[1]\ldots M[m]$ conv. K and $K^*M[1]\ldots M[m]$ conv. K^* conv. $PM[1]\ldots M[m]$, and this contradicts the choice of path as being a shortest one. Thus for $p = n + 3$, by Jacopini

$$S \vdash K = P[1]M \ \&\ P[1]N = P[2]M \ \&\ \ldots\ \&\ P[p]N = K^*$$

in other words

$$K = x[1]M,\ x[1]N = x[2]M, \ldots\ldots, x[p]N = K^*$$

has a solution over S. ■

The following corollary follows from the proof.

Corollary: If S is a set of combinator equations of finite degree then there exists a functional equation

$$Pxyz = Qxyz$$

such that $M = N$ is inconsistent with some consistent extension of $S \Leftrightarrow PMNz = QMNz$ is consistently solvable over S.

Remark: For the case that S is empty the construction in the proof of Theorem 1 does not work. This is verified in [6].

However, it is still the case that the theorem is true (the best proof comes from [4]).

Theorem 2: There exist S of every finite degree.

We shall present a proof of this theorem elsewhere; the theorem is not used below.

Theorem 3: There exists an inevitably consistent combinator equation.

Proof: For the proof we need to recall a result of [5].

As usual, $W[e]$ is the eth recursively enumerable set. For each RE set S of natural numbers there exists a combinator H such that if S satisfies

S is not empty
if e belongs to S then $W[e]$ is non-empty
if i and j both belong to S and $W[i]$ intersect $W[j]$ is non-empty then $W[i] = W[j]$
if e belongs to S, $\#M$ belongs to $W[e]$, and M conv. N then $\#N$ belongs to $W[e]$ then HM conv. $HN \Leftrightarrow M$ conv. N or there exists e in S s.t. M and N both belong to $W[e]$. In addition, the construction of H is uniform in S i.e., there exists a combinator G such that if e is an RE index for S we have Ge conv. H.

To apply this result consider a fixed enumeration of the finite sequences of combinators. We let $p(i)$ be the number of combinators in the ith sequence and we let $P[i,j]$ be the jth combinator in the ith sequence where j is between 1 and $p(i)$. Given combinators N and M we define two RE sets of combinators.

1. $\{\langle\langle P[i,1],\ldots,P[i,p(i)]\rangle,\ \langle K, P[i,1]N,\ldots,P[i,p(i)]N\rangle\rangle : i = 0,1,\ldots\}$
2. $\{\langle\langle P[i,1],\ldots,P[i,p(i)]\rangle,\ \langle P[i,1]M,\ldots,P[i,p(i)]M, K^*\rangle\rangle : i = 0,1,\ldots\}$

where $\langle X[1],\ldots,X[n]\rangle$ is the usual sequencing combinator $\lambda x.\ xX[1]\ldots X[n]$. If these two sets intersect modulo beta conversion (i.e., if their beta conversion closures intersect) then for some i we have

$$
\begin{array}{ll}
K & \text{conv. } P[i,1]M \\
P[i,1]N & \text{conv. } P[i,2]M \\
P[i,2]N & \text{conv. } P[i,3]M \\
\cdot & \\
\cdot & \\
\cdot & \\
P[i,p(i)]N & \text{conv.} K^*
\end{array}
$$

and the equation $M = N$ is inconsistent with beta conversion.

Conversely, if $M = N$ is inconsistent with beta conversion then by Jacopini's theorem the two sets intersect modulo beta conversion. Now let k be an RE index for the beta conversion closure of the first set and let ℓ be an RE index for the beta conversion closure of the second set. Apply the above theorem to $S = \{k, \ell\}$ to obtain H. Now the constant F such that $F\ \#\langle N, M\rangle$ conv. H. Now let

$$L = \lambda x.\ \langle\langle P[1,1],\ldots,P[1,p(1)]\rangle,\ \langle K, P[1,1]x,\ldots,P[1,p(1)]x\rangle\rangle$$
$$J = \lambda x.\ \langle\langle P[1,1],\ldots,P[1,p(1)]\rangle,\ \langle P[1,1]x,\ldots,P[1,p(1)]x, K^*\rangle\rangle.$$

Then by the fixed point theorem [1] there exists a pair $\langle N, M\rangle$ such that

$$\langle N, M\rangle \quad \text{conv.} \quad \langle F\#\langle N, M\rangle(LN), F\#\langle N, M\rangle(JM)\rangle$$

We claim that the equation $M = N$ is inevitably consistent.

First suppose that M conv. N. Then $H(LN)$ conv. $H(JM)$ and so by the above theorem S must fail to satisfy one of the stated conditions. This can only be that $W[k]$ intersect $W[\ell]$ is non-empty and thus $M = N$ is inconsistent with beta conversion. We conclude that M does not beta convert to N. Next suppose that S is a consistent set of combinator equations such that $M = N$ is inconsistent with S. By Jacopini's theorem there exist $P[i,1],\ldots,P[i,p(i)]$ such that

$$
\begin{array}{ll}
S \vdash & K = P[i,1]M \\
S \vdash & P[i,1]N = P[i,2]M \\
. & \\
. & \\
. & \\
\multicolumn{2}{l}{S \vdash P[i,p(i)]N = K^*}
\end{array}
$$

that is

$$
\begin{array}{l}
S \vdash \langle\langle P[i,1],\ldots,P[i,p(i)]\rangle,\ \langle K, P[i,1]N,\ldots,P[i,p(i)]N\rangle\rangle = \\
\quad \langle\langle P[i,1],\ldots,P[i,p(i)]\rangle, \langle P[i,1]M,\ldots,P[i,p(i)]M, K^*\rangle\rangle \text{ and} \\
S \vdash H\langle\langle P[i,1],\ldots,P[i,p(i)]\rangle,\ \langle K, P[i,1]N,\ldots,P[i,p(i)]N\rangle\rangle = \\
\quad H\langle\langle P[i,1],\ldots,P[i,p(i)]\rangle\langle P[i,1]M,\ldots P[i,p(i)]M, K^*\rangle\rangle.
\end{array}
$$

However,

$$
\begin{array}{l}
S \vdash H\langle\langle P[1,1],\ldots,P[1,p(1)]\rangle,\ \langle K, P[1,1]N,\ldots,P[1,p(1)N\rangle\rangle = \\
\quad H\langle\langle [i,1],\ldots,P[i,p(i)]\rangle\langle K, P[i,1]N,\ldots,P[i,p(i)]N\rangle\rangle \text{ and} \\
S \vdash H\langle\langle P[1,1],\ldots,P[1,p(1)]\rangle,\ \langle P[1,1]M,\ldots,P[1,p(1)]M, K^*\rangle\rangle = \\
\quad H\langle\langle P[i,1],\ldots,P[i,p(i)]\rangle\langle P[i,1]M,\ldots,P[i,p(i)]M, K^*\rangle\rangle \text{ thus} \\
S \vdash F\#\langle N, M\rangle \text{ (LN) } = \mathrm{F\#\#\langle N,M\rangle\ (JM)} \text{ and} \\
S \vdash M = N
\end{array}
$$

contradicting the choice of S. Thus $M = N$ is inevitably consistent. ■

Remark: It can be proved from Mitschke's theorem ([1]) that any inevitably consistent equation must contain a universal generator. This is indeed the case for our example.

The following theorem follows from theorem 3; however, it has a simpler proof.

Theorem 4: There exist inevitably consistently solvable functional equations.

Proof: We can restate Jacopini's theorem for the empty S as follows. $M = N$ is inconsistent with beta conversion $\Leftrightarrow$ there exists a combinator P of the form $\lambda a.\ aP[1]\ldots P[p]$ such that $B(C^*K^*)(PM)$ beta converts to $B(PN)(C^*K)$. Now by [4] there exists a combinator R such that RP beta converts to R if and only if P beta converts to the form $\lambda a.\ aP[1]\ldots P[p]$ for combinators $P[1],\ldots,P[p]$.

Thus the equations

$$(^*) \qquad Rx = R, \quad B(C^*K^*)(xM) = B(xN)(C^*K)$$

have a solution modulo beta conversion if and only if $M = N$ is inconsistent with beta conversion. Moreover, if S is a consistent set of combinator equations then $(^*)$ has a solution over S if $M = N$ is inconsistent with S. Hence the equations

$$(^{**}) \qquad Rx = R, \quad B(C^*K^*)(x(Omega)) = B(xy)(C^*K),$$

once the two variables are replaced by one variable through pairing, are inevitable. For , since Omega is easy there is no solution to the given equations in beta conversion alone. However, for each S there is an extension with a solution for either Omega is already inconsistent with each solvable term, or for one of them, say N, Omega $= N$ is consistent with S. In the extension S U $\{\text{Omega} = N\}$ then Omega is inconsistent in S U $\{\text{Omega} = N\}$ is inconsistent with some other solvable M by Bohm's theorem. In addition, if $N(N[1])\ldots(N[n])$ converts to I then set $L = Y(\lambda x.\ xN[1]\ldots N[n]M)$. Then Omega is inconsistent in S U $\{\text{Omega} = N\}$ with the unsolvable L. ■

4 Consistency Sets

Clearly if S is RE then the set of consistent solutions to $Mx = Nx$ is a co-Visseral ([5]) set. It is natural to ask if every co-Visseral set is representable in this manner as a "consistency set". By [5] it suffices to consider only co-Visseral sets of the form $\{P : P$ does not beta convert to $Q\ \}$.

Theorem 5: Let $Mx = Nx$ be given. Then there exists a combinator P not beta convertible to Omega such that either $M(\text{Omega}) = N(\text{Omega})$ is consistent or $MP = NP \Rightarrow M(\text{Omega}) = N(\text{Omega})$.

Proof: Suppose that $Mx = Nx$ is given and $M(\text{Omega}) = N(\text{Omega})$ is inconsistent. Then $M(\text{Omega})$ amd $N(\text{Omega})$ have beta eta distinct Bohm trees ([1]) page 504 and page 244). Without loss of generality we may assume that $M(\text{Omega})$ and $N(\text{Omega})$ are not separable. Thus $M(\text{Omega})$ and $N(\text{Omega})$ have reducts with equivalent subterms one of which is unsolvable and the other of which has a head normal form. Symmetrically assume that the unsolvable one is in a reduct of M. By the Bohm-out technique there exists a possibly open term X such that

$$X(Mx) \text{ beta converts to } \begin{cases} \text{no head normal form} \\ \lambda y[1]\ldots y[r].\ xY[1]\ldots Y[s] \end{cases}$$
$$X(Nx) \text{ beta converts to } \quad y.$$

Clearly we may assume that the second alternative for $X(Mx)$ does not occur. In addition we can arrange it so that $X(Mx)$ does not occur. In addition, we can arrange it so that $X(Mx)$ has the property

$X(Mx)$ either has infinite order or order zero.

By the fixed point theorem there exists a combinator P such that P beta converts to $(\lambda z.\ z((\lambda y.\ X(MP))z))(\lambda z.\ zz)$. By the standardization theorem P does not beta convert to Omega. However, whenever $MP = NP$ we have $P =$ Omega. This completes the proof.

Corollary: The set $\{P : P$ does not beta convert to Omega$\}$ is not a consistency set.

5 Finitely Consistently Solvable Sets of Equations

Definition: If S is a set of functional equations then S is said to be (effectively) finitely consistently solvable if there is a partial (recursive) function f defined on exactly the finite subsets of S such that if F is a finite subset of S then

$$\{M(f(F)) = N(f(F)) : Mx = Nx \text{ in } F\}$$

is consistent with beta conversion.

Remark: The effectiveness condition in the definition really has two parts

(a) S is RE
(b) constistent solutions can be computed for finite subsets. Next we show that neither of these restrictions can be relaxed.

Theorem 6: There exists a finitely consistently solvable set which is not consistently solvable.

Proof: We shall actually build two variations on the same example only one of which is RE. The RE example goes as follows,

For each combinator M we shall use two "local" variables y and z which actually depend on M. For each such combinator we take the equations

$$zx = K, \quad zM = yM, \quad yx = K^*.$$

Our example consists of all these equations with all the local variables replaces by one global variable through pairing. Clearly this set is not consistently solvable. However, for any finite subset corresponding to the combinators $M[1], \ldots, M[m]$ we can find a consistent solution as follows. Let M have a head normal form distinct from the head normal forms of the solvable members of $\{M[1], \ldots, M[m]\}$.

Let $N[1], \ldots, N[n]$ be such that

$$MN[1] \ldots N[n] \twoheadrightarrow I,$$

then for each of the sets

$$zx = K, \ zM[i] = yM[i], \ \ yx = K^*$$

we have the solution of M for x and

if $M[i]$ is solvable then there exists a Bohm-out term P such that PM beta converts to $K^*$and $PM[i]$ beta converts to K and put P for y and KK for z.

if $M[i]$ is unsolvable then there exists a fixed point P without head normal form such that P beta converts to $PN[1] \ldots N[n]K^*$. Put $\lambda x.\, xN[1] \ldots N[n]$ for y and I for z. This works in the Bohm tree model where all the unsolvable are equal; in particular $M[i] = P$.

Clearly computing the finite consistent solution requires determining the solvability of $M[i]$. Computing a finite consistent solution can be simplified by passing to a non-RE example. We keep the above equations for those terms M which are unsolvable and add the following for terms N in head normal form

$$yN = K, \ \ yx = K^*.$$

It should be clear how to solve for the variables in any finite subset of these equations. This completes the construction.

Theorem 7: If S is effectively finitely consistently solvable then S is consistently solvable.

Proof: Suppose that S is effectively finitely consistently solvable and the function f is as above. For each finite subset F of S define $T(F) = \{Mf(F) = Nf(F) : Mx = Nx$ belongs to $F\}$. By Visser's theorem 3.8 ([7]) there exists a combinator P such that for each finite subset F of S

$$T(F) \; U \; \{P = f(F)\} \text{ is consistent.}$$

Thus by the compactness theorem the set

$$\{MP = NP : Mx = Nx \text{ belongs to } S\}$$

is consistent. This completes the proof.

References

1. Barendregt, The Lambda Calculus, North Holland, 1984.
2. Berrarducci and Intrigila, Some new results on easy lambda terms in Dezani-Ciancaglini, Ronchi Della Rocha, and Venturini Zilli, A Collection of Contributions in Honor Corrado Bohm, Elsevier, 1993.
3. Jacopini, A condition for identifying two elements of whatever model of combinatory logic in Bohm, ed. *Lecture Notes in Computer Science*, **37**, Springer-Verlag, 1975.
4. Statman, On sets of solutions to combinator equations, *TCS*, **66**, p. 99-104.
5. Statman, Morphism and partitions of V-sets, Carnegie Mellon University, Department of Mathematical Sciences Research Report 97-210, September, 1997.
6. Statman, On the existence of n but not $n+1$ easy combinators, Carnegie Mellon University, Department of Mathematical Sciences Research Report 94-165, March, 1994.
7. Visser, Numerations, lambda calculus, and arithmetic in Hindley and Seldin, eds., To H. B. Curry: Essays on Combinatory Logic, Lambda Calculus, and Formalism, Academic Press, 1980.

Strong Normalisation of Cut-Elimination in Classical Logic

C. Urban and G.M. Bierman

University of Cambridge Computer Laboratory
{cu200,gmb}@cl.cam.ac.uk

Abstract. In this paper a strongly normalizing cut-elimination procedure is presented for classical logic. The procedure adapts the standard cut transformations, see for example [12]. In particular our cut-elimination procedure requires no special annotations on formulae. We design a term calculus for a variant of Kleene's sequent calculus G3 via the Curry-Howard correspondence and the cut-elimination steps are given as rewrite rules. In the strong normalization proof we adapt the symmetric reducibility candidates developed by Barbanera and Berardi.

1 Introduction

Gentzen has shown in his seminal paper [10] that all cuts can be eliminated from proofs in LK and LJ. Since then many *Hauptsätze* (cut-elimination theorems) have appeared for various sequent calculus formulations. Most of them, including Gentzen's original, provide a cut-elimination procedure which is weakly normalising, i.e., they employ a particular reduction strategy (for example an inner-most reduction strategy or the elimination of the cut with the highest rank). Besides these weakly normalising methods a few strongly normalising cut-elimination procedures have been developed; for example in [4–7, 13, 14]. However, all those methods impose some form of restriction on the reduction rules to ensure strong normalisation. A common restriction is to not allow a cut-rule to pass over another cut-rule (exceptions are [6, 13]). However this limits, in the intuitionistic case, the correspondence between cut-elimination and beta-reduction [8, 14]. Therefore in this paper we develop a strongly normalising cut-elimination procedure adapting the standard cut-elimination steps for *logical cuts* and allowing *commuting cuts* to pass over other cuts. (A cut-rule is said to be a logical cut when both cut-formulae are introduced by axioms or logical inference rules; otherwise the cut is said to be a commuting cut.) Our method is closely related to the cut-elimination procedure developed for LK^{tq} [6, 15]. However we do not need their colour annotations.

The problem of non-termination of cut-elimination occurs in both intuitionistic logic and classical logic. One example of a non-terminating reduction sequence in intuitionistic logic is given in [20]; for classical logic [6] and [9] give the following example:

$$\dfrac{\dfrac{\dfrac{\overline{A\vdash A}\quad\overline{A\vdash A}}{A\vee A\vdash A,A}\,\vee_L}{A\vee A\vdash A}\,Contr_R \qquad \dfrac{\dfrac{\overline{A\vdash A}\quad\overline{A\vdash A}}{A,A\vdash A\wedge A}\,\wedge_R}{A\vdash A\wedge A}\,Contr_L}{A\vee A\vdash A\wedge A}\,Cut$$

where a commuting cut needs to be eliminated. There are two possible reductions: either the cut can be permuted upwards in the left proof branch or in the right proof branch. If one is not careful, applying these reductions in alternation can lead to arbitrary big normal forms and to non-termination. This is remedied in [6] by devising a specific protocol for cut-elimination, which depends on additional information ('colours') attached to every cut-formula. For this cut-elimination procedure strong normalisation and confluence has been proved; the colours are used to ingeniously map every LK^{tq}-proof to a corresponding proof-net in linear logic and every cut-elimination step to a series of reductions on proof-nets (strong normalisation for proof-nets has been proved in [11]).

We shall consider a sequent calculus formulation very similar to Kleene's G3 [16] and G3c of [18], where the structural rules are completely implicit in the form of the logical rules. Another feature of our work is that we shall annotate proofs with terms and term rewrite rules will describe the cut-elimination steps. In our approach no additional information is required to guide the cut-elimination process. The rest of the paper is organised as follows: §2 contains various notational conventions and definitions; §3 contains a detailed proof of strong normalisation for the rewrite system. The proof adapts the technique of symmetric reducibility candidates [1]; §4 concludes and gives suggestions for further work.

2 Terms, Judgements, Rewrite Rules and Substitution

The main idea behind the cut-elimination procedure presented in this paper is to transport one subderivation of a commuting cut to the place(s) where the cut-formula is introduced. Consider the following proof in G3c:

$$\dfrac{\pi_1\left\{\dfrac{\overline{A\vdash B\supset C, A^\bullet} \quad \dfrac{\overline{A,B\vdash C,A^\bullet}}{A\vdash B\supset C,A}\supset_R}{A\vee A\vdash B\supset C,A}\vee_L\right. \qquad \left.\dfrac{\dfrac{\overline{A^\star\vdash D,A}\quad\overline{A^\star\vdash D,A}}{A\vdash D,A\wedge A}\wedge_R \quad \dfrac{\overline{A^\star,E\vdash A}\quad\overline{A^\star,E\vdash A}}{A,E\vdash A\wedge A}\wedge_R}{A,D\supset E\vdash A\wedge A}\supset_L\right\}\pi_2}{A\vee A, D\supset E\vdash B\supset C, A\wedge A}\mathit{Cut}$$

The cut-formula A is neither a main formula in the inference rule $\vee_L$, nor in $\supset_L$. Therefore the cut is a commuting cut. In π_1 the cut-formula is a main formula in the axioms marked with a bullet; in π_2, respectively, in the axioms marked with a star. Eliminating the cut in the proof above means to either transport the derivation π_2 to the places marked with a bullet and 'cut it against' the corresponding axioms, or to transport π_1 and 'cut it against' the axioms marked with a star. In both cases the derivation being transported is duplicated.

In the remainder of this section we shall annotate proofs, via the Curry-Howard correspondence, with terms and present a rewrite system for cut-elimination. The raw terms are defined in Figure 1 using *names* and *co-names* as binders. Besides the terms, which are going to be used as annotations for proofs, there are two other syntactic categories which play an important rle in the definition of substitution and in the strong normalisation proof. Let M and N be terms, then $(x{:}B)M$ and $\langle a{:}B\rangle N$ are called *named terms* and *co-named terms*,

Raw Terms: $M, N ::=$	$\mathsf{Ax}(x, a)$	Axiom	
|	$\mathsf{Cut}(\langle a{:}B\rangle M, (x{:}B)N)$	Cut	
|	$\mathsf{And}_R(\langle a{:}B\rangle M, \langle b{:}C\rangle N, c)$	And-R	
|	$\mathsf{And}_L^i((x{:}B)M, y)$	And-L$_i$	$(i = 1, 2)$
|	$\mathsf{Or}_R^i(\langle a{:}B\rangle M, b)$	Or-R$_i$	$(i = 1, 2)$
|	$\mathsf{Or}_L((x{:}B)M, (y{:}C)N, z)$	Or-L	
|	$\mathsf{Imp}_R((x{:}B)\langle a{:}C\rangle M, b)$	Imp-R	
|	$\mathsf{Imp}_L(\langle a{:}B\rangle M, (x{:}C)N, y)$	Imp-L	

Fig. 1. The grammar for the raw terms where B and C are are types; x, y, z are taken from a set of *names* and a, b, c from a set of *co-names*.

respectively. We use round brackets to signify that a name becomes bound in a term and angle brackets that a co-name becomes bound in a term. Analogous to the Church-style formation rules for the λ-calculus, all binders are explicitly typed (types are defined as normal). However in what follows we will omit these typings when they are clear from the context. Given a term M, its set of free names is written as $FN(M)$ and its set of free co-names is written as $FC(M)$ (similarly for named and co-named terms) – their routine definitions are omitted. We assume that the three types of terms are equal up to α-conversion and that a Barendregt-style naming convention holds for names and co-names (see 2.1.13 in [2]). Rewriting a name x to y in M is written as $M\{x \mapsto y\}$ (respectively $M\{a \mapsto b\}$ for co-names). The routine formalisation of the rewriting operation is omitted.

In the following we are only concerned with terms which can be well-typed by the inference system given in Figure 2. The typing judgements are of the form $\Gamma \triangleright M \triangleright \Delta$ where Γ is a set of name-type pairs and Δ is a set of co-name-type pairs. The reader will see that this system is the term system for a variant of Kleene's G3 formulation via the Curry-Howard correspondence. Our $\wedge_L$ and $\vee_R$ rules differ slightly from the G3 and G3c of [18]: they provide more convenience in the strong normalisation proof, but the original rules could be used as well (see Section 4). There are no primitive rules for contraction and weakening: they are completely implicit in the form of the logical rules. However, special care needs to be taken with implicit contractions. Consider the proof fragment:

$$\frac{x:B, \Gamma \triangleright M \triangleright \Delta, b:B\supset C, a:C}{\Gamma \triangleright \mathsf{Imp}_R((x)\langle a\rangle M, b) \triangleright \Delta, b:B\supset C}\ \supset_R \qquad (1)$$

The typing rule introduces the co-name-type pair $b : B\supset C$ in the conclusion. However it is allowed that this pair can already be present in the premise. On the other hand, the name-type pair $x:B$ and the co-name-type pair $a:C$ in the premise are not allowed to be in the conclusion: they become bound in the term.

The following definition corresponds to the traditional notion of what the main formula of a inference rule is.

$$x{:}B,\Gamma \triangleright \mathsf{Ax}(x,a) \triangleright \Delta, a{:}B$$

$$\frac{x{:}B_i,\Gamma \triangleright M \triangleright \Delta}{y{:}B_1\wedge B_2,\Gamma \triangleright \mathsf{And}^i_L((x) M,y) \triangleright \Delta}\ \wedge_{L_i} \qquad \frac{\Gamma \triangleright M \triangleright \Delta, a{:}B \quad \Gamma \triangleright N \triangleright \Delta, b{:}C}{\Gamma \triangleright \mathsf{And}_R(\langle a\rangle M,\langle b\rangle N,c) \triangleright \Delta, c{:}B\wedge C}\ \wedge_R$$

$$\frac{x{:}B,\Gamma \triangleright M \triangleright \Delta \quad y{:}C,\Gamma \triangleright N \triangleright \Delta}{z{:}B\vee C,\Gamma \triangleright \mathsf{Or}_L((x)M,(y)N,z) \triangleright \Delta}\ \vee_L \qquad \frac{\Gamma \triangleright M \triangleright \Delta, a{:}B_i}{\Gamma \triangleright \mathsf{Or}^i_R(\langle a\rangle M,b) \triangleright \Delta, b{:}B_1\vee B_2}\ \vee_{R_i}$$

$$\frac{\Gamma \triangleright M \triangleright \Delta, a{:}B \quad x{:}C,\Gamma \triangleright N \triangleright \Delta}{y{:}B\supset C,\Gamma \triangleright \mathsf{Imp}_L(\langle a\rangle M,(x)N,y) \triangleright \Delta}\ \supset_L \qquad \frac{x{:}B,\Gamma \triangleright M \triangleright \Delta, a{:}C}{\Gamma \triangleright \mathsf{Imp}_R((x)\langle a\rangle M,b) \triangleright \Delta, b{:}B\supset C}\ \supset_R$$

$$\frac{\Gamma_1 \triangleright M \triangleright \Delta_1, a{:}B \quad x{:}B,\Gamma_2 \triangleright N \triangleright \Delta_2}{\Gamma_1,\Gamma_2 \triangleright \mathsf{Cut}(\langle a\rangle M,(x)N) \triangleright \Delta_1,\Delta_2}\ Cut$$

Fig. 2. The typing rules for the propositional fragment.

Definition 1.
A term M *introduces* the name z or co-name c if M is of the form:

for z: $\mathsf{Ax}(z,c)$	for c: $\mathsf{Ax}(z,c)$
$\mathsf{And}^i_L((x)S,z)$	$\mathsf{And}_R(\langle a\rangle S,\langle b\rangle T,c)$
$\mathsf{Or}_L((x)S,(y)T,z)$	$\mathsf{Or}^i_R(\langle a\rangle S,c)$
$\mathsf{Imp}_L(\langle a\rangle S,(x)T,z)$	$\mathsf{Imp}_R((x)\langle a\rangle S,c)$

Recall our example from the beginning of this section where a commuting cut can be permuted in two different directions. Therefore the rewrite system for our cut-elimination procedure is defined using two, symmetric forms of substitution, which are written as $P[x := \langle a\rangle Q]$ and $S[b := (y)T]$. These substitutions are used when the inference rules directly above the cut do not introduce the cut-formula. In these cases the cuts can permute, or 'jump' directly to the place(s) where the cut-formula is introduced (i.e., is a main formula). Whenever a substitution 'hits' a term where the cut-formula is introduced the substitution 'expands' to a cut. Two examples are as follows:

$$\mathsf{And}_R(\langle a\rangle M,\langle b\rangle N,c)[c := (x)P] \stackrel{\text{def}}{=} \mathsf{Cut}(\langle c\rangle \mathsf{And}_R(\langle a\rangle M,\langle b\rangle N,c),(x)P)$$
$$\mathsf{Ax}(x,a)[x := \langle b\rangle Q] \stackrel{\text{def}}{=} \mathsf{Cut}(\langle b\rangle Q,(x)\mathsf{Ax}(x,a))$$

In the first term the formula labelled with c is the main formula and in the second the formula labelled with x is a main formula. So in both cases the substitution expands to a cut. In the other cases where the name or co-name that is substituted is not a label for the main formula, then the substitution is pushed into the subterms or vanishes in case of the axioms. Two examples are as follows (assume the substitution $[\sigma]$ is *not* of the form $[z := \ldots]$ or $[a := \ldots]$):

$$\mathsf{Or}_L((x)M,(y)N,z)[\sigma] \stackrel{\text{def}}{=} \mathsf{Or}_L((x)\ M[\sigma],(y)\ N[\sigma],z)$$
$$\mathsf{Ax}(z,a)[\sigma] \stackrel{\text{def}}{=} \mathsf{Ax}(z,a)$$

However, special care needs to be taken for axioms, because they have two main formulae. For technical reasons in the strong normalisation proof we need the following property:

$$M[x := \langle a \rangle P][b := (y)Q] \equiv M[b := (y)Q][x := \langle a \rangle P] \tag{2}$$

if $b \notin FC(\langle a \rangle P)$ and $x \notin FN((y)Q)$. The nave definition outlined above does not satisfy this property: in case M is of the form $\mathsf{Ax}(x, b)$ we get two different terms:

$$\begin{aligned}
\mathsf{Ax}(x,b)[x := \langle a \rangle P][b := (y)Q] &\stackrel{\text{def}}{=} \mathsf{Cut}(\langle a \rangle P, (x)\mathsf{Cut}(\langle b \rangle \mathsf{Ax}(x,b), (y)Q)) \\
\mathsf{Ax}(x,b)[b := (y)Q][x := \langle a \rangle P] &\stackrel{\text{def}}{=} \mathsf{Cut}(\langle b \rangle \mathsf{Cut}(\langle a \rangle P, (x)\mathsf{Ax}(x,b)), (y)Q)
\end{aligned}$$

Furthermore the nested cuts with an axiom as an immediate subterm could be a source for non-termination as noted in [6]. Therefore we use a more subtle definition of substitution and introduce two special clauses to handle the problematic example above.

Definition 2. Substitution

$$\begin{aligned}
\mathsf{Cut}(\langle a \rangle \mathsf{Ax}(x,a), (y)M)[x := \langle b \rangle P] &\stackrel{\text{def}}{=} \mathsf{Cut}(\langle b \rangle P, (x)M\{y \mapsto x\}) \\
\mathsf{Cut}(\langle a \rangle M, (x)\mathsf{Ax}(x,b))[b := (y)P] &\stackrel{\text{def}}{=} \mathsf{Cut}(\langle b \rangle M\{a \mapsto b\}, (y)P) \\
M[c := (y)P] &\stackrel{\text{def}}{=} \mathsf{Cut}(\langle c \rangle M, (y)P) \quad \text{if } M \text{ introduces } c \\
M[y := \langle c \rangle P] &\stackrel{\text{def}}{=} \mathsf{Cut}(\langle c \rangle P, (y)M) \quad \text{if } M \text{ introduces } y
\end{aligned}$$

otherwise

$$\begin{aligned}
\mathsf{Ax}(x,a)[\sigma] &\stackrel{\text{def}}{=} \mathsf{Ax}(x,a) \\
\mathsf{Cut}(\langle a \rangle M, (x)N)[\sigma] &\stackrel{\text{def}}{=} \mathsf{Cut}(\langle a \rangle\ M[\sigma], (x)\ N[\sigma]) \\
\mathsf{And}_R(\langle a \rangle M, \langle b \rangle N, c)[\sigma] &\stackrel{\text{def}}{=} \mathsf{And}_R(\langle a \rangle\ M[\sigma], \langle b \rangle\ N[\sigma], c) \\
\mathsf{And}_L^i((x)M, y)[\sigma] &\stackrel{\text{def}}{=} \mathsf{And}_L^i((x)\ M[\sigma], y) \\
\mathsf{Or}_R^i(\langle a \rangle M, b)[\sigma] &\stackrel{\text{def}}{=} \mathsf{Or}_R^i(\langle a \rangle\ M[\sigma], b) \\
\mathsf{Or}_L((x)M, (y)N, z)[\sigma] &\stackrel{\text{def}}{=} \mathsf{Or}_L((x)\ M[\sigma], (y)\ N[\sigma], z) \\
\mathsf{Imp}_R((x)\langle a \rangle M, b)[\sigma] &\stackrel{\text{def}}{=} \mathsf{Imp}_R((x)\langle a \rangle\ M[\sigma], b) \\
\mathsf{Imp}_L(\langle a \rangle M, (x)N, y)[\sigma] &\stackrel{\text{def}}{=} \mathsf{Imp}_L(\langle a \rangle\ M[\sigma], (x)\ N[\sigma], y)
\end{aligned}$$

Recall that we assumed a Barendregt-style naming condition for (co-)names. A substitution $M[a := (x{:}B)N]$ is said to be *well-formed*, iff $\mathsf{Cut}(\langle a{:}B \rangle M, (x{:}B)N)$ is well-typed. In the following we shall consider only well-formed substitutions.

A nave translation of the traditional, logical cut-elimination rules into our term calculus is, for example, as follows ($\wedge_1$ case):

$$\mathsf{Cut}(\langle c \rangle \mathsf{And}_R(\langle a \rangle M, \langle b \rangle N, c), (y)\mathsf{And}_L^1((x)P, y)) \longrightarrow \mathsf{Cut}(\langle a \rangle M, (x)P)$$

However, there is a problem with this reduction rule. In our sequent calculus, the structural rules are implicit (see the discussion of proof (1)). This makes the calculus smaller, and more importantly it provides a very convenient way to define substitution (no explicit contractions are required when a term is duplicated). Unfortunately, we have to pay a price for this in the logical cut-elimination rules. Consider the following instance of the redex above:

$$\dfrac{\dfrac{\Gamma_1 \triangleright M \triangleright \Delta_1, c\!:\!B\wedge C, a\!:\!B \quad \Gamma_1 \triangleright N \triangleright \Delta_1, b\!:\!C}{\Gamma_1 \triangleright \mathsf{And}_R(\langle a\rangle M, \langle b\rangle N, c) \triangleright \Delta_1, c\!:\!B\wedge C}\wedge_R \quad \dfrac{x\!:\!B, \Gamma_2 \triangleright P \triangleright \Delta_2}{y\!:\!B\wedge C, \Gamma_2 \triangleright \mathsf{And}^1_L(\langle x\rangle P, y) \triangleright \Delta_2}\wedge_{L_1}}{\Gamma_1, \Gamma_2 \triangleright \mathsf{Cut}(\langle c\rangle \mathsf{And}_R(\langle a\rangle M, \langle b\rangle N, c), \langle y\rangle \mathsf{And}^1_L(\langle x\rangle P, y)) \triangleright \Delta_1, \Delta_2}\,Cut$$

where $c\!:B\wedge C \in FC(M)$. The nave reduction rule given above would (incorrectly!) reduce this proof to the following:

$$\dfrac{\Gamma_1 \triangleright M \triangleright \Delta_1, c\!:\!B\wedge C, a\!:\!B \quad x\!:\!B, \Gamma_2 \triangleright P \triangleright \Delta_2}{\Gamma_1, \Gamma_2 \triangleright \mathsf{Cut}(\langle a\rangle M, \langle x\rangle P) \triangleright \Delta_1, \Delta_2, c\!:\!B\wedge C}\,Cut$$

Unfortunately c has now become free! In order to obtain a subject reduction property for the rewrite system we have to include in every logical reduction step extra substitutions (the main formula of the conclusion could potentially be in every subterm). These substitutions ensure that no bound (co-)name becomes free. In effect the logical reduction rules look slightly complicated, but that is the price we have to pay for the convenience of not having explicit structural rules. The cut-elimination procedure is defined (in its entirety) as follows:

Definition 3. Cut-Elimination

Logical Cuts ($i = 1, 2$)

1. $\mathsf{Cut}(\langle b\rangle \mathsf{And}_R(\langle a_1\rangle M_1, \langle a_2\rangle M_2, b), \langle y\rangle \mathsf{And}^i_L(\langle x\rangle N, y))$
   $\longrightarrow \mathsf{Cut}(\langle a_i\rangle M_i[b := \langle y\rangle \mathsf{And}^i_L(\langle x\rangle N, y)], \langle x\rangle N[y := \langle b\rangle \mathsf{And}_R(\langle a_1\rangle M_1, \langle a_2\rangle M_2, b)])$
2. $\mathsf{Cut}(\langle b\rangle \mathsf{Or}^i_R(\langle a\rangle M, b), \langle y\rangle \mathsf{Or}_L(\langle x_1\rangle N_1, \langle x_2\rangle N_2, y))$
   $\longrightarrow \mathsf{Cut}(\langle a\rangle M[b := \langle y\rangle \mathsf{Or}_L(\langle x_1\rangle N_1, \langle x_2\rangle N_2, y)], \langle x_i\rangle N_i[y := \langle b\rangle \mathsf{Or}^i_R(\langle a\rangle M, b)])$
3. $\mathsf{Cut}(\langle b\rangle \mathsf{Imp}_R(\langle x\rangle\langle a\rangle M, b), \langle z\rangle \mathsf{Imp}_L(\langle c\rangle N, \langle y\rangle P, z))$
   $\longrightarrow \mathsf{Cut}(\langle a\rangle \mathsf{Cut}(\langle c\rangle N[z := \langle b\rangle S], \langle x\rangle M[b := \langle z\rangle T]), \langle y\rangle P[z := \langle b\rangle S])$ or
   $\longrightarrow \mathsf{Cut}(\langle c\rangle N[z := \langle b\rangle S], \langle x\rangle \mathsf{Cut}(\langle a\rangle M[b := \langle z\rangle T], \langle y\rangle P[z := \langle b\rangle S]))$
   where $S \equiv \mathsf{Imp}_R(\langle x\rangle\langle a\rangle M, b)$ and $T \equiv \mathsf{Imp}_L(\langle c\rangle N, \langle y\rangle P, z)$
4. $\mathsf{Cut}(\langle a\rangle M, \langle x\rangle \mathsf{Ax}(x, b)) \longrightarrow M\{a \mapsto b\}$ if M introduces a
5. $\mathsf{Cut}(\langle a\rangle \mathsf{Ax}(y, a), \langle x\rangle M) \longrightarrow M\{x \mapsto y\}$ if M introduces x

Commuting Steps (otherwise)

6. $\mathsf{Cut}(\langle a\rangle M, \langle x\rangle N) \longrightarrow M[a := \langle x\rangle N]$ if M does not introduce a or
   $\longrightarrow N[x := \langle a\rangle M]$ if N does not introduce x

There are a few subtleties in the reduction rule for the third case. Firstly, there are two ways to reduce a cut-rule having an implication as the cut-formula. Therefore we have included two reductions for this case. Secondly, special care needs to be taken that there is no clash between bound and free (co-)names. In the first reduction rule we need to ensure that a is not a free co-name in N; in the second rule that x is not free in P. This can always be achieved by renaming a and x appropriately (they are binders in $\mathsf{Imp}_R(\langle x\rangle\langle a\rangle M, b)$). We assume that the renaming is done implicitly in the cut-elimination procedure.

The main difference between our rules and the cut-elimination procedure defined for LK^{tq} is the inclusion of non-determinism. Recall our example from the beginning of this section where a commuting cut can move in two directions. Let $\mathsf{Cut}(\langle a\rangle M, (x)N)$ be the term annotation for this commuting cut where M and N are the corresponding term annotations for proofs π_1 and π_2, respectively. According to our last rule, this term can reduce to either $M[a := (x)N]$ or $N[x := \langle a\rangle M]$. The choice to which term it reduces is not specified (similarly for the reduction of the logical cut in the third case). In contrast, in LK^{tq} this choice is completely determined by the colour annotation. In general the colour annotation reduces the number of normal forms (cut-free proofs) reachable from a proof containing cuts (see §4 for an example). For the substitution we have the following lemmas:

Lemma 1.
(i) $M[x := \langle a\rangle \mathsf{Ax}(y,a)] \longrightarrow^{+} M\{x \mapsto y\}$ *or* $M[x := \langle a\rangle \mathsf{Ax}(y,a)] \equiv M$
(ii) $M[a := (x)\mathsf{Ax}(x,b)] \longrightarrow^{+} M\{a \mapsto b\}$ *or* $M[a := (x)\mathsf{Ax}(x,b)] \equiv M$

Proof. Routine induction on the structure of M.

Lemma 2. *For any arbitrary substitution* $[\sigma]$

if $M \longrightarrow M'$, *then* $M[\sigma] \longrightarrow M'[\sigma]$ *or* $M[\sigma] \equiv M'[\sigma]$

Proof. Induction on the structure of M. One interesting case is where $M[\sigma] \equiv M'[\sigma]$; it is as follows:

Case $M \equiv \mathsf{Cut}(\langle a\rangle \mathsf{Ax}(y,a), (x)P)$: Let P introduce x, then $M \longrightarrow M'$ with $M' \equiv P\{x \mapsto y\}$. Let $[\sigma]$ be $[y := \langle c\rangle Q]$. We have:

$$M[\sigma] \equiv \mathsf{Cut}(\langle a\rangle \mathsf{Ax}(y,a), (x)P)[y := \langle c\rangle Q] \stackrel{\text{def}}{=} \mathsf{Cut}(\langle c\rangle Q, (y)P\{x \mapsto y\})$$
$$M'[\sigma] \equiv P\{x \mapsto y\}[y := \langle c\rangle Q] \stackrel{\text{def}}{=} \mathsf{Cut}(\langle c\rangle Q, (y)P\{x \mapsto y\})$$

3 Proof of Strong Normalisation

We give in this section a detailed proof of strong normalisation for the reduction system developed in the previous section. To save space only details for the $\wedge$-fragment are presented, but some pointers are given at the end of this section for the other connectives. The proof uses the notion of symmetric reducibility candidates from [1]. The proof proceeds as follows:

1. Define the sets of candidates over types using a fixed point construction.
2. Prove that candidates are closed under reduction.
3. Show that a named or co-named term in a candidate implies strong normalisation for the corresponding term.
4. Prove that all terms are strongly normalising.

The set SN denotes the set of strongly normalising terms. The candidates are defined only for named and co-named terms. We say that $\langle B\rangle$ is the *type* of co-named terms of the form $\langle a{:}B\rangle M$; similarly (B) is the type of named terms of the form $(x{:}B)M$. We define:

1. $CT_{\langle B\rangle}$ is the set of co-named terms of type $\langle B\rangle$,
2. $NT_{(B)}$ is the set of named terms of type (B).

In the following we define for every type $\langle B\rangle$ and (B) the candidates, written as $\mathrm{J}\langle B\rangle\mathrm{K}$ and $\mathrm{J}(B)\mathrm{K}$; they are subsets of $CT_{\langle B\rangle}$ and $NT_{(B)}$, respectively. The definition of the candidates uses set operators for which we define the types as follows (where the set of all subsets of a given set S will be denoted as $\mathcal{P}(S)$):

$$\begin{array}{l}
\textsc{andright}_{\langle B\wedge C\rangle} : \mathcal{P}(CT_{\langle B\rangle}) \times \mathcal{P}(CT_{\langle C\rangle}) \times \mathcal{P}(NT_{(B\wedge C)}) \rightarrow \mathcal{P}(CT_{\langle B\wedge C\rangle})\\
\textsc{andleft}^{i}_{(B_1\wedge B_2)} : \mathcal{P}(NT_{(B_i)}) \times \mathcal{P}(CT_{\langle B_1\wedge B_2\rangle}) \rightarrow \mathcal{P}(NT_{(B_1\wedge B_2)})\\
\textsc{binding}_{(B)} : \mathcal{P}(CT_{\langle B\rangle}) \rightarrow \mathcal{P}(NT_{(B)})\\
\textsc{binding}_{\langle B\rangle} : \mathcal{P}(NT_{(B)}) \rightarrow \mathcal{P}(CT_{\langle B\rangle})\\
\textsc{neg}_{(B)} : \mathcal{P}(CT_{\langle B\rangle}) \rightarrow \mathcal{P}(NT_{(B)})\\
\textsc{neg}_{(B)} : \mathcal{P}(CT_{\langle B\rangle}) \rightarrow \mathcal{P}(NT_{(B)})
\end{array}$$

The operators are indexed on types. When defining the set operators we use the following two sets of named and co-named axioms:

$$\begin{array}{l}
\textsc{axioms}_{(B)} \stackrel{\text{def}}{=} \{(x{:}B)\mathsf{Ax}(y,b) \mid \text{for all } \mathsf{Ax}(y,b)\} \subseteq NT_{(B)}\\
\textsc{axioms}_{\langle B\rangle} \stackrel{\text{def}}{=} \{\langle a{:}B\rangle\mathsf{Ax}(y,b) \mid \text{for all } \mathsf{Ax}(y,b)\} \subseteq CT_{\langle B\rangle}
\end{array}$$

The set operators $\textsc{andright}$, $\textsc{andleft}^{i}$ and $\textsc{binding}$ are defined as follows:

$$\begin{array}{rl}
\textsc{andright}_{\langle B\wedge C\rangle}(X,Y,Z) \stackrel{\text{def}}{=} & \{\langle c{:}B\wedge C\rangle\mathsf{And}_R(\langle a{:}B\rangle M, \langle b{:}C\rangle N, c) \mid\\
& \forall\,(x{:}B\wedge C)P \in Z.\ \langle a\rangle\ M[c := (x)P] \in X \text{ and } \langle b\rangle\ N[c := (x)P] \in Y\}\\
\textsc{andleft}^{i}_{(B_1\wedge B_2)}(X,Y) \stackrel{\text{def}}{=} & \{(y{:}B_1\wedge B_2)\mathsf{And}^{i}_{L}((x{:}B_i)M, y) \mid\\
& \forall\,\langle a{:}B_1\wedge B_2\rangle P \in Y.\ (x)\ M[y := \langle a\rangle P] \in X\}\\
\textsc{binding}_{(B)}(X) \stackrel{\text{def}}{=} & \{(x{:}B)M \mid \forall\langle a{:}B\rangle P \in X.\ M[x := \langle a{:}B\rangle P] \in SN\}\\
\textsc{binding}_{\langle B\rangle}(Y) \stackrel{\text{def}}{=} & \{\langle a{:}B\rangle M \mid \forall(x{:}B)P \in Y.\ M[a := (x{:}B)P] \in SN\}
\end{array}$$

The set operator $\textsc{neg}$ and the candidates $\mathrm{J}(B)\mathrm{K}$ and $\mathrm{J}\langle B\rangle\mathrm{K}$ are defined simultaneously over types:

$$\begin{array}{lll}
\textsc{neg}_{\langle B\rangle}(X) & \stackrel{\text{def}}{=} \textsc{axioms}_{\langle B\rangle} \cup \textsc{binding}_{\langle B\rangle}(X) & \langle B\rangle \text{ atomic}\\
& \stackrel{\text{def}}{=} \textsc{axioms}_{\langle C\wedge D\rangle} \cup \textsc{binding}_{\langle C\wedge D\rangle}(X) \cup & \langle B\rangle \equiv \langle C\wedge D\rangle\\
& \quad \textsc{andright}_{\langle C\wedge D\rangle}(\mathrm{J}\langle C\rangle\mathrm{K}, \mathrm{J}\langle D\rangle\mathrm{K}, X) & \\
\textsc{neg}_{(B)}(Y) & \stackrel{\text{def}}{=} \textsc{axioms}_{(B)} \cup \textsc{binding}_{(B)}(Y) & (B) \text{ atomic}\\
& \stackrel{\text{def}}{=} \textsc{axioms}_{(C\wedge D)} \cup \textsc{binding}_{(C\wedge D)}(Y) \cup & (B) \equiv (C\wedge D)\\
& \quad \textsc{andleft}^{1}_{(C\wedge D)}(\mathrm{J}(C)\mathrm{K}, Y) \cup \textsc{andleft}^{2}_{(C\wedge D)}(\mathrm{J}(D)\mathrm{K}, Y) &
\end{array}$$

For the definition of the candidates we use fixed points of an increasing set operator. A set operator *op* is said to be:

$$\begin{array}{ll} \text{increasing, iff} & S \subseteq S' \Rightarrow op(S) \subseteq op(S'), \text{ and} \\ \text{decreasing, iff} & S \subseteq S' \Rightarrow op(S) \supseteq op(S'). \end{array}$$

The candidates are defined as follows:

$$\mathrm{J}(B)\mathrm{K} \stackrel{\text{def}}{=} X_0 \text{ and } \mathrm{J}\langle B\rangle\mathrm{K} \stackrel{\text{def}}{=} \mathrm{NEG}_{\langle B\rangle}(\mathrm{J}(B)\mathrm{K})$$

where X_0 is the least fixed point of the operator $\mathrm{NEG}_{(B)} \circ \mathrm{NEG}_{\langle B\rangle}$.[1] We have that $\mathrm{BINDING}_{\langle B\rangle}$ and $\mathrm{ANDRIGHT}_{\langle C\wedge D\rangle}$ (i.e., $X \mapsto \mathrm{ANDRIGHT}_{\langle C\wedge D\rangle}(\mathrm{J}\langle C\rangle\mathrm{K}, \mathrm{J}\langle D\rangle\mathrm{K}, X)$) are decreasing operators. But then $\mathrm{NEG}_{\langle B\rangle}$ must be a decreasing operator (similarly $\mathrm{NEG}_{(B)}$ must be decreasing). If both $\mathrm{NEG}_{\langle B\rangle}$ and $\mathrm{NEG}_{(B)}$ are decreasing, then the operator $\mathrm{NEG}_{(B)} \circ \mathrm{NEG}_{\langle B\rangle}$ is increasing and the least fixed point X_0 exists according to Tarski's fixed point theorem. For the candidates we have:

$$\mathrm{J}(B)\mathrm{K} = \mathrm{NEG}_{(B)}(\mathrm{J}\langle B\rangle\mathrm{K}) \text{ and } \mathrm{J}\langle B\rangle\mathrm{K} = \mathrm{NEG}_{\langle B\rangle}(\mathrm{J}(B)\mathrm{K}).$$

Since NEG is closed under AXIOMS we also have have:

$$\mathrm{AXIOMS}_{(B)} \subseteq \mathrm{J}(B)\mathrm{K} \text{ and } \mathrm{AXIOMS}_{\langle B\rangle} \subseteq \mathrm{J}\langle B\rangle\mathrm{K}. \tag{3}$$

Lemma 3.

(i) If $\langle a{:}B\rangle M \in \mathrm{J}\langle B\rangle\mathrm{K}$ *and* $M \longrightarrow M'$ *then* $\langle a{:}B\rangle M' \in \mathrm{J}\langle B\rangle\mathrm{K}$.
(ii) If $(x{:}B)M \in \mathrm{J}(B)\mathrm{K}$ *and* $M \longrightarrow M'$ *then* $(x{:}B)M' \in \mathrm{J}(B)\mathrm{K}$.

Proof. We prove both cases simultaneously by induction on $\langle B\rangle$ and (B).

Case $\langle B\rangle$ atomic: For (i) we have $\mathrm{J}\langle B\rangle\mathrm{K} = \mathrm{NEG}_{\langle B\rangle}(\mathrm{J}(B)\mathrm{K})$; therefore $\langle a{:}B\rangle M \in \mathrm{AXIOMS}_{\langle B\rangle} \cup \mathrm{BINDING}_{\langle B\rangle}(\mathrm{J}(B)\mathrm{K})$. M cannot be an axiom (because axioms do not reduce), therefore $\langle a{:}B\rangle M \in \mathrm{BINDING}_{\langle B\rangle}(\mathrm{J}(B)\mathrm{K}) \stackrel{\text{def}}{=} \{\langle a{:}B\rangle S \mid \forall (x{:}B)T \in \mathrm{J}(B)\mathrm{K}. S[a := (x{:}B)T] \in SN\}$. For $\langle a{:}B\rangle M$ we have $M[a := (x{:}B)P] \in SN$ for all $(x{:}B)P \in \mathrm{J}(B)\mathrm{K}$ and since $M \longrightarrow M'$ we know by Lemma 2 that either $M[a := (x)P] \longrightarrow M'[a := (x)P]$ or $M[a := (x)P] \equiv M'[a := (x)P]$. In both cases we have $M'[a := (x{:}B)P] \in SN$ for all $(x{:}B)P \in \mathrm{J}(B)\mathrm{K}$. This implies that $\langle a{:}B\rangle M' \in \mathrm{BINDING}_{\langle B\rangle}(\mathrm{J}(B)\mathrm{K})$ and hence $\langle a{:}B\rangle M' \in \mathrm{NEG}_{\langle B\rangle}(\mathrm{J}(B)\mathrm{K})$. Therefore $\langle a{:}B\rangle M' \in \mathrm{J}\langle B\rangle\mathrm{K}$. Similarly for (ii).

Case $\langle B\rangle \equiv \langle C\wedge D\rangle$: $\langle a{:}C\wedge D\rangle M$ is element of $\mathrm{J}\langle C\wedge D\rangle\mathrm{K} = \mathrm{NEG}_{\langle C\wedge D\rangle}(\mathrm{J}(C\wedge D)\mathrm{K}) \stackrel{\text{def}}{=} \mathrm{AXIOMS}_{\langle C\wedge D\rangle} \cup \mathrm{BINDING}_{\langle C\wedge D\rangle}(\mathrm{J}(C\wedge D)\mathrm{K}) \cup \mathrm{ANDRIGHT}_{\langle C\wedge D\rangle}(\mathrm{J}\langle C\rangle\mathrm{K}, \mathrm{J}\langle D\rangle\mathrm{K}, \mathrm{J}(D\wedge C)\mathrm{K})$. $\langle a{:}C\wedge D\rangle M \notin \mathrm{AXIOMS}_{\langle C\wedge D\rangle}$, because axioms do not reduce. Therefore we have that $\langle a{:}C\wedge D\rangle M \in \mathrm{ANDRIGHT}_{\langle C\wedge D\rangle}(\mathrm{J}\langle C\rangle\mathrm{K}, \mathrm{J}\langle D\rangle\mathrm{K}, \mathrm{J}(C\wedge D)\mathrm{K})$ or that $\langle a{:}C\wedge D\rangle M \in \mathrm{BINDING}_{\langle C\wedge D\rangle}(\mathrm{J}(C\wedge D)\mathrm{K})$. In the second case we reason as in the atomic case. In the first case we know that $\langle a\rangle M$ is of the form

[1] In all rigour we also have to assume that the candidates are closed under α-conversion.

$\langle c{:}C\wedge D\rangle \mathsf{And}_R(\langle d\rangle S, \langle e\rangle T, c)$ and $\langle a\rangle M' \equiv \langle c{:}C\wedge D\rangle \mathsf{And}_R(\langle d\rangle S', \langle e\rangle T', c)$ where either $S \longrightarrow S'$ and $T \equiv T'$ or $S \equiv S'$ and $T \longrightarrow T'$. Assume the former case (the other case being similar). We have that $\langle d{:}C\rangle S[c := (x)P] \in \mathrm{J}\langle C\rangle\mathrm{K}$ for all $(x{:}C\wedge D)P \in \mathrm{J}(C\wedge D)\mathrm{K}$. Since $S \longrightarrow S'$ we know by Lemma 2 that either $S[c := (x)P] \equiv S'[c := (x)P]$ or $S[c := (x)P] \longrightarrow S'[c := (x)P]$. In both cases (in the second by IH) we can infer that $\langle d\rangle S'[c := (x)P] \in \mathrm{J}(C)\mathrm{K}$ for all $(x{:}C\wedge D)P \in \mathrm{J}(C\wedge D)\mathrm{K}$. Therefore we know that $\langle a{:}C\wedge D\rangle M'$ must be in $\text{ANDRIGHT}_{\langle C\wedge D\rangle}(\mathrm{J}\langle C\rangle\mathrm{K}, \mathrm{J}\langle D\rangle\mathrm{K}, \mathrm{J}(C\wedge D)\mathrm{K})$ and we can conclude that $\langle a{:}C\wedge D\rangle M' \in \mathrm{J}\langle C\wedge D\rangle\mathrm{K}$. Similarly for (ii).

Lemma 4.

(i) If $\langle a{:}B\rangle M \in \mathrm{J}\langle B\rangle\mathrm{K}$, *then* $M \in SN$.
(ii) If $(x{:}B)M \in \mathrm{J}(B)\mathrm{K}$, *then* $M \in SN$.

Proof. Simultaneous induction on the types $\langle B\rangle$ and (B).

Case $\langle B\rangle$ atomic: Since $\mathrm{J}\langle B\rangle\mathrm{K} = \text{NEG}_{\langle B\rangle}(\mathrm{J}(B)\mathrm{K})$ we have $\langle a{:}B\rangle M \in \text{AXIOMS}_{\langle B\rangle}$ or $\langle a{:}B\rangle M \in \text{BINDING}_{\langle B\rangle}(\mathrm{J}(B)\mathrm{K})$. In the first case M is an axiom and therefore strongly normalising. In the second case we know that $M[a := (x{:}B)P] \in SN$ for all $(x{:}B)P \in \mathrm{J}(B)\mathrm{K}$. By (3) we have $(x{:}B)\mathsf{Ax}(x,a) \in \mathrm{J}(B)\mathrm{K}$ and therefore $M[a := (x)\mathsf{Ax}(x,a)] \in SN$. Furthermore we know by Lemma 2 that either $M[a := (x)\mathsf{Ax}(x,a)] \equiv M$ or $M[a := (x)\mathsf{Ax}(x,a)] \longrightarrow^+ M$. Therefore $M \in SN$. Similarly for (ii).

Case $\langle B\rangle \equiv \langle C\wedge D\rangle$: By $\mathrm{J}\langle C\wedge D\rangle\mathrm{K} = \text{NEG}_{\langle C\wedge D\rangle}(\mathrm{J}(C\wedge D)\mathrm{K})$ we have that:

$$\langle a{:}C\wedge D\rangle M \in \text{AXIOMS}_{\langle C\wedge D\rangle} \cup \text{BINDING}_{\langle C\wedge D\rangle}(\mathrm{J}(C\wedge D)\mathrm{K}) \cup \text{ANDRIGHT}_{\langle C\wedge D\rangle}(\mathrm{J}\langle C\rangle\mathrm{K}, \mathrm{J}\langle D\rangle\mathrm{K}, \mathrm{J}(C\wedge D)\mathrm{K})$$

If $\langle a{:}C\wedge D\rangle M$ is element of the first two sets we reason as in the atomic case. Left to show is that $M \in SN$ if $\langle a\rangle M \in \text{ANDRIGHT}_{\langle C\wedge D\rangle}(\mathrm{J}\langle C\rangle\mathrm{K}, \mathrm{J}\langle D\rangle\mathrm{K}, \mathrm{J}(C\wedge D)\mathrm{K})$. In this case $\langle a\rangle M$ is of the form $\langle c\rangle \mathsf{And}_R(\langle d\rangle S, \langle e\rangle T, c)$ where $\langle d\rangle S[c := (x)P] \in \mathrm{J}\langle C\rangle\mathrm{K}$ and $\langle e\rangle T[c := (x)P] \in \mathrm{J}\langle D\rangle\mathrm{K}$ for all $(x{:}C\wedge D)P \in \mathrm{J}(C\wedge D)\mathrm{K}$. By (3) we know that $(x{:}C\wedge D)\mathsf{Ax}(x,c) \in \mathrm{J}(C\wedge D)\mathrm{K}$ and we have $\langle d\rangle S[c := (x)\mathsf{Ax}(x,c)] \in \mathrm{J}\langle C\rangle\mathrm{K}$ and $\langle e\rangle T[c := (x)\mathsf{Ax}(x,c)] \in \mathrm{J}\langle D\rangle\mathrm{K}$. By IH we can infer that $S[c := (x)\mathsf{Ax}(x,c)] \in SN$ and $T[c := (x)\mathsf{Ax}(x,c)] \in SN$. From Lemma 1 we can infer that $S[c := (x)\mathsf{Ax}(x,c)] \equiv S$ or $S[c := (x)\mathsf{Ax}(x,c)] \longrightarrow^+ S$. In both cases we know that $S \in SN$ (similarly $T \in SN$). But then $\mathsf{And}_R(\langle d\rangle S, \langle e\rangle T, c)$ must be strongly normalising too. Similarly for (ii).

Lemma 5. *If* $M, N \in SN$ *and* $\langle a{:}B\rangle M \in \mathrm{J}\langle B\rangle\mathrm{K}$, $(x{:}B)N \in \mathrm{J}(B)\mathrm{K}$ *then* $\mathsf{Cut}(\langle a{:}B\rangle M, (x{:}B)N) \in SN$.

Proof. We assign to each term of the form $\mathsf{Cut}(\langle a{:}B\rangle M, (x{:}B)N)$ a lexicographically ordered induction value of the form $(\delta, l(M), l(N))$ where δ is the degree of the cut-formula B; $l(M)$ and $l(N)$ are the lengths of the maximal reduction sequences starting from M and N, respectively. By assumption both $l(M)$ and $l(N)$ are finite. We prove that all terms to which $\mathsf{Cut}(\langle a\rangle M, (x)N)$ reduces are strongly normalising.

Inner Reduction: $\mathsf{Cut}(\langle a\rangle M, (x)N) \longrightarrow \mathsf{Cut}(\langle a\rangle M', (x)N')$ where either $M \equiv M'$ and $N \longrightarrow N'$ or $M \longrightarrow M'$ and $N \equiv N'$. Assume the later case (the other case being similar). We have to prove that $\mathsf{Cut}(\langle a\rangle M', (x)N) \in SN$. From $\langle a{:}B\rangle M \in \mathrm{J}\langle B\rangle\mathrm{K}$ we can infer by Lemmas 3 and 4 that $\langle a{:}B\rangle M' \in \mathrm{J}\langle B\rangle\mathrm{K}$ and $M' \in SN$. We know that the degree of the cut-formula is in both terms equal, but $l(M') < l(M)$. Therefore we can apply the IH and infer that $\mathsf{Cut}(\langle a\rangle M', (x)N) \in SN$.

Commuting Reduction: $\mathsf{Cut}(\langle a\rangle M, (x)N) \longrightarrow M[a := (x)N]$. By assumption we have $\langle a{:}B\rangle M \in \mathrm{J}\langle B\rangle\mathrm{K} = \textsc{neg}_{\langle B\rangle}(\mathrm{J}(B)\mathrm{K})$. We know that the commuting reduction is only applicable if M does not introduce a; therefore we have that $\langle a{:}C{\wedge}D\rangle M \notin \textsc{andright}_{\langle C\wedge D\rangle}(\mathrm{J}\langle C\rangle\mathrm{K}, \mathrm{J}\langle D\rangle\mathrm{K}, \mathrm{J}(C{\wedge}D)\mathrm{K})$ (where $B \equiv C{\wedge}D$). That means that $\langle a{:}B\rangle M \in \textsc{axioms}_{\langle B\rangle}$ or $\langle a{:}B\rangle M \in \textsc{binding}_{\langle B\rangle}(\mathrm{J}(B)\mathrm{K})$. In the first case we have $\mathsf{Cut}(\langle a\rangle M, (x)N) \longrightarrow M[a := (x)N] \equiv M$ (because M is an axiom and does not introduce a); M is strongly normalising by assumption.
In the second case we have that $M[a := (y{:}B)P] \in SN$ for all $(y{:}B)P \in \mathrm{J}(B)\mathrm{K}$. Set $(y{:}B)P$ to $(x{:}B)N$ which is in $\mathrm{J}(B)\mathrm{K}$ by assumption. Symmetric case is similar.

Case Logical Reduction I: $\mathsf{Cut}(\langle a\rangle \mathsf{Ax}(y,a), (x)N) \longrightarrow N\{x \mapsto y\}$. By assumption we know that $N \in SN$. This implies that $N\{x \mapsto y\} \in SN$. Symmetric case is similar.

Case Logical Reduction II: $\mathsf{Cut}\langle c\rangle \mathsf{And}_R(\langle a\rangle S, \langle b\rangle T, c), (y)$ $\mathsf{And}^1_L((x)U, y)$, *where* $B \equiv C{\wedge}D$. For more clarity we set $\langle c\rangle M \equiv \langle c{:}C{\wedge}D\rangle \mathsf{And}_R(\langle a\rangle S, \langle b\rangle T, c)$ and $(y)N \equiv (y{:}C{\wedge}D)\mathsf{And}^1_L((x)U, y)$.

$$\begin{array}{l}\mathsf{Cut}(\langle c\rangle \mathsf{And}_R(\langle a\rangle S, \langle b\rangle T, c), (y)\mathsf{And}^1_L((x)U, y)) \\ \longrightarrow \mathsf{Cut}(\langle a\rangle S[c := (y)N], (x)U[y := \langle c\rangle M]).\end{array}$$

By assumption we know that $\langle c{:}C{\wedge}D\rangle M \in \mathrm{J}\langle C{\wedge}D\rangle\mathrm{K}$ and $(y{:}C{\wedge}D)N \in \mathrm{J}(C{\wedge}D)\mathrm{K}$. We have to show that $\mathsf{Cut}(\langle a: C\rangle S[c := (y)N], (x: C)U[y := \langle c\rangle M]) \in SN$. Since $\langle c\rangle M \in \mathrm{J}\langle C{\wedge}D\rangle\mathrm{K} = \textsc{neg}_{\langle C\wedge D\rangle}(\mathrm{J}(C{\wedge}D)\mathrm{K})$ and $\langle c\rangle M \notin \textsc{axioms}_{\langle C\wedge D\rangle}$ we know that:

$\langle c{:}C{\wedge}D\rangle M \in \textsc{binding}_{\langle C\wedge D\rangle}(\mathrm{J}(C{\wedge}D)\mathrm{K})$ or
$\langle c{:}C{\wedge}D\rangle M \in \textsc{andright}_{\langle C\wedge D\rangle}(\mathrm{J}\langle C\rangle\mathrm{K}, \mathrm{J}\langle D\rangle\mathrm{K}, \mathrm{J}(C{\wedge}D)\mathrm{K})$.

Similarly

$(y{:}C{\wedge}D)N \in \textsc{binding}_{(C\wedge D)}(\mathrm{J}\langle C{\wedge}D\rangle\mathrm{K})$ or
$(y{:}C{\wedge}D)N \in \textsc{andleft}^1_{(C\wedge D)}(\mathrm{J}(C)\mathrm{K}, \mathrm{J}\langle C{\wedge}D\rangle\mathrm{K})$.

If $\langle c{:}C{\wedge}D\rangle M \in \textsc{binding}_{\langle C\wedge D\rangle}(\mathrm{J}(C{\wedge}D)\mathrm{K})$ we know that $M[c := (z)P] \in SN$ for all $(z{:}C{\wedge}D)P \in \mathrm{J}(C{\wedge}D)\mathrm{K}$. By assumption $(y{:}C{\wedge}D)N \in \mathrm{J}(C{\wedge}D)\mathrm{K}$ and therefore $M[c := (y)N] \equiv \mathsf{Cut}(\langle c\rangle M, (y)N) \in SN$. But then we also have that its reduct $\mathsf{Cut}(\langle a\rangle S[c := (y)N], (x)U[y := \langle c\rangle M]) \in SN$. Similarly for the case $(y{:}C{\wedge}D)N \in \textsc{binding}_{(C\wedge D)}(\mathrm{J}\langle C{\wedge}D\rangle\mathrm{K})$. It is left to show strong normalisation in the case where $\langle c : C{\wedge}D\rangle M \in \textsc{andright}_{(C\wedge D)}(J\langle C\rangle K, J\langle D\rangle K, J(C{\wedge}D)K)$ and $(y{:}C{\wedge}D)N \in \textsc{andleft}^1_{(C\wedge D)}(\mathrm{J}(C)\mathrm{K}, \mathrm{J}\langle C{\wedge}D\rangle\mathrm{K})$. We have $\langle a\rangle\ S[c := (y)P] \in \mathrm{J}\langle C\rangle\mathrm{K}$ and $(x)\ U[y := \langle c\rangle Q] \in \mathrm{J}(C)\mathrm{K}$ for all terms $(y{:}C{\wedge}D)P \in \mathrm{J}(C{\wedge}D)\mathrm{K}$ and $\langle c{:}C{\wedge}D\rangle Q \in \mathrm{J}\langle C{\wedge}D\rangle\mathrm{K}$. By assumption we know that $\langle c{:}C\wedge D\rangle M \in \mathrm{J}\langle C\wedge D\rangle\mathrm{K}$ and $(y{:}C\wedge D)N \in \mathrm{J}(C\wedge D)\mathrm{K}$; set $\langle c\rangle M$ for $\langle c\rangle Q$ and $(y)N$ for $(y)P$ respectively. Therefore we know that $\langle a\rangle\, S[c := (y)N] \in \mathrm{J}\langle C\rangle\mathrm{K}$ and $(x)\, U[y := \langle c\rangle M] \in \mathrm{J}(C)\mathrm{K}$. Furthermore, by Lemma 4 we have $S[c := (y)N] \in SN$ and $U[y := \langle c\rangle M] \in SN$. Because the degree of the cut-formula decreased we can apply the IH and infer that

$$\mathsf{Cut}(\langle a\rangle S[c := (y)N], (x)U[y := \langle c\rangle M]) \in SN.$$

We have shown that all immediate reducts of $\mathsf{Cut}(\langle a\rangle M, (x)N)$ are strongly normalising. Consequently $\mathsf{Cut}(\langle a\rangle M, (x)N)$ must be strongly normalising.

It is left to show that all well-typed terms are strongly normalising. To do so, we shall consider a special class of simultaneous substitutions, which are called safe. The principal property of safe substitutions $[\sigma_1]$ and $[\sigma_2]$ is that they can be commuted, i.e. $M[\sigma_1][\sigma_2] \equiv M[\sigma_2][\sigma_1]$.

Let $\hat{\sigma}$ be a set of substitutions of the form $[x := \langle a\rangle P]$ and $[b := (y)Q]$. Let us call the set of the x's and b's the domain of $\hat{\sigma}$ (written as $dom(\hat{\sigma})$); the set of named terms $(y)Q$ and co-named terms $\langle a\rangle P$ is called the co-domain of $\hat{\sigma}$ (written as $codom(\hat{\sigma})$). A *safe simultaneous substitution* (sss) is a set of substitutions where no variable clash between the domain and co-domain occurs (this can always be achieved by appropriate α-conversions, however, we omit a precise definition).

The next lemma shows that a specific type of simultaneous substitutions is safe.

Lemma 6. *Let $\hat{\sigma}$ be of the form:*

$$\left\{ \bigcup_{i=0,\ldots,n} [x_i := \langle c\rangle \mathsf{Ax}(x_i, c)] \right\} \cup \left\{ \bigcup_{j=0,\ldots,m} [a_j := (y)\mathsf{Ax}(y, a_j)] \right\}$$

where the x_i's and a_i's are distinct names and co-names, respectively. Substitution $\hat{\sigma}$ is a sss.

Proof. Induction on the length of $\hat{\sigma}$.

Lemma 7. *For every term M (not necessarily strongly normalising) and for every sss $\hat{\sigma}$, such that $FN(M) \cup FC(M) \subseteq dom(\hat{\sigma})$ (i.e., $\hat{\sigma}$ is a closing substitution[2]) and for every $(x{:}B)P \in codom(\hat{\sigma})$ $(x{:}B)P \in \mathrm{J}(B)\mathrm{K}$ and every $\langle a{:}C\rangle Q \in codom(\hat{\sigma})$ $\langle a{:}C\rangle Q \in \mathrm{J}\langle C\rangle\mathrm{K}$, we have $M\hat{\sigma} \in SN$.*

Proof. We proceed by induction over the structure of M. We write $\hat{\sigma}, [\sigma]$ for the set $\hat{\sigma} \cup [\sigma]$ where $[\sigma] \notin \hat{\sigma}$.

Case $\mathsf{Ax}(x, a)$: We have to prove that: $\mathsf{Ax}(x, a)\ \hat{\sigma}, [x := \langle b\rangle P], [a := (y)Q] \in SN$. By definition of substitution $\mathsf{Ax}(x, a)\ \hat{\sigma}, [x := \langle b\rangle P], [a := (y)Q] \equiv \mathsf{Cut}(\langle b\rangle P, (y)Q)$. By assumption $\langle b{:}B\rangle P \in \mathrm{J}\langle B\rangle\mathrm{K}$ and $(y{:}B)Q \in \mathrm{J}(B)\mathrm{K}$. By Lemma 4 we know that $P \in SN$ and $Q \in SN$. Therefore we can apply Lemma 5 and can infer that $\mathsf{Cut}(\langle b\rangle P, (y)Q) \in SN$. Therefore $\mathsf{Ax}(x, a)\hat{\sigma}, [x := \langle b\rangle P], [a := (y)Q] \in SN$.

[2] All free names and co-names of M are amongst the domain of $\hat{\sigma}$.

Case $\mathsf{And}_R(\langle a\rangle M, \langle b\rangle N, c)$: We prove that $\mathsf{And}_R(\langle a\rangle M, \langle b\rangle N, c)\ \hat{\sigma}, [c := (z)R] \in SN$ where $(z{:}B\wedge C)R$ is an arbitrary named term in $\mathrm{J}(B\wedge C)\mathrm{K}$. We can infer that $\mathsf{And}_R(\langle a\rangle M, \langle b\rangle N, c)\,\hat{\sigma}, [c := (z)R] \equiv \mathsf{Cut}(\langle c\rangle \mathsf{And}_R(\langle a\rangle M\hat{\sigma}, \langle b\rangle N\hat{\sigma}, c), (z)R)$. By IH we know that $M\ \hat{\sigma}, [c := (x)S], [a := (y)P] \in SN$ and $N\ \hat{\sigma}, [c := (x)S], [b := (v)Q] \in SN$ for arbitrary $(y{:}B)P \in \mathrm{J}\langle B\rangle\mathrm{K}$, $(v{:}C)Q \in \mathrm{J}\langle C\rangle\mathrm{K}$ and $(x{:}B\wedge C)S \in \mathrm{J}(B\wedge C)\mathrm{K}$.
Making appropriate α-conversions we have $(M\hat{\sigma})[c := (x)S][a := (y)P] \in SN$ and $(N\hat{\sigma})[c := (x)S][b := (v)Q] \in SN$. By definition of BINDING we have $\langle a{:}B\rangle\ (M\hat{\sigma})[c := (x)S] \in \mathrm{J}\langle B\rangle\mathrm{K}$ and $\langle b{:}C\rangle(N\hat{\sigma})[c := (x)S] \in \mathrm{J}\langle C\rangle\mathrm{K}$. Because $(x{:}B\wedge C)S$ is an arbitrary named term in the candidate $\mathrm{J}(B\wedge C)\mathrm{K}$ we have by definition of $\text{ANDRIGHT}_{\langle B\wedge C\rangle}$ that $\langle c{:}B\wedge C\rangle \mathsf{And}_R(\langle a\rangle M\hat{\sigma}, \langle b\rangle N\hat{\sigma}, c) \in \mathrm{J}\langle B\wedge C\rangle\mathrm{K}$. Furthermore we know by Lemma 4 that $\mathsf{And}_R(\langle a\rangle M\hat{\sigma}, \langle b\rangle N\hat{\sigma}, c) \in SN$.
For $(z{:}B\wedge C)R \in \mathrm{J}(B\wedge C)\mathrm{K}$ we have by Lemma 4 that $R \in SN$. We can apply Lemma 5 and have $\mathsf{Cut}(\langle c\rangle \mathsf{And}_R(\langle a\rangle M\hat{\sigma}, \langle b\rangle N\hat{\sigma}, c), (z)R) \in SN$ and therefore $\mathsf{And}_R(\langle a\rangle M, \langle b\rangle N, c)\ \hat{\sigma}, [c := (z)R] \in SN$.

Case $\mathsf{And}^i_L((x)M, y)\ (i = 1, 2)$: We have to prove that $\mathsf{And}^i_L((x)M, y)\ \hat{\sigma}, [y := \langle c\rangle R] \in SN$ where $\langle c{:}B_1\wedge B_2\rangle R$ is an arbitrary co-named term in $\mathrm{J}\langle B_1\wedge B_2\rangle\mathrm{K}$. We have $\mathsf{And}^i_L((x)M, y)\ \hat{\sigma}, [y := \langle c\rangle R] \equiv \mathsf{Cut}(\langle c\rangle R, (y)\mathsf{And}^i_L((x)M\hat{\sigma}, y))$ by definition of substitution. By IH we know that $M\ \hat{\sigma}, [y := \langle a\rangle S], [x := \langle b\rangle T] \in SN$ for arbitrary $\langle a{:}B_1\wedge B_2\rangle S \in \mathrm{J}\langle B_1\wedge B_2\rangle\mathrm{K}$, and arbitrary $\langle b{:}B_i\rangle T \in \mathrm{J}\langle B_i\rangle\mathrm{K}$. Making appropriate α-conversions we have $(M\hat{\sigma})[y := \langle a\rangle S][x := \langle b\rangle T] \in SN$. By definition of BINDING we have $(x{:}B_i)\ (M\hat{\sigma})[y := \langle a\rangle S] \in \mathrm{J}(B_i)\mathrm{K}$. Since $\langle a{:}B_1\wedge B_2\rangle S$ is an arbitrary co-named term in $\mathrm{J}\langle B_1\wedge B_2\rangle\mathrm{K}$ we have by definition of $\text{ANDLEFT}^i_{(B_1\wedge B_2)}$ that $(y{:}B_1\wedge B_2)\mathsf{And}^i_L((x)M\hat{\sigma}, y) \in \mathrm{J}(B_1\wedge B_2)\mathrm{K}$. By Lemma 4 we can infer that $\mathsf{And}^i_L((x)M\hat{\sigma}, y) \in SN$. For $(c{:}B_1\wedge B_2)R \in \mathrm{J}(B_1\wedge B_2)\mathrm{K}$ we have by Lemma 4 that $R \in SN$. We can apply Lemma 5 and have $\mathsf{Cut}(\langle c\rangle R, (y)\mathsf{And}^i_L((x)M\hat{\sigma}, y)) \in SN$. Therefore $\mathsf{And}^i_L((x)M, y)\ \hat{\sigma}, [y := \langle c\rangle R] \in SN$.

Case $\mathsf{Cut}(\langle a\rangle M, (x)N)$:

Subcase I: M is an axiom (case N being an axiom is similar). We have to show that $\mathsf{Cut}(\langle a\rangle \mathsf{Ax}(x, a), (y)N)\ [x := \langle b\rangle S], \hat{\sigma} \in SN$. By definition of substitution $\mathsf{Cut}(\langle a\rangle \mathsf{Ax}(x, a), (y)N)\ [x := \langle b\rangle S], \hat{\sigma} \equiv \mathsf{Cut}(\langle b\rangle S, (x)\ N\{x\mapsto y\}\hat{\sigma})$. By assumption we know that $\langle b{:}B\rangle S \in \mathrm{J}\langle B\rangle\mathrm{K}$; using Lemma 4 we know that $S \in SN$. By assumption we know that $N\ \hat{\sigma}, [x := \langle b\rangle S], [y := \langle b\rangle S] \in SN$ for arbitrary $\langle b{:}B\rangle S \in \mathrm{J}\langle B\rangle\mathrm{K}$. Because $\hat{\sigma}, [x := \langle b\rangle S], [y := \langle b\rangle S]$ is a safe simultaneous substitution we have (making appropriate α-conversions) $N\ \hat{\sigma}, [x := \langle b\rangle S], [y := \langle b\rangle S] \equiv (N\{y\mapsto x\}\hat{\sigma})\ [x := \langle b\rangle S]$. By definition of BINDING we know that $(x{:}B)\ N\{y\mapsto x\}\hat{\sigma} \in \mathrm{J}(B)\mathrm{K}$. By Lemma 4 we can infer that $N\{y\mapsto x\}\hat{\sigma} \in SN$. Then we can apply Lemma 5 and can show that $\mathsf{Cut}(\langle b\rangle S, (x)\ N\{y\mapsto x\}\hat{\sigma}) \in SN$. Therefore $\mathsf{Cut}(\langle a\rangle \mathsf{Ax}(x, a), (y)N)\ \hat{\sigma}, [x := \langle b\rangle S] \in SN$.

Subcase II: M and N are not axioms. We prove that $\mathsf{Cut}(\langle a\rangle M, (x)N)\ \hat{\sigma} \in SN$. By IH we know that $M\ \hat{\sigma}, [a := (y)S] \in SN$ and $N\ \hat{\sigma}, [x := \langle b\rangle T] \in SN$ for arbitrary $(y{:}B)S \in \mathrm{J}(B)\mathrm{K}$ and $\langle b{:}B\rangle T \in \mathrm{J}\langle B\rangle\mathrm{K}$. Making appropriate α-conversions we know that $(M\hat{\sigma})[a := (y)S] \in SN$ and $(N\hat{\sigma})[x := \langle b\rangle T] \in$

SN. By definition of BINDING we can infer that $\langle a{:}B\rangle\, M\hat{\sigma} \in \mathrm{J}\langle B\rangle\mathrm{K}$ and $(x{:}B)\, N\hat{\sigma} \in \mathrm{J}(B)\mathrm{K}$. By Lemma 4 we have that $M\hat{\sigma} \in SN$ and $N\hat{\sigma} \in SN$. Therefore we can apply Lemma 5 and infer $\mathsf{Cut}(\langle a\rangle\, M\hat{\sigma}, (x)\, N\hat{\sigma}) \equiv \mathsf{Cut}(\langle a\rangle M, (x)N)\,\hat{\sigma} \in SN$.

We can now prove our main theorem.

Theorem 1. *All well-typed terms are strongly normalising.*

Proof. We know by Lemma 7 that for arbitrary well-typed terms M and arbitrary safe simultaneous substitution $\hat{\sigma}$, we have $M\hat{\sigma} \in SN$. Let $\hat{\sigma}$ be the safe simultaneous substitution from Lemma 6. Using Lemma 1 we can infer that either $M\hat{\sigma} \longrightarrow^{+} M$ or $M\hat{\sigma} \equiv M$. From this we have $M \in SN$.

This theorem can be extended to the full classical logic. To save space we give only the definitions for the set operators with implicational type:

$$\begin{array}{l}\text{IMPLEFT}_{(B\supset C)} : \mathcal{P}(CT_{\langle B\rangle}) \times \mathcal{P}(NT_{(C)}) \times \mathcal{P}(CT_{\langle B\supset C\rangle}) \to \mathcal{P}(NT_{(B\supset C)})\\ \text{IMPRIGHT}_{\langle B\supset C\rangle} : \mathcal{P}(NT_{(B)}) \times \mathcal{P}(CT_{\langle C\rangle}) \times \mathcal{P}(NT_{(B\supset C)}) \to \mathcal{P}(CT_{\langle B\supset C\rangle})\end{array}$$

$$\begin{array}{rl}\text{IMPLEFT}_{(B\supset C)}(X,Y,Z) \stackrel{\text{def}}{=} & \{(z{:}B\supset C)\mathsf{Imp}_L(\langle a{:}B\rangle M, (x{:}C)N, z) \mid \\ & \forall\, \langle c{:}B\supset C\rangle P \in Z.\langle a\rangle\, M[z := \langle c\rangle P] \in X \text{ and } (x)\, N[z := \langle c\rangle P] \in Y\}\\ \text{IMPRIGHT}_{\langle B\supset C\rangle}(X,Y,Z) \stackrel{\text{def}}{=} & \{\langle b{:}B\supset C\rangle\mathsf{Imp}_R((x{:}B)\langle a{:}C\rangle M, b) \mid \\ & \forall\, (z{:}B\supset C)P \in Z, \forall\, \langle c{:}B\rangle S \in X.\langle a\rangle\, M[z := \langle c\rangle P][x := \langle c\rangle S] \in Y \text{ and}\\ & \forall\, (z{:}B\supset C)P \in Z, \forall\, (y{:}C)T \in Y.(x)\, M[z := \langle c\rangle P][a := (y)T] \in X\}\end{array}$$

$$\begin{array}{rl}\text{NEG}_{\langle B\supset C\rangle}(X) \stackrel{\text{def}}{=} & \text{AXIOMS}_{\langle B\supset C\rangle} \cup \text{BINDING}_{\langle B\supset C\rangle}(X) \cup \\ & \text{IMPRIGHT}_{\langle B\supset C\rangle}(\mathrm{J}(B)\mathrm{K}, \mathrm{J}\langle C\rangle\mathrm{K}, X)\end{array}$$

$$\begin{array}{rl}\text{NEG}_{(B\supset C)}(X) \stackrel{\text{def}}{=} & \text{AXIOMS}_{(B\supset C)} \cup \text{BINDING}_{(B\supset C)}(X) \cup \\ & \text{IMPLEFT}_{(B\supset C)}(\mathrm{J}\langle B\rangle\mathrm{K}, \mathrm{J}(C)\mathrm{K}, X)\end{array}$$

The strong normalisation proof can be easily extended using the definitions above. The only difficulty arises in Lemma 5 for the cut-elimination reduction for the connective $\supset$. The reduct of such a cut contains two nested cuts. Although the degree of the cut-formula decreases for the outer cut, the IH is not immediately applicable. In order to apply the induction hypothesis for the outer cut one has to show for the inner cut that:

$$\langle a\rangle\mathsf{Cut}(\langle c\rangle N[z := \langle b\rangle\mathsf{Imp}_R((x)\langle a\rangle M, b)], (x)M[b := (z)\mathsf{Imp}_L(\langle c\rangle N, (y)P, z)]) \in \mathrm{J}\langle C\rangle\mathrm{K}$$

and

$$(x)\mathsf{Cut}(\langle a\rangle M[b := (z)\mathsf{Imp}_L(\langle c\rangle N, (y)P, z)], (y)P[z := \langle b\rangle\mathsf{Imp}_R((x)\langle a\rangle M, b)]) \in \mathrm{J}(B)\mathrm{K}$$

In the first case (the other being similar) one has to show that:

$$\mathsf{Cut}(\langle c\rangle N[z := \langle b\rangle\mathsf{Imp}_R((x)\langle a\rangle M, b)], (x)M[b := (z)\mathsf{Imp}_L(\langle c\rangle N, (y)P, z)])\\ [a := (v)T] \in SN.$$

To infer this it is essential to know that a is not a free name in N and P (requirement of the reduction rule which can always be achieved by renaming a appropriately).

$$
\cfrac{
\cfrac{\overline{A\vee A\vdash A\vee A}\quad\overline{A\vee A\vdash A\vee A}}{(A\vee A)\vee(A\vee A)\vdash A\vee A}\vee_L
\quad
\cfrac{\cfrac{\overline{A\vdash A}\quad\overline{A\vdash A}}{A\vee A\vdash A}\vee_L \quad \cfrac{\overline{A\vdash A}\quad\overline{A\vdash A}}{A\vdash A\wedge A}\wedge_R}{A\vee A\vdash A\wedge A}Cut
}{(A\vee A)\vee(A\vee A)\vdash A\wedge A}Cut
$$

$$
\cfrac{
\cfrac{\cfrac{\overline{A\vdash A}\quad\overline{A\vdash A}}{A\vdash A\wedge A}\wedge_R \quad \cfrac{\overline{A\vdash A}\quad\overline{A\vdash A}}{A\vdash A\wedge A}\wedge_R}{A\vee A\vdash A\wedge A}\vee_L
\quad
\cfrac{\cfrac{\overline{A\vdash A}\quad\overline{A\vdash A}}{A\vee A\vdash A}\vee_L \quad \cfrac{\overline{A\vdash A}\quad\overline{A\vdash A}}{A\vee A\vdash A}\vee_L}{A\vee A\vdash A\wedge A}\wedge_R
}{(A\vee A)\vee(A\vee A)\vdash A\wedge A}\vee_L
$$

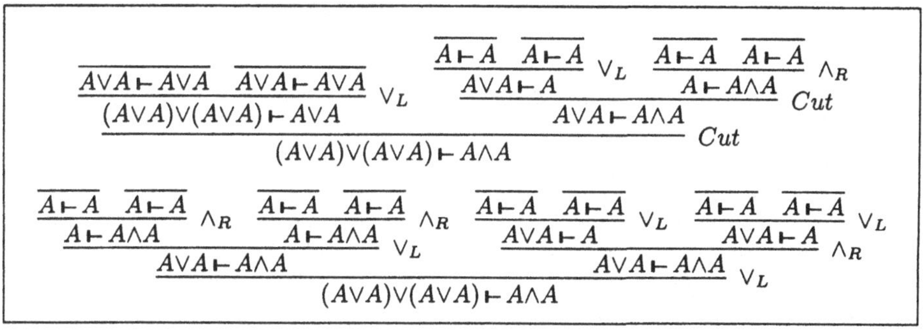

Fig. 3. A proof in G3c and a cut-free normalform which is not reachable by a cut-elimination procedure using colours as in LK^{tq}.

4 Conclusion

In this paper we presented a reduction system for cut-elimination in classical logic. One feature of the reduction system is to permute a subderivation of a commuting cut directly to the place(s) where the cut-formula is a main formula. This is an idea taken from the work in LK^{tq} [6]. However we do not require their colour annotations on the cut-formulae (in fact no additional information is required at all). One consequence is that, in general, more normal forms can be reached from a given proof containing cuts (see Figure 3 for an example). Because of the fewer constraints on our reduction system strong normalisation cannot be proved by translating every reduction to a series of reductions in proof-nets as done for LK^{tq}. The use of a term calculus for sequent derivations allowed us to use directly proof techniques from the λ^{Sym}-calculus [1] to prove strong normalisation. This use of syntax to study proof structures is part of a on-going research project [3, 19].

The result presented in this paper can be extended to the first-order calculus and can be adapted to LK or free-style LK^{tq}. There are many directions for further work. For example what is the precise correspondence in the intuitionistic case between normalisation and our strongly normalising cut-elimination procedure? For classical logic the correspondence between our cut-elimination procedure and normalisation in, for example, Parigot's $\lambda\mu$ [17] is another interesting question.

References

1. F. Barbanera and S. Berardi. A Symmetric Lambda Calculus for "Classical" Program Extraction. In *Proc. of Theoretical Aspects of Computer Software*, volume 789 of *LNCS*, pages 495–515. Springer Verlag, 1994.
2. H. Barendregt. *The Lambda Calculus – Its Syntax and Semantics*, volume 103 of *Studies in Logic and the Foundations of Mathematics*. North-Holland Publishing Company, 1981.

3. G. Bierman. Some Lectures on Proof Theory. Notes from course given at PUC-Rio, November 1997.
4. E. T. Bittar. Strong Normalisation Proofs for Cut Elimination in Gentzen's Sequent Calculi. In *Proc. of the Symposium: Logic, Algebra and Computer Science, Warsaw*, December 1996.
5. E. A. Cichon, M. Rusinowitch, and S. Selhab. Cut Elimination and Rewriting: Termination Proofs. *Theoretical Computer Science*, 1996. (to appear).
6. V. Danos, J.-B. Joinet, and H. Schellinx. A New Deconstructive Logic: Linear Logic. *Journal of Symbolic Logic*, 62(3):755–807, Sept. 1997.
7. A. G. Dragalin. *Mathematical Intuitionism*, volume 67. American Mathematical Society, Providence, Rhode Island, 1988.
8. R. Dyckhoff and L. Pinto. Cut-Elimination and a Permutation-Free Sequent Calculus for Intuitionistic Logic. *Studia Logica*, 60:107–118, 1998.
9. J. Gallier. Constructive Logics. Part I: A Tutorial on Proof Systems and Typed λ-Calculi. *Theoretical Computer Science*, 110(2):249–239, 1993.
10. G. Gentzen. Untersuchungen über das logische Schließen I and II. *Mathematische Zeitschrift*, 39:176–210, 405–431, 1935.
11. J.-Y. Girard. Linear Logic. *Theoretical Computer Science*, 50:1–102, 1987.
12. J.-Y. Girard, Y. Lafont, and P. Taylor. *Proofs and Types*. Cambridge Tracts in Theoretical Computer Science. Cambridge University Press, 1989.
13. E. H. Hauesler and L. C. Pereira. Gentzen's Second Consistency Proof and Strong Cut-Elimination. *Logique and Analyse*, 154:95–111, 1996.
14. H. Herbelin. A λ-Calculus Structure Isomorphic to Sequent Calculus Structure. In *Proc. of the Conference on Computer Science Logic*, volume 933 of *LNCS*, pages 67–75. Springer Verlag, 1995.
15. J.-B. Joinet, H. Schellinx, and L. Tortora de Falco. SN and CR for Free-Style LK^{tq}: Linear Decorations and Simulation of Normalisation. Technical Report, May 1998.
16. S. C. Kleene. *Introduction to Metamathematics*. North-Holland Publishing Company, Amsterdam, 1952.
17. M. Parigot. $\lambda\mu$-Calculus: An Algorithmic Interpretation of Classical Logic. In *Proc. of the Int. Conference on Logic Programming and Automated Deduction*, volume 624 of *LNCS*, pages 190–201. Springer Verlag, 1992.
18. A. S. Troelstra and H. Schwichtenberg. *Basic Proof Theory*. Cambridge Tracts in Theoretical Computer Science. Cambridge University Press, 1996.
19. C. Urban. First-Year Report: Computational Content of Classical Proofs. Technical Report, 1997.
20. J. Zucker. The Correspondance Between Cut-Elimination and Normalisation. *Annals of Mathematical Logic*, 7:1–112, 1974.

Pure Type Systems with Subtyping
(Extended Abstract)

Jan Zwanenburg

Eindhoven University of Technology, The Netherlands
janz@@win.tue.nl

Abstract. We extend the framework of Pure Type Systems with subtyping, as found in $F^{\omega}_{\leq}$. This leads to a concise description of many existing systems with subtyping, and also to some new interesting systems. We develop the meta-theory for this framework, including Subject Reduction and Minimal Typing.
The main problem was how to formulate the rules of the framework in such a way that we avoid circularities between theory about typing and theory about subtyping. We solve this problem by a simple but rigorous design decision: the subtyping rules do not depend on the typing rules.

1 Introduction

The Pure Type Systems (*PTSs*, see [Bar92]) provide a framework of type systems, in which many particular systems, such as F, F^{ω}, λP and the Calculus of Constructions can be concisely expressed and easily compared. Furthermore, the *PTSs* also include many new interesting systems.

We introduce a framework of Pure Type Systems with Subtyping ($PTS^{\leq}s$), which includes a number of *PTSs* extended with subtyping, e.g. $\lambda{\rightarrow}^{\leq}$ [Car88], $F_{\leq}$ [CG92], $F^{\omega}_{\leq}$ [PS94], $\lambda P_{\leq}$ [AC96] and $\lambda C_{\leq}$ [Che97]. This framework also yields new systems, e.g. the Calculus of Constructions with subtyping.

The main problem is how to define it in such a way that we can develop the meta-theory. The most straightforward approach seems to be the combination of the rules of *PTSs* with subtyping rules found in systems like $F^{\omega}_{\leq}$. Some of these subtyping rules have typing judgments as premises. This is very awkward for the meta-theory, since results about the subtyping judgment cannot be proved independently of results about the typing judgments: soon one gets circular dependencies of lemmas about subtyping and lemmas about typing. Each particular system with subtyping in the literature avoids or solves this problem by exploiting the particular nature of that system, and none of these solutions work also for $PTS^{\leq}s$ (see section 2.3).

This leads us to consider a reformulation of the definition of the $PTS^{\leq}s$, where we conform to the following major design decision:

The subtyping rules do not depend on the typing rules.

In other words, we define the subtyping relation on pseudoterms rather than only on well-typed terms. Now we can develop the theory for subtyping first, and then proceed to the typing judgment.

It turns out to be hard to prove some essential properties about the subtyping judgment. We solve this by considering an equivalent reformulation of the subtyping rules, roughly similar to the subtyping algorithms proposed in the literature. A surprising element in this reformulation is a subtyping rule that relates terms that are per definition untypable.

Furthermore, the proof of Uniqueness of Typing for ordinary *PTSs* couldn't be easily extended to a proof of Minimal Typing for $PTS^{\leq}s$. We solved this by proving a weak form of Minimal Typing, and by introducing another form of reduction.

In section 2 we define the syntax of $PTS^{\leq}s$, and give the typing and subtyping rules. We also relate to subtyping systems in the literature. Section 3 gives the meta-theory including Subject Reduction and Minimal Typing. Section 4 gives the conclusions.

2 Syntax and Typing Rules

We specify the syntax of $PTS^{\leq}s$ in section 2.1, and the typing and subtyping rules in section 2.2. Section 2.3 shows how many existing systems with subtyping can be considered as a $PTS^{\leq}$. Section 2.4 show a number of alternatives and extensions for our rules.

2.1 Syntax

Three constructs are new in $PTS^{\leq}s$ (compared to ordinary *PTSs*). We have bounded abstractions $\lambda x \leq a : A.\ b$, bounded quantifications $\Pi x \leq a : A.\ B$, and bounded declarations $\Gamma, x \leq a : A$. These constructs are important in the explanation of specifications of $PTS^{\leq}s$.

Definition 1. *A* specification *of a* $PTS^{\leq}$ *is a 5-tuple* $(\mathcal{S}, \mathcal{A}, \mathcal{R}, \mathcal{S}^{\leq}, \mathcal{R}^{\leq})$, *with the following properties:*

1. $\mathcal{S}$ *is a set of symbols called the* sorts.
2. $\mathcal{A} \subseteq \mathcal{S} \times \mathcal{S}$, *a set of* axioms *of the form* $(s : s')$.
3. $\mathcal{R} \subseteq \mathcal{S} \times \mathcal{S} \times \mathcal{S}$, *a set of* rules *of the form* (s_1, s_2, s_3).
4. $\mathcal{S}^{\leq} \subseteq \mathcal{S}$ *is a set of* subtyping sorts.
5. $\mathcal{R}^{\leq} \subseteq \mathcal{S}^{\leq} \times \mathcal{S} \times \mathcal{S}$, *a set of* bounded rules.

We write (s_1, s_2) for a (bounded) rule, as abbreviation for (s_1, s_2, s_2). The first three elements of the tuple serve exactly the same purpose as in *PTSs* [Bar92]. The subset of sorts $\mathcal{S}^{\leq}$ controls on which levels we can introduce subtyping. We can make a bounded declaration $x \leq a : A$, which declares variable x as a subtype of a, if $a : A$ and $A : s$ and $s \in \mathcal{S}^{\leq}$. Intuitively, in the system $\lambda\omega^{\leq}$ where $\mathcal{S}^{\leq} = \{\Box\}$, we admit $\mathtt{Nat} \leq \mathtt{Int} : *$, since $* : \Box$ and $\Box \in \mathcal{S}^{\leq}$, and we admit $\mathtt{CarI} \leq \mathtt{VehicleI} : * \to *$, since $* \to * : \Box$ and $\Box \in \mathcal{S}^{\leq}$, but we do not admit $\mathtt{x} \leq \mathtt{true} : \mathtt{Bool}$ since $\mathtt{Bool} : *$ and $* \notin \mathcal{S}^{\leq}$.

Just as $\mathcal{R}$ controls which Π-types (quantifications) we can form, $\mathcal{R}^{\leq}$ controls which bounded Π-types (bounded quantifications) we can make, and hence also which bounded abstractions we can make. For example, in $\lambda 2$ the rule $(\square, *) \in \mathcal{R}$ makes the Π-type $\Pi \mathtt{X}:*. \mathtt{X} \to \mathtt{X}$ possible, and similarly, in $\lambda 2^{\leq}$ the bounded rule $(\square, *) \in \mathcal{R}^{\leq}$ permits the bounded quantification $\Pi \mathtt{X} \leq \mathtt{Int} : *. \mathtt{X} \to \mathtt{X}$. Typically, $\mathcal{R}^{\leq}$ is a subset of $\mathcal{R}$.

Definition 2 (Pseudoterms).
The set of pseudoterms T *of a* $PTS^{\leq}$ $\lambda(\mathcal{S}, \mathcal{A}, \mathcal{R}, \mathcal{S}^{\leq}, \mathcal{R}^{\leq})$ *is defined by*

$$T ::= V \mid \mathcal{S} \mid (T\, T) \mid (\lambda V : T.\, T) \mid (\Pi V : T.\, T) \mid (\lambda V \leq T : T.\, T) \mid (\Pi V \leq T : T.\, T)$$

where V *is the set of variables.*

In a pseudoterm $\lambda x \leq a : A.\, b$ the λ binds occurrences of x in b, and similarly for Π. The notions of free and bound variables are defined accordingly. As usual, we write $A \to B$ for $\Pi x : A.\, B$ when $x \notin \mathrm{FV}(B)$.

Definition 3 (Pseudocontexts).
The set of pseudocontexts $\mathcal{C}$ *of a* $PTS^{\leq}$ $\lambda(\mathcal{S}, \mathcal{A}, \mathcal{R}, \mathcal{S}^{\leq}, \mathcal{R}^{\leq})$ *is defined by*

- $\epsilon \in \mathcal{C}$
- $\Gamma, x : A \in \mathcal{C}$ *if* $\Gamma \in \mathcal{C}, A \in T, x \in V$ *is* Γ*-fresh and* $x \notin \mathrm{FV}(A)$.
- $\Gamma, x \leq a : A \in \mathcal{C}$ *if* $\Gamma \in \mathcal{C}, a, A \in T, x \in V$ *is* Γ*-fresh and* $x \notin \mathrm{FV}(a) \cup \mathrm{FV}(A)$.

Here ϵ *denotes the empty context, and a variable* x *is called* Γ-fresh *if* $x \notin \{y\} \cup \mathrm{FV}(B)$ *for all* $y : B$ *occurring in* Γ*, and* $x \notin \{y\} \cup \mathrm{FV}(b) \cup \mathrm{FV}(B)$ *for all* $y \leq b : B$ *occurring in* Γ.

Definition 4 (Reduction). *The* β*-reduction relation* $\rhd_\beta \subseteq T \times T$ *is defined by*

$$(\lambda x : A.\, b)\, a \rhd_\beta b[x := a]$$
$$(\lambda x \leq a' : A.\, b)\, a \rhd_\beta b[x := a]$$

and all the compatibility rules. The relation $\rhd\!\!\rhd_\beta$ *is the reflexive and transitive closure of* $\rhd_\beta$*, and* $=_\beta$ *is the reflexive, symmetric and transitive closure of* $\rhd\!\!\rhd_\beta$.

2.2 Typing Rules

We have three kinds of judgments: $\Gamma \vdash ok$ for Γ is well-formed, $\Gamma \vdash a : A$ for term a has type A in Γ, and $\Gamma \vdash A \leq B$ for A is a subtype of B in Γ.

Definition 5 (Well-formedness of contexts).

(C-empty) $$\frac{}{\epsilon \vdash ok}$$

(C-var) $$\frac{\Gamma \vdash A : s}{\Gamma, x : A \vdash ok}$$

(C-Bvar) $$\frac{\Gamma \vdash a : A \quad \Gamma \vdash A : s \quad s \in \mathcal{S}^{\leq}}{\Gamma, x \leq a : A \vdash ok}$$

By our definition of pseudocontext, x must be Γ-fresh in rules (C-var) and (C-Bvar). The (C-Bvar) rule formalizes that the set $\mathcal{S}^{\leq}$ controls on which levels we may introduce subtyping, as explained above. E.g. if $\mathcal{S} = \{*, \Box\}$, $\mathcal{A} = \{(*:\Box)\}$ and $\mathcal{S}^{\leq} = \{\Box\}$ the context $\Gamma_{\text{init}} \equiv \texttt{Int}:*, \texttt{Nat}\leq\texttt{Int}:*$ is well-formed.

Definition 6 (Unbounded typing rules). *These rules are a slight reformulation of the rules for PTSs [Bar92], except for the absence of the conversion rule.*

$$\text{(axiom)} \quad \frac{\Gamma \vdash ok \quad (s_1 : s_2) \in \mathcal{A}}{\Gamma \vdash s_1 \,:\, s_2}$$

$$\text{(var)} \quad \frac{\Gamma \vdash ok \quad x : A \in \Gamma}{\Gamma \vdash x \,:\, A}$$

$$(\Pi\text{-form}) \quad \frac{\Gamma \vdash A \,:\, s_1 \quad \Gamma, x : A \vdash B \,:\, s_2 \quad (s_1, s_2, s_3) \in \mathcal{R}}{\Gamma \vdash (\Pi x : A.\, B) \,:\, s_3}$$

$$(\Pi\text{-intro}) \quad \frac{\Gamma, x : A \vdash b \,:\, B \quad \Gamma \vdash (\Pi x : A.\, B) \,:\, s}{\Gamma \vdash (\lambda x : A.\, b) \,:\, (\Pi x : A.\, B)}$$

$$(\Pi\text{-elim}) \quad \frac{\Gamma \vdash b \,:\, (\Pi x : A.\, B) \quad \Gamma \vdash a \,:\, A}{\Gamma \vdash b\, a \,:\, B[x := a]}$$

Definition 7 (Bounded typing rules).

$$\text{(subsum)} \quad \frac{\Gamma \vdash b \,:\, B \quad \Gamma \vdash B' \,:\, s \quad \Gamma \vdash B \leq B'}{\Gamma \vdash b \,:\, B'}$$

$$\text{(Bvar)} \quad \frac{\Gamma \vdash ok \quad x \leq a : A \in \Gamma}{\Gamma \vdash x \,:\, A}$$

$$(\text{B}\Pi\text{-form}) \quad \frac{\Gamma \vdash A \,:\, s_1 \quad \Gamma, x \leq a : A \vdash B \,:\, s_2 \quad (s_1, s_2, s_3) \in \mathcal{R}^{\leq}}{\Gamma \vdash (\Pi x \leq a : A.\ B) \,:\, s_3}$$

$$(\text{B}\Pi\text{-intro}) \quad \frac{\Gamma, x \leq a : A \vdash b \,:\, B \quad \Gamma \vdash (\Pi x \leq a : A.\ B) \,:\, s}{\Gamma \vdash (\lambda x \leq a : A.\ b) \,:\, (\Pi x \leq a : A.\ B)}$$

$$(\text{B}\Pi\text{-elim}) \quad \frac{\Gamma \vdash b \,:\, (\Pi x \leq a : A.\ B) \quad \Gamma \vdash a' \,:\, A \quad \Gamma \vdash a' \leq a}{\Gamma \vdash b\, a' \,:\, B[x := a']}$$

The (subsum) rule is the usual rule for subtyping, with the additional premise $\Gamma \vdash B' \,:\, s$. This is necessary to ensure B' is not an ill-behaved pseudoterm (recall that subtyping is possible on all pseudo-terms). This rule is similar to the conversion rule in ordinary *PTSs*. Instead of demanding $B =_\beta B'$ we have $\Gamma \vdash B \leq B'$. By looking ahead to the ($\leq$-conv) rule in definition 8, we see that (subsum) generalizes the conversion rule.

Rules (Bvar) through (BΠ-elim) are the Bounded analogues of rules (var) through (Π-elim). The rule (BΠ-elim) expresses what a bounded quantification means; if b has type $\Pi x \leq a \,:\, A.\ B$, then it may only be applied to terms a' which are a subtype of a (and also have type A).

Definition 8 (Subtyping rules).

$$(\leq\text{-conv}) \qquad \frac{a =_\beta b}{\Gamma \vdash a \leq b}$$

$$(\leq\text{-trans}) \qquad \frac{\Gamma \vdash a \leq b \quad \Gamma \vdash b \leq c}{\Gamma \vdash a \leq c}$$

$$(\leq\text{-var}) \qquad \frac{x \leq a : A \in \Gamma}{\Gamma \vdash x \leq a}$$

$$(\leq\text{-}\Pi) \qquad \frac{\Gamma \vdash A' \leq A \quad \Gamma, x : A' \vdash B \leq B'}{\Gamma \vdash (\Pi x : A.\, B) \leq (\Pi x : A'.\, B')}$$

$$(\leq\text{-B}\Pi) \qquad \frac{\Gamma, x \leq a : A \vdash B \leq B'}{\Gamma \vdash (\Pi x \leq a : A.\ B) \leq (\Pi x \leq a : A.\ B')}$$

$$(\leq\text{-}\lambda) \qquad \frac{\Gamma, x : A \vdash b \leq b'}{\Gamma \vdash (\lambda x : A.\, b) \leq (\lambda x : A.\, b')}$$

$$(\leq\text{-app}) \qquad \frac{\Gamma \vdash b \leq b'}{\Gamma \vdash b\, a \leq b'\, a}$$

No subtyping rule depends on a typing judgment. As a consequence, $\leq$ is a relation on pseudo-terms. The rules ($\leq$-conv), ($\leq$-trans), ($\leq$-var), ($\leq$-λ) and ($\leq$-app) are formulated as usual. Rule ($\leq$-Π) is a general formulation of the usual subtyping rule for $\rightarrow$-types. We can use this rule even if there is no interesting subtyping on B (or A). E.g. $\Gamma_{\text{init}} \vdash \texttt{Int} \rightarrow * \leq \texttt{Nat} \rightarrow *$.

Rule ($\leq$-BΠ) is called the kernel-Fun rule, since it appears in Cardelli and Wegner's original Fun calculus [CW85]. There are alternatives for this rule, but ($\leq$-BΠ) has the best meta-theoretical properties [CG92,Pie94,CP94].

2.3 Examples of $PTS^{\leq}$s

We show how examples of systems of the λ-cube [Bar92] extended with subtyping fit in our framework. These systems have $\mathcal{S} = \{*, \Box\}$, $\mathcal{A} = \{(* : \Box)\}$ and $\mathcal{R}$ consists only of pairs. The systems are extended with subtyping by choosing $\mathcal{S}^{\leq} = \{\Box\}$, and taking for $\mathcal{R}^{\leq}$ a subset of rules $(\Box, s_2)$ from $\mathcal{R}$. We do not repeat these common properties. We also briefly discuss some approaches to the meta-theories, and why these approaches fail to work for $PTS^{\leq}s$.

The $PTS^{\leq}$ $\lambda{\rightarrow}^{\leq}$ is specified by $\mathcal{R} = \{(*, *)\}$ and $\mathcal{R}^{\leq} = \emptyset$. Since $\Box \in \mathcal{S}^{\leq}$ we can make and use subtyping declarations. The system $\lambda{\rightarrow}^{\leq}$ is the standard extension of $\lambda{\rightarrow}$ with subtyping, e.g. defined in [Com95], and is the basis of [Car88].

The system $\lambda 2^{\leq}$ is specified by $\mathcal{R} = \{(*, *), (\Box, *)\}$ and $\mathcal{R}^{\leq} = \{(\Box, *)\}$. Since $(\Box, *) \in \mathcal{R}^{\leq}$, we can make bounded quantifications. The system $\lambda 2^{\leq}$ is equal to kernel-Fun [CW85], except for their *Top* type. The subtyping rules for the *Top* type in $\lambda 2^{\leq}$ would be:

$$(\leq\text{-}Top)\ \frac{\Gamma \vdash A : *}{\Gamma \vdash A \leq Top}$$

We didn't include *Top*, since this subtyping rule essentially depends on a typing judgment. This is incompatible with our approach, where subtyping does not depend on typing. The absence of *Top* types in $PTS^{\leq}s$ is not as bad as it seems, since we also have ordinary quantifications. The system $F_{\leq}$ (e.g. [CG92]) is equal to $\lambda 2^{\leq}$ except for a *Top*-type and a more liberal ($\leq$-BΠ) rule.

The system $\lambda\omega^{\leq}$ is specified by $\mathcal{R} = \{(*,*),(\Box,*),(\Box,\Box)\}$, $\mathcal{R}^{\leq} = \{(\Box,*)\}$. The difference with $\lambda 2^{\leq}$ is that $(\Box,\Box) \in \mathcal{R}$, resulting in type-constructors. This has two effects on subtyping. First, we have bounded quantifications were the bound is a type-constructor. Second, we have lifted subtyping on type-constructors by rules ($\leq$-app) and ($\leq$-λ). The system $\lambda\omega^{\leq}$ is equal to $F^{\omega}_{\leq}$ [PS94], except that $F^{\omega}_{\leq}$ has a family of *Top*-types. In [Com95] a further extension of $F^{\omega}_{\leq}$ is given. The meta-theory is developped in three stages: first the theory about typing type-constructors, then about the subtyping judgment and finally about typing programs. This cannot be done in general for $PTS^{\leq}s$, since typing for the various categories of terms is mutually dependent. In [PS94] the Minimal Typing property is proved using the typing algorithm, whereas we prove this property separately.

The system $\lambda\omega^{\leq^{+}}$ is specified by $\mathcal{R} = \{(*,*),(\Box,*),(\Box,\Box)\}$ and $\mathcal{R}^{\leq} = \{(\Box,*),(\Box,\Box)\}$. The difference with $\lambda\omega^{\leq}$ is that we have $(\Box,\Box) \in \mathcal{R}^{\leq}$. With this rule, we can type bounded constructor abstractions, i.e. terms like $\lambda X \leq \texttt{Int} : *.\ X \to X$. The system $\lambda\omega^{\leq^{+}}$ corresponds with the system $\mathcal{F}^{\omega}_{\leq}$ defined in [CG97]. There are two differences. First, we have no *Top*-types. Second, we do not have subtyping on these bounded abstractions, because it destroys the property we formulated in lemma 5. The meta-theory developed in [CG97] follows a quite different approach than works mentioned above and our work; by giving a typed operational semantics they solve the mutual dependence between the typing and subtyping judgments occurring in $\mathcal{F}^{\omega}_{\leq}$. We don't know whether this approach is applicable to $PTS^{\leq}s$.

The system $\lambda P^{\leq}$ is specified by $\mathcal{R} = \{(*,*),(*,\Box)\}$ and $\mathcal{R}^{\leq} = \emptyset$. The rule $(*,\Box) \in \mathcal{R}$ gives types depending on programs and corresponding type-constructors, for which lifted subtyping is possible. The system $\lambda P_{\leq}$ as described in [AC96] is roughly the same as this $PTS^{\leq}$, and typing on programs in both systems is exactly equivalent. This system is the first calculus discussed here with mutual dependency between programs and type-constructors. They avoid circularities between lemmas about typing and lemmas about subtyping by syntactically distinguishing β-reduction on programs and on type-constructors. This syntactical distinction is impossible in $PTS^{\leq}s$. Just as in [PS94] the Minimal Typing property is proved using the typing algorithm.

The $PTS^{\leq}$ $\lambda C^{\leq}$ is specified by $\mathcal{R} = \mathcal{S}^2$ and $\mathcal{R}^{\leq} = \{(\Box,*),(\Box,\Box)\}$. This is the Calculus of Constructions [CH88], the most powerful system in the λ-cube, extended with subtyping and bounded quantifications. It includes all systems given above. The system $\lambda C^{\leq}$ hasn't come up in the literature.

The $PTS^{\leq}$ $\lambda C^{\leq^-}$ is specified by $\mathcal{R} = \mathcal{S}^2$ and $\mathcal{R}^{\leq} = \emptyset$. This is the subsystem of $\lambda C^{\leq}$ where bounded quantifications have been left out. The $PTS^{\leq}$ $\lambda C^{\leq^-}$ is exactly the same as the system defined in [Che97]. Here, programs and type-constructors are also mutually dependent, but the typing judgments occurring in subtyping rules all have the simple form $\Gamma \vdash A : s$. Using this in combination with the specific rules $\mathcal{R}$ of $\lambda C^{\leq^-}$, enough meta-theory for typing can be proved before subtyping is examined. This method does not work for $PTS^{\leq}s$, since terms involved in the subtype relation are not always typable with a sort.

2.4 Alternatives for Rules

In this section we discuss two alternatives for our rules, and why we have rejected these alternatives. Some other alternatives were given in section 2.3.

The ($\leq$-λ) rule can be generalized, so that $\Gamma \vdash A' \leq A$ and $\Gamma, x : A' \vdash B \leq B'$ imply $\Gamma \vdash \lambda x : A.\, B \leq \lambda x : A'.\, B'$. We have chosen for ($\leq$-$\lambda$) because it is simpler and the generalization does not have any effect in most $PTS^{\leq}s$.

We first considered a more constrained version of the (subsum) rule:

$$\text{(subsum')} \quad \frac{\Gamma \vdash b : B \quad \Gamma \vdash B' : s \quad \Gamma \vdash s : s' \quad s' \in \mathcal{S}^{\leq} \quad \Gamma \vdash B \leq B'}{\Gamma \vdash b : B'}$$

We did so, because we believed the meta-theory would be easier because of the additional constraints on s. It turned out, however, that the meta-theory was more difficult, so we rejected this rule.

3 Meta-theory

In this section we develop the meta-theory for $PTS^{\leq}s$. First we establish a number of properties of the subtyping judgment (section 3.1). We are able to do so, because subtyping does not depend on typing. Using these properties, we prove that the Subject Reduction property holds for all $PTS^{\leq}s$ (section 3.2). In functional $PTSs$ (without subtyping), we have that every term has a unique type (modulo β-conversion). Subtyping destroys this property, but every term has a so-called minimal type. Section 3.3 shows this for functional $PTS^{\leq}s$. First, we mention the Church-Rosser property for β-reduction.

Theorem 1. *If $a =_\beta b$ then there is a c with $a \rhd\!\!\!> _\beta c$ and $b \rhd\!\!\!>_\beta c$.*

3.1 Properties of Subtyping

Unfortunately, the subtyping rules given in definition 8 are quite intractable; it is hard to prove properties about them. They are so intractable, because there is a lot of redundancy in the subtyping rules; there can be several quite different derivations of the same subtyping judgment. Therefore we introduce a set of more restricted rules, equivalent to the set in definition 8, but with only a little redundancy. This set of restricted rules behaves much better, and in particular has the following crucial property: a subtype derivation using the restricted rules does not *introduce* untypable terms. To be more precise, if the terms in the conclusion of such a subtyping judgment are typable, then all terms occurring in the derivation of this judgment are typable. We will only be able to show this at the end of section 3.2, in lemma 6. The original rules do not have this property.

Two rules are responsible for the intractibility of the set of original rules, namely ($\leq$-trans) and ($\leq$-app). We discuss for both rules which problems they cause, and which restricted rules (in definition 10) below replace them.

The ($\leq$-trans) rule. As in most systems with subtyping, this rule is the most responsible for the intractibility of the original subtyping rules, and we have a similar solution. Recall that ($\leq$-trans) allows deriving $\Gamma \vdash a \leq c$ from $\Gamma \vdash a \leq b$ and $\Gamma \vdash b \leq c$.

It can be used at any moment in a derivation, since there are no restrictions on the form of the conclusions a and c. Even worse, the term b in the premises cannot be determined from a and c, and even when a and c are typable terms, b can be a non-typable term. It is essential in two situations. First, using ($\leq$-trans) and ($\leq$-conv) we can derive $\Gamma \vdash a \leq b$ from $\Gamma \vdash a' \leq b'$ whenever $a \rhd\!\!\!>_\beta a'$ and $b \rhd\!\!\!>_\beta b'$. This use of the ($\leq$-trans) rule is taken over by the more direct ($\leq$-red) rule (in definition 10). As a side-effect, the ($\leq$-conv) rule can be simplified to ($\leq$-refl).

Second, the ($\leq$-trans) rule is necessary when the term a is a variable x, and c is not convertible to a. This use of transitivity is taken over by the ($\leq$-transvar) rule.

In all other cases, the ($\leq$-trans) rule is not essential, because it can be "pushed" upwards through the derivation, ending only in one of the situations sketched above. This property, sometimes called "Transitivity Elimination" [Com95,Che97], is formally proved in lemma 4.

The ($\leq$-app) rule. Another source of intractability is the ($\leq$-app) rule, which says that $\Gamma \vdash b\ a \leq b'\ a$ is derivable from $\Gamma \vdash b \leq b'$. It is not apparent that this rule gives problems, but consider the case when b is an abstraction: instead of using ($\leq$-app), we could also reduce $b\ a$ using the ($\leq$-trans) and ($\leq$-conv) rules (in the same way as above), and proceed from there. If a judgment of this form holds, it can always be derived without ($\leq$-app). So for this kind of judgments, we do not need the ($\leq$-app) rule, and we would like to remove it, to have less redundancy.

However, it is essential in two situations: First, if term b (in judgment $b\ a \leq b'\ a$) is a variable. This is catered for by the ($\leq$-transvar) rule given in definition 10, where ($\leq$-app) is combined with ($\leq$-trans) and ($\leq$-var). Second, if the term b is a (bounded or unbounded) Π-type. For example we can derive with ($\leq$-app) that $\Gamma_{\text{init}}, \mathtt{B}\!:\!* \vdash (\Pi x\!:\!\mathtt{B}.\,\mathtt{Nat})\ \mathtt{B} \leq (\Pi x\!:\!\mathtt{B}.\,\mathtt{Int})\ \mathtt{B}$. The reader might reject this situation by saying that $b\ a$ is never typable if b is a Π-type. This is true, however subtyping is defined on pseudo-terms, rather than (typable) terms, so we cannot ignore this situation here. This is a consequence of the decision that the subtyping rules do not depend on typing judgments.

This situation is catered for by the ($\leq$-Πapp) rule below. In the end, when we have shown Subject Reduction, we will see we do not need ($\leq$-Πapp) rule after all (lemma 6). This seems to be a contradiction with the statement that ($\leq$-app) is essential in this situation. But it is not a contradiction: for *pseudoterms* the ($\leq$-Πapp) rule is essential, and for *typable terms* it is redundant. In other words, the rule is needed only as a catalyst, in order to prove the meta-theory for subtyping as smooth as possible.

In other situations, the ($\leq$-app) rule is not essential, which is proved in lemma 3.

Definition 9. *A term a is a Π-type if $a \equiv \Pi x{:}B.\,C$ or $a \equiv \Pi x{\leq}b : B.\,C$ for some b, B and C.*

Definition 10 (Subtyping rules).

($\leq$-refl) $$\frac{}{\Gamma \vdash b \leq b}$$

($\leq$-red) $$\frac{a \rhd\!\!>_\beta a' \quad b \rhd\!\!>_\beta b' \quad \Gamma \vdash a' \leq b'}{\Gamma \vdash a \leq b}$$

($\leq$-transvar) $$\frac{x{\leq}a : A \in \Gamma \quad \Gamma \vdash a\, c_1\, c_2\, \ldots\, c_n \leq b \quad n \geqslant 0}{\Gamma \vdash x\, c_1\, c_2\, \ldots\, c_n \leq b}$$

($\leq$-Πapp) $$\frac{\Gamma \vdash a \leq b \quad a \text{ is a } \Pi\text{-type} \quad b \text{ is a } \Pi\text{-type} \quad n \geqslant 1}{\Gamma \vdash a\, c_1\, c_2\, \ldots\, c_n \leq b\, c_1\, c_2\, \ldots\, c_n}$$

($\leq$-Π) $$\frac{\Gamma \vdash A' \leq A \quad \Gamma, x : A' \vdash B \leq B'}{\Gamma \vdash (\Pi x{:}A.\,B) \leq (\Pi x{:}A'.\,B')}$$

($\leq$-BΠ) $$\frac{\Gamma, x{\leq}a : A \vdash B \leq B'}{\Gamma \vdash (\Pi x{\leq}a : A.\,B) \leq (\Pi x{\leq}a : A.\,B')}$$

($\leq$-λ) $$\frac{\Gamma, x : A \vdash b \leq b'}{\Gamma \vdash (\lambda x{:}A.\,b) \leq (\lambda x{:}A.\,b')}$$

Convention. From this point onwards, we will always use the subtyping rules of definition 10. We will refer to the original, liberal rules (definition 8) using $\Gamma \vdash_l a \leq b$. Note that the typing rules use the liberal rules.

Most other works [PS94,AC96,CG97,Che97] also have a set of alternative typing rules, roughly similar to ours, i.e. with ($\leq$-app) and ($\leq$-trans) replaced by ($\leq$-transvar). There are two important differences. First, none of the sets of alternative rules in the literature have the rule ($\leq$-Πapp). Second, the alternative rules differ considerably in the approach of reduction in subtyping judgments (our rule ($\leq$-red)).

In the rest of this section we show some properties of the new subtyping rules, including some Generation properties, equivalence with the liberal rules, and the Substitution lemmas. But first we show that subtyping is closed under β-conversion, i.e. β-converting a term in a subtyping judgment keeps it derivable (theorem 2). Even the number of interesting steps — i.e. not ($\leq$-red) — in the derivation for this judgment stays the same.

Convention. We use the letter Υ as meta-variable for derivation trees, because it resembles a willow.

Definition 11. *The NR-height of a subtyping derivation Υ, written as NR-height(Υ), is the height of Υ, not counting applications of the ($\leq$-red) rule. NR stands for "Not counting Reductions". We write $\Upsilon < \Upsilon'$ as shorthand for NR-height(Υ) $<$ NR-height(Υ'), and similarly for $\leq$.*

Definition 12. $\rhd_\beta$, $\rhd\!\!\!\rhd_\beta$ *and* $=_\beta$ *are extended to contexts in the usual way.*

Theorem 2 ($\leq$-Conversion-closed). *Suppose $\Gamma =_\beta \Gamma'$ and $a =_\beta a'$ and $b =_\beta b'$. If Υ derives $\Gamma \vdash a \leq b$ then there is a $\Upsilon' \leq \Upsilon$ such that Υ' derives $\Gamma' \vdash a' \leq b'$.*

This property is very important, since it allows us to convert terms in a subtyping judgment without increasing the NR-height. This makes the NR-height a very useful induction measure. An example of use of this lemma is in the proof of lemma 4.

Now we prove Generation properties for subtyping. We mention only

Lemma 1 ($\leq$-Generation).

1. *If $\Gamma \vdash \Pi x{:}A_1.\, B_1 \leq \Pi x{:}A_2.\, B_2$ then $\Gamma \vdash A_2 \leq A_1$ and $\Gamma, x{:}A_2 \vdash B_1 \leq B_2$.*
2. *If $\Gamma \vdash (\Pi x{\leq}a_1 : A_1.\, B_1) \leq (\Pi x{\leq}a_2 : A_2.\, B_2)$ then $a_1 =_\beta a_2$ and $A_1 =_\beta A_2$ and $\Gamma, x{\leq}a_1 : A_1 \vdash B_1 \leq B_2$.*

One of the reasons for introducing the restricted subtyping rules is that generation properties like these are very hard to prove for the original subtyping rules.

The first Substitution property for subtyping allows us to replace an unbounded variable y with any term c. Note that c does not have to be typable.

Lemma 2 ($\leq$-Substitution). *If $\Gamma, y : C, \Gamma' \vdash a \leq b$ and $\Gamma, \Gamma'[y := c]$ is a pseudocontext then $\Gamma, \Gamma'[y := c] \vdash a[y := c] \leq b[y := c]$.*

The other part of the Substitution property — replacing a *bounded* variable — is proved at the end of this section. We use the $\leq$-Substitution property just given for showing admissibility of the ($\leq$-app) rule.

Lemma 3 (App-admissible). *If $\Gamma \vdash a \leq b$ then $\Gamma \vdash a\, c \leq b\, c$.*

Proof. By induction on the NR-height of the derivation, and by case distinction to the last derivation rule other than ($\leq$-red). □

Note that this property depends essentialy on the ($\leq$-Πapp) rule; for an example see the discussion about the ($\leq$-app) rule at the start of this section.

Lemma 4 (Trans-admissible). *If $\Gamma \vdash a \leq b$ and $\Gamma \vdash b \leq c$ then $\Gamma \vdash a \leq c$.*

Proof. By induction on the sum of the NR-heights of the derivations. Consider in each derivation the last rule other than ($\leq$-red). So we have $a \rhd\!\!>_\beta a'$, $b \rhd\!\!>_\beta b'$, $\Gamma \vdash a' \leq b'$, $b \rhd\!\!>_\beta b''$, $c \rhd\!\!>_\beta c'$ and $\Gamma \vdash b'' \leq c'$, where the last rule in both derivation is not ($\leq$-red). Now we make a case distinction to the last rule used for deriving $\Gamma \vdash a' \leq b'$. Cases ($\leq$-refl) and ($\leq$-transvar) are straightforward, using theorem 2 (and the IH for ($\leq$-transvar)). For the other cases, we make case distinction to the last rule used for deriving $\Gamma \vdash b'' \leq c'$. Since $b' =_\beta b''$, it is easy to see this derivation uses either ($\leq$-refl), then we are done by theorem 2), or the same rule as for $\Gamma \vdash a' \leq b'$, then we finish the proof using theorem 2 and the IH. □

Now all liberal rules (definition 8) have been shown to be admissible, so every subtyping judgment derivable with the liberal rules is also derivable with the restricted rules.

Theorem 3 (Equivalence). $\Gamma \vdash_l a \leq b \Longleftrightarrow \Gamma \vdash a \leq b$.

Proof. Soundness ($\Longleftarrow$) is easy. Completeness ($\Longrightarrow$) follows from the admissibility lemmas 3 and 4; admissibility of ($\leq$-conv) and ($\leq$-var) is simple. □

Soundness only holds because the liberal rules do not have typing judgments as premise. The equivalence allows replacing each premise of the form $\Gamma \vdash_l a \leq b$ in typing rules by the premise $\Gamma \vdash a \leq b$ without changing the set of derivable typing judgments, so we can use properties like lemma 1 when proving properties about the typing judgment.

Using the admissibility of ($\leq$-trans) and ($\leq$-app), we prove the other Substitution property.

Lemma 5 ($\leq$-Substitution). *If $\Gamma \vdash c \leq c'$ and $\Gamma, y \leq c' : C, \Gamma' \vdash a \leq b$ and $\Gamma, \Gamma'[y := c]$ is a pseudocontext then $\Gamma, \Gamma'[y := c] \vdash a[y := c] \leq b[y := c]$.*

The $\leq$-Substitution properties are essential to prove Subject Reduction, via the Substitution properties for typing.

3.2 Subject Reduction

The proof of Subject Reduction goes along the same lines as in ordinary *PTSs*, and is longer but not more complicated. We first have to prove the usual Substitution properties, using $\leq$-Substitution, then prove the Generation and Correctness of Types lemmas and then proceed to Subject Reduction.

Theorem 4 (Subject Reduction). *If $\Gamma \vdash a : A$ and $a \rhd_\beta b$ then $\Gamma \vdash b : A$.*

Proof. By strengthening the IH as usual, and by induction on the derivation. All cases go straightforward by IH, except in the first clauses, when a is a redex. These cases are proved in a similar way as in [Bar92], now using lemma 1. □

Subject Reduction has an important consequence on subtyping derivations: the subtyping rules do not introduce untypable terms. In other words, if the terms in the conclusion are typable, then all terms in the derivation are typable.

Lemma 6. *Suppose Υ derives $\Gamma \vdash a \leq b$, and $\Gamma \vdash a : A$ and $\Gamma \vdash b : B$ hold. Then for all subderivations Υ' of Υ, where Υ' shows $\Gamma' \vdash c \leq d$ for some c and d, there are C, D such that $\Gamma' \vdash c : C$ and $\Gamma' \vdash d : D$.*

This lemma has the following consequence. The ($\leq$-Πapp) rule is not used in subtyping derivations with typable terms in the conclusion. This follows from the fact that the conclusion of the ($\leq$-Πapp) rule contains two terms that are never typable, since the terms consist of an application of a Π-type to one or more arguments.

3.3 Minimal Typing

For ordinary functional *PTSs*, we have Uniqueness of Typing, which says that a term has only one type, modulo β-conversion. We do not have unique types in $PTS^{\leq}s$, since by the subsumption rule a term can have different types. We will show that we do have a weaker property, *Minimal Typing*. This means that every typable term has a minimal type.

Definition 13. *Term a has* minimal *type A in Γ, notated as $\Gamma \vdash_m a : A$, if $\Gamma \vdash a : A$ and for all B $\Gamma \vdash a : B \implies \Gamma \vdash A \leq B$.*

Minimal Typing is important for type-checking, since the problem "does term a have type B" can then be split into the simpler problems "compute a minimal type A for a" and "is A a subtype of B".

Minimal Typing holds only for functional $PTS^{\leq}s$:

Definition 14. *A $PTS^{\leq}$ $\lambda(\mathcal{S}, \mathcal{A}, \mathcal{R}, \mathcal{S}^{\leq}, \mathcal{R}^{\leq})$ is* functional *if*

$$\begin{aligned} (s : s') \in \mathcal{A} \text{ and } (s : s'') \in \mathcal{A} &\implies s' \equiv s'' \\ (s_1, s_2, s_3) \in \mathcal{R} \text{ and } (s_1, s_2, s_3') \in \mathcal{R} &\implies s_3 \equiv s_3' \\ (s_1, s_2, s_3) \in \mathcal{R}^{\leq} \text{ and } (s_1, s_2, s_3') \in \mathcal{R}^{\leq} &\implies s_3 \equiv s_3' \end{aligned}$$

However, Minimal Typing is not easily proved. A direct proof by induction on the structure of the term (say a) fails, because of two problems.

First, we sometimes need the induction hypothesis for a *type* of a, instead of a subterm. We solve this problem by first proving a property called Weak Minimal Typing (lemma 8), which is strong enough to replace the IH for the type of a.

Second, if a is an application $b\ c$, we get by the IH a minimal type B of b. But we are not interested in B itself, but in B', the least supertype of B that is a Π-type; only from B' we can calculate a minimal type of the application $b\ c$. We obtain this B' by introducing a new kind of reduction $\rhd_{\mathsf{wh}\beta\sigma}$ (weak head β σ), which reduces B to B'.

Convention. In this section we consider only functional $PTS^{\leq}s$.

First we show the Weak Minimal Typing property that says that common types of a term have a (common) lower bound, and then we define $\rhd_{wh\beta\sigma}$ reduction.

Definition 15. *Terms a and b have a* lower bound *in Γ, $\Gamma \vdash a \sqcup b$, if there is a c such that $\Gamma \vdash c \leq a$ and $\Gamma \vdash c \leq b$.*

Lemma 7.

- *If $\Gamma \vdash s_1 \sqcup s_2$ then $s_1 \equiv s_2$.*
- *If $\Gamma \vdash (\Pi x{:}A_1.\, B_1) \sqcup (\Pi x{:}A_2.\, B_2)$ then $\Gamma, x{:}A_1 \vdash B_1 \sqcup B_2$.*
- *If $\Gamma \vdash (\Pi x{\leq}a_1 : A_1.\, B_1) \sqcup (\Pi x{\leq}a_2 : A_2.\, B_2)$ then $\Gamma, x{\leq}a_1 : A_1 \vdash B_1 \sqcup B_2$.*
- *Not $\Gamma \vdash (\Pi x{:}A_1.\, B_1) \sqcup (\Pi x{\leq}a_2 : A_2.\, B_2)$.*
- *If $\Gamma \vdash A \sqcup s$ then $\Gamma \vdash A \leq s$.*

Lemma 8 (Weak Minimal Typing).
If $\Gamma \vdash a : A$ and $\Gamma \vdash a : B$ then $\Gamma \vdash A \sqcup B$.

Proof. By induction on the structure of a. Apply the Generation lemma to $\Gamma \vdash a : A$ and $\Gamma \vdash a : B$. For some cases, we need lemmas 7, 4, 2 and 5. □

Note that if we read "$=_\beta$" for "$\leq$", then "$\sqcup$" is equal to "$=_\beta$", and we have the Uniqueness of Types property. Using Weak Minimal Typing, we prove the following lemma that relates the types of two terms that are in the subtype relation. We need also this lemma in the proof of Minimal Typing.

Lemma 9. *If $\Gamma \vdash a \leq b$ and $\Gamma \vdash a : A$ and $\Gamma \vdash b : B$ then $\Gamma \vdash A \sqcup B$.*

Proof. By induction on the subtyping derivation, using lemmas 7 and 6, and Weak Minimal Typing. □

We define $\rhd_{wh\beta\sigma}$ reduction as the union of $\rhd_{wh\beta}$ and $\rhd_{wh\sigma}$, where $\rhd_{wh\beta}$ is the usual weak head restriction of $\rhd_\beta$, and $\rhd_{wh\sigma}$ reduces a term $x\, b_1 \ldots b_n$ to $c\, b_1 \ldots b_n$ if c is the bound of x. This reduction is the weak head restriction of the so-called Γ-reduction found in [PS94,Che97,CG97].

Definition 16. *The relation $\_ \vdash \_ \rhd_{wh\beta\sigma} \_$ is defined as follows:*

$$
\begin{array}{rcl}
 & & \Gamma \vdash (\lambda x{:}A.\, b)\, a \rhd_{wh\beta\sigma} b[x := a] \\
 & & \Gamma \vdash (\lambda x{\leq}a' : A.\, b)\, a \rhd_{wh\beta\sigma} b[x := a] \\
x{\leq}a : A \in \Gamma & \Longrightarrow & \Gamma \vdash x \rhd_{wh\beta\sigma} a \\
\Gamma \vdash a \rhd_{wh\beta\sigma} a' & \Longrightarrow & \Gamma \vdash a\, b \rhd_{wh\beta\sigma} a'\, b
\end{array}
$$

$\Gamma \vdash \_ \rhd\!\!\!\rhd_{wh\beta\sigma} \_$ is the reflexive and transitive closure of $\Gamma \vdash \_ \rhd_{wh\beta\sigma} \_$.

Lemma 10.

- *If* $\Gamma \vdash A \leq s$ *then* $\Gamma \vdash A \rhd\!\!>_{wh\beta\sigma} s$.
- *If* $\Gamma \vdash A \leq (\Pi x{:}B.\,C)$ *then* $\Gamma \vdash A \rhd\!\!>_{wh\beta\sigma} (\Pi x{:}B'.\,C')$ *and* $\Gamma \vdash (\Pi x{:}B'.C') \leq (\Pi x{:}B.\,C)$.
- *If* $\Gamma \vdash A \leq (\Pi x{\leq}b : B.\ C)$ *then* $\Gamma \vdash A \rhd\!\!>_{wh\beta\sigma} (\Pi x{\leq}b' : B'.\ C')$ *and* $\Gamma \vdash (\Pi x{\leq}b' : B'.\ C') \leq (\Pi x{\leq}b : B.\ C)$.

Theorem 5 (Minimal Typing for functional $PTS^{\leq}s$). *If* a *is typable in* Γ, *there is a type* M *with* $\Gamma \vdash_m a : M$.

Proof. By induction on the structure of a. Use the Generation lemma. For every case we have two parts. First, find an M such that $\Gamma \vdash a : M$. Second, show that this M is minimal. We sketch the proof of two cases.

If $a \equiv b\,c$ then $\Gamma \vdash b : \Pi x{:}C.D$ or $\Gamma \vdash b : \Pi x{\leq}c' : C.\ D$. Assume we are in the first case, and $\Gamma \vdash c : C$. By IH $\Gamma \vdash_m b : M_1$, so $\Gamma \vdash M_1 \leq \Pi x{:}C.D$. By lemma 10 $\Gamma \vdash M_1 \rhd\!\!>_{\mathrm{wh}\beta\sigma} \Pi x : C'.\,D'$ and $\Gamma \vdash \Pi x{:}C'.\,D' \leq \Pi x{:}C.D$. A minimal type of a is now $D'[x := c]$.

If $a \equiv \lambda x : C.\,d$ then $\Gamma, x{:}C \vdash d : D$ and $\Gamma \vdash \Pi x{:}C.D : s_3$, which gives $\Gamma \vdash C : s_1$ and $\Gamma, x{:}C \vdash D : s_2$. By IH $\Gamma, x{:}C \vdash_m d : M_1$, so $\Gamma, x{:}C \vdash M_1 \leq D$. By Correctness of Types $\Gamma, x{:}C \vdash M_1 : s_2'$, and by lemma 9 $\Gamma, x : C \vdash s_2 \sqcup s_2'$ and hence by lemma 7 $s_2 \equiv s_2'$. So $\Gamma \vdash \Pi x{:}C.M_1 : s_3$ and we can derive $\Gamma \vdash a : \Pi x{:}C.M_1$. It is easy to show that this type is minimal. □

Finally, each PTS can be seen as a $PTS^{\leq}$ with $\mathcal{S}^{\leq}$ and $\mathcal{R}^{\leq}$ empty.

Theorem 6. *Take the PTS* P *with specification* $(\mathcal{S}, \mathcal{A}, \mathcal{R})$ *and the* $PTS^{\leq}$ S *with specification* $(\mathcal{S}, \mathcal{A}, \mathcal{R}, \emptyset, \emptyset)$. *Then* $\Gamma \vdash_P a : A \Longleftrightarrow \Gamma \vdash_S a : A$.

4 Conclusions

In this paper we defined the framework of Pure Type Systems with Subtyping, an extension of the PTSs with subtyping, bounded quantification and lifted subtyping. We do not have subtyping on sorts (e.g. as in [Luo89]), or coercive subtyping, which means that subtyping between existing types can be defined with coercions [Bar96]. Many existing type systems with subtyping can be seen as members of our framework, viz. $\lambda{\to}^{\leq}$, $F_{\leq}$, $F^{\omega}_{\leq}$, $\mathcal{F}^{\omega}_{\leq}$, $\lambda P_{\leq}$ and the calculus of [Che97]. Other members, like $\lambda C^{\leq}$, are new systems which have promising features, both applicable in programming languages and in theorem proving.

We developed the meta-theory for $PTS^{\leq}s$, including Subject Reduction and Minimal Typing. In order to prove these properties, we adopted the design decision that the subtyping rules do not depend on the typing rules. This allows us to develop the meta-theory for the subtyping judgment before the theory of the typing judgment. However, this decision alone was not sufficient: we had to give a reformulation of the subtyping rules (definition 10), that behaved better. In particular, the reformulated rules do not introduce untypable terms: if the terms in the conclusion are typable, so are all terms in the premises of the rule (lemma 6). This property is important in our proof of Minimal Typing.

The decision has two drawbacks. First, $\leq$ is defined for all pseudoterms instead of only for (typable) terms. Similarly, the meta-theory for the subtyping judgment is done for pseudoterms. This forced us to introduce a weird rule, ($\leq$-Πapp), to have equivalence between the original rules and the reformulated rules on pseudoterms. This rule is weird, since it only relates untypable terms, but we showed in lemma 6 that ($\leq$-Πapp) is never used in sensible subtyping derivations. Second, the design decision makes it hard to extend the $PTS^{\leq}s$ with some features, like *Top*-types or subtyping on bounded operator abstractions (section 2.3). For many systems, these extensions make little sense and this drawback has no effect.

A type-checking algorithm and decidability of typing for a range of $PTS^{\leq}s$ is beyond the scope of this paper, but will appear elsewhere [Zwa99].

Acknowledgments

The author wishes to thank Adriana Compagnoni, Kees Hemerik, Kruseman Aretz, Erik Poll and an anonymous referee for suggestions and corrections.

References

[AC96] David Aspinall and Adriana Compagnoni. Subtyping dependent types. In *11th Annual IEEE Symposium on Logic in Computer Science, New Brunswick, New Jersey, USA*, 1996.

[Bar92] H. Barendregt. Lambda calculi with types. In S. Abramsky, D. M. Gabbai, and T. S. E. Maibaum, editors, *Handbook of Logic in Computer Science*, volume 2. Oxford University Press, 1992.

[Bar96] Gilles Barthe. Implicit coercions in type systems. In *Proceedings of the International Workshop Types '95, Torino, Italy*, volume 1158 of *Lecture Notes in Computer Science*, 1996.

[Car88] Luca Cardelli. A semantics of multiple inheritance. *Information and Computation*, 176:138–164, 1988.

[CG92] Pierre-Louis Curien and Giorgio Ghelli. Coherence of subsumption, minimum typing and the type checking in $F_{\leq}$. *Mathematical Structures in Computer Science*, 2(1):55–91, 1992.

[CG97] Adriana Compagnoni and Healfdene Goguen. Typed operational semantics for higher order subtyping. Technical Report ECS-LFCS-97-361, University of Edinburgh, 1997.

[CH88] Thierry Coquand and Gérard P. Huet. The calculus of constructions. *Information and Computation*, 76:95–120, 1988.

[Che97] Gang Chen. Subtyping calculus of construction. In *Mathematical Foundations of Computer Science (MFCS'97), Bratislava, Slovakia*, volume 1295, pages 189–198. Springer-Verlag, August 1997.

[Com95] Adriana B. Compagnoni. *Higher-Order Subtyping with Intersection Types.* PhD thesis, University of Nijmegen, The Netherlands, 1995.

[CP94] Giuseppe Castagna and Benjamin C. Pierce. Decidable bounded quantification. In *Proceedings of the 21st ACM Symposium on Principles of Programming Languages (POPL), Portland, Oregon*. ACM, 1994.

[CW85] Luca Cardelli and Peter Wegner. On understanding types, data abstraction, and polymorphism. *Computing Surveys*, 17(4):471–522, 1985.

[Luo89] Zhaohui Luo. ECC: an Extended Calculus of Constructions. In *Proceedings of IEEE 4th Annual Symposium on Logic in Computer Science (LICS'89), Asilomar, California*, 1989.

[Pie94] Benjamin C. Pierce. Bounded quantification is undecidable. *Information and Computation*, 112(1):131–165, 1994.

[PS94] Benjamin Pierce and Martin Steffen. Higher-order subtyping. In *IFIP Working Conference on Programming Concepts, Methods and Calculi (PROCOMET)*, 1994.

[Zwa99] Jan Zwanenburg. *An Object Oriented Programming Logic Based on Type Theory*. PhD thesis, Eindhoven University of Technology, 1999. To appear.

Author Index

Lecture Notes in Computer Science

For information about Vols. 1–1495
please contact your bookseller or Springer-Verlag

Vol. 1496: W.M. Wells, A. Colchester, S. Delp (Eds.), Medical Image Computing and Computer-Assisted Intervention – MICCAI'98. Proceedings, 1998. XXII, 1256 pages. 1998.

Vol. 1497: V. Alexandrov, J. Dongarra (Eds.), Recent Advances in Parallel Virtual Machine and Message Passing Interface. Proceedings, 1998. XII, 412 pages. 1998.

Vol. 1498: A.E. Eiben, T. Bäck, M. Schoenauer, H.-P. Schwefel (Eds.), Parallel Problem Solving from Nature – PPSN V. Proceedings, 1998. XXIII, 1041 pages. 1998.

Vol. 1499: S. Kutten (Ed.), Distributed Computing. Proceedings, 1998. XII, 419 pages. 1998.

Vol. 1500: J.-C. Derniame, B.A. Kaba, D. Wastell (Eds.), Software Process: Principles, Methodology, and Technology. XIII, 307 pages. 1999.

Vol. 1501: M.M. Richter, C.H. Smith, R. Wiehagen, T. Zeugmann (Eds.), Algorithmic Learning Theory. Proceedings, 1998. XI, 439 pages. 1998. (Subseries LNAI).

Vol. 1502: G. Antoniou, J. Slaney (Eds.), Advanced Topics in Artificial Intelligence. Proceedings, 1998. XI, 333 pages. 1998. (Subseries LNAI).

Vol. 1503: G. Levi (Ed.), Static Analysis. Proceedings, 1998. IX, 383 pages. 1998.

Vol. 1504: O. Herzog, A. Günter (Eds.), KI-98: Advances in Artificial Intelligence. Proceedings, 1998. XI, 355 pages. 1998. (Subseries LNAI).

Vol. 1505: D. Caromel, R.R. Oldehoeft, M. Tholburn (Eds.), Computing in Object-Oriented Parallel Environments. Proceedings, 1998. XI, 243 pages. 1998.

Vol. 1506: R. Koch, L. Van Gool (Eds.), 3D Structure from Multiple Images of Large-Scale Environments. Proceedings, 1998. VIII, 347 pages. 1998.

Vol. 1507: T.W. Ling, S. Ram, M.L. Lee (Eds.), Conceptual Modeling – ER '98. Proceedings, 1998. XVI, 482 pages. 1998.

Vol. 1508: S. Jajodia, M.T. Özsu, A. Dogac (Eds.), Advances in Multimedia Information Systems. Proceedings, 1998. VIII, 207 pages. 1998.

Vol. 1510: J.M. Zytkow, M. Quafafou (Eds.), Principles of Data Mining and Knowledge Discovery. Proceedings, 1998. XI, 482 pages. 1998. (Subseries LNAI).

Vol. 1511: D. O'Hallaron (Ed.), Languages, Compilers, and Run-Time Systems for Scalable Computers. Proceedings, 1998. IX, 412 pages. 1998.

Vol. 1512: E. Giménez, C. Paulin-Mohring (Eds.), Types for Proofs and Programs. Proceedings, 1996. VIII, 373 pages. 1998.

Vol. 1513: C. Nikolaou, C. Stephanidis (Eds.), Research and Advanced Technology for Digital Libraries. Proceedings, 1998. XV, 912 pages. 1998.

Vol. 1514: K. Ohta, D. Pei (Eds.), Advances in Cryptology – ASIACRYPT'98. Proceedings, 1998. XII, 436 pages. 1998.

Vol. 1515: F. Moreira de Oliveira (Ed.), Advances in Artificial Intelligence. Proceedings, 1998. X, 259 pages. 1998. (Subseries LNAI).

Vol. 1516: W. Ehrenberger (Ed.), Computer Safety, Reliability and Security. Proceedings, 1998. XVI, 392 pages. 1998.

Vol. 1517: J. Hromkovič, O. Sýkora (Eds.), Graph-Theoretic Concepts in Computer Science. Proceedings, 1998. X, 385 pages. 1998.

Vol. 1518: M. Luby, J. Rolim, M. Serna (Eds.), Randomization and Approximation Techniques in Computer Science. Proceedings, 1998. IX, 385 pages. 1998.

1519: T. Ishida (Ed.), Community Computing and Support Systems. VIII, 393 pages. 1998.

Vol. 1520: M. Maher, J.-F. Puget (Eds.), Principles and Practice of Constraint Programming - CP98. Proceedings, 1998. XI, 482 pages. 1998.

Vol. 1521: B. Rovan (Ed.), SOFSEM'98: Theory and Practice of Informatics. Proceedings, 1998. XI, 453 pages. 1998.

Vol. 1522: G. Gopalakrishnan, P. Windley (Eds.), Formal Methods in Computer-Aided Design. Proceedings, 1998. IX, 529 pages. 1998.

Vol. 1524: G.B. Orr, K.-R. Müller (Eds.), Neural Networks: Tricks of the Trade. VI, 432 pages. 1998.

Vol. 1525: D. Aucsmith (Ed.), Information Hiding. Proceedings, 1998. IX, 369 pages. 1998.

Vol. 1526: M. Broy, B. Rumpe (Eds.), Requirements Targeting Software and Systems Engineering. Proceedings, 1997. VIII, 357 pages. 1998.

Vol. 1527: P. Baumgartner, Theory Reasoning in Connection Calculi. IX, 283. 1999. (Subseries LNAI).

Vol. 1528: B. Preneel, V. Rijmen (Eds.), State of the Art in Applied Cryptography. Revised Lectures, 1997. VIII, 395 pages. 1998.

Vol. 1529: D. Farwell, L. Gerber, E. Hovy (Eds.), Machine Translation and the Information Soup. Proceedings, 1998. XIX, 532 pages. 1998. (Subseries LNAI).

Vol. 1530: V. Arvind, R. Ramanujam (Eds.), Foundations of Software Technology and Theoretical Computer Science. XII, 369 pages. 1998.

Vol. 1531: H.-Y. Lee, H. Motoda (Eds.), PRICAI'98: Topics in Artificial Intelligence. XIX, 646 pages. 1998. (Subseries LNAI).

Vol. 1096: T. Schael, Workflow Management Systems for Process Organisations. Second Edition. XII, 229 pages. 1998.

Vol. 1532: S. Arikawa, H. Motoda (Eds.), Discovery Science. Proceedings, 1998. XI, 456 pages. 1998. (Subseries LNAI).

Vol. 1533: K.-Y. Chwa, O.H. Ibarra (Eds.), Algorithms and Computation. Proceedings, 1998. XIII, 478 pages. 1998.

Vol. 1534: J.S. Sichman, R. Conte, N. Gilbert (Eds.), Multi-Agent Systems and Agent-Based Simulation. Proceedings, 1998. VIII, 237 pages. 1998. (Subseries LNAI).

Vol. 1535: S. Ossowski, Co-ordination in Artificial Agent Societies. XV; 221 pages. 1999. (Subseries LNAI).

Vol. 1536: W.-P. de Roever, H. Langmaack, A. Pnueli (Eds.), Compositionality: The Significant Difference. Proceedings, 1997. VIII, 647 pages. 1998.

Vol. 1537: N. Magnenat-Thalmann, D. Thalmann (Eds.), Modelling and Motion Capture Techniques for Virtual Environments. Proceedings, 1998. IX, 273 pages. 1998. (Subseries LNAI).

Vol. 1538: J. Hsiang, A. Ohori (Eds.), Advances in Computing Science – ASIAN'98. Proceedings, 1998. X, 305 pages. 1998.

Vol. 1539: O. Rüthing, Interacting Code Motion Transformations: Their Impact and Their Complexity. XXI,225 pages. 1998.

Vol. 1540: C. Beeri, P. Buneman (Eds.), Database Theory – ICDT'99. Proceedings, 1999. XI, 489 pages. 1999.

Vol. 1541: B. Kågström, J. Dongarra, E. Elmroth, J. Waśniewski (Eds.), Applied Parallel Computing. Proceedings, 1998. XIV, 586 pages. 1998.

Vol. 1542: H.I. Christensen (Ed.), Computer Vision Systems. Proceedings, 1999. XI, 554 pages. 1999.

Vol. 1543: S. Demeyer, J. Bosch (Eds.), Object-Oriented Technology ECOOP'98 Workshop Reader. 1998. XXII, 573 pages. 1998.

Vol. 1544: C. Zhang, D. Lukose (Eds.), Multi-Agent Systems. Proceedings, 1998. VII, 195 pages. 1998. (Subseries LNAI).

Vol. 1545: A. Birk, J. Demiris (Eds.), Learning Robots. Proceedings, 1996. IX, 188 pages. 1998. (Subseries LNAI).

Vol. 1546: B. Möller, J.V. Tucker (Eds.), Prospects for Hardware Foundations. Survey Chapters, 1998. X, 468 pages. 1998.

Vol. 1547: S.H. Whitesides (Ed.), Graph Drawing. Proceedings 1998. XII, 468 pages. 1998.

Vol. 1548: A.M. Haeberer (Ed.), Algebraic Methodology and Software Technology. Proceedings, 1999. XI, 531 pages. 1999.

Vol. 1550: B. Christianson, B. Crispo, W.S. Harbison, M. Roe (Eds.), Security Protocols. Proceedings, 1998. VIII, 241 pages. 1999.

Vol. 1551: G. Gupta (Ed.), Practical Aspects of Declarative Languages. Proceedings, 1999. VIII, 367 pgages. 1999.

Vol. 1552: Y. Kambayashi, D.L. Lee, E.-P. Lim, M.K. Mohania, Y. Masunaga (Eds.), Advances in Database Technologies. Proceedings, 1998. XIX, 592 pages. 1999.

Vol. 1553: S.F. Andler, J. Hansson (Eds.), Active, Real-Time, and Temporal Database Systems. Proceedings, 1997. VIII, 245 pages. 1998.

Vol. 1554: S. Nishio, F. Kishino (Eds.), Advanced Multimedia Content Processing. Proceedings, 1998. XIV, 454 pages. 1999.

Vol. 1555: J.P. Müller, M.P. Singh, A.S. Rao (Eds.), Intelligent Agents V. Proceedings, 1998. XXIV, 455 pages. 1999. (Subseries LNAI).

Vol. 1557: P. Zinterhof, M. Vajteršic, A. Uhl (Eds.), Parallel Computation. Proceedings, 1999. XV, 604 pages. 1999.

Vol. 1558: H. J.v.d. Herik, H. Iida (Eds.), Computers and Games. Proceedings, 1998. XVIII, 337 pages. 1999.

Vol. 1559: P. Flener (Ed.), Logic-Based Program Synthesis and Transformation. Proceedings, 1998. X, 331 pages. 1999.

Vol. 1560: K. Imai, Y. Zheng (Eds.), Public Key Cryptography. Proceedings, 1999. IX, 327 pages. 1999.

Vol. 1561: I. Damgård (Ed.), Lectures on Data Security.VII, 250 pages. 1999.

Vol. 1563: Ch. Meinel, S. Tison (Eds.), STACS 99. Proceedings, 1999. XIV, 582 pages. 1999.

Vol. 1567: P. Antsaklis, W. Kohn, M. Lemmon, A. Nerode, S. Sastry (Eds.), Hybrid Systems V. X, 445 pages. 1999.

Vol. 1568: G. Bertrand, M. Couprie, L. Perroton (Eds.), Discrete Geometry for Computer Imagery. Proceedings, 1999. XI, 459 pages. 1999.

Vol. 1569: F.W. Vaandrager, J.H. van Schuppen (Eds.), Hybrid Systems: Computation and Control. Proceedings, 1999. X, 271 pages. 1999.

Vol. 1570: F. Puppe (Ed.), XPS-99: Knowledge-Based Systems. VIII, 227 pages. 1999. (Subseries LNAI).

Vol. 1572: P. Fischer, H.U. Simon (Eds.), Computational Learning Theory. Proceedings, 1999. X, 301 pages. 1999. (Subseries LNAI).

Vol. 1575: S. Jähnichen (Ed.), Compiler Construction. Proceedings, 1999. X, 301 pages. 1999.

Vol. 1576: S.D. Swierstra (Ed.), Programming Languages and Systems. Proceedings, 1999. X, 307 pages. 1999.

Vol. 1577: J.-P. Finance (Ed.), Fundamental Approaches to Software Engineering. Proceedings, 1999. X, 245 pages. 1999.

Vol. 1578: W. Thomas (Ed.), Foundations of Software Science and Computation Structures. Proceedings, 1999. X, 323 pages. 1999.

Vol. 1579: W.R. Cleaveland (Ed.), Tools and Algorithms for the Construction and Analysis of Systems. Proceedings, 1999. XI, 445 pages. 1999.

Vol. 1580: A. Včkovski, K.E. Brassel, H.-J. Schek (Eds.), Interoperating Geographic Information Systems. Proceedings, 1999. XI, 329 pages. 1999.

Vol. 1581: J.-Y. Girard (Ed.), Typed Lambda Calculi and Applications. Proceedings, 1999. VIII, 397 pages. 1999.

Vol. 1582: A. Lecomte, F. Lamarche, G. Perrier (Eds.), Logical Aspects of Computational Linguistics. Proceedings, 1997. XI, 251 pages. 1999. (Subseries LNAI).

Vol. 1587: J. Pieprzyk, R. Safavi-Naini, J. Seberry (Eds.), Information Security and Privacy. Proceedings, 1999. XI, 327 pages. 1999.